Springer Series in
SOLID-STATE SCIENCES 157

Springer Series in
SOLID-STATE SCIENCES

Series Editors:
M. Cardona   P. Fulde   K. von Klitzing   R. Merlin   H.-J. Queisser   H. Störmer

The Springer Series in Solid-State Sciences consists of fundamental scientific books prepared by leading researchers in the field. They strive to communicate, in a systematic and comprehensive way, the basic principles as well as new developments in theoretical and experimental solid-state physics.

147 **Electron Scattering in Solid Matter**
A Theoretical
and Computational Treatise
By J. Zabloudil, R. Hammerling,
L. Szunyogh, and P. Weinberger

148 **Physical Acoustics in the Solid State**
By B. Lüthi

149 **Solitary Waves
in Complex Dispersive Media**
Theory · Simulation · Applications
By V.Yu. Belashov and
S.V. Vladimirov

150 **Topology in Condensed Matter**
Editor: M.I. Monastyrsky

151 **Particle Penetration
and Radiation Effects**
By P. Sigmund

152 **Magnetism**
From Fundamentals
to Nanoscale Dynamics
By J. Stöhr and H.C. Siegmann

153 **Quantum Chemistry of Solids**
The LCAO First Principles Treatment
of Crystals
By R.A. Evarestov

154 **Low-Dimensional Molecular Metals**
By N. Toyota, M. Lang, and J. Müller

155 **Diffusion in Solids**
Fundamentals, Methods, Materials,
Diffusion-Controlled Processes
By H. Mehrer

156 **Physics
of Zero- and One-Dimensional
Nanoscopic Systems**
Editors: S.N. Karmakar, S.K. Maiti,
and C. Jayeeta

157 **Spin Physics in Semiconductors**
Editor: M.I. Dyakonov

Volumes 100–146 are listed at the end of the book.

M.I. Dyakonov
(Ed.)

Spin Physics in Semiconductors

With 176 Figures

 Springer

Professor Michel I. Dyakonov

Laboratoire de Physique Théorique et Astroparticules
cc 070 Université Montpellier II, 34095 Montpellier, France
E-mail: Michel.DYAKONOV@LPTA,univ-montp2.fr

Series Editors:

Professor Dr., Dres. h. c. Manuel Cardona
Professor Dr., Dres. h. c. Peter Fulde*
Professor Dr., Dres. h. c. Klaus von Klitzing
Professor Dr., Dres. h. c. Hans-Joachim Queisser

Max-Planck-Institut für Festkörperforschung, Heisenbergstrasse 1, 70569 Stuttgart, Germany
*Max-Planck-Institut für Physik komplexer Systeme, Nöthnitzer Strasse 38
  01187 Dresden, Germany

Professor Dr. Roberto Merlin

Department of Physics, University of Michigan
450 Church Street, Ann Arbor, MI 48109-1040, USA

Professor Dr. Horst Störmer

Dept. Phys. and Dept. Appl. Physics, Columbia University, New York, NY 10027 and
Bell Labs., Lucent Technologies, Murray Hill, NJ 07974, USA

Springer Series in Solid-State Sciences   ISSN 0171-1873

ISBN 978-3-540-78819-5   e-ISBN 978-3-540-78820-1

Library of Congress Control Number: 2008926904

© Springer-Verlag Berlin Heidelberg 2008

Typesetting by the author and Vtex using a Springer LaTeX macro
Cover: eStudio Calamar Steinen

SPIN: 12037452   57/3100/Vtex
Printed on acid-free paper

9 8 7 6 5 4 3 2 1

springeronline.com

To the memory of Vladimir Idelevich Perel (1928–2007)

Preface

The purpose of this collective book is to present a non-exhaustive survey of spin-related phenomena in semiconductors with a focus on recent research. In some sense it may be regarded as an updated version of the *Optical Orientation* book, which was entirely devoted to spin physics in bulk semiconductors.

During the 24 years that have elapsed, we have witnessed, on the one hand, an extraordinary development in the wonderful semiconductor physics in two dimensions with the accompanying revolutionary applications. On the other hand, during the last maybe 15 years there was a strong revival in the interest in spin phenomena, in particular in low-dimensional semiconductor structures. While in the 1970s and 1980s the entire world population of researchers in the field never exceeded 20 persons, now it can be counted by the hundreds and the number of publications by the thousands. This explosive growth is stimulated, to a large extent, by the hopes that the electron and/or nuclear spins in a semiconductor will help to accomplish the dream of factorizing large numbers by quantum computing and eventually to develop a new spin-based electronics, or "spintronics". Whether any of this will happen or not, still remains to be seen. Anyway, these ideas have resulted in a large body of interesting and exciting research, which is a good thing by itself.

The field of spin physics in semiconductors is extremely rich and interesting with many spectacular effects in optics and transport. We believe that a representative part of them is reviewed in this book. We have tried to make the presentation accessible to graduate students and to researchers new to the field.

Montpellier,   *Michel Dyakonov*
May 2008

Contents

Preface vii

List of Contributors xvii

1 Basics of Semiconductor and Spin Physics 1
M.I. Dyakonov 1
1.1   Historical Background 1
1.2   Spin Interactions 2
     1.2.1   The Pauli Principle 2
     1.2.2   Exchange Interaction 3
     1.2.3   Spin–Orbit Interaction 3
     1.2.4   Hyperfine Interaction with Nuclear Spins 4
     1.2.5   Magnetic Interaction 5
1.3   Basics of Semiconductor Physics 5
     1.3.1   Electron Energy Spectrum in a Crystal 5
     1.3.2   Effective Masses of Electrons and Holes 5
     1.3.3   The Effective Mass Approximation 6
     1.3.4   Role of Impurities 7
     1.3.5   Excitons 8
     1.3.6   The Structure of the Valence Band. Light and Heavy Holes 8
     1.3.7   Band Structure of GaAs 11
     1.3.8   Photo-generation of Carriers and Luminescence 11
     1.3.9   Angular Momentum Conservation in Optical Transitions 12
     1.3.10 Low Dimensional Semiconductor Structures 13
1.4   Overview of Spin Physics in Semiconductors 15
     1.4.1   Optical Spin Orientation and Detection 15
     1.4.2   Spin Relaxation 16
     1.4.3   Hanle Effect 21
     1.4.4   Mutual Transformations of Spin and Charge Currents 22
     1.4.5   Interaction between the Electron and Nuclear Spin Systems 23
1.5   Overview of the Book Content 25
     References 26

2 Spin Dynamics of Free Carriers in Quantum Wells 29
R.T. Harley 29
2.1   Introduction 29
2.2   Optical Measurements of Spin Dynamics 29
2.3   Mechanisms of Spin Relaxation of Free Electrons................. 32
2.4   Electron Spin Relaxation in Bulk Semiconductors 35
2.5   Electron Spin Relaxation in [001]-Oriented Quantum Wells 37
      2.5.1   Symmetrical [001]-Oriented Quantum Wells................ 37
      2.5.2   Structural Inversion Asymmetry in [001]-Oriented Quantum
              Wells... 40
      2.5.3   Natural Interface Asymmetry in Quantum Wells 42
      2.5.4   Oscillatory Spin-Dynamics in Two-dimensional Electron
              Gases 45
2.6   Spin Dynamics of Free Holes in Bulk Material and Quantum Wells 47
2.7   Engineering and Controlling the Spin Dynamics in Quantum Wells 49
2.8   Conclusions 51
      References 52

3 Exciton Spin Dynamics in Semiconductor Quantum Wells 55
T. Amand and X. Marie 55
3.1   Two-dimensional Exciton Fine Structure 55
      3.1.1   Short-Range Electron–Hole Exchange 56
      3.1.2   Long-Range Electron–Hole Exchange 57
3.2   Optical Orientation of Exciton Spin in Quantum Wells............ 58
3.3   Exciton Spin Dynamics in Quantum Wells......................... 60
      3.3.1   Exciton Formation in Quantum Wells 60
      3.3.2   Spin Relaxation of Exciton-Bound Hole 62
      3.3.3   Spin Relaxation of Exciton-Bound Electron 65
      3.3.4   Exciton Spin Relaxation Mechanism 66
3.4   Exciton Exchange Energy and g-Factor in Quantum Wells 72
      3.4.1   Exchange Interaction of Excitons and g-Factor Measured with
              cw Magneto-Photoluminescence Spectroscopy 73
      3.4.2   Exciton Spin Quantum Beats Spectroscopy 76
3.5   Exciton Spin Dynamics in Type II Quantum Wells 81
3.6   Spin Dynamics in Dense Excitonic Systems....................... 83
      References 86

4 Exciton Spin Dynamics in Semiconductor Quantum Dots 91
X. Marie, B. Urbaszek, O. Krebs and T. Amand 91
4.1   Introduction 91
4.2   Electron–Hole Complexes in Quantum Dots 92
      4.2.1   Coulomb Corrections to the Single Particle Picture 93
      4.2.2   Fine Structure of Neutral Excitons 93
4.3   Exciton Spin Dynamics in Neutral Quantum Dots without Applied
      Magnetic Fields 95

4.3.1   Exciton Spin Dynamics under Resonant Excitation   95
4.3.2   Exciton Spin Quantum Beats: The Role of Anisotropic
        Exchange   97
4.4   Exciton Spin Dynamics in Neutral Quantum Dots in External Magnetic
      Fields.   98
      4.4.1   Zeeman Effect Versus Anisotropic Exchange Splittings in Single
              Dot Spectroscopy   98
      4.4.2   Exciton Spin Quantum Beats in Applied Magnetic Fields   100
4.5   Charged Exciton Complexes: Spin Dynamics without Applied Magnetic
      Fields.   101
      4.5.1   Formation of Trions: Doped and Charge Tuneable Structures   102
      4.5.2   Fine Structure and Polarization of X^+ and X^- Excitons   103
      4.5.3   Spin Dynamics in Negatively Charged Exciton
              Complexes X^{n-}   104
      4.5.4   Spin Memory of Trapped Electrons   106
4.6   Charged Exciton Complexes: Spin Dynamics in Applied Magnetic
      Fields.   106
      4.6.1   Electron Spin Polarization in Positively Charged Excitons in
              Longitudinal Magnetic Fields   107
      4.6.2   Electron Spin Coherence in Positively Charged Excitons in
              Transverse Magnetic Fields   109
4.7   Conclusions   110
      References   110

5 Time-Resolved Spin Dynamics and Spin Noise Spectroscopy   115
   J. Hübner and M. Oestreich   115
5.1   Introduction   115
5.2   Time- and Polarization-Resolved Photoluminescence   116
      5.2.1   Experimental Technique.   117
      5.2.2   Experimental Example I: Spin Relaxation in (110) Oriented
              Quantum Wells   119
      5.2.3   Experimental Example II: Coherent Dynamics of Coupled
              Electron and Hole Spins in Semiconductors   122
      5.2.4   Photoluminescence and Spin-Optoelectronic Devices   123
5.3   Time-Resolved Faraday/Kerr Rotation   123
      5.3.1   Experimental Set-Up   125
      5.3.2   Experimental Example: Spin Amplification.   127
5.4   Spin Noise Spectroscopy   129
      5.4.1   Experimental Realization   129
5.5   Spin Noise Measurements in n-GaAs   131
5.6   Conclusions   132
      References   133

6 Coherent Spin Dynamics of Carriers 135
D.R. Yakovlev and M. Bayer 135
6.1   Introduction 135
       6.1.1   Spin Coherence and Spin Dephasing Times 136
       6.1.2   Optical Generation of Spin Coherent Carriers 137
       6.1.3   Experimental Technique 138
6.2   Spin Coherence in Quantum Wells 140
       6.2.1   Electron Spin Coherence 141
       6.2.2   Hole Spin Coherence 151
6.3   Spin Coherence in Singly Charged Quantum Dots 153
       6.3.1   Exciton and Electron Spin Beats Probed by Faraday Rotation ... 155
       6.3.2   Generation of Electron Spin Coherence 157
       6.3.3   Mode Locking of Spin Coherence in an Ensemble of Quantum
                Dots 160
       6.3.4   Nuclei Induced Frequency Focusing of Spin Coherence 169
6.4   Conclusions 174
       References 175

7 Spin Properties of Confined Electrons in Si 179
W. Jantsch and Z. Wilamowski 179
7.1   Introduction 179
7.2   Spin–Orbit Effects in Si Quantum Wells 182
       7.2.1   The Bychkov–Rashba Field 182
7.3   Spin Relaxation of Conduction Electrons in Si/SiGe Quantum Wells 186
       7.3.1   Mechanisms of Spin Relaxation of Conduction Electrons 186
       7.3.2   Linewidth and the Longitudinal Relaxation Time of the
                Two-dimensional Electron Gas in Si/SiGe 187
       7.3.3   Dephasing and Longitudinal Spin Relaxation 191
       7.3.4   Comparison with Experiment 194
7.4   Current Induced Spin–Orbit Field 195
7.5   ESR Excited by an ac Current 197
       7.5.1   Electric Dipole vs. Magnetic Dipole Spin Excitation 197
       7.5.2   The ESR Signal Strength in Two-dimensional Si/SiGe
                Structures—Experimental Results 198
       7.5.3   Modeling the Current Induced Excitation and Detection
                of ESR 199
       7.5.4   Power Absorption, Line Shape 201
7.6   Spin Relaxation under Lateral Confinement 201
       7.6.1   Shallow Donors 202
       7.6.2   From the Two-dimensional Electron Gas to Quantum Dots 204
       7.6.3   Spin Relaxation and Dephasing in Si Quantum Dots 205
7.7   Conclusions 206
       References 207

8 Spin Hall Effect 211
M.I. Dyakonov and A.V. Khaetskii 211
8.1   Background: Magnetotransport in Molecular Gases 211
8.2   Phenomenology (with Inversion Symmetry)......................... 213
       8.2.1   Preliminaries 213
       8.2.2   Spin and Charge Current Coupling 213
       8.2.3   Phenomenological Equations................................ 214
       8.2.4   Physical Consequences of Spin–Charge Coupling 215
       8.2.5   Related Problems 218
       8.2.6   Electrical Effects of Second Order in Spin–Orbit Interaction 219
8.3   Phenomenology (without Inversion Symmetry) 222
8.4   Microscopic Mechanisms 223
       8.4.1   Spin Asymmetry in Electron Scattering 223
       8.4.2   The Side Jump Mechanism 226
       8.4.3   Intrinsic Mechanism 231
8.5   Experiments 235
8.6   Conclusion 239
       Appendix A: The Generalized Kinetic Equation 239
       References 241

9 Spin–Photogalvanics... 245
E.L. Ivchenko and S. Ganichev 245
9.1   Introduction. Phenomenological Description 245
9.2   Circular Photogalvanic Effect................................... 247
       9.2.1   Historical Background 247
       9.2.2   Basic Experiments 248
       9.2.3   Microscopic Model for Inter-Sub-Band Transitions 251
       9.2.4   Relation to k-Linear Terms 251
       9.2.5   Circular PGE Due to Inter-Sub-Band Transitions 251
       9.2.6   Interband Optical Transitions 253
       9.2.7   Spin-Sensitive Bleaching 254
9.3   Spin–Galvanic Effect... 256
       9.3.1   Microscopic Mechanisms 257
       9.3.2   Spin–Galvanic Photocurrent Induced by the Hanle Effect 259
       9.3.3   Spin–Galvanic Effect at Zero Magnetic Field 261
       9.3.4   Determination of the Rashba/Dresselhaus Spin Splitting Ratio ... 262
9.4   Inverse Spin–Galvanic Effect 263
       9.4.1   Spin-Flip Mediated Current-Induced Polarization 264
       9.4.2   Precessional Mechanism 265
       9.4.3   Current Induced Spin Faraday Rotation 266
       9.4.4   Current Induced Polarization of Photoluminescence 267
9.5   Pure Spin Currents... 268
       9.5.1   Pure Spin Current Injected by a Linearly Polarized Beam 269
       9.5.2   Pure Spin Currents Due to Spin-Dependent Scattering.......... 271

9.6   Concluding Remarks 274
      References 274

10 Spin Injection 279
M. Johnson 279
10.1 Introduction 279
      10.1.1 History 279
10.2 Theoretical Models of Spin Injection and Spin Accumulation 281
      10.2.1 Heuristic Introduction 281
      10.2.2 Microscopic Transport Model 285
      10.2.3 Thermodynamic Theory of Spin Transport 286
      10.2.4 Hanle Effect 292
10.3 Spin Injection Experiments in Metals 292
10.4 Spin Injection in Semiconductors 295
      10.4.1 Optical Experiments 297
      10.4.2 Transport Experiments 301
10.5 Related Topics 305
      References 306

11 Dynamic Nuclear Polarization and Nuclear Fields 309
V.K. Kalevich, K.V. Kavokin and I.A. Merkulov 309
11.1 Electron–Nuclear Spin System of the Semiconductor: Characteristic
      Values of Effective Fields and Spin Precession Frequencies 310
      11.1.1 Zeeman Splitting of Spin Levels 310
      11.1.2 Quadrupole Interaction 311
      11.1.3 Hyperfine Interaction 311
      11.1.4 Nuclear Dipole–Dipole Interaction 313
11.2 Electron Spin Relaxation by Nuclei: from Short to Long Correlation
      Time 314
11.3 Dynamic Polarization of Nuclear Spins 316
      11.3.1 Electron Spin Splitting in the Overhauser Field 317
      11.3.2 Stationary States of the Electron–Nuclear Spin System in
              Faraday Geometry 319
      11.3.3 Dynamic Polarization by Localized Electrons 320
      11.3.4 Cooling of the Nuclear Spin System 322
      11.3.5 Polarization of Nuclei by Excitons in Neutral Quantum Dots 324
      11.3.6 Current-Induced Dynamic Polarization in Tunnel-Coupled
              Quantum Dots 325
      11.3.7 Self-Polarization of Nuclear Spins 325
11.4 Dynamic Nuclear Polarization in Oblique Magnetic Field 326
      11.4.1 Larmor Electron Spin Precession 327
      11.4.2 Polarization of Electron–Nuclear Spin-System in an Oblique
              Magnetic Field 329
      11.4.3 Bistability of the Electron–Nuclear Spin System in Structures
              with Anisotropic Electron g-Factor and Spin Relaxation Time ... 331

11.5 Optically Detected and Optically Induced Nuclear Magnetic
        Resonances 333
        11.5.1 Optically Detected Nuclear Magnetic Resonance 333
        11.5.2 Multispin and Multiquantum NMR 333
        11.5.3 Optically Induced NMR 335
11.6 Spin Conservation in the Electron–Nuclear Spin System of a Quantum
        Dot 337
        11.6.1 Time Scales for Preservation of Spin Direction and Spin
                Temperature 337
        11.6.2 A Guide to Interpretation of Experiments on "Spin Memory".... 338
11.7 Conclusions 342
        References 343

12 Nuclear–Electron Spin Interactions in the Quantum Hall Regime 347
Y.Q. Li and J.H. Smet 347
12.1 Introduction 348
        12.1.1 The Quantum Hall Effects in a Nutshell 348
        12.1.2 Electron Spin Phenomena in the Quantum Hall Effects 353
        12.1.3 Nuclear Spins in GaAs-Based 2D Electron Systems 356
12.2 Experimental Techniques 360
12.3 Nuclear Spin Phenomena in the Quantum Hall Regime 362
        12.3.1 The Role of Disorder 362
        12.3.2 Edge Channel Scattering 364
        12.3.3 Skyrmions 367
        12.3.4 Nuclear–Electron Spin Interactions at $\nu = 2/3$ 369
        12.3.5 Resistively Detected NMR at $\nu = 2/3$ 371
        12.3.6 Composite Fermion Fermi Sea at $\nu = 1/2$ 379
        12.3.7 Other Cases 382
12.4 Summary and Outlook 384
        References 384

13 Diluted Magnetic Semiconductors:
Basic Physics and Optical Properties 389
J. Cibert and D. Scalbert 389
13.1 Introduction 389
13.2 Band Structure of II–VI and III–V DMS 390
13.3 Exchange Interactions in DMS 392
        13.3.1 $s, p–d$ Exchange Interaction 392
        13.3.2 $d–d$ Exchange Interactions 394
13.4 Magnetic Properties 396
        13.4.1 Undoped DMS 396
        13.4.2 Carrier-Induced Ferromagnetism 399
13.5 Basic Optical Properties 402
        13.5.1 Giant Zeeman Effect 402
        13.5.2 Optically Detected Ferromagnetism in II–VI DMS 408

13.5.3 Quantum Dots 410
13.5.4 Spin-Light Emitting Diodes 412
13.5.5 III–V Diluted Magnetic Semiconductors 412
13.6 Spin Dynamics 414
13.6.1 Electron Spin Relaxation Induced by $s–d$ Exchange 415
13.6.2 Mn Spin Relaxation 415
13.6.3 Collective Spin Excitations in CdMnTe Quantum Wells 419
13.7 Advanced Time-Resolved Optical Experiments 422
13.7.1 Carrier Spin Dynamics 423
13.7.2 Magnetization Dynamics 424
References 427

Index 433

List of Contributors

Thierry Amand
Laboratoire de Physique et Chimie
  des Nano-Objets
Université de Toulouse
INSA-CNRS-UPS
135 avenue de Rangueil
31077 Toulouse cedex 4, France
amand@insa-toulouse.fr

Manfred Bayer
Experimentelle Physik 2
Technische Universität Dortmund
D-44221 Dortmund, Germany
manfred.bayer@
physik.uni-dortmund.de

Joël Cibert
Institut Néel, CNRS-UJF, BP166
38042 Grenoble, France
joel.cibert@
grenoble.cnrs.fr

Michel Dyakonov
Laboratoire de Physique Théorique
  et Astroparticules
Université Montpellier II, CNRS
Place E. Bataillon
34090 Montpellier, France
dyakonov@
lpta.univ-montp2.fr

Sergey Ganichev
University of Regensburg
D-93040 Regensburg, Germany
sergey.ganichev@
physik.uni-regensburg.de

Richard Harley
School of Physics & Astronomy
University of Southampton
Southampton, SO17 1BJ, UK
R.T.Harley@soton.ac.uk

Jens Hübner
Institute for Solid State Physics
Leibniz
Universität Hannover, Appelstr. 2
D-30167 Hannover, Germany
jhuebner@
nano.uni-hannover.de

E.L. Ivchenko
A.F. Ioffe Physico-Technical Institute
Russian Academy of Sciences
194021 St. Petersburg, Russia
ivchenko@coherent.ioffe.ru

Wolfgang Jantsch
Johannes Kepler Universität
Altenbergerstrasse 69
A-4040 Linz, Austria
Wolfgang.Jantsch@jku.at

Mark Johnson
Naval Research Laboratory
Washington, DC 20375, USA
mark.b.johnson@nrl.navy.mil

Vladimir Kalevich
Ioffe Physico-Technical Institute
194021 St. Petersburg, Russia
kalevich@solid.ioffe.ru

Kirill Kavokin
Ioffe Physico-Technical Institute
194021 St. Petersburg, Russia
kidd.orient@mail.ioffe.ru

Alexander Khaetskii
Institute of Microelectronics
   Technology
Russian Academy of Sciences
142432 Chernogolovka, Moscow
District, Russia
khaetski@ipmt-hpm.ac.ru

Olivier Krebs
CNRS-Laboratoire de
   Photonique et de Nanostructures
route de Nozay
91460 Marcoussis, France
Olivier.Krebs@lpn.cnrs.fr

Yongqing Li
Max-Planck-Institute
   for Solid State Research
Heisenbergstraße 1
D-70569 Stuttgart, Germany
y.li@fkf.mpg.de

Xavier Marie
Laboratoire de Physique et Chimie
   des Nano-Objets
Université de Toulouse
INSA-CNRS-UPS
135 avenue de Rangueil
31077 Toulouse cedex 4, France
marie@insa-toulouse.fr

Igor Merkulov
Ioffe Physico-Technical Institute
194021 St. Petersburg, Russia
merkulov@orient.ioffe.ru

Michael Oestreich
Institute for Solid State Physics
Leibniz
Universität Hannover, Appelstr. 2
D-30167 Hannover, Germany
oest@nano.uni-hannover.de

Denis Scalbert
Université Montpellier 2
Place Eugène
Bataillon, 34095 Montpellier, France
scalbert@
GES.univ-montp2.fr

Jurgen Smet
Max-Planck-Institute
   for Solid State Research
Heisenbergstraße 1
D-70569 Stuttgart, Germany
j.smet@fkf.mpg.de

Bernhard Urbaszek
Laboratoire de Physique et Chimie
   des Nano-Objets
Université de Toulouse
INSA-CNRS-UPS
135 avenue de Rangueil
31077 Toulouse cedex 4, France
urbaszek@insa-toulouse.fr

Zbysław Wilamowski
Institute of Physics Polish Academy
   of Sciences, Alea Lotnikow 32/46
02-668 Warsaw, Poland
and
University of Warmia and Mazury,
Żołnierska 14, 10-561 Olsztyn, Poland
wilamz@ifpan.edu.pl

Dmitri Yakovlev
Experimentelle Physik 2
Technische Universität Dortmund
D-44221 Dortmund, Germany
and
Ioffe Physico-Technical Institute
Russian Academy of Sciences
194021 St. Petersburg, Russia
dmitri.yakovlev@
physik.uni-dortmund.de

1

Basics of Semiconductor and Spin Physics

M.I. Dyakonov

This introductory chapter is mainly addressed to readers new to the field. In Sect. 1.1 a brief review of the historical roots of the current research is given. Section 1.2 describes various spin interactions. Section 1.3 is a mini textbook on semiconductor physics designed for beginners. A short overview of spin phenomena in semiconductors is given in Sect. 1.4. Finally, Sect. 1.5 presents the topics discussed in the chapters to follow.

1.1 Historical Background

The first step towards today's activity was made by Robert Wood in 1923/1924 when even the notion of electron spin was not yet introduced. In a charming paper [1] Wood and Ellett describe how the initially observed high degree of polarization of mercury vapor fluorescence (resonantly excited by polarized light) was found to diminish significantly in later experiments. "It was then observed that the apparatus was oriented in a different direction from that which obtained in earlier work, and on turning the table on which everything was mounted through ninety degrees, bringing the observation direction East and West, we at once obtained a much higher value of the polarization." In this way Wood and Ellett discovered what we now know as the Hanle effect, i.e., depolarization of luminescence by transverse magnetic field (the Earth's field in their case). It was Hanle [2] who carried out detailed studies of this phenomenon and provided the physical interpretation.

The subject did not receive much attention until 1949 when Brossel and Kastler [3] initiated profound studies of optical pumping in atoms, which were conducted by Kastler and his school in Paris in the 1950s and 1960s. (See Kastler's Nobel Prize award lecture [4].) The basic physical ideas and the experimental technique of today's "spintronic" research originate from these seminal papers: creation of a non-equilibrium distribution of atomic angular moments by optical excitation, manipulating this distribution by applying *dc* or *ac* fields, and detecting the result by studying the luminescence polarization. The relaxation times for the decay of atomic angular moments can be quite long, especially when hyperfine splitting due to the nuclear spin is involved.

A number of important applications have emerged from these studies, such as gyroscopes and hypersensitive magnetometers, but in my opinion, the knowledge obtained is even more valuable. The detailed understanding of various atomic processes and of many aspects of the interaction between light and matter was pertinent to the future developments, e.g., for laser physics.

The first experiment on the optical spin orientation of electrons in a semiconductor (Si) was done by Georges Lampel [5] in 1968, as a direct application of the ideas of optical pumping in atomic physics. The greatest difference, which has important consequences, is that now these are the free conduction band electrons (or holes) that get spin-polarized, rather than electrons bound in an atom. This pioneering work was followed by extensive experimental and theoretical studies mostly performed by small research groups at Ioffe Institute in St. Petersburg (Leningrad) and at Ecole Polytéchnique in Paris in the 1970s and early 1980s. At the time this research met with almost total indifference by the rest of the physics community.

1.2 Spin Interactions

This section serves to enumerate the possible types of spin interactions that can be encountered in a semiconductor.

The existence of an electron spin, $s = 1/2$, and the associated magnetic moment of the electron, $\mu = e\hbar/2mc$, has many consequences, some of which are very important and define the very structure of our world, while others are more subtle, but still quite interesting. Below is a list of these consequences in the order of decreasing importance.

1.2.1 The Pauli Principle

Because of $s = 1/2$, the electrons are fermions, and so no more than one electron per quantum state is allowed. Together with Coulomb law and the Schrödinger equation, it is this principle that is responsible for the structure of atoms, chemical properties, and the physics of condensed matter, biology included. It is interesting to speculate what would our world look like without the Pauli principle and whether any kind of life would be possible in such a world! Probably, only properties of the high-temperature, fully ionized plasma would remain unchanged. Note that the Pauli exclusion principle is not related to any interaction: if we could switch off the Coulomb repulsion between electrons (but leave intact their attraction to the nuclei), no serious changes in atomic physics would occur, although some revision of the Periodic Table would be needed.

Other manifestations of the electronic spin are due to interactions, either electric (the Coulomb law) or magnetic (related to the electron magnetic moment μ_B).

1.2.2 Exchange Interaction

It is, in fact, the result of the electrostatic Coulomb interaction between electrons, which becomes spin-dependent because of the requirement that the wave function of a pair of electrons be anti-symmetric with respect to the interchange of electron coordinates and spins. If the electron spins are parallel, the coordinate part of the wave function should be antisymmetric: $\psi_{\uparrow\uparrow}(r_2, r_1) = -\psi_{\uparrow\uparrow}(r_1, r_2)$, which means that the probability that two electrons are very close to each other is small compared to the opposite case, when the spins are antiparallel and accordingly their coordinate wave function is symmetric. Electrons with parallel spins are then better separated in space, so that their repulsion is less and consequently the energy of the electrostatic interaction for parallel spins is lower.

The exchange interaction is responsible for ferromagnetism. In semiconductors, it is normally not of major importance, except for magnetic semiconductors (like CdMnTe) and for the semiconductor-ferromagnet interface.

1.2.3 Spin–Orbit Interaction

If an observer moves with a velocity v in an external electric field E, he will see a magnetic field $B = (1/c)E \times v$, where c is the velocity of light. This magnetic field acts on the electron magnetic moment. This is the physical origin of the spin–orbit interaction,[1] the role of which strongly increases for heavy atoms (with large Z). The reason is that there is a certain probability for the outer electron to approach the nucleus and thus to see the very strong electric field produced by the unscreened nuclear charge $+Ze$ at the center. Due to the spin–orbit interaction, any electric field acts on the spin of a moving electron.

Being perpendicular both to E and v, in an atom the vector B is normal to the plane of the orbit, thus it is parallel to the orbital angular momentum L. The energy of the electron magnetic moment in this magnetic field is $\pm\mu_B B$ depending on the orientation of the electron spin (and hence its magnetic moment) with respect to B (or to L).[2]

---

[1] It is often stated that the origin of the spin–orbit interaction is relativistic and quantum-mechanical. This is true in the sense that it can be derived from the relativistic Dirac equation by keeping terms on the order of $1/c^2$. However, the above formula $B = (1/c)E \times v$ is *not* relativistic: one does not need the theory of relativity to understand that, when moving with respect to a stationary charge, a current, and hence a magnetic field will be seen. Given that the electron has a magnetic moment, the spin–orbit interaction follows directly. It is also not really quantum-mechanical: a classical object having a magnetic moment would experience the same interaction. The only place where quantum mechanics enters is the value of the electron magnetic moment and, of course, the fact that the electron spin is $1/2$.

[2] In fact the interaction energy derived in this simple-minded way should be cut in half (the "Thomas's one half" [6]) if one takes properly into account that, because of the electron acceleration in the electric field of the nucleus, its moving frame is not inertial. This finding, made in 1926, resolved the factor of 2 discrepancy between the measured and previously calculated fine structure splittings.

Thus the spin–orbit interaction can be written as $A(\boldsymbol{LS})$, the constant A depending on the electron state in an atom. This interaction results in a splitting of atomic levels (the fine structure), which strongly increases for heavy atoms.[3]

In semiconductors, the spin–orbit interaction depends not only on the velocity of the electron (or its quasi-momentum), but also on the structure of the Bloch functions defining the motion on the atomic scale. As in isolated atoms, it defines the values of the electron g-factors. More details can be found in [7].

Spin–orbit interaction is key to the subject of this book as it enables optical spin orientation and detection (the electrical field of the light wave does not interact directly with the electron spin). It is (in most cases) responsible for spin relaxation. And finally, it makes the transport and spin phenomena inter-dependent.

1.2.4 Hyperfine Interaction with Nuclear Spins

This is the magnetic interaction between the electron and nuclear spins, which may be quite important if the lattice nuclei in a semiconductor have non-zero spin (like in GaAs). If the nuclei get polarized, this interaction is equivalent to the existence of an effective nuclear magnetic field acting on electron spins. The effective field of 100% polarized nuclei in GaAs would be several Tesla!

Because the nuclear magnetic moment is so small (2 000 times less than that of the electron) the equilibrium nuclear polarization at the (experimentally inaccessible) magnetic field of 100 T and a temperature of 1 K would be only about 1%. However, much higher degrees of polarization may be easily achieved through *dynamic nuclear polarization* due to a hyperfine interaction with non-equilibrium electrons.

Experimentally, non-equilibrium nuclear polarization of several percent is easily achieved, recently values up to 50% were observed (see Chap. 11).

Similar to the spin–orbit interaction, the hyperfine interaction may be expressed in the form $A(\boldsymbol{IS})$ (the Fermi contact interaction), where $\boldsymbol{I}$ is the nuclear spin, $\boldsymbol{S}$ is the electron spin, and the *hyperfine constant* A is proportional to $|\psi(0)|^2$, the square of the electron wave function at the location of the nucleus.

Like spin–orbit interaction, the hyperfine interaction strongly increases in atoms with large Z, and for the same reason. An s-electron in an outer shell has a certain probability to be at the center of the atom, where the nucleus is located, and the nearer it is to the center, the less the nucleus is shielded by the inner electrons. Thus the electron wave function of an s-electron will have a sharp spike in the vicinity of the nucleus. For example, for the In atom the value of $|\psi(0)|^2$ is 6 000 times larger than in the hydrogen atom.

For p-states, and generally for states with $l \neq 0$, the Fermi interaction does not work, since $\psi(0) = 0$, and the electron and nuclear spins are coupled by the much weaker dipole–dipole interaction.

---

[3] Interestingly, general relativity predicts spin–orbit effects (on the order of $(v/c)^2$) in the motion of planets. Thus the "spin" of the Earth should make a slow precession around its orbital angular momentum.

1.2.5 Magnetic Interaction

This is the direct dipole–dipole interaction between the magnetic moments of a pair of electrons. For two electrons located at neighboring sites in a crystal lattice this gives an energy on the order of 1 K. This interaction is normally too weak to be of any importance in semiconductors.

1.3 Basics of Semiconductor Physics

A semiconductor is an insulator with a relatively small forbidden gap and shallow energy levels of electrons bound to impurities. The main feature of a semiconductor is its extreme sensitivity to impurities: a concentration of impurities like one per million of host atoms may determine the electrical conductivity and its temperature dependence.

1.3.1 Electron Energy Spectrum in a Crystal

The potential energy of an electron in a crystal is periodic in space. The most important consequence of this is that the energy spectrum consists of allowed and forbidden energy bands, and that the electron states can be characterized by its quasi-momentum p (or quasi-wave vector $k = p/\hbar$). The energy in an allowed band is a periodic function of k, so it may be considered only in a certain region of k-space called the first Brillouin zone. The number of states in an allowed band is equal to twice the number of elementary cells in the crystal (the doubling is due to spin). Thus the energy spectrum is given by the dependence of energy on quasi-momentum, $E(p)$, for all the allowed bands.

In insulators and pure semiconductors at zero temperature a certain number of the lowest allowed bands are completely filled with electrons (according to the Pauli principle), while the higher bands are empty. In most cases only the upper filled band (valence band) and the first empty band (conduction band) are of interest. The conduction and valence bands are separated by a forbidden energy gap of width E_g. In semiconductors the value of E_g may vary from zero (so-called gapless semiconductors, like HgTe) to about 2–3 eV. For Si $E_g \approx 1.1\,\text{eV}$, for GaAs $E_g \approx 1.5\,\text{eV}$.

1.3.2 Effective Masses of Electrons and Holes

The important property of semiconductors is that the number of free carriers (electrons in the conduction band or holes in the valence band) is always small compared to the number of atoms. The carriers are produced either by thermal excitation, in which case one has an equal number of electrons and holes, or by doping (see Sect. 1.3.4). Whatever the case, the carrier concentration never exceeds $10^{20}\,\text{cm}^{-3}$ (normally much less than that), while the number of states per cm^3 in a given band is on the order of 10^{22}, which is also a typical electron concentration in a metal. This means that electrons occupy only a very small fraction of the conduction band where their energy is lowest (and holes occupy only a very small fraction of the valence band). Consequently, when dealing with a semiconductor, we should be mostly interested in the properties of the energy spectrum in the vicinity of the minimum of the function $E(p)$ for the conduction band and in the vicinity of its maximum for the valence band. If these extrema correspond to the center of the Brillouin zone ($p = 0$), as it is the case for GaAs and many other materials, then for small p the function $E(p)$ should be parabolic:

$$E_c = \frac{p^2}{2m_c} \quad \text{for the conduction band,}$$

$$E_v = -\frac{p^2}{2m_v} \quad \text{for the valence band.}$$

Here m_c and m_v are the effective masses of electrons and holes, respectively. The effective masses may differ considerably from the free electron mass m_0, for example in GaAs $m_c = 0.067m_0$. Generally, the extrema of $E(p)$ do not necessarily occur at the center of the Brillouin zone, also the effective mass may be anisotropic, i.e., have different values for different directions in the crystal.

1.3.3 The Effective Mass Approximation

The effective masses were initially introduced just as convenient parameters to describe the curvature of the $E(p)$ parabolic dependence in the vicinity of its minimum or maximum. However this concept has a more profound meaning. In many cases we are interested in what happens to an electron, or a hole, under the action of some external forces due to, for example, electric and magnetic fields, deformation of the crystal, etc.

It can be shown, that if the spatial variation of these forces is much slower than that of the periodic crystal potential and if the carrier energy remains small compared to the forbidden gap, E_g, we can forget about the existence of the periodic potential and consider our electrons (or holes) as free particles moving in this external field. The only difference is that they have an *effective mass*, not the free electron mass. Thus the classical motion of a conduction electron in an electric field E and a magnetic field B is described by the conventional Newton's law: $m_c \, d^2r/dt^2 = -eE - (e/c)v \times B$. In particular, the cyclotron frequency of an electron rotating in a magnetic field is determined by the effective mass m_c, and this gives a valuable method of determining the effective masses experimentally (the cyclotron resonance).

If quantum treatment is needed, one can use the Schrödinger equation for an electron in the external field with its effective mass, *forgetting* about the existence of the crystal periodic potential.

Clearly, the validity of the effective mass approximation simplifies enormously the understanding of various physical phenomena in semiconductors.

1.3.4 Role of Impurities

Consider a crystal of germanium in which each atom is linked to its first neighbors by 4 tetrahedral bonds (Ge is an element of column IV of the Periodic Table, it has four electrons to form bonds). Replace one of the host atoms by an atom of As, which belongs to column V. Arsenic will give four of its valence electrons to participate in bonding, and give its remaining fifth electron to the conduction band of the crystal. Thus, arsenic is a *donor* for germanium. The extra electron can travel far away from the donor, which then has a positive charge. Alternatively, the electron may be bound by the positive charge of the donor forming a hydrogen-like "atom".

If the binding energy is small compared to E_g, and if the effective Bohr radius a_B^* is large compared to the lattice constant, this bound state can be studied using the effective mass approximation described in the previous section. This means that we can use the theory of the hydrogen atom and simply replace in all final formulas the free electron mass m_0 by the effective mass m_c. There is also another simple modification, which takes into account the static dielectric constant of the material, ϵ. The Coulomb potential energy of two opposite charges in vacuum is $-e^2/r$, while inside a polarizable medium it should be replaced by $-e^2/(\epsilon r)$. The ionization energy and the Bohr radius for the hydrogen atom are, respectively: $E_0 = m_0 e^4/(2\hbar) = 13.6\,\text{eV}$, $a_B = \hbar^2/(m_0 e^2) \sim 10^{-8}\,\text{cm}$. To obtain the corresponding values for an electron bound to a donor in a semiconductor, we make the replacements: $m_0 \rightarrow m_c$, $e^2 \rightarrow e^2/\epsilon$.

Suppose, for example, that $m_c = 0.1 m_0$ and $\epsilon = 10$, which are typical values for a semiconductor. Then our electron bound to a donor will have an ionization energy smaller by a factor of 1000 ($E_0^* \sim 10\,\text{meV}$) and an effective Bohr radius larger by a factor of 100 ($a_B^* \sim 10\,\text{nm}$) than the corresponding values for a hydrogen atom. This justifies the validity of the effective mass approximation. It is interesting that within the electron orbit there are roughly 10^5 host atoms! The electron simply does not see these atoms, their only role being to change the free electron mass to m_c. Because of the small value of the binding energy E_0^*, the donor is very easily ionized at moderate temperatures.

Conversely, if we replace the Ge atom by a group III impurity, like gallium, which has three valence electrons, it will take the fourth electron, needed to form the tetrahedral bonds, from the Ge valence band. Then the Ga *acceptor* will become a negatively charged center and a positively charged hole will appear in the valence band. Now the same story applies to the hole: it can either be free, or it may be bound to the negative acceptor forming a hydrogen-like state. It is the effective mass of the hole, m_v, which will now define the ionization energy and the effective Bohr radius. Since in most cases $m_v > m_c$, the acceptor radius is normally smaller that the donor radius, and the ionization of acceptors occurs at higher temperatures. Some complications of this simple picture arise if the effective mass is anisotropic.

Semiconductors are always, either intentionally or non-intentionally, *doped* by impurities and may be n-type or p-type depending on the dominant impurity type.

1.3.5 Excitons

An exciton in a semiconductor is a bound state of an electron and hole. It is again a hydrogen-like system with properties similar to an electron bound to a donor impurity. The important difference is that an exciton as a whole can move inside the crystal. Another difference is that excitons practically never exist in conditions of equilibrium. Usually they are created by optical excitation. Excitons have a certain lifetime with respect to recombination, during which the bound electron–hole pair annihilates. They can be seen as an absorption line somewhat below E_g.

1.3.6 The Structure of the Valence Band. Light and Heavy Holes

The allowed bands in crystals may be thought of as originating from discrete atomic levels, which are split to form a band when isolated atoms become close to each other. However atomic levels are generally degenerate, i.e., there may be several distinct states having the same energy. This degeneracy may have important consequences for the band energy spectrum of a crystal.

Neglecting Spin–Orbit Interaction

We now restrict the discussion to cubic semiconductors and at first do not consider spin effects. The $p = 0$ conduction band state is s-type ($l = 0$), the corresponding valence band state is p-type ($l = 1$) and is triply degenerate ($m_l = 0, \pm 1$). Here l is the atomic orbital angular momentum, and m_l is its projection on an arbitrary axis. The problem is to construct an effective mass description of the valence band structure taking into account this threefold degeneracy. This may be done using symmetry considerations: we have a vector $\boldsymbol{p}$ and a pseudo-vector of angular momentum $\boldsymbol{L}$ (which is a set of 3×3 matrices L_x, L_y, and L_z, corresponding to $l = 1$, L_z is a diagonal matrix with eigenvalues 1, 0, and -1), and a scalar Hamiltonian should be constructed, which must be quadratic in $\boldsymbol{p}$.

If we require invariance under rotations, the only possibility is the Luttinger Hamiltonian [8]:

$$H = Ap^2\mathcal{I} + B(\boldsymbol{p}\boldsymbol{L})^2, \qquad (1.1)$$

where A and B are arbitrary constants, $\mathcal{I}$ is a unit 3×3 matrix.

Thus the Hamiltonian H is also a 3×3 matrix, and the energy spectrum in the valence band should be found by diagonalizing this matrix. We can greatly simplify this procedure by noting that the choice of the axes x, y, z is arbitrary. Accordingly, we can choose the direction of the z-axis along the vector $\boldsymbol{p}$ (naturally, the final result does not depend on how the axes are chosen). Then $(\boldsymbol{p}\boldsymbol{L})^2 = p^2 L_z^2$, so that H becomes diagonal with eigenvalues

$$E_h(p) = (A + B)p^2 \quad \text{for } L_z = \pm 1, \qquad E_l(p) = Ap^2 \quad \text{for } L_z = 0.$$

Thus the valence band energy spectrum has two parabolic branches, $E_h(p)$ and $E_l(p)$, the first one being two-fold degenerate. We can now introduce two effective masses, m_h and m_l, by the relations: $A + B = 1/(2m_h)$ and $A = 1/(2m_l)$ and say that we have two types of holes in the valence band, the *light* and *heavy* holes (usually $B < 0$, but $A + B > 0$). The difference between these particles is that the heavy hole has a projection of its orbital momentum L on the direction of p (*helicity*) equal to ± 1, while the light hole has a projection equal to 0.

Effects of Spin–Orbit Interaction

If we now include spin but do not take into account the spin–orbit interaction, this will simply double all the states, both in the conduction band and in the valence band. However the spin–orbit interaction essentially changes the energy spectrum of the valence band.

We start again with the atomic states from which the bands originate. The spin–orbit interaction results in an additional energy proportional to (LS) (see Sect. 1.2.3). Because of this, L and S are no longer conserved separately, but only the total angular momentum $J = L + S$.

The eigenvalues of J^2 are $j(j+1)$ with $|l - s| \leq j \leq l + s$. Thus the state with $l = 0$ (from which the conduction band is built) is not affected ($j = s = 1/2$), while the state with $l = 1$ (from which the valence band is built) is split into two states with $j = 3/2$ and $j = 1/2$. In atomic physics this splitting leads to the fine structure of spectral lines.

The symmetry properties of band states at $p = 0$ are completely similar to those of the corresponding atomic states. Thus for $p = 0$ we must have a four-fold degenerate state ($j = 3/2$, $J_z = +3/2, +1/2, -1/2, -3/2$), which is separated by an energy distance Δ, the *spin–orbit splitting*, from a doubly degenerate state ($j = 1/2$, $J_z = +1/2, -1/2$). The conduction band remains doubly degenerate. The value of Δ is small for materials with light atoms, like Si, and may be quite large (comparable to E_g) in semiconductors composed of heavy atoms, like InSb (see Sect. 1.2.3). In GaAs $\Delta \approx 0.3$ eV.

To see what happens to the $j = 3/2$ state for $p \neq 0$ for energies $E(p) \ll \Delta$ we construct the Luttinger Hamiltonian in a way quite similar to the procedure in the previous section. The only difference is that the 3×3 matrices L_x, L_y, and L_z, corresponding to $l = 1$, should now be replaced by 4×4 matrices J_x, J_y, and J_z, corresponding to $j = 3/2$:

$$H = Ap^2 \mathcal{I} + B(pJ)^2, \tag{1.2}$$

where now $\mathcal{I}$ is a unit 4×4 matrix, the matrix J_z is diagonal with eigenvalues $3/2$, $1/2$, $-1/2$, and $-3/2$.

Proceeding as above, we obtain the spectrum of the heavy and light holes, which is valid for energies much less than Δ:

$$E_{\mathrm{h}}(p) = \left(A + \frac{9B}{4} \right) p^2 = \frac{p^2}{2m_{\mathrm{h}}} \quad (J_z = \pm 3/2) \quad \text{heavy hole band;}$$

$$E_{\mathrm{l}}(p) = \left(A + \frac{B}{4} \right) p^2 = \frac{p^2}{2m_{\mathrm{l}}} \quad (J_z = \pm 1/2) \quad \text{light hole band.}$$

Both bands are doubly degenerate. Heavy holes have projection of the angular momentum J on the direction of p (or helicity) equal to $\pm 3/2$, while for light holes the helicity is $\pm 1/2$. Normally $B < 0$, but $A + 9B/4 > 0$, so that both masses are positive.

The combined description of all three bands (light, heavy, and split-off) on the energy scale $\Delta \sim E(p) \ll E_g$, including effects of non-parabolicity, can be found in [9].

Gapless Semiconductors

Interestingly, the signs of the expressions $A + 9B/4$ and $A + B/4$ may be opposite, which is the case of the so-called gapless semiconductors, like HgTe. In these materials the light hole mass becomes negative, so that this band becomes a conduction band. The conduction band and the valence band (which now consists of heavy holes only) are degenerate at $p = 0$, so that the energy gap is absent.

Warping of the Iso-energetic Surfaces

Also, it should be noted that the Luttinger Hamiltonian (1.2) presents the so-called spherical approximation: it is invariant under arbitrary rotations. In a cubic crystal the symmetry is generally lower. Thus the true Luttinger Hamiltonian should have a more general form:

$$H = Ap^2\mathcal{I} + B(pJ)^2 + C\left(J_x^2 p_x^2 + J_y^2 p_y^2 + J_z^2 p_z^2\right), \qquad (1.3)$$

where now the x, y, z axes are not arbitrary, they coincide with the crystallographic axes. The last term makes the iso-energetic surfaces of light and heavy holes anisotropic, so that the energy branches $E_h(p)$ and $E_l(p)$ will not have the simple parabolic form given above. (A similar term should be added to (1.1).)

Oddities in the Behavior of Light and Heavy Holes

In the valence band the "spin" of light and heavy holes is tightly bound to their momentum, and this has many interesting consequences. If some external forces exist, the light and heavy hole states generally become mixed. A simple example is the reflection from an interface.

Suppose that a heavy hole is incident on an ideal flat potential wall. If the incidence is normal, nothing very interesting happens, except that the initial state with helicity $+3/2$ (angular momentum J parallel to p) will be transformed after reflection into a state with opposite helicity: $-3/2$. This can be explained by noting that while the initial momentum p changes sign under reflection, the internal angular momentum remains unchanged.

However for an arbitrary angle of incidence the same reasoning tells us that the reflected heavy hole will have a certain arbitrary angle between J and p. But such

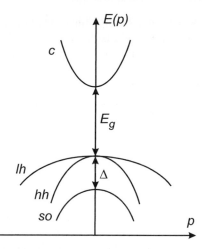

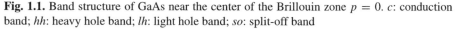

Fig. 1.1. Band structure of GaAs near the center of the Brillouin zone $p = 0$. c: conduction band; hh: heavy hole band; lh: light hole band; so: split-off band

free states do not exist! This means that the incident heavy hole will be partly transformed into the light hole. (A similar phenomenon of transformation between ordinary and the extraordinary waves during reflection is known in optics of uniaxial crystals.)

One can reconsider all the textbook problems of quantum mechanics (potential well, tunnel effect, the hydrogen problem, movement in magnetic field, etc.) for a particle, described by the Luttinger Hamiltonian; and these exercises reveal the rather bizarre physics of light and heavy holes in a semiconductor.

1.3.7 Band Structure of GaAs

The above considerations lead to the band structure presented in Fig. 1.1. Near the center of the Brillouin zone there is a simple isotropic conduction band, which is doubly degenerate in spin (for the moment we neglect the spin splitting, see Sect. 1.4.2). The valence band, consists of the sub-bands of light and heavy holes, which are anisotropic (see Sect. 1.3.6), and the isotropic split-off band, which are all doubly degenerate.

1.3.8 Photo-generation of Carriers and Luminescence

In the process of interband absorption of a photon with energy $\hbar\omega > E_g$ in a semiconductor, an electron in the conduction band and a hole in the valence band are generated. During the process the (quasi)momentum is conserved, however the photon momentum $\hbar k = 2\pi\hbar/\lambda$, where λ is the photon wavelength, is very small (compared, for example, to the electron thermal momentum) and normally may be neglected.

In this approximation the optical transitions are *vertical*: to see what happens, we must simply apply a vertical arrow of length $\hbar\omega$ to Fig. 1.1, so that the arrow touches one of the valence bands and the conduction band. The ends of the arrow will give us the initial energies of the generated electrons and holes. An electron may be created in company with a heavy hole, or a light hole; for $\hbar\omega > E_g + \Delta$ the electron–hole pair can also involve a hole in the split-off band. Note, that for a given photon energy the initial electron energy will be different for these three processes.

The photoexcited carriers live some time τ before recombination, which may be radiative (i.e., accompanied by photon emission, which results in luminescence), or non-radiative. In direct-band semiconductors, like GaAs, the recombination is predominantly radiative with a lifetime on the order of 1 ns.

It is important to realize that this time is normally very long compared to the carriers thermalization time. Thermalization means energy relaxation of carriers in their respective bands due to phonon emission and absorption, which results in an equilibrium Boltzmann (or Fermi, depending on temperature and concentration) distribution function of electrons and holes. Thermal equilibrium *between* electrons and holes is established by recombination, on the time scale τ.

Because the recombination time τ is so long compared to the energy relaxation time, the luminescence is produced mostly by thermalized carriers and the emitted photons have energies close to the value of E_g, irrespective of the energy of exciting photons.[4]

It should be noted that semiconductors are normally either intentionally, or non-intentionally doped by impurities. In a *p*-type semiconductor at moderate excitation power the number of photo-generated holes is small compared to the number of equilibrium holes, so that the photo-created electron will recombine with these equilibrium holes, rather than with photo-generated ones.

1.3.9 Angular Momentum Conservation in Optical Transitions

This section is most important for our subject. Along with energy and momentum conservation, the conservation of the angular momentum is a fundamental law of physics. Just like particles, electromagnetic waves have angular momentum. Photons of right or left polarized light have a projection of the angular momentum on the direction of their propagation (helicity) equal to $+1$ or -1, respectively (in units of $\hbar$). Linearly polarized photons are in a superposition of these two states.

When a circularly polarized photon is absorbed, this angular momentum is distributed between the photo-excited electron and hole according to the selection rules determined by the band structure of the semiconductor. Because of the complex nature of the valence band, this distribution depends on the value of the momentum of the created electron–hole pair (p and $-p$). However, it can be shown that if we take the average over the directions of p, the result is the same as in optical transitions

---

[4] A small part of the excited electrons can emit photons *before* losing their energy by thermalization. The studies of the spectrum and polarization properties of this so-called *hot luminescence* reveal interesting and unusual physics, see [10, 11].

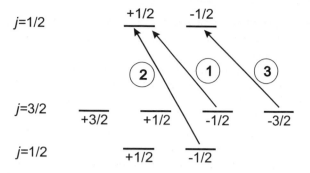

Fig. 1.2. Optical transitions between levels with $j = 3/2$ and $j = 1/2$ (the bands of light and heavy holes, and the split-off band) and the levels with $j = 1/2$ (the conduction band) during an absorption of a right-polarized photon. The probability ratio for the three transitions is 3:2:1

between atomic states with $j = 3/2, m_j = -3/2, -1/2, +1/2, +3/2$ (corresponding to bands of light and heavy holes) and $j = 1/2, m_j = -1/2, +1/2$ (corresponding to the conduction band), see Sect. 1.4.1 below.

Possible transitions between these states, as well as between states in the split-off band and the conduction band, for absorption of a right circularly polarized photon with corresponding relative probabilities are presented in Fig. 1.2. Note, that if we add up *all* transitions, which is the correct thing to do if the photon energy sufficiently exceeds $E_g + \Delta$ the two spin states in the conduction band will be populated equally. This demonstrates the role of spin–orbit interaction for optical spin pumping, see [9, 14] for the details of photon energy dependence of the spin polarization.

1.3.10 Low Dimensional Semiconductor Structures

The development of semiconductor physics in the last two decades is mainly related to studies of artificially engineered low dimensional semiconductor structures, two-dimensional (quantum wells), one-dimensional (quantum wires), and zero-dimensional (quantum dots). By growing a structure consisting of a thin semiconductor layer, for example GaAs, surrounded by material with a larger band gap, for example a solid solution GaAlAs, one obtains a potential well for electrons (and for holes) with a typical width of 20–200 Å.

Thus the first problem in quantum mechanics courses, a particle in a one-dimensional rectangular potential well, which since 1926 was tackled by generations of students as the simplest training exercise, has finally become relevant to some reality!

Energy Spectrum of Electrons and Holes in a Quantum Well

The motion in the direction perpendicular to the layer (the *growth direction*), z, is quantized in accordance with textbooks, while the motion in the plane of the layer xy is unrestrained. Thus the energy spectrum of an electron in a quantum well consists

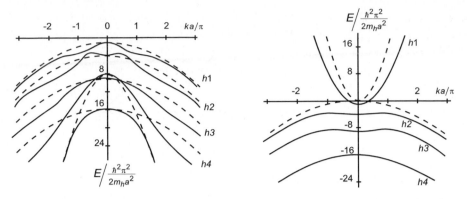

Fig. 1.3. The energy spectrum $E(k)$ of holes (*left*) and of carriers in a gapless semiconductor (*right*) in an infinite rectangular quantum well within the spherical approximation [12]. *Dashed lines* represent the spectrum that would exist if the two types of carriers were independent particles

of two-dimensional *sub-bands*: $E_n(\boldsymbol{p}) = E_n^0 + p^2/(2m)$, where E_n^0 are the energy levels for the one-dimensional motion in the z direction, $\boldsymbol{p}$ is the two-dimensional (quasi)momentum in the xy plane, and m is the electron effective mass.

In most cases the electron concentration in the well is such that only the lowest sub-band is occupied. The motion of such electrons is purely two-dimensional (2D). One important consequence is that in an applied magnetic field perpendicular to the 2D plane the spectrum becomes discrete: it consists of Landau levels. A magnetic field parallel to the 2D plane has no effect on the orbital motion of electrons, however it has the usual influence on their spins.

For the case of holes in a quantum well, the problem is not so simple. For $\boldsymbol{p} = 0$ one has two independent ladders of levels for heavy and light holes, given (for an infinite well) by the textbook formula $E_n^0 = (\pi n \hbar)^2/(2ma^2)$, where m is the respective effective mass, a is the well width, and $n = 1, 2, 3, \ldots$. However, for $\boldsymbol{p} \neq 0$ the spectrum is determined by the mutual transformations of light and heavy holes during reflections from the potential walls, see Sect. 1.3.6.

Figure 1.3 shows the spectrum of holes and of carriers in a gapless semiconductor in an infinite quantum well calculated in [12] within the spherical approximation (1.2).[5]

Especially interesting is the case of a gapless semiconductor. In a quantum well, a gap will obviously appear due to quantization of the transverse motion. Naively, one would expect this gap to be $E_{e1}^0 - E_{h1}^0 = (1/2)(\pi \hbar/a)^2/(1/m_e - 1/m_h)$, i.e., mostly determined by the small electron mass. In fact, this is not true, because the

---

[5] More accurately, one should use the Hamiltonian in (1.3), which takes care of the warping of iso-energetic surfaces. In fact, the energy spectrum depends on the growth direction, and on the orientation of the vector $\boldsymbol{p}$ in the xy plane with respect to the crystal axes. However the general properties of the spectrum are the same.

$h1$ sub-band, originating from the first hole level at $p = 0$, becomes *electronic* (see Fig. 1.3). Thus the gap is $\approx m_e/m_h \sim 1/10$ times smaller than expected.

For $pa/\hbar \gg 1$ the first electronic sub-band $h1$ corresponds to *surface states* localized near the well boundaries. Such states should exist also near the surface of a bulk gapless semiconductor [13].

In fact, it is not even necessary to have a sandwich structure to obtain 2D electrons. A simple interface between two different materials plus an electric field of ionized donors gives the same effect, except that now the quantum well is not rectangular, but more like triangular, and that its shape depends on the electron concentration.

The heterostructure design allows to accomplish what was impossible in bulk semiconductors: a spatial separation of the electrons and the donors, from which they originate. The technique of *delta doping* provides a 2D electron gas with previously unimaginable mobilities on the order of 10^7 V/cm^2 s.

Quantum Dots

Quantum dots are zero-dimensional structures, a sort of large artificial atoms. Under certain growth conditions, *self-assembled* quantum dots appear spontaneously. Typically, they have the form of a flat cake with a hight ~ 30 Å and a base diameter of ~ 300 Å. They are embedded in a different material, so that there is a large potential barrier at the interface.

Normally, samples contain an ensemble of many quantum dots with varying parameters, however special techniques allow us to deal with individual dots. Like in an atom, the energy spectrum is discrete. A quantum dot may contain a few electrons or holes.

1.4 Overview of Spin Physics in Semiconductors

The basic ideas concerning spin phenomena in semiconductors were developed both theoretically and experimentally more than 30 years ago. Some of these ideas have been rediscovered only recently. A review of non-equilibrium spin physics in bulk semiconductors can be found in [14], as well as in other chapters of the *Optical Orientation* book.

1.4.1 Optical Spin Orientation and Detection

To date, the most efficient way of creating non-equilibrium spin orientation in a semiconductor is provided by an interband absorption of circularly polarized light.

It can be seen from Fig. 1.2 that for $E_g < \hbar\omega < E_g + \Delta$ absorption produces an average electron spin along the direction of excitation equal to $(-1/2)(3/4) + (+1/2)(1/4) = -1/4$ and an average hole spin equal to $+5/4$, with a sum $+1$, equal to the angular momentum of the absorbed right circularly polarized photon. Thus in a *p*-type semiconductor the degree of spin polarization of the photo-excited electrons will be −50%; the minus sign indicating that the spin orientation is opposite to the angular momentum of incident photons.

If our electron immediately recombines with its partner hole, a 100% circularly polarized photon will be emitted. However in a *p*-type semiconductor electrons will predominantly recombine with the majority holes, which are not polarized. Then the same selection rules show that the circular polarization of luminescence should be $\mathcal{P}_0 = 25\%$, if the holes are not polarized, and if no electron spin relaxation occurs during the electron lifetime τ, i.e., if $\tau_s \gg \tau$. Generally, the degree $\mathcal{P}$ of circular polarization of the luminescence excited by circularly polarized light is less than $\mathcal{P}_0$:

$$\mathcal{P} = \frac{\mathcal{P}_0}{1 + \tau/\tau_s}. \tag{1.4}$$

In an optical spin orientation experiment a semiconductor (usually *p*-type) is excited by circularly polarized light with $\hbar\omega > E_g$. The circular polarization of the luminescence is analyzed, which gives a direct measure of the electron spin polarization. Actually, the degree of circular polarization is simply equal to the average electron spin. Thus various spin interactions can be studied by simple experimental means. The electron polarization will be measured provided the spin relaxation time τ_s is not very short compared to the recombination time τ, a condition, which often can be achieved even at room temperature.

1.4.2 Spin Relaxation

Spin relaxation, i.e., disappearance of initial non-equilibrium spin polarization, is the central issue for all spin phenomena. Spin relaxation can be generally understood as a result of the action of fluctuating in time magnetic fields. In most cases, these are not real magnetic fields, but rather "effective" magnetic fields originating from the spin–orbit, or, sometimes, exchange interactions, see Sect. 1.2.

Generalities

A randomly fluctuating magnetic field is characterized by two important parameters: its amplitude (or, more precisely, its rms value), and its correlation time, τ_c, i.e., the time during which the field may be roughly considered as constant. Instead of the amplitude, it is convenient to use the rms value of the spin precession frequency in this random field, ω.

Thus we have the following physical picture of spin relaxation: the spin makes a precession around the (random) direction of the effective magnetic field with a typical frequency ω and during a typical time τ_c. After a time τ_c the direction and the absolute value of the field change randomly, and the spin starts its precession around the new direction of the field. After a certain number of such steps the initial spin direction will be completely forgotten.

How this happens depends on the value of the dimensionless parameter $\omega\tau_c$, which is the typical angle of spin precession during the correlation time. Two limiting cases may be considered:

$\omega\tau_c \ll 1$ *(Most Frequent Case)*

The precession angle is small, so that the spin vector experiences a slow angular diffusion. During a time t, the number of random steps is t/τ_c, for each step the squared precession angle is $(\omega\tau_c)^2$. These steps are not correlated, so that the total squared angle after a time t is $(\omega\tau_c)^2(t/\tau_c)$. The spin relaxation time may be defined as the time at which this angle becomes of the order of 1. Hence,

$$\frac{1}{\tau_s} \sim \omega^2\tau_c. \tag{1.5}$$

This is essentially a *classical* formula (the Planck constant does not enter), although certainly it can be also derived quantum-mechanically. Note, that in this case $\tau_s \gg \tau_c$.

$\omega\tau_c \gg 1$

This means that during the correlation time the spin will make many rotations around the direction of the magnetic field. During the time on the order of $1/\omega$ the spin projection transverse to the random magnetic field is (on the average) completely destroyed, while its projection along the direction of the field is conserved. At this stage the spin projection on its initial direction will diminish three times. [Let the random magnetic field have an angle θ with the initial spin direction. After many rotations the projection of the spin on the initial direction will diminish as $(\cos\theta)^2$. In three dimensions, the average of this value over the possible orientations of the random field yields $1/3$.]

After time τ_c the magnetic field changes its direction, and the initial spin polarization will finally disappear. Thus in the case $\omega\tau_c \gg 1$ the time decay of spin polarization is not exponential, and the process has two distinct stages: the first one has a duration $1/\omega$, and the second one has a duration τ_c. The overall result is $\tau_s \sim \tau_c$.

This consideration is quite general and applies to any mechanism of spin relaxation. We have only to understand the values of the relevant parameters ω and τ_c for a given mechanism.

Spin Relaxation Mechanisms

There are several possible mechanisms providing the fluctuating magnetic fields responsible for spin relaxation.

Elliott–Yafet Mechanism [15, 16]

The electrical field, accompanying lattice vibrations, or the electric field of charged impurities is transformed to an effective magnetic field through a spin–orbit interaction. Thus momentum relaxation should be accompanied by spin relaxation.

For phonons, the correlation time is on the order of the inverse frequency of a typical thermal phonon. Spin relaxation by phonons is normally rather weak, especially at low temperatures.

For scattering by impurities, the direction and the value of the random magnetic field depends on the geometry of the individual collision (the impact parameter). This random field cannot be characterized by a single correlation time, since it exists only *during* the brief act of collision and is zero between collisions. In each act of scattering the electron spin rotates by some small angle ϕ. These rotations are uncorrelated for consequent collisions, so the average square of spin rotation angle during time t is on the order of $\langle\phi^2\rangle(t/\tau_p)$, where τ_p is the time between collisions and $\langle\phi^2\rangle$ is the average of ϕ^2 over the scattering geometry.

Thus $1/\tau_s \sim \langle(\phi)^2\rangle/\tau_p$. The relaxation rate is obviously proportional to the impurity concentration.

Dyakonov–Perel Mechanism [9, 17]

This one is related to the spin–orbit splitting of the conduction band in non-centrosymmetric semiconductors like GaAs (but not Si or Ge, which are centrosymmetric). For bulk semiconductors, this splitting was first pointed out by Dresselhaus [18]. The additional spin-dependent term in the electron Hamiltonian can be presented as

$$\hbar\boldsymbol{\Omega}(\boldsymbol{p})\boldsymbol{S}, \tag{1.6}$$

which can be viewed as the energy of a spin in an effective magnetic field. Here $\boldsymbol{\Omega}(\boldsymbol{p})$ is a vector depending on orientation of the electron momentum with respect to the crystal axes (xyz), such that

$$\Omega_x \sim p_x(p_y^2 - p_z^2), \qquad \Omega_y \sim p_y(p_z^2 - p_x^2), \qquad \Omega_z \sim p_z(p_x^2 - p_y^2). \tag{1.7}$$

For a given $\boldsymbol{p}$, $\boldsymbol{\Omega}(\boldsymbol{p})$ is the spin precession frequency in this field. This frequency is proportional to $p^3 \sim E^{3/2}$. The effective magnetic field changes in time because the direction of $\boldsymbol{p}$ varies due to electron collisions. Thus the correlation time is on the order of the momentum relaxation time, τ_p, and if $\Omega\tau_p$ is small, which is normally the case, we get

$$\frac{1}{\tau_s} \sim \Omega^2\tau_p. \tag{1.8}$$

In contrast to the Elliott–Yafet mechanism, now the spin rotates not during, but *between* the collisions. Accordingly, the relaxation rate *increases* when the impurity concentration decreases (i.e., when τ_p becomes longer). It happens that this mechanism is often the dominant one, both in bulk $A^{III}B^V$ and $A^{II}B^{VI}$ semiconductors, like GaAs and in 2D structures (where $\boldsymbol{\Omega}(\boldsymbol{p}) \sim p$, see below).

Bir–Aronov–Pikus Mechanism [19]

This is a mechanism of spin relaxation of non-equilibrium electrons in p-type semiconductors due to the exchange interaction between the electron and hole spins (or, expressing it otherwise, exchange interaction between an electron in the conduction band and all the electrons in the valence band). This spin relaxation rate, being proportional to the number of holes, may become the dominant one in heavily p-doped semiconductors.

Relaxation via Hyperfine Interaction with Nuclear Spins

The electron spin interacts with the spins of the lattice nuclei (see Sect. 1.4.5 below), which are normally in a disordered state. Thus the nuclei provide a random effective magnetic field, acting on the electron spin. The corresponding relaxation rate is rather weak, but may become important for localized electrons, when other mechanisms, associated with electron motion, do not work.

Spin Relaxation of Holes in the Valence Band

The origin of this relaxation is in the splitting of the valence band into sub-bands of light and heavy holes. In this case, $\hbar\Omega(p)$ is equal to the energy difference between light and heavy holes for a given p and the correlation time is again τ_p. However, in contrast to the situation for electrons in the conduction band, we have now the opposite limiting case: $\Omega(p)\tau_p \gg 1$. So, the hole spin relaxation time is on the order of τ_p, which is very short. One can say that the hole "spin" J is rigidly fixed with respect to its momentum p, and because of this, momentum relaxation leads automatically to spin relaxation.

For this reason, normally it is virtually impossible to maintain an appreciable non-equilibrium polarization of bulk holes. However, Hilton and Tang [20] have managed to observe the spin relaxation (on the femtosecond time scale) of both light and heavy holes in undoped bulk GaAs. The general theory of the relaxation of spin, as well as helicity and other correlations between J and p, for holes in the valence band was given in [21].

Influence of Magnetic Field on Spin Relaxation

In the presence of an external magnetic field B, the spins perform a regular precession with a frequency $\Omega = g\mu B/\hbar$, and one should distinguish between relaxation of the spin component along B and the relaxation, or *dephasing*, of the perpendicular components. In the magnetic resonance literature it is customary to denote the corresponding longitudinal and transverse times as T_1 and T_2, respectively.

To understand what happens, it is useful to go to a frame rotating around B with the spin precession frequency Ω. In the absence of random fields, the spin vector would remain constant in the rotating frame. Relaxation is due to random fields in the rotating frame, and obviously these fields now rotate around B with the same frequency Ω.

Thus random fields directed along B are the same as in the rest frame, and cause the same relaxation of the perpendicular spin components with a characteristic time $T_2 \sim \tau_s$. However the perpendicular components of the random field, which are responsible for the relaxation of the spin component along B, do rotate. The importance of this rotation is determined by the parameter $\Omega\tau_c$, the angle of rotation of the random field during the correlation time.

If $\Omega\tau_c \ll 1$, then rotation is of no importance, since the random field will anyway change its direction after a time τ_c. However, for $\Omega\tau_c \gg 1$ the rotating random field will effectively average out during the correlation time, resulting in a decrease of the longitudinal spin relaxation rate.

A simple calculation gives

$$\frac{1}{T_1} = \frac{1}{\tau_s} \frac{1}{1 + (\Omega \tau_c)^2} = \frac{\omega^2 \tau_c}{1 + (\Omega \tau_c)^2}. \tag{1.9}$$

Interestingly, with increasing magnetic field the longitudinal spin relaxation rate changes from being proportional to τ_c to becoming proportional to $1/\tau_c$.

Again, the classical formula (1.9) can be derived quantum mechanically. From the quantum point of view the longitudinal relaxation is due to flips of the spin projection on $\boldsymbol{B}$, which requires an energy $g\mu B$. Since the energy spectrum of the random field has a width $\hbar/\tau_c$ the process becomes ineffective when $g\mu B \gg \hbar/\tau_c$, or equivalently, when $\Omega \tau_c \gg 1$.

Ivchenko [22] has calculated the influence of magnetic field on the Dyakonov–Perel spin relaxation. The result coincides with (1.9) with $\tau_c = \tau_p$, except that the spin precession frequency Ω is replaced by the (greater) electron cyclotron frequency, ω_c. The reason is that for this case the rotation of the vector $\boldsymbol{\Omega}(\boldsymbol{p})$ is primarily due to the rotation of the electron momentum $\boldsymbol{p}$ in the magnetic field.

Spin Relaxation of Two-dimensional Electrons and Holes

Usually the Dyakonov–Perel mechanism is the dominant one. However, the momentum dependence of the effective magnetic field, or the vector $\boldsymbol{\Omega}(\boldsymbol{p})$, is quite different.

First, because the projection of momentum perpendicular to the 2D plane is quantized and fixed, and because it is usually much greater than the in-plane projections, the spin splitting defined by (1.6) becomes *linear* in the in-plane momentum [23].

For the simplest case when the growth direction is (001), we must replace p_z and p_z^2 in (1.7) by their quantum-mechanical average values in the lowest sub-band, which are equal to 0 and $\langle p_z^2 \rangle$, respectively (for a deep rectangular well of width a, $\langle p_z^2 \rangle = (\pi \hbar/a)^2$). These considerations give

$$\Omega_x \sim -p_x \langle p_z^2 \rangle, \qquad \Omega_y \sim p_y \langle p_z^2 \rangle, \qquad \Omega_z = 0. \tag{1.10}$$

We see that the effective magnetic field is linear in $\boldsymbol{p}$ and lies in the 2D plane. As a consequence, the spin relaxation is anisotropic: the spin component perpendicular to the plane decays two times faster than the spin in-plane components.[6]

Thus the spin relaxation of 2D electrons is generally anisotropic and depends on the growth direction [23]. An interesting case is when the growth direction corresponds to (110). If we now take *this* direction as the z axis, and take x and y axes along the in-plane $(1\bar{1}0)$ and (001) directions, respectively, in the same manner as above we obtain

$$\Omega_x = 0, \qquad \Omega_y = 0, \qquad \Omega_z \sim p_x. \tag{1.11}$$

---

[6] The reason is that the z projection of the spin is rotated by both x and y components of the random field, while the x spin projection is influenced only by the y component, since the z component of the random field is zero.

The random effective magnetic field is now always perpendicular to the 2D plane! Its value and sign depend only on the projection of electron momentum on the $(1\bar{1}0)$ direction. This means that now the relaxation times for both in-plane components of the spin are equal, however the normal to the plane spin component does not relax at all.[7]

Second, if the quantum well is asymmetric, e.g., the triangular well in a heterostructure, there is another source of effective magnetic field, besides that originating from the Dresselhaus term, (1.7) and (1.6). This is due to the Bychkov–Rashba splitting [25, 26], which has the form (1.6) with

$$\boldsymbol{\Omega}(\boldsymbol{p}) \sim \boldsymbol{E}^R \times \boldsymbol{p}, \tag{1.12}$$

where $\boldsymbol{E}^R$ is the so-called "Rashba field", a built-in vector oriented along the growth direction and defined by the asymmetry of the quantum well.[8] For this case $\boldsymbol{\Omega}$ also lies in the 2D plane and is perpendicular to $\boldsymbol{p}$.

Although the $\boldsymbol{\Omega}(\boldsymbol{p})$ dependence is different from the one considered above for the (001) growth direction, the relaxation process is quite similar. However, if both types of interactions coexist and are of the same order of magnitude, a specific anisotropy of relaxation in the xy plane arises due to a kind of interference between the two terms [28].

The spin structure of holes in a quantum well is also completely different that in the bulk. More details on spin–orbit interaction in two-dimensional systems can be found in Winkler's book [29].

1.4.3 Hanle Effect

Depolarization of luminescence by a transverse magnetic field (first discovered by Wood and Ellett, as described in Sect. 1.1) is effectively employed in experiments on spin orientation in semiconductors.

The reason for this effect is the precession of electron spins around the direction of the magnetic field. Under continuous illumination, this precession leads to the decrease of the average projection of the electron spin on the direction of observation, which defines the degree of circular polarization of the luminescence. Thus the degree of polarization decreases as a function of the transverse magnetic field. Measuring this dependence under steady state conditions makes it possible to determine both the spin relaxation time and the recombination time.

This effect is due to the precession of electron spins in a magnetic field $\boldsymbol{B}$ with the Larmor frequency $\boldsymbol{\Omega}$. This precession, along with spin pumping, spin relaxation, and recombination is described by the following simple equation of motion of the

---

[7] In fact, the normal spin component *will* slowly decay because of the small cubic in $\boldsymbol{p}$ terms, which were neglected in deriving (1.10) and (1.11). Experimentally, a $\sim$20 times suppression of spin relaxation in (110) quantum wells is observed.

[8] The corresponding term in the Hamiltonian of 2D electrons was previously derived by Vasko [27].

average spin vector S:

$$\frac{\mathrm{d}S}{\mathrm{d}t} = \mathit{\Omega} \times S - \frac{S}{\tau_\mathrm{s}} - \frac{S - S_0}{\tau}, \tag{1.13}$$

where the first term on the rhs describes spin precession in a magnetic field ($\mathit{\Omega} = g\mu B/\hbar$), the second term describes spin relaxation, and the third one describes generation of spin by optical excitation (S_0/τ) and recombination ($-S/\tau$). The vector S_0 is directed along the exciting light beam, its absolute value is equal to the initial average spin of photo-created electrons.

In the stationary state ($\mathrm{d}S/\mathrm{d}t = 0$) and in the absence of a magnetic field, one finds

$$S_z(0) = \frac{S_0}{1 + \tau/\tau_\mathrm{s}}, \tag{1.14}$$

where $S_z(0)$ is the projection of the spin on the direction of S_0 (z-axis). Since $S_z(0)$ is equal to the degree of polarization of the luminescence (Sect. 1.4.1), this formula is equivalent to the expression for $\mathcal{P}$ in (1.4). In the presence of magnetic field transverse to S_0 we obtain

$$S_z(B) = \frac{S_z(0)}{1 + (\Omega\tau^*)^2}, \qquad \frac{1}{\tau^*} = \frac{1}{\tau} + \frac{1}{\tau_\mathrm{s}}. \tag{1.15}$$

The effective time τ^* defines the width of the depolarization curve. Thus the spin projection S_z (and hence the degree of circular polarization of the luminescence) decreases as a function of the transverse magnetic field. Combining the measurements of the zero-field value $\mathcal{P} = S_z(0)$ and of the magnetic field dependence in the Hanle effect, we can find the two essential parameters: the electron lifetime, τ, and the spin relaxation time, τ_s, under steady-state conditions.

If polarized electrons are created by a short pulse, time-resolved measurements reveal, very impressively, the damped spin precession around the direction of magnetic field [30], which follows from (1.13) for a given initial spin value.

1.4.4 Mutual Transformations of Spin and Charge Currents

Because of spin–orbit interaction, charge and spin transport are interconnected: an electrical current produces a transverse spin current and vice versa [31, 32]. In recent years this has become a subject of considerable interest and intense research, both experimental and theoretical, see Chap. 8.

One of the new phenomena, predicted in [31, 32] and now called the Spin Hall Effect, consists of the current-induced spin accumulation at the boundaries of a conductor. The spins are perpendicular to the direction of the electric current and have opposite signs on the opposing boundaries.[9] Accumulation occurs on the spin diffusion length $L_\mathrm{s} = \sqrt{D\tau_\mathrm{s}}$, where D is the diffusion coefficient. Typically L_s is on the order of 1 μm.

---

[9] This is reminiscent of what happens in the normal Hall effect, where *charges* of opposite sign accumulate at the boundaries because of the Lorentz force.

Inversely, a spin current, due for example to the inhomogenuity of the spin density, generates an electric current. More precisely, there is an electric current proportional to curl S (the Inverse Spin Hall Effect). This effect was found experimentally for the first time by Bakun et al. [33].

In gyrotropic crystals a current can be induced by a *homogeneous* non-equilibrium spin density, as it was shown theoretically by Ivchenko and Pikus [34] and by Belinicher [35]. The first experimental demonstration of this effect was reported in [36]. Inversely, an electric current will generate a uniform spin polarization.

Thus, generally, an electric current can induce spin accumulation at the boundaries, or a uniform spin polarization, or both effects simultaneously. Reciprocal effects exist too.

Phenomenologically, all these effects (including the well-known anomalous Hall effect [37]) can be derived from pure symmetry considerations, according to the general principle: *everything, that is not forbidden by symmetry or conservation laws, will happen.* In an isotropic media with inversion symmetry, the only building block is the unit antisymmetric tensor ϵ_{ijk}. If the symmetry is lower, there will be other tensors, that the theory may use. The microscopic theory should provide the physical mechanism of the phenomenon under consideration, as well as the values of the observable quantities. More details can be found in Chaps. 8 and 9.

1.4.5 Interaction between the Electron and Nuclear Spin Systems

The non-equilibrium spin-oriented electrons can easily transmit their polarization to the lattice nuclei, thus creating an effective magnetic field. This field will, in turn, influence the spin of electrons (but not their orbital motion). For example, it can strongly influence the electron polarization via the Hanle effect [38]. Thus the spin-oriented electrons and the polarized lattice nuclei form a strongly coupled system, in which spectacular non-linear phenomena, like self-sustained slow oscillations and hysteresis are observed by simply looking at the circular polarization of the luminescence [14, 39]. Optical detection of the nuclear magnetic resonance in a semiconductor was demonstrated for the first time by Ekimov and Safarov [40].

The physics of these phenomena are governed by three basic interactions:

Hyperfine Interaction between Electron and Nuclear Spins

The interaction has the form $A(IS)$, where I is the nuclear spin and S is the electron spin. If the electrons are in equilibrium this interaction provides a mechanism for nuclear spin relaxation. If the electron spin system is out of equilibrium, it leads to dynamic nuclear polarization. These processes are very slow compared to the characteristic electron time scale. On the other hand, if the nuclei are polarized, this interaction is equivalent to the existence of an effective nuclear magnetic field. The field of 100% polarized nuclei in GaAs would be about 6 T. Experimentally, nuclear polarization of several percent is easily achieved.

The time of build-up of nuclear polarization due to interaction with electrons is given by the general formula (1.5), where ω should be understood as the precession frequency of the nuclear spin in the effective electron magnetic field due to hyperfine interaction, and the correlation time τ_c depends on the electron state. For mobile electrons this time is extremely short: $\tau_c \sim \hbar/E$, where E is the electron energy. As first pointed out by Bloembergen [43], nuclear polarization (or depolarization) by electrons is much more effective when the electrons are localized, for example, bound to donors, or confined in a quantum dot. In this case τ_c is generally much longer than for mobile carriers. It is defined by the shortest of processes like recombination, hopping to another donor site, thermal ionization, or spin relaxation.

Dipole–Dipole Interaction between Nuclear Spins

This interaction can be characterized by the local magnetic field, B_L, on the order of several Gauss, which is created at a given nuclear site by the neighboring nuclei.[10] The precession period of a nuclear spin in the local field, on the order of $T_2 \sim 10^{-4}$ s, gives a typical intrinsic time scale for the nuclear spin system. During this time, thermal equilibrium within this system is established, with a *nuclear spin temperature* Θ_N, which may be very different from the crystal temperature T, for example, something like 10^{-6} K.

Since the times characterizing the interaction of the nuclear spin system with the outside world (electrons, or lattice) is much greater than T_2, the nuclear spin system can be considered as always being in a state of internal thermal equilibrium with a nuclear spin temperature defined by the energy exchange with the electrons and/or the lattice. Accordingly, the nuclear polarization *is always given by the thermodynamic formula* $\mathcal{P} \sim \mu_N B/(k\Theta_N)$, where μ_N is the nuclear magnetic moment. The most important concept of the nuclear spin temperature was introduced by Redfield [41], see also [42].

The dipole–dipole interaction is also responsible for the nuclear spin diffusion [43]—a process that tends to make the nuclear polarization uniform in space. The nuclear spin diffusion coefficient can be estimated as $D_N \sim a_0^2/T_2 \sim 10^{-12}$ cm^2/s, where a_0 is the distance between the neighboring nuclei. Thus it takes about 1 s to spread out the nuclear polarization on a distance of 100 Å, and several hours for a distance of 1 μm.

Zeeman Interaction of Electron and Nuclear Spins

The energy of a nuclear magnetic moment in an external magnetic field is roughly 2 000 times smaller than that for the electron. However, it becomes important in magnetic fields exceeding the local field $B_L \sim 3$ G. Accordingly, the behavior of the nuclear spin system in small fields, less than B_L, is quite different than in larger

---

[10] As was pointed out in Sect. 1.2, the magnetic dipole–dipole interaction between electron spins can be usually neglected. Given that a similar interaction between nuclear spins is about a million times smaller, it may seem strange that this interaction may be of any importance. The answer comes when we consider the extremely long time scale in the nuclear spin system (seconds or more) compared to the characteristic times for the electron spin system (nanoseconds or less).

fields. At zero magnetic field the nuclear spins can not be polarized (the Zeeman energy is zero, while Θ_N remains finite, see the thermodynamic formula above).

Also, as the magnetic field increases, the time of polarization will increase according to (1.9), where Ω is the electron spin precession frequency. Quantum mechanically, this increase is the result of the strong mismatch between the electron and nuclear Zeeman energies. Because of this mismatch the electron–nucleus flip–flop transitions would violate energy conservation. They can occur, however, because of the energy uncertainty $\Delta E \sim \hbar/\tau_c$.

The interplay of these interactions under various experimental conditions accounts for the extremely rich and interesting experimental findings in this domain, see Chap. 11.

1.5 Overview of the Book Content

Within the scope of this introductory chapter it is only possible to briefly outline the main directions of the current research.

Time-Resolved Optical Techniques. The innovative time resolved optical techniques, based on Faraday or Kerr polarization rotation, were developed by Awschalom's group in Santa Barbara [45] and by Harley's group in Southampton [46]. These techniques opened a new era in experimental spin physics. They have allowed for the visualization of spin dynamics on the sub-picosecond time scale and study of the intimate details of various spin processes in a semiconductor. This book presents several subjects, where most of the experimental results are obtained by using these optical techniques.

Spin Dynamics in Quantum Wells and Quantum Dots. The spin dynamics of carriers in quantum wells is discussed in Chap. 2. Exciton spin dynamics and the fine structure of neutral and charged excitons are presented in Chaps. 3 (quantum wells) and 4 (quantum dots). The interplay between carrier exchange and confinement leads to quite a number of interesting and subtle effects, that are now well understood. These chapters show how many important parameters, like spin splittings and relaxation times, can be accurately determined.

Spin Noise Spectroscopy. Chapter 5 gives a general introduction to experimental time-resolved techniques. It also presents quite a new way of research in spin physics, where the methods of noise spectroscopy, known in other domains, are applied to the spin system in a semiconductor. Unlike other techniques, this allows for the study of spin dynamics without perturbing the system by an external excitation.

Coherent Spin Dynamics in Quantum Dots. This topic is covered in Chap. 6. It contains extraordinarily interesting and surprising new results on "mode-locking" of spin coherence in an ensemble of quantum dots excited by a periodic sequence of laser pulses and, in particular, on spin precession "focusing" induced by the hyperfine interaction with the nuclear spins.

Spin Properties of Confined Electrons in Silicon. Spin-related studies in silicon were somewhat neglected in recent years, because it practically does not give photoluminescence, has a weak spin–orbit interaction, and contains few nuclear spins. However, Chap. 7 demonstrates interesting new spin physics in Si-based quantum wells and quantum dots, studied mostly by the electron spin resonance, which may have extremely small line-widths.

Coupling of Spin and Charge Currents. Chapter 8 is devoted to the coupling between the spin and charge currents due to spin–orbit interaction and the Spin Hall Effect, which was observed only recently and caused widespread interest. A related subject is treated in Chap. 9 describing spin-related photocurrents, or circular photo-galvanic effect, in two-dimensional structures. There are a variety of interesting experiments, which reveal subtle physics.

Spin Injection. Spin injection from a ferromagnet to a normal metal, originally proposed by Aronov [47], and spin detection using a ferromagnet, originally proposed by Silsbee [48], was first observed by Johnson and Silsbee [49]. Injection through a ferromagnet/semiconductor junction has been investigated in many recent works. Chapter 10 describes these and related phenomena, which have some promising applications.

Nuclear Spin Effects in Optics and Electron Transport. Chapter 11 discusses electron-nuclear spin systems formed by the hyperfine interaction in quantum wells and quantum dots. Nuclear spin polarization results in spectacular optical effects, including unusual magnetic resonances and hysteretic behavior.

Chapter 12 describes some astonishing manifestations of nuclear spins in low temperature magneto-transport in two dimensions, first observed by Dobers et al. [50]. Strong changes of the magnetoresistance in the Quantum Hall Effect regime are observed and shown to be caused by the dynamic nuclear spin polarization. Such studies yield unique insights into the properties of fragile quantum Hall states, which only exist at ultra-low temperatures and in the highest mobility samples. Some of the experimental results still remain to be understood.

Spin Dynamics in Diluted Magnetic Semiconductors. Mn doped III–V and II–VI systems, both bulk and two-dimensional, have attracted intense interest. The giant Zeeman splitting due to exchange interaction with Mn, combination of ferromagnetic and semiconductor properties, and the possibility of making a junction between a ferromagnetic and a normal semiconductor have been the focus of numerous studies. The basic physics, the magnetic and optical properties are reviewed in Chap. 13.

References

[1] R.W. Wood, A. Ellett, Phys. Rev. **24**, 243 (1924)
[2] W. Hanle, Z. Phys. **30**, 93 (1924)
[3] J. Brossel, A. Kastler, C. R. Hebd. Acad. Sci. **229**, 1213 (1949)
[4] A. Kastler, Science **158**, 214 (1967)
[5] G. Lampel, Phys. Rev. Lett. **20**, 491 (1968)

[6] L.H. Thomas, Nature **117**, 514 (1926)
[7] P.Y. Yu, M. Cardona, *Fundamental of Semiconductors*, 3rd edn. (Springer, Berlin, 2001)
[8] J.M. Luttinger, Phys. Rev. **102**, 1030 (1956)
[9] M.I. Dyakonov, V.I. Perel, Z. Eksp. Teor. Fiz. **60**, 1954 (1971); Sov. Phys. JETP **33**, 1053 (1971)
[10] V.D. Dymnikov, M.I. Dyakonov, V.I. Perel, Z. Eksp. Teor. Fiz. **71**, 2373 (1976); Sov. Phys. JETP **44**, 1252 (1976)
[11] D.N. Mirlin, in *Optical Orientation*, ed. by F. Meier, B.P. Zakharchenya (North-Holland, Amsterdam, 1984), p. 133
[12] M.I. Dyakonov, A.V. Khaetskii, Z. Eksp. Teor. Fiz. **82**, 1584 (1982); Sov. Phys. JETP **55**, 917 (1982)
[13] M.I. Dyakonov, A.V. Khaetskii, Pis'ma Z. Eksp. Teor. Fiz. **33**, 110 (1981); Sov. Phys. JETP Lett. **33**, 115 (1981)
[14] M.I. Dyakonov, V.I. Perel, in *Optical Orientation*, ed. by F. Meier, B.P. Zakharchenya (North-Holland, Amsterdam, 1984), p. 15
[15] R.J. Elliott, Phys. Rev. **96**, 266 (1954)
[16] Y. Yafet, in *Solid State Physics*, vol. 14, ed. by F. Seits, D. Turnbull (Academic, New York, 1963), p. 1
[17] M.I. Dyakonov, V.I. Perel, Fiz. Tverd. Tela **13**, 3581 (1971); Sov. Phys. Solid State **13**, 3023 (1972)
[18] G. Dresselhaus, Phys. Rev. **100**, 580 (1955)
[19] G.I. Bir, A.G. Aronov, G.E. Pikus, Z. Eksp. Teor. Fiz. **69**, 1382 (1975); Sov. Phys. JETP **42**, 705 (1976)
[20] D.J. Hilton, C.L. Tang, Phys. Rev. Lett. **89**, 146601 (2002)
[21] M.I. Dyakonov, A.V. Khaetskii, Z. Eksp. Teor. Fiz. **86**, 1843 (1984); Sov. Phys. JETP **59**, 1072 (1984)
[22] E.L. Ivchenko, Fiz. Tverd. Tela **15**, 1566 (1973); Sov. Phys. Solid State **15**, 1048 (1973)
[23] M.I. Dyakonov, V.Yu. Kachorovskii, Fiz. Techn. Poluprov. **20**, 178 (1986); Sov. Phys. Semicond. **20**, 110 (1986)
[24] Y. Ohno, R. Terauchi, T. Adachi, F. Matsukura, H. Ohno, Phys. Rev. Lett. **83**, 4196 (1999)
[25] Y.A. Bychkov, E.I. Rashba, J. Phys. C **17**, 6039 (1984)
[26] Y.A. Bychkov, E.I. Rashba, Z. Eksp. Teor. Fiz. Pis'ma, **39**, 66 (1984); Sov. Phys. JETP Lett. **39**, 78 (1984)
[27] F.T. Vasko, Z. Eksp. Teor. Fiz. Pis'ma, **30**, 574 (1979); Sov. Phys. JETP Lett. **30**, 541 (1979)
[28] N.S. Averkiev, L.E. Golub, Phys. Rev. B **60**, 15582 (1999)
[29] R. Winkler, *Spin–Orbit Coupling Effects in Two-dimensional Electron and Hole Systems* (Springer, Berlin, 2003)
[30] J.A. Gupta, X. Peng, A.P. Alivisatos, D.D. Awschalom, Phys. Rev. B **59**, 10421 (1999)
[31] M.I. Dyakonov, V.I. Perel, Pis'ma Z. Eksp. Teor. Fiz. **13**, 657 (1971); Sov. Phys. JETP Lett. **13**, 467 (1971)
[32] M.I. Dyakonov, V.I. Perel, Phys. Lett. A **35**, 459 (1971)
[33] A.A. Bakun, B.P. Zakharchenya, A.A. Rogachev, M.N. Tkachuk, V.G. Fleisher, Pis'ma Z. Eksp. Teor. Fiz. **40**, 464 (1984); Sov. Phys. JETP Lett. **40**, 1293 (1984)
[34] E.L. Ivchenko, G.E. Pikus, Pis'ma Z. Eksp. Teor. Fiz. **27**, 640 (1978); Sov. Phys. JETP Lett. **27**, 604 (1978)
[35] V.I. Belinicher, Phys. Lett. A **66**, 213 (1978)
[36] V.M. Asnin, A.A. Bakun, A.M. Danishevskii, E.L. Ivchenko, G.E. Pikus, A.A. Rogachev, Solid State Commun. **30**, 565 (1979)

[37] R. Karplus, J.M. Luttinger, Phys. Rev. **95**, 1154 (1954)
[38] M.I. Dyakonov, V.I. Perel, V.I. Berkovits, V.I. Safarov, Z. Eksp. Teor. Fiz. **67**, 1912 (1974); Sov. Phys. JETP **40**, 950 (1975)
[39] V.G. Fleisher, I.A. Merkulov, in *Optical Orientation*, ed. by F. Meier, B.P. Zakharchenya (North-Holland, Amsterdam, 1984), p. 173
[40] A.I. Ekimov, V.I. Safarov, Pis'ma Z. Eksp. Teor. Fiz. **15**, 453 (1972); Sov. Phys. JETP Lett. **15**, 179 (1972)
[41] A.G. Redfield, Phys. Rev. **98**, 1787 (1955)
[42] A. Abragam, *The Principles of Nuclear Magnetism* (Oxford University Press, Oxford, 1983)
[43] N. Bloembergen, Physica **20**, 1130 (1954)
[44] N. Bloembergen, Physica **25**, 386 (1949)
[45] J.J. Baumberg, D.D. Awschalom, N. Samarth, J. Appl. Phys. **75**, 6199 (1994)
[46] N.I. Zheludev, M.A. Brummell, A. Malinowski, S.V. Popov, R.T. Harley, Solid State Commun. **89**, 823 (1994)
[47] A.G. Aronov, Pis'ma Z. Eksp. Teor. Fiz. **24**, 37 (1976); Sov. Phys. JETP Lett. **24**, 32 (1976)
[48] R.H. Silsbee, Bull. Mag. Res. **2**, 284 (1980)
[49] M. Johnson, R.H. Silsbee, Phys. Rev. Lett. **55**, 1790 (1985)
[50] M. Dobers, K. von Klitzing, J. Schneider, G. Weinmann, K. Ploog, Phys. Rev. Lett. **61**, 1650 (1988)

2

Spin Dynamics of Free Carriers in Quantum Wells

R.T. Harley

2.1 Introduction

In this chapter we describe the spin-dynamics of free carriers in quantum wells, concentrating on zinc-blende structure semiconductors. The spin properties of electrons in silicon are discussed in Chap. 7. The basic question in this chapter is: how does a spin polarized carrier population return to equilibrium? We shall also specialize it to exclude effects of spatial nonuniformity; so we are not talking of spin diffusion or transport, an exciting field covered in Chaps. 8, 9, and 10.

The most important feature of the quantum wells to be discussed here is that their constituents have direct band gaps so that optical excitation and probing directly reveal spin phenomena. In Sect. 2.2 we approach the topic from an experimental viewpoint discussing optical measurements and what they can be expected to reveal. The sections which follow describe experiments on the spin-dynamics of electrons and of holes relating them to theoretical mechanisms. This leads to a discussion of methods for engineering and controlling the spin phenomena in quantum wells and finally we look to the future of research in the area.

Apart from the intrinsic interest of this small piece of condensed matter physics, one of its remarkable features is the way that it continues to throw surprises at its aficionados. Perhaps the most obvious is how, under the frequently dominant Dyakonov–Perel spin relaxation mechanism, strong scattering of the electron momenta equates to weak spin relaxation; but there are other surprises.

2.2 Optical Measurements of Spin Dynamics

Almost all of the experiments on spin dynamics in semiconductors described in the *Optical Orientation* book [1] were based on measurements of photoluminescence under continuous wave excitation and these techniques are still useful, not least because they are relatively simple. More recent experiments have used time-resolved techniques, based on mode-locked lasers, as are described in some detail in Chap. 5.

In the continuous wave methods the sample is excited with circularly polarized light which, due to the optical selection rules (see Chap. 1), results in generation of spin-polarized populations of electrons and holes with spins oriented along the exciting beam. The photoluminescence is measured in the backward direction, again along the axis of the incident beam. A dynamic equilibrium is set up between excitation, spin evolution or relaxation and recombination and the degree of circular polarization of the photoluminescence is a direct measure of the degree of spin polarization of the photo-excited populations. On the assumption of exponential relaxation times for the spins (τ_s) and for recombination (τ_r), the degree of polarization of photoluminescence is determined by the ratio τ_s/τ_r. Thus if the recombination time is known independently, the spin relaxation time can be extracted. An example of such measurements is discussed in Sect. 2.6.

The Hanle effect is a development of this method [1]. A magnetic field is applied perpendicular to the exciting beam, at right angles to the initial direction of the spins, causing the spins to precess at the Larmor frequency in a plane perpendicular to the field axis. This produces a new dynamic equilibrium in which excitation, recombination, spin relaxation and spin precession are balanced. The measured polarization now depends on three times; τ_s, τ_r and the Larmor precession period. The precession reduces the average spin component along the excitation direction, so reducing the polarization of the photoluminescence. In the simplest case, the polarization is theoretically a Lorentzian function of field. Such a curve is specified by two experimentally accessible parameters, width and height. Thus it is necessary to know one of the three times in order to determine the other two. As an example, the Hanle effect, in combination with independent determination of τ_r, was used to make the first measurements of electron Landé g-factor in quantum wells [2]. Other examples are discussed in Sects. 2.4 and 2.5.2.

In time-resolved techniques, pulsed excitation followed by measurement of the intensity and polarization of the transient photoluminescence gives directly the spin evolution of the photo-excited populations. Time-resolved pump-and-probe methods involve pulsed excitation followed by the measurement of the pump-induced change of transmission (known as time-resolved Faraday effect) or reflection (Kerr effect) using a weak, delayed probe pulse [3]. These measurements are indirect in the sense that some nonlinear-optical modification of the optical constants determines the signal. Fortunately, among the possible nonlinear mechanisms the most important in a quantum well is 'phase-space-filling'; as a result of the Pauli principle, states in the conduction or valence bands which are occupied by photo-excited carriers are not available for optical transitions so that the techniques are effectively sensitive directly to the photo-excited populations [4, 5]. In some experiments two separate measurements are made, one with the probe having the same circular polarization as the pump and the other the opposite polarization. A more sensitive technique is to use a linearly polarized probe in which case the rotation of its plane measures the difference of pump-induced populations.

Interpretation of the signals from any of the experiments is quite subtle and often rapidly glossed over in the literature. Interband excitation may produce excitons or

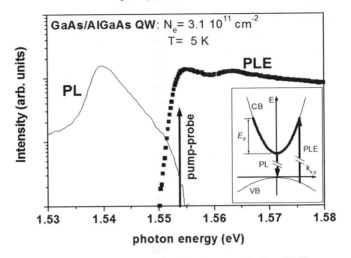

Fig. 2.1. Photoluminescence (PL) and photoluminescence excitation (PLE) spectra of a degenerate n-type quantum well and (inset) the conduction and valence band structure with arrows indicating the optical transitions. The arrow marked pump–probe shows the photon energy of a typical time-resolved Faraday or Kerr effect measurement [6]

free holes and electrons, and their contributions will depend on many factors such as temperature, excitation energy and the sample—whether intrinsic, n-type or p-type, etc. In Sect. 2.5.1 we discuss the interplay of excitons and free carriers in measurements on intrinsic quantum wells.

Figure 2.1 illustrates the subtleties for one particular situation; a degenerate n-type quantum well [6]. It shows the absorption (PLE) and emission (PL) spectra and (inset) the band structure with arrows indicating the peak of the photoluminescence and the onset of photoluminescence excitation which are separated by a 'Stokes' shift determined by the depth of the Fermi sea (E_F). A pump–probe measurement with low intensity, circular polarized optical excitation at the onset of photoluminescence excitation, as indicated, will measure the phase-space-filling effect of the photo-excited electrons and holes. This will be dominated by spin-evolution of the electrons near the Fermi level because the holes will rapidly thermalize into states near to the zone-center, under the 'umbrella' of the Fermi sea of filled electron states, leaving the initially-occupied valence band states empty. Time-resolved or continuous wave photoluminescence measurements will involve polarized excitation at (or above) the photoluminescence excitation onset and detection of the emission near the maximum of the photoluminescence spectrum. The degree of polarization of the photoluminescence depends in a quite complicated way on the polarization of the photo-excited holes and the net polarization of the electron Fermi sea when recombination takes place [1, 7–9].

2.3 Mechanisms of Spin Relaxation of Free Electrons

Three mechanisms are known for spin relaxation of free conduction electrons in non-magnetic zinc-blende semiconductors [1]. Two usually make minor contributions and consequently have not been extensively investigated; they are spin–flips accompanying electron scattering as a result of spin–orbit coupling, known as the Elliott–Yafet (EY) mechanism [10, 11] and spin–flips induced by exchange interaction with an unpolarized population of holes, known as the Bir, Aronov, and Pikus (BAP) [12] mechanism. In both these mechanisms the spin relaxation rate is proportional to the electron scattering rate. The third mechanism, the Dyakonov–Perel (DP) mechanism [13], is the most important. In bulk material there is clear evidence for involvement of both Bir–Aronov–Pikus and Elliott–Yafet mechanisms, as discussed in Sect. 2.4, but in quantum wells there is very little evidence that mechanisms other than Dyakonov–Perel are important.

In the Dyakonov–Perel mechanism, as described in Chap. 1, the driving force for spin reorientation is the spin–orbit splitting of the conduction band which gives an intrinsic tendency of electron spins to precess in flight between scattering events with an effective, momentum-dependent, Larmor vector Ω_p whose magnitude corresponds to the conduction band spin-splitting. Scattering of the momentum of the electron randomizes the precession and, in the strong scattering limit, spin reorientation proceeds as a series of random fractional rotations. The spin relaxation rate along a particular axis, i, turns out to be

$$1/\tau_{s,i} = \langle \Omega_\perp^2 \rangle \tau_p^*, \qquad (2.1)$$

where $\langle \Omega_\perp^2 \rangle$ is the mean square precession vector in the plane perpendicular to i and τ_p^* is the momentum relaxation time of an individual electron [13, 14]. This formula incorporates the counterintuitive, 'motional slowing' feature of the Dyakonov–Perel mechanism whereby the spin relaxation rate is inversely proportional to the electron scattering rate.

The presence of the two factors in (2.1) complicates the interpretation of experimental measurements. The spin splitting can be calculated theoretically; its dependence on momentum is determined to a large extent by symmetry considerations and its magnitude can also be calculated for example by the $\mathbf{k} \cdot \mathbf{p}$ method [15]. On the other hand the scattering time tends to be sample-dependent, being determined in part by the degree of perfection of the structure. Alas, all too few experiments have characterized the scattering time well enough for really hard conclusions to be drawn.

The scattering represented by the rate, $1/\tau_p^*$, randomizes the momentum of an individual electron and is the sum of two rates

$$1/\tau_p^* = 1/\tau_p + 1/\tau_{ee}, \qquad (2.2)$$

where $1/\tau_{ee}$ is due to electron–electron scattering and $1/\tau_p$ is due to all other processes such as scattering from imperfections and phonons. The latter processes exchange momentum between the electrons and the crystal lattice and so determine the mobility of an ensemble of electrons. Electron–electron scattering becomes significant in n-type semiconductors. It redistributes momentum within an ensemble of electrons but does not contribute directly to the mobility.

For years it was stated in the literature that $1/\tau_p^*$ was the same as $1/\tau_p$ [14] and could therefore be determined experimentally from the mobility; no one seemed to notice the potential importance of electron–electron scattering. So it was a surprise when recent theoretical papers and experiments in two-dimensional electron gases, described in Sect. 2.5.4, highlighted the importance of electron–electron scattering [16–19].

The vector Ω_p arises from combined effects of spin–orbit coupling and inversion asymmetry. In bulk material there is just one contribution to Ω_p, the Dresselhaus or bulk inversion asymmetry (referred to in the literature as BIA) which arises because the zinc-blende structure lacks inversion symmetry. What makes spin dynamics in quantum wells much richer is the existence of two other contributions together with possibilities for playing one off against the others. The extra contributions are, the Rashba or structural inversion asymmetry (SIA) due to inversion asymmetry of the quantum well potential including electric fields, built-in or externally applied, and the natural interface asymmetry (NIA) due to asymmetry associated with the chemical bonding within interfaces.[1]

The various contributions to Ω_p have different dependences on the components of momentum as was mentioned in Chap. 1. In bulk material, where only the bulk inversion asymmetry exists,

$$\underline{\Omega}^{\text{BIA}} = \frac{\gamma}{\hbar^3}\left\{p_x\left(p_y^2 - p_z^2\right), \, p_y\left(p_z^2 - p_x^2\right), \, p_z\left(p_x^2 - p_y^2\right)\right\}, \qquad (2.3)$$

where γ is known as the Dresselhaus coefficient and x, y, and z are the cubic crystal axes. Thus $|\Omega|^2 \sim |p|^6$ and since the kinetic energy of an electron $\sim p^2$, when the average in (2.1) is taken at temperature T we get

$$1/\tau_s \sim (k_B T)^3 \tau_p^* \qquad (2.4)$$

if the electrons are nondegenerate, where k_B is Boltzmann's constant, or

$$1/\tau_s \sim (E_F)^3 \tau_p^* \qquad (2.5)$$

for a degenerate electron gas.

In a quantum well the momentum component along the growth axis is quantized and the z-components are replaced by their expectation values $\langle p_z \rangle = 0$ and $\langle p_z^2 \rangle$. The latter is proportional to the electron confinement energy E_{1e}; the leading term is obtained from (2.3) by assuming $\langle p_z^2 \rangle$ is much greater than p_x^2 and p_y^2 (see Chap. 1). Then if we transform to 'laboratory' axes [20] in which the growth axis is z and the principal axes in the quantum well plane are x and y, the leading terms are

---

[1] The different contributions to Ω_p are well reviewed by Flatté et al. in [14].

$$\underline{\Omega}^{\text{BIA}} = \frac{\gamma}{\hbar^3}\langle p_z^2\rangle\{-p_x, p_y, 0\} \quad \text{for [001] growth axis,} \tag{2.6}$$

$$\underline{\Omega}^{\text{BIA}} = \frac{\gamma}{2\hbar^3}\langle p_z^2\rangle\{0, 0, p_y\} \quad \text{for [110] growth axis,} \tag{2.7}$$

$$\underline{\Omega}^{\text{BIA}} = \frac{2\gamma}{\hbar^3\sqrt{3}}\langle p_z^2\rangle\{p_y, -p_x, 0\} \quad \text{for [111] growth axis.} \tag{2.8}$$

Notice that for both [001] and [111] growth Ω_p is oriented in the quantum well plane so that the bulk inversion asymmetry will cause Dyakonov–Perel spin relaxation along the growth axis. For [110] grown wells Ω_p is along z so that there is no Dyakonov–Perel spin relaxation along the growth axis. We shall return to these features in Sect. 2.5.

Since $\langle p_z^2\rangle$ is proportional to the quantum confinement energy E_{1e}, we obtain, for $E_{1e} > k_B T$,

$$1/\tau_s \sim (k_B T)E_{1e}^2\tau_p^* \tag{2.9}$$

for non-degenerate electrons and, assuming $E_{1e} \gg E_F$,

$$1/\tau_s \sim (E_F)E_{1e}^2\tau_p^* \tag{2.10}$$

for a degenerate Fermi sea.

In a wide quantum well at relatively high temperature we may have the condition $E_{1e} \leq k_B T$. This means that the thermal average kinetic energy of an electron in the plane becomes comparable to or greater than the electron confinement energy; in other words $\langle p_x^2\rangle$ and $\langle p_y^2\rangle$ become comparable to $\langle p_z^2\rangle$. Referring to (2.3) we see that (2.9) must now be replaced by

$$1/\tau_s \sim \left[(k_B T)E_{1e}^2 + (k_B T)^3\right]\tau_p^*. \tag{2.11}$$

Measurements which show this effect are described in Sect. 2.5.1.

The structural inversion asymmetry arises from inversion asymmetry of the quantum well potential and vanishes in a symmetrical quantum well. It is typically produced by application of an external electric field or by asymmetry of doping which generates a built-in electric field. The spin splitting is an example of the Rashba effect for which $\Omega = \beta F \times p$ where F is the electric field and β is known as the Rashba coefficient. For field along the z axis this becomes, for all growth orientations,

$$\underline{\Omega}^{\text{SIA}} = \frac{\beta F}{\hbar}\{p_y, -p_x, 0\}. \tag{2.12}$$

Asymmetry of the potential may also be generated by asymmetric variation of the alloy composition in the wells or barriers. The simultaneous presence of both bulk inversion asymmetry and structural inversion asymmetry can produce interesting interference effects, as described in Sects. 2.5.2 and 2.7, but if the structural inversion asymmetry dominates, the corresponding spin relaxation rates are

$$1/\tau_s \sim (k_B T)\tau_p^* \tag{2.13}$$

for non-degenerate electrons and

$$1/\tau_s \sim (E_F)\tau_p^* \qquad (2.14)$$

for degenerate electrons.

The natural interface asymmetry component is discussed in more detail in Sect. 2.5.3. It is zero if the well and barrier have anions or cations in common and for some crystallographic orientations, notably [110].

It is important to note that the expressions for Ω_p quoted here are leading terms in a perturbation expansion. Lau et al. [21] have carried out non-perturbative calculations for a number of heterostructure systems and shown that, particularly at high temperatures the simple perturbation theory is inadequate. We shall see an example of this in Sect. 2.5.1.

2.4 Electron Spin Relaxation in Bulk Semiconductors

Although this chapter is mainly about quantum wells it is appropriate to include a short discussion of recent studies of electron spin dynamics in bulk material because it is in bulk that interplay of the Dyakonov–Perel with the Elliott–Yafet and Bir–Aronov–Pikus mechanisms has been observed. Early work was reviewed in the *Optical Orientation* book [1].

Both the Dyakonov–Perel and the Elliott–Yafet mechanisms rely on spin orbit coupling. The Elliott–Yafet is, in general, not only intrinsically weaker than the Dyakonov–Perel mechanism but since it gives spin-relaxation rate $1/\tau_s$ proportional to the electron momentum scattering rate, $1/\tau_p^*$, it becomes less significant where scattering is minimized for example in high quality samples. On the other hand, spin reorientation via precession, as in the Dyakonov–Perel mechanism, becomes more efficient for weak momentum scattering.

The Elliott–Yafet mechanism can become important in particular circumstances for example in narrow gap bulk material where the spin–orbit interaction is large; Murzyn et al. [22] found that in bulk InAs the spin relaxation time measured at 300 K increased from 20 ps to 1600 ps in going from intrinsic to weakly degenerate n-type material, while in InSb there was a corresponding increase from 16 ps to 600 ps. The interpretation of this surprising increase is that the Dyakonov–Perel mechanism is dominant and the increased spin relaxation time, in going from intrinsic to n-type material, corresponds to an increased momentum scattering rate of the electrons. The increase is presumably due to electron–electron scattering. Furthermore in the n-type samples, the Dyakonov–Perel spin relaxation is so completely suppressed that the rate is actually limited by the Elliott–Yafet mechanism.

The Bir–Aronov–Pikus mechanism for electron spin relaxation requires a significant concentration of holes. The main studies of it have been made in p-type bulk material; Fig. 2.2 shows the results of a calculation of the relative importance of Bir–Aronov–Pikus and Dyakonov–Perel mechanisms in GaAs and GaSb [1]. This summarizes the basic result that the Bir–Aronov–Pikus mechanism comes into play

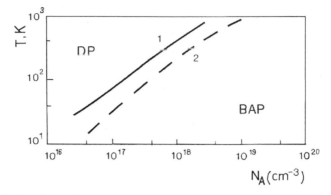

Fig. 2.2. Calculated boundaries of temperature and acceptor (hole) concentration where Dyakonov–Perel (DP) and Bir–Aronov–Pikus (BAP) spin relaxation rates are equal for (*1*) bulk GaAs and (*2*) bulk GaSb [1]

most strongly at low temperatures and at high hole concentrations; the holes may originate from doping or from interband optical excitation.

A remarkable recent observation in *n*-type InSb at 300 K by Murdin et al. [23] demonstrated a competition between the Bir–Aronov–Pikus and Dyakonov–Perel mechanisms. Scattering of electrons from holes might increase the spin relaxation rate if the Bir–Aronov–Pikus mechanism is dominant or reduce it if the Dyakonov–Perel mechanism is more important. They measured an electron spin relaxation rate $1/\tau_s = 26\,\text{ns}^{-1}$ by means of interband excitation, in which holes are photocreated with the electrons and in the same sample they obtained $1/\tau_s = 71\,\text{ns}^{-1}$ using an intraband excitation technique in which no holes are excited. As the presence of holes reduced the relaxation rate, it can be deduced that the Dyakonov–Perel mechanism dominates over the Bir–Aronov–Pikus.

Among recent measurements of electron spin relaxation in bulk materials is the important discovery of a maximum in the spin relaxation time in *n*-type material at low temperatures at a concentration corresponding to the metal insulator transition [24]. Figure 2.3 shows the results from Dzhioev et al. for GaAs [9], similar data have been obtained for GaN [25]. At very low concentrations the electrons are localized and their spin relaxation is determined by a variety of processes, for example fluctuating hyperfine fields, which we do not consider here. At the metal-insulator transition the spin relaxation time reaches a value ∼200 ns and then falls rapidly as $\sim N_D^{-1.7}$. This is as expected for the Dyakonov–Perel mechanism. At the peak, the relaxation time approaches that expected for the Elliott–Yafet mechanism. As temperature is increased the spin relaxation time becomes shorter in a way consistent with the Dyakonov–Perel mechanism [24].

The very long spin relaxation time indicated in Fig. 2.3 has enabled beautiful studies of electron spin transport and manipulation in bulk GaAs using optical techniques particularly by Crooker and coworkers [26].

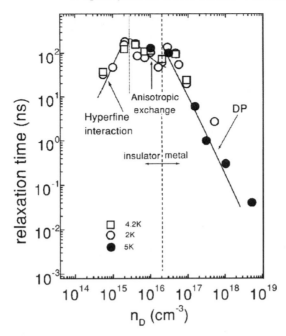

Fig. 2.3. Spin relaxation time at low temperatures for electrons in bulk GaAs. For concentrations below the metal-insulator transition electrons are localized. At higher concentrations the Dyakonov–Perel (DP) relaxation mechanism dominates [9, 24]

2.5 Electron Spin Relaxation in [001]-Oriented Quantum Wells

The standard crystallographic orientation of substrate for growth of quantum wells is [001] and many studies of spin dynamics have used this orientation. However it was first realized theoretically by Dyakonov and Kachorovskii [27] that, for studies of spin, other orientations, for example [110] and [111] are of great interest. Over the past decade experiments have started to catch up with theory and ideas for 'engineering' spin properties of quantum wells through the crystallography are being investigated. Some of these developments are described in Sect. 2.7. In this section we describe measurements on [001]-oriented wells.

2.5.1 Symmetrical [001]-Oriented Quantum Wells

We start by considering spin dynamics in symmetrical [001]-oriented quantum wells. This means structures in which the only asymmetry is the bulk inversion asymmetry and the precession vector is oriented in the plane of the wells (see (2.6)), giving spin-relaxation along the growth axis. Firstly this requires structural inversion symmetry, i.e., there must be no built-in or applied electric field (or other odd-parity external perturbation of the layer structure); and it requires that spatial variations of alloy composition in wells and barriers have inversion symmetry. Secondly the natural

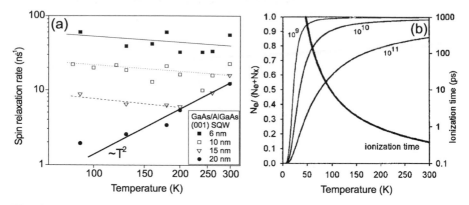

Fig. 2.4. (a) Temperature dependence of electron spin relaxation rate in undoped GaAs/AlGaAs quantum wells of different widths grown sequentially on the same substrate. The lines are to guide the eye. **(b)** Calculations of thermodynamic equilibrium ratio of electron (N_e) and electron plus exciton ($N_e + N_x$) concentrations in GaAs quantum wells for excitation densities of 10^9, 10^{10}, and 10^{11} cm^{-2} and of ionization time for excitons [28]

interface asymmetry must also be absent; we discuss this requirement in more detail in Sect. 2.5.3 but natural interface asymmetry is absent when the well and barrier material share a common atom, as for example in GaAs/AlGaAs structures.

Some representative measurements of electron spin relaxation rates in such samples appear in Fig. 2.4(a) [28] which shows time resolved Kerr measurements of spin relaxation along the growth axis in a sample containing a series of single, symmetrical and undoped GaAs/AlGaAs quantum wells of different widths. The excitation energy corresponds to resonant creation of heavy-hole excitons but, even so, the measured spin evolution in this temperature range is controlled by free electrons for two reasons. First the photo-excited excitons dissociate on a much faster timescale into free holes and electrons and second the photo-excited holes lose their spin memory very fast in this temperature range, in 1 ps or less.

We shall see the justification for the second assertion in Sect. 2.6 and to justify the first Fig. 2.4(b) shows the calculated temperature dependences of the thermodynamic equilibrium between excitons and free carriers and of the dissociation time of excitons. The interesting point from these two calculation is that, although for typical experimental excitation densities, $\sim 10^9$ cm^{-2}, thermodynamic equilibrium favors free carriers at all temperatures above ~ 20 K, the thermal dissociation of the excitons, which is controlled by phonon interactions, is on a nanosecond time scale at this temperature and only drops below 10 ps at about 80 K. Therefore up to this temperature the spin evolution will still be dominated by excitonic effects because thermal equilibrium is not established on the timescale of the spin evolution. Above 80 K the observed spin dynamics are dominated by free electrons.

The temperature dependence of the spin relaxation time Fig. 2.4(a) and its dependence on electron confinement energy at 300 K (Fig. 2.5) indicate the dominance of the Dyakonov–Perel mechanism. For narrow quantum wells where $k_B T < E_{1e}$, we

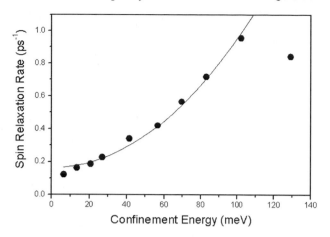

Fig. 2.5. Dependence of electron spin relaxation rate on confinement energy at 300 K for a series of undoped GaAs/AlGaAs quantum wells grown sequentially on the same substrate [28]

expect $1/\tau_s \sim E_{1e}^2 (k_B T)\tau_p^*$ (cf. (2.9)) whereas for wider wells and high temperatures with $k_B T > E_{1e}$ we expect $1/\tau_s \sim [k_B T E_{1e}^2 + (k_B T)^3]\tau_p^*$ (cf. (2.11)).

Figure 2.4(a) shows that, for the narrower wells, the experimental spin relaxation rate is approximately temperature-independent. This is consistent with (2.9) provided that τ_p^* varies as T^{-1}. Indeed this is the variation expected for a non-degenerate electron gas with dominant phonon scattering in a 2D system where the density of states is constant [29]. For the wider wells at higher temperatures the variation becomes approximately T^2, consistent with (2.11) with the same behavior of τ_p^*. Although τ_p^* was not determined in this study, since all the wells were grown sequentially in the same sample, it is likely that the temperature dependence and magnitude is constant. At room temperature (Fig. 2.5) the spin relaxation rate shows an approximate (constant $+ E_{1e}^2$)-dependence again consistent with (2.11). At the highest value of confinement energy the spin relaxation rate falls sharply. This is probably a result of strong interface scattering of the electrons.

Although the temperature and confinement energy dependencies just described support the Dyakonov–Perel mechanism, it turns out that there is a discrepancy between the measured magnitudes of the spin relaxation rate and predictions based on the leading term analysis of (2.9). This was revealed in work by Terauchi et al. [30] who investigated the dependence of electron spin relaxation time in GaAs/AlGaAs quantum wells on both mobility and confinement energy at 300 K. Their results are shown in Fig. 2.6. These samples have comparatively low mobility and electron concentration so electron–electron scattering should be weak making $\tau_p^* = \tau_p$ (cf. (2.2)). The calculated relaxation times are about a factor ten too short and the dependences on both mobility and confinement are not as predicted. These discrepancies have been explained by Lau et al. [21] (see also [14]) on the basis of a more refined version of the Dyakonov–Perel mechanism; they used a nonperturbative method to calculate $\langle \Omega^2 \rangle$ which removed the order of magnitude discrepancy and reproduced

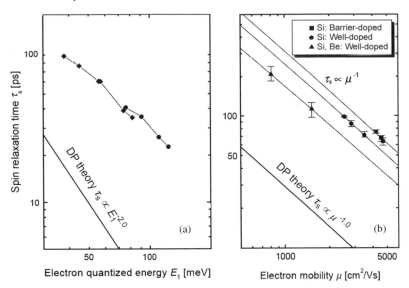

Fig. 2.6. Electron spin relaxation times at 300 K as a function of **(a)** confinement energy and **(b)** mobility in a variety of GaAs/AlGaAs quantum well samples. The Dyakonov–Perel theory calculation is based on the simple, leading-term perturbation calculation outlined in (2.9) [30]

the observed dependence on confinement energy (Fig. 2.6(a)). They were also able to interpret the departures from strict μ^{-1} behavior in Fig. 2.6(b) in terms of different angular dependences for different scattering mechanisms in the samples [14].

2.5.2 Structural Inversion Asymmetry in [001]-Oriented Quantum Wells

In a quantum well grown on a [001]-oriented substrate the bulk inversion asymmetry component of Ω_p is oriented in the well plane and its magnitude is isotropic to lowest order (see (2.6)). This not only produces Dyakonov–Perel relaxation for spins along the growth direction but also isotropic spin-relaxation for electron spins oriented in the plane.[2] If there is a structural inversion asymmetry component it also is oriented in-plane but it has different momentum dependence from the bulk inversion asymmetry component (see (2.6) and (2.12)); the resultant precession vector is anisotropic. Figure 2.7(a) illustrates, in a radial plot, the effect of adding (2.6) and (2.12), with the arrows indicating the orientation of the precession vector. The anisotropy of the plot reflects the fact that the additional asymmetry of the quantum well potential removes the fourfold rotation/reflection symmetry (S_4) of the quantum well reducing the point group from D_{2d} to C_{2v}.

The anisotropy of the conduction band spin splitting in a quantum well was first measured directly by Raman scattering by Jusserand et al. [31] and the anisotropy

---

[2] If terms in p_x^2 and p_y^2 are retained in (3.6a) Ω_{BIA} becomes anisotropic with four-fold symmetry in the plane.

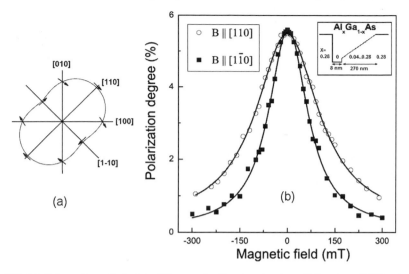

Fig. 2.7. (a) Schematic radial plot of the conduction band spin splitting in a [001]-oriented quantum well with combined bulk inversion and structural inversion asymmetry terms. The axes are p_x and p_y and the curve represents the splitting for a given total in-plane momentum. The arrows indicate the direction of the precession vector. When $\frac{\gamma}{\hbar^2}\langle p_z^2\rangle = -\beta F$ (see (2.6) and (2.12)) the splitting for momentum in the $(1\bar{1}0)$ direction vanishes and the precession vector lies along that axis for all p_x and p_y. (b) Hanle depolarization curves measured at 80 K in an asymmetrically grown GaAs/AlGaAs quantum well as shown in the right inset. The different widths of the curves indicate different spin relaxation times for spins along the [110] and $(1\bar{1}0)$ directions [34]

can also be observed in transport measurements (see Chap. 9). Averkiev and Golub [32] pointed out that since the components of Ω_p along the [110] and $[\bar{1}10]$ directions are different (see arrows in Fig. 2.7(a)) the anisotropy also implies that the spin relaxation rates along these two directions should be different. Furthermore, if the electric field is set so that $\frac{\gamma}{\hbar^2}\langle p_z^2\rangle = -\beta F$ the [110] component of Ω_p would vanish for all values of p_x and p_y and therefore the spin relaxation rate along $[\bar{1}10]$ would vanish.

To measure the in-plane spin relaxation rate requires a spin polarization in the plane. This can be achieved by the standard circular polarized optical excitation normal to the plane with magnetic field applied in the plane. The spins are thus first oriented along the growth direction, then precess through the in-plane direction and consequently sense the in-plane relaxation rate for 50% of their evolution.

Two schemes for achieving the possible cancellation have been proposed. The first involves using degenerate n-type modulation doping of the well [32]. The doping asymmetry produces a built-in electric field which determines the structural inversion asymmetry (SIA) term (βF) while the quantum well width and doping density determine the bulk inversion asymmetry (BIA) contribution ($\frac{\gamma}{\hbar^2}\langle p_z^2\rangle$). Very recently progress on this has been reported by Stich et al. [33] using the time-resolved Faraday

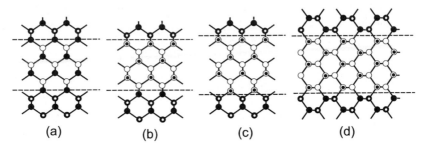

Fig. 2.8. Illustrating the origin of the natural interface asymmetry in quantum wells grown on [001] substrates (**a**)–(**c**) and on a [011] substrate (**d**). (**a**) represents a common-atom system and (**b**)–(**d**) no-common-atom systems. (**a**) and (**d**) have no natural interface asymmetry because the top and bottom interfaces are equivalent. After [37]

rotation technique, showing an SIA/BIA ratio of 0.65 in a GaAs/AlGaAs modulation doped quantum well. The second involves built-in asymmetry of the quantum well by spatial grading of the alloy composition at one interface while maintaining the other interface abrupt [34]. Figure 2.7(b) indicates the conduction band profile for the sample (inset) and (main figure) the Hanle depolarization curves for magnetic field oriented in the plane along [110] and [1$\bar{1}$0]. For these two field directions the electron spins precess in orthogonal planes and thereby sample the two perpendicular in-plane spin relaxation rates. There is a clear difference indicating that the SIA/BIA ratio is $\sim$4 [34]. (Although it seems obvious that the source of asymmetry in the experiment is the graded interface, analysis by Winkler [15, 35] appears to contradict this interpretation, suggesting that the structural inversion asymmetry spin splitting associated with alloy variations alone is very small.)

2.5.3 Natural Interface Asymmetry in Quantum Wells

The natural interface anisotropy (NIA) is a form of inversion asymmetry which results from the structure of chemical bonding at the interfaces of quantum wells. It was first noted by Krebs and Voisin [36] and has received relatively low-level attention since, mainly because the 'fruit fly' material systems such as GaAs/AlGaAs, do not show the effect. As with the structural inversion asymmetry, the effect of natural interface asymmetry is to reduce the quantum well symmetry to C_{2v} and to make the precession vector anisotropic, as in Fig. 2.7(a). Here we outline the basic idea and describe recent experiments which demonstrate natural interface asymmetry particularly clearly.

Figure 2.8 illustrates schematically the atomic structure of various zinc blend quantum wells. In Figs. 2.8(a)–(c) we are looking along the [110] axis of a quantum well, grown from bottom to top of the diagram, on a [001]-oriented substrate. In Fig. 2.8(b) and (c) the well and barrier are depicted with no common atom; barriers are black and black with white-dot, wells are white with black-dot and white. In Fig. 2.8(b) the lower barrier is terminated with a layer of black atoms and the well has been terminated with a layer of black-dot atoms. The interfaces are indicated by

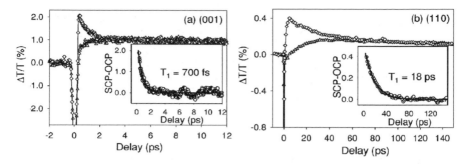

Fig. 2.9. Time-resolved Faraday signals at 115 K for InAs/GaSb superlattices grown on (**a**) [001]- and (**b**) [110]-oriented substrates. The insets show the difference of measurements with same-circular and opposite-circular probe polarization with respect to the pump (SCP-OCP) and indicate the decay of electron spin polarization. The 25-fold increase of spin relaxation time for [110] is due to the absence of natural interface asymmetry for that growth orientation [37]. Copyright by American Physical Society (2003)

dashed lines. The growth in Fig. 2.8(c) is the same except that the lower barrier has been terminated with a layer of white-dot atoms; again the interfaces are indicated with dashed lines. In both these examples it can be seen that the interface structure removes the S_4 symmetry operation and so the natural interface asymmetry will be important. In Fig. 2.8(a) the barriers are again black and white-dot but now the well is black and white giving the structure a common atom (black). In this case each interface coincides with a layer of black (i.e., the common) atoms and the structure retains the S_4 symmetry and has no natural interface asymmetry. Figure 2.8(d) shows the case of a quantum well grown along the [110] direction and viewed along [1$\bar{1}$0]. In this case the interfaces contain both anions and cations and, even when there is no common atom no natural interface asymmetry is expected.

Of course, in practice interfaces between layers cannot be grown as perfectly as depicted in Fig. 2.8 and there will always be a degree of disorder which tends to reduce the natural interface asymmetry effect. Thus reliable calculations of the magnitude of the effect are difficult. Another point is that, since it is an interface effect, whereas the bulk inversion asymmetry is a bulk effect, the natural interface asymmetry will be most significant in narrow quantum wells.

One particularly clear experimental demonstration of natural interface asymmetry is illustrated in Fig. 2.9 which depicts time-resolved circular-polarized transmission measurements at 115 K on InAs/GaSb superlattice samples grown on both [001] and [110] oriented substrates [37]. In these measurements the sample is excited with a circularly polarized pump pulse and transmission of a delayed probe pulse is measured either having the same or opposite circular polarization as the pump. The difference of these two signals (inset) represents the decay of electron spin polarization in the sample. For the [001]-grown sample the spin relaxation time is 700 fs; this very short time can only be understood in terms of a dominant contribution from the natural interface asymmetry in this sample. The interpretation is supported by the 25

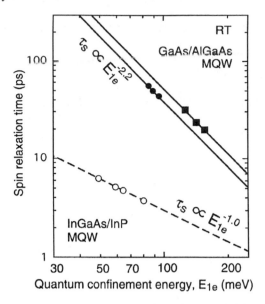

Fig. 2.10. Comparison of electron spin relaxation times for GaAs/AlGaAs and InGaAs/GaAs quantum wells at 300 K. The large decrease in relaxation time in the latter system may be due to natural interface asymmetry [38]

times enhancement of the spin relaxation time to, 18 ps, in the [110] sample where natural interface asymmetry will be absent. Actually for a [110] sample with perfect interfaces an enhancement to 600 ps was expected and the lesser observed enhancement was interpreted as due to interface roughness on the scale of one monolayer fluctuations [37].

The original paper by Krebs and Voisin [36] considered [001]-grown InGaAs/InP quantum wells. There should be a natural interface asymmetry contribution to the spin splitting of the conduction band because, although both well and barrier contain indium, the 'average' atom (InGa) is, for this purpose, different from (In). Figure 2.10 shows spin relaxation times for GaAs/AlGaAs and (InGa)As/InP undoped quantum wells at 300 K, measured as functions of the electron confinement energy E_{1e} [38]. The variation for GaAs/AlGaAs is $E_{1e}^{-2.2}$, very close to that expected for the Dyakonov–Perel mechanism (2.9). The spin relaxation times for (InGa)As/InP are at least a factor ten shorter and show a weaker dependence on confinement energy. Taking into account the differences of effective mass and band gap between the two material systems, without natural interface asymmetry the relaxation time in (InGa)As/InP should be shorter than in GaAs/AlGaAs by a factor $\sim$2.5 at a given confinement energy and its dependence on confinement energy should be the same [38]. Thus it seems possible (but not conclusive) that the extra reduction is a result of natural interface asymmetry in (InGa)As/InP [14].

2.5.4 Oscillatory Spin-Dynamics in Two-dimensional Electron Gases

It is almost always safe to assume that the Dyakonov–Perel spin dynamics is in a strong scattering regime, characterized by $\Omega_\perp \tau_p^* \ll 1$ (2.1) where the electron spins precess through small angles between momentum scattering events and the resulting spin evolution is an exponential decay. This turns out not to be true in two dimensional electron gases at low temperatures allowing investigation of the weak scattering regime as well as the transition between weak and strong regimes [6, 19, 39].

In a degenerate electron gas the Pauli principle inhibits electron–electron scattering and at absolute zero the electron–electron contribution to $1/\tau_p^*$, $1/\tau_{ee}$ (2.2) vanishes. Furthermore the ensemble momentum scattering contribution, $1/\tau_p^*$, can be made small at low temperatures by modulation doping which physically separates the electron gas from the donors. For example in a 10 nm GaAs/AlGaAs quantum well, mobility $\sim 10^6$ cm^2 V^{-1} s^{-1} is possible, corresponding to $\tau_p = 38$ ps. If the two dimensional electron gas has a typical density, N_s, say $\sim 3 \times 10^{11}$ cm^{-2}, we can estimate (2.6) the magnitude of the conduction band spin splitting at the Fermi momentum to be ~ 0.2 ra ps^{-1}, giving $|\Omega|\tau_p \sim 7.6$ which is comfortably in the weak scattering regime.

Thus we expect that at very low temperatures the electron spin will, on average, precess through several full cycles before being scattered and therefore that such a two dimensional electron gas will show oscillatory spin evolution with decay time $\sim \tau_p$ (Sect. 1.4.2). As the temperature is increased, $1/\tau_p^*$ will increase due to increases of both electron-phonon and electron–electron scattering, the evolution will become exponential and the decay time will increase corresponding to (2.1).

Figure 2.11(a) shows experimental time-resolved Kerr data for a two dimensional electron gas in a quantum well with nominal width 10.2 nm and mobility $\sim 0.34 \times 10^6$ cm^2 V^{-1} s^{-1} [6]. In these measurements the photo-excited electron population is very low, about one percent of the density of the two dimensional electron gas. At 5 K (inset) the spin evolution is a damped oscillation and for higher temperatures it becomes exponential with an increasing decay time. The degeneracy temperature of the two dimensional electron gas is ~ 129 K and it is possible to estimate that the value of $\langle \Omega^2 \rangle$ *increases* slightly over this range of temperatures. Thus the oscillatory behavior at low temperatures and the increase of spin relaxation time with temperature is a clear example of the 'motional slowing' which is integral to the Dyakonov–Perel mechanism. Calculation shows [6] that it is electron–electron scattering which produces the rapid increase of spin relaxation time with temperature which actually reaches a peak near the Fermi temperature.

Figure 2.11(b) illustrates the oscillatory spin evolution for a range of quantum well widths in the weak-scattering regime. The oscillation frequency increases as the width is reduced, corresponding to increasing quantum confinement energy, E_{1e}. The data were analyzed to give the Dresselhaus coefficient, γ, (2.6) as a function of the confinement energy (see inset).

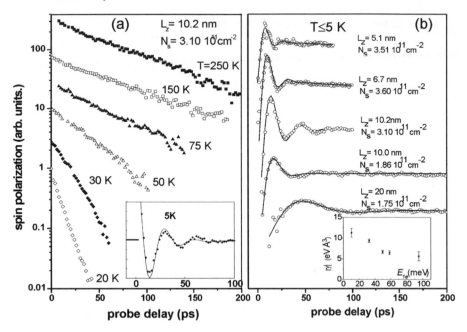

Fig. 2.11. (a) Spin relaxation at a range of temperatures for a small fraction of spin polarized electrons injected at the Fermi energy in a GaAs/AlGaAs two dimensional electron gas in a quantum well with nominal width 10.2 nm. The inset shows the oscillatory spin evolution observed at 5 K in the weak scattering regime [6]. (b) Spin evolution in the weak-scattering regime for five samples with different quantum well width (L_Z) and electron concentration (N_S). From top to bottom the τ_p values given by the mobility are: 13 ps, 13 ps, 13 ps, 10 ps, and 27 ps. Inset: Dresselhaus coefficient γ (2.6) vs. electron confinement energy E_{1e} [39]

The photo-excited population in the experiments just described was deliberately much less than the two-dimensional electron gas density so that the initial spin-polarization of the two-dimensional electron gas was ∼1%. Remarkable many-body effects occur when the initial polarization of the two-dimensional electron gas is comparable to the density [40, 41]. Figure 2.12(a) shows the experimental time-resolved Faraday signals obtained at 4.2 K by Stich et al. [40, 41] from a two-dimensional electron gas sample with well width 20 nm, two-dimensional electron gas density 2.1×10^{11} cm^{-2} and electron mobility 1.6×10^6 cm^2 V^{-1} s^{-1}. As the initial spin polarization is increased up to 30%, not only are the oscillations completely suppressed but there is also a dramatic increase in spin decay time. Figure 2.12(b) shows values of the spin relaxation time compared with calculations performed with and without Hartree-Fock contribution to the Coulomb interaction. The latter serves as an effective magnetic field along the z-axis and therefore strongly modifies the spin precession driven by the bulk inversion asymmetry field, which is perpendicular to z.

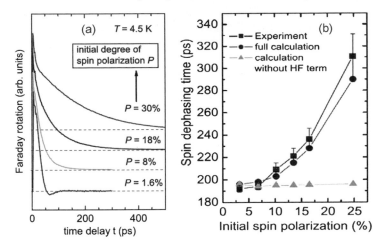

Fig. 2.12. (**a**) Spin evolution in a two dimensional electron gas in a GaAs/AlGaAs quantum well at 4.2 K as a function of the injected polarization P. (**b**) Comparison of experimental spin relaxation (*spin dephasing*) times as function of initial polarization at 4.2 K with calculations including (*circles*) and not including (*triangles*) the Hartree–Fock (HF) term which acts as an effective internal magnetic field stabilizing the spin polarization [40, 41]. Copyright by American Physical Society (2007)

2.6 Spin Dynamics of Free Holes in Bulk Material and Quantum Wells

The spin relaxation of holes in bulk material is expected to be much more rapid than for electrons because of the strong mixing of light and heavy hole valence bands, as explained in Sect. 1.4. This was verified experimentally in GaAs by Titkov et al. in 1978 [1, 42]; using Hanle measurements on weakly n-type material (10^{15} cm^{-3}) they measured 4 ps at 1.5 K. They also found that application of uniaxial stress along [100] increased the spin relaxation time to $\sim$100 ps. The latter effect is consistent with stress-induced splitting, and consequent decoupling of the heavy and light hole bands at the center of the Brillouin zone (zero momentum).

In early experimental studies on quantum wells it was also assumed [43] that the spin-relaxation of holes would be fast. It was therefore a surprise to find that, at low temperatures, the circular polarization of photoluminescence from n-type GaAs/AlGaAs quantum wells under circularly polarized optical excitation could approach $\sim$100% [7, 8, 43]. In such a measurement the polarization should be controlled by the photo-excited holes, implying a very long hole spin relaxation time. Furthermore, time-resolved photoluminescence measurements gave apparently contradictory values for the heavy hole spin relaxation time in GaAs/AlGaAs quantum wells; $\sim$1 ns at 4 K [44] and $\sim$4 ps at 10 K [45].

It was eventually pointed out theoretically in about 1990 by Uenoyama and Sham [46] and by Ferreira and Bastard [47] that spin relaxation of heavy holes in quantum wells should be very dependent on momentum. The quantum well potential splits

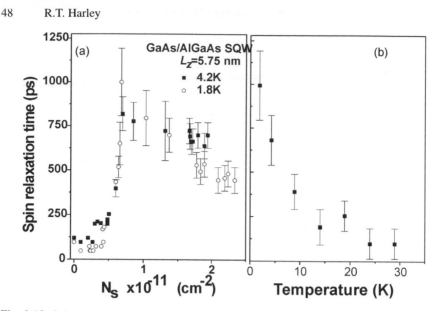

Fig. 2.13. Spin relaxation time for a GaAs/AlGaAs quantum well as a function of electron concentration obtained from measurements of polarization of photoluminescence under polarized continuous wave excitation. (**b**) Temperature dependence of the spin relaxation time at a concentration $\sim 10^{11}$ cm^{-2} [7, 8]

the light and heavy hole bands and decouples them at the zone center. Therefore, at the zone center the spin-relaxation time should be extremely long and limited by processes similar to those of conduction electrons, while at finite momenta it would fall very rapidly due to mixing of light and heavy hole bands. At low temperatures, a thermalized population of holes would occupy states very close to the zone center and therefore have long spin relaxation time. With increasing temperature the average spin relaxation time should fall due to thermal excitation of the holes to higher momenta. (The situation is similar to that of bulk material under uniaxial strain but, remarkably, it took more than a decade after Titkov's work for this theoretical picture to emerge for quantum wells.)

An example of continuous wave polarization measurements is shown in Fig. 2.13(a) which plots the spin relaxation time as a function of electron concentration in a 5.75 nm GaAs/AlGaAs quantum well. (The spin relaxation time was deduced from the polarization using a separate measurement of the photoluminescence decay time [7, 8].) The electron concentration was varied by means of externally applied bias voltage between the substrate and an electrode on top of the structure. At very low concentrations the electrons and holes form excitons which have spin-relaxation times below 100 ps. Above a density $\sim 5 \times 10^{10}$ cm^{-2}, corresponding to screening-out of excitonic binding, the relaxation time increases dramatically. Figure 2.13(b) shows the temperature dependence of the spin relaxation time at a concentration $\sim 10^{11}$ cm^{-2}. This can be interpreted as the spin relaxation time of the heavy holes showing the expected strong temperature dependence.

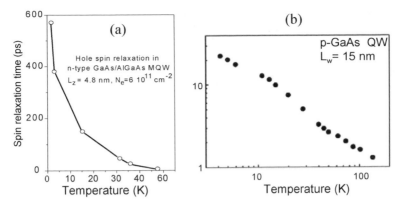

Fig. 2.14. Spin relaxation time as a function of temperature for holes in GaAs/AlGaAs quantum wells of different width. The data in (**a**) are obtained in an *n*-type sample with width 4.8 nm by interband time-resolved photoluminescence measurements [48]. The data in (**b**) are obtain by infrared intraband excitation in a *p*-type well with width 15 nm [49]

Figure 2.14(a) shows the results of a definitive time-resolved photoluminescence investigation of hole spin relaxation in 4.8 nm wide GaAs/AlGaAs quantum wells by Baylac et al. [48] confirming the strong temperature dependence. Figure 2.14(b) shows data for 15 nm GaAs/AlGaAs quantum wells obtained using intraband infrared excitation and thereby avoiding photo-excitation of electrons [49].

The details of the spin relaxation mechanism for the holes in these experiments is not clear. The continuous wave and time-resolved photoluminescence measurements were both performed on relatively narrow quantum wells and give spin relaxation times of order 500 ps to 800 ps at 4.2 K. The intraband measurement was made on a much wider well and gave a spin-relaxation time more than an order of magnitude faster, between 20 ps and 30 ps at 4.2 K. In order to distinguish the relative importance of precessional, Dyakonov–Perel-like, from spin–flip scattering mechanisms, it would be necessary to make measurements on samples with known hole scattering times. It would also be necessary to investigate the possibility that localization of the holes is affecting the results. Such systematic experiments have yet to be made.

2.7 Engineering and Controlling the Spin Dynamics in Quantum Wells

A number of possibilities for manipulating the Dyakonov–Perel spin dynamics in quantum wells is embodied in (2.6)–(2.8) and (2.12). Averkiev and Golub [32] considered the possible interference between the bulk inversion asymmetry and structural inversion asymmetry components of Ω_p in [100] oriented quantum wells, as described above in Sect. 2.5.2. Recently Cartoixa et al. [20] have pointed out that for the case of [111] oriented quantum wells the bulk inversion asymmetry and structural inversion asymmetry terms, (2.8) and (2.12), have the same momentum dependence

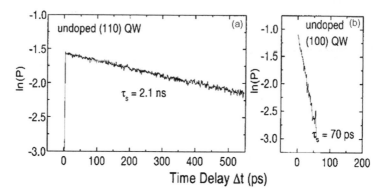

Fig. 2.15. Electron spin relaxation signals in (**a**) [110]-oriented and (**b**) [100]-oriented GaAs/AaGlAs quantum well samples at 300 K, demonstrating the dramatic enhancement of spin relaxation time for spins along the growth axis in [110]-grown wells [50]. Copyright by American Physical Society (1999)

but their coefficients may be varied independently. One possibility is to engineer the structure so that $(2/\sqrt{3})\frac{\gamma\langle p_z^2\rangle}{\hbar^2} = -\beta F$. In this case, all components of Ω_p vanish and therefore the Dyakonov–Perel spin relaxation rate will 'vanish' for all spin orientations. In principle $\langle p_z^2\rangle$ can be adjusted by varying the electron confinement via the well width and βF can be varied by built-in asymmetry and/or by applied electric field. Actually (2.6)–(2.8) and (2.12) represent only the lowest order term in perturbation expansions of the components of p, so the cancellations of the Dyakonov–Perel mechanism will be only partial. To date no experimental measurements of spin dynamics in [111] oriented quantum wells have been reported.

Dyakonov and Kachorovskii [27] were the first to point out that for [110]-oriented wells the bulk inversion asymmetry vector is oriented along the growth axis for all electron momenta and therefore spin relaxation along that axis vanishes. The first experiments on electron spin dynamics in [110] oriented quantum wells were carried out by Ohno et al. [50] They showed that the spin relaxation time of undoped 7.5 nm GaAs/AlGaAs quantum wells at 300 K increased from ∼70 ps for [100] orientation to ∼2.1 ns for [110] orientation; the data are illustrated in Fig. 2.15. This was a beautiful verification of the theoretical prediction but, in spite of systematic measurements as a function of well width, temperature, and electron mobility, it was not possible to determine the mechanism which limited the spin relaxation time in the [110] case.

Lau and Flatté [51] pointed out that application of electric field along the growth axis in a [110] oriented quantum well would induce an in-plane structural inversion asymmetry component of Ω_p (2.12) and thereby switch the Dyakonov–Perel mechanism back on for the z-component of spin. The effect of applied electric field on the spin relaxation rate in undoped [110] oriented GaAs/AlGaAs quantum wells at 170 K is shown in Fig. 2.16 for a well width of 7.5 nm [52]. The quantum wells form the insulating region of a *pin* structure and the electric field is the sum of the built-in

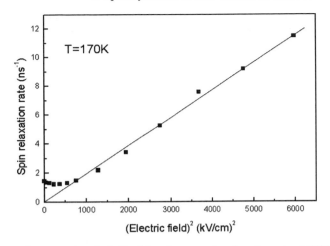

Fig. 2.16. Dependence on electric field of the electron spin relaxation rate at 170 K in a 7.5 nm wide [110]-oriented GaAs/AlGaAs quantum well sample. The data extrapolate through the origin from high fields indicating no significant contribution of non-field-dependent terms in the Dyakonov–Perel relaxation process. After [52]

field and that due to external reverse bias voltage. The data are interesting in several ways. First, for sufficiently high fields the variation is accurately quadratic in field, as expected for the Dyakonov–Perel relaxation mechanism (3.1) with $\langle\Omega^2\rangle$ from the structural inversion asymmetry (2.12). Secondly, the variation extrapolates through the origin, which appears to rule out any field-independent contributions to the spin splitting in the Dyakonov–Perel spin relaxation. This contradicts a suggestion by Karimov et al. [52] that the spin relaxation in zero field for [110]-oriented wells is due to interface roughness and leaves the mechanism of spin relaxation in zero electric field undetermined.

Although the spin relaxation rate along the growth axis in a [110] oriented quantum well is very low, spins oriented in the plane will be subject to the component of Ω along the growth axis and consequently should show much higher rates. This was verified by Döhrmann et al. [53] by measuring the decay rate of spin beats for magnetic fields applied in-plane as described in Chap. 5.

2.8 Conclusions

In this chapter we have made a survey of spin dynamics of free carriers in quantum wells, selecting a few pieces of data to discuss from the very large literature. The emphasis has been on experiments which reveal the mechanisms of spin dynamics and test our understanding of them. In quantum wells this turns out to mean investigating the Dyakonov–Perel mechanism because in almost all experiments to date it has been the dominant mechanism.

What are the exciting and important areas for future research?

Most of the work we have discussed has been on GaAs/AlGaAs because it is the best developed material system and has its band gap in a particularly convenient wavelength range for optical investigations. Investigation of other material systems is obviously desirable; for example narrow gap semiconductors, which have larger spin-orbit interactions should enable the investigation of the alternatives to the Dyakonov–Perel mechanism. Not very much is known about the spin dynamics of holes in quantum wells, as reflected in the relatively short section in this chapter. Systematic measurements on hole spin dynamics in well characterized samples are necessary. Another important area involves engineering and controlling the spin dynamics of the carriers in quantum wells as described in Sect. 2.5. We discussed the possibilities of different crystallographic growth directions for quantum wells and of applied electric fields. Measurements to test the predictions for [111] growth will be very interesting as will studies of the interplay of applied electric fields and other perturbations for example due to alloy asymmetry; can one design and realize an asymmetric [111]-oriented quantum well sample in which the Dyakonov–Perel mechanism of spin relaxation is canceled for all spin orientations? Other possibilities include reduction of the dimensionality of the structure to one dimension [54, 55]. Finally spin diffusion and spin transport are receiving increasing attention; the so-called spin grating technique, a powerful optical method for probing the dynamics of spatially varying spin populations, which has been dormant since its first development in 1996 [56], has come into use and is producing extremely interesting results [57].

Acknowledgements. I wish to thank all authors who have kindly given permission for me to use figures from their published papers.

References

[1] F. Meier, B.P. Zakharchenya (eds.), *Optical Orientation Modern Problems in Condensed Matter Science* (North-Holland, Amsterdam, 1984)

[2] M.J. Snelling, G.P. Flinn, A.S. Plaut, R.T. Harley, A.C. Tropper, R. Eccleston, C.C. Phillips, Phys. Rev. B **44**, 11345 (1991)

[3] D.D. Awschalom, N. Samarth, Optical manipulation, transport and storage of spin coherence in semiconductors, in *Semiconductor Spintronics and Quantum Computation*, ed. by D.D. Awschalom et al. (Springer, Berlin, 2002), Chap. 5

[4] M.J. Snelling, P. Perozzo, D.C. Hutchings, I. Galbraith, A. Miller, Phys. Rev. B **49**, 17160 (1994)

[5] S. Schmitt-Rink, D.S. Chemla, D.A. Miller, B Phys. Rev. B **32**, 6601 (1985)

[6] W.J.H. Leyland, G.H. John, R.T. Harley, M.M. Glazov, E.L. Ivchenko, D.A. Ritchie, I. Farrer, A.J. Shields, M. Henini, Phys. Rev. B **75**, 165309 (2007)

[7] M.J. Snelling, Optical orientation in quantum wells, PhD Thesis, Southampton (1991)

[8] M.J. Snelling, A.S. Plaut, G.P. Flinn, R.T. Harley, A.C. Tropper, T.M. Kerr, J. Lumin. **45**, 208 (1990)

[9] R.I. Dzhioev, K.V. Kavokin, V.L. Korenev, M.V. Lazarev, B.Ya. Meltser, M.N. Stepanova, B.P. Zakharchenya, D. Gammon1, D.S. Katzer, Phys. Rev. B. **66**, 245204 (2002)

[10] R.J. Elliott, Phys. Rev. **96**, 266 (1954)
[11] Y. Yafet, in *Solid State Physics*, vol. 14, ed. by F. Seitz, D. Turnbull, (Academic, New York, 1963), p. 2
[12] G.L. Bir, A.G. Aronov, G.E. Pikus, Sov. Phys. JETP **42**, 705 (1976)
[13] M.I. D'yakonov, V.I. Perel', Sov. Phys. JETP **33**, 1053 (1971)
[14] M.E. Flatté, J.M. Byers, W.H. Lau, Spin dynamics in semiconductors, in *Semiconductor Spintronics and Quantum Computation*, ed. by D.D. Awschalom et al. (Springer, Berlin, 2002), Chap. 4
[15] R. Winkler, *Spin Orbit Coupling Effects in Two-dimensional Electron and Hole Systems*. Springer Tracts in Modern Physics, vol. 191 (2003)
[16] M.W. Wu, C.Z. Nong, Eur. Phys J. B **18**, 373 (2000)
[17] M.M. Glazov, E.L. Ivchenko, JETP Lett. **75**, 403 (2002)
[18] M.M. Glazov, E.L. Ivchenko, JETP **99**, 1279 (2004)
[19] M.A. Brand, A. Malinowski, O.Z. Karimov, P.A. Marsden, R.T. Harley, A.J. Shields, D. Sanvitto, D.A. Ritchie, M.Y. Simmons, Phys. Rev. Lett. **89**, 236601 (2002)
[20] X. Cartoixà, D.Z.-Y. Ting, Y.-C. Chang, Phys. Rev. B **71**, 045313 (2005)
[21] W.H. Lau, J.T. Olesberg, M.E. Flatté, Phys. Rev. B **64**, 161301 (2001) (R)
[22] P. Murzyn, C.R. Pidgeon, P.J. Phillips, M. Merrick, K.L. Litvinenko, J. Allam, B.N. Murdin, T. Ashley, J.H. Jefferson, A. Miller, L.F. Cohen, Appl. Phys. Lett. **83**, 5220 (2003)
[23] B.N. Murdin, K. Litvinenko, D.G. Clarke, C.R. Pidgeon, P. Murzyn, P.J. Phillips, D. Carder, G. Berden, B. Redlich, A.F.G. van der Meer, S. Clowes, J.J. Harris, L.F. Cohen, T. Ashley, L. Buckle, Phys. Rev. Lett. **96**, 096603 (2006)
[24] J.M. Kikkawa, D.D. Awschalom, Phys. Rev. Lett. **80**, 4313 (1998)
[25] B. Beschoten, E. Johnston-Halperin, D.K. Young, M. Poggio, J.E. Grimaldi, S. Keller, S.P. DenBaars, U.K. Mishra, E.L. Hu, D.D. Awschalom, Phys. Rev. B **63**, R121202 (2001)
[26] S.A. Crooker, D.L. Smith, Phys. Rev. Lett. **94**, 236601 (2005)
[27] M.I. D'yakonov, V.Yu. Kachorovskii, Sov. Phys. Semicond. **20**, 110 (1986)
[28] A. Malinowski, R.S. Britton, T. Grevatt, R.T. Harley, D.A. Ritchie, M.Y. Simmons, Phys. Rev. B **62**, 13034 (2000)
[29] J.M. Ziman, *Electrons and Phonons* (Oxford University Press, London, 1972), Chap. 10
[30] R. Terauchi, Y. Ohno, T. Adachi, A. Sato, F. Matsukura, A. Tackeuchi, H. Ohno, Jpn. J. Appl. Phys. **38**, 2549 (1999)
[31] B. Jusserand, D. Richards, G. Allan, C. Priester, B. Etienne, Phys. Rev. B **51**, 707 (1995)
[32] N.S. Averkiev, L.E. Golub, Phys. Rev. **60**, 15582 (1999)
[33] D. Stich, J.H. Jiang, T. Korn, R. Schulz, W. Wegscheider, M.W. Wu, C. Schüller, Phys. Rev. B **76**, 073309 (2007)
[34] N.S. Averkiev, L.E. Golub, A.S. Gurevich, V.P. Evtikhiev, V.P. Kochereshko, A.V. Platonov, A.S. Shkolnik, Yu.P. Efimov, Phys. Rev. B **74**, 033305 (2006)
[35] R. Winkler, Physica E **22**, 450 (2004)
[36] O. Krebs, P. Voisin, Phys. Rev. Lett. **77**, 1829 (1996)
[37] K.C. Hall, K. Gründoğdu, E. Altunkaya, W.H. Lau, M.E. Flatté, T.F. Boggess, J.J. Zinck, W.B. Barvosa-Carter, S.L. Skeith, Phys. Rev. B **68**, 115311 (2003)
[38] A. Tackeuchi, O. Wada, Y. Nishikawa, Appl. Phys. Lett. **70**, 1131 (1997)
[39] W.J.H. Leyland, R.T. Harley, M. Henini, D. Taylor, A.J. Shields, I. Farrer, D.A. Ritchie, cond-mat/0707.4180 (2007)
[40] D. Stich, J. Zhou, T. Korn, R. Schulz, D. Schuh, W. Wegscheider, M.W. Wu, C. Schüller, Phys. Rev. Lett. **98**, 176401 (2007)
[41] D. Stich, J. Zhou, T. Korn, R. Schulz, D. Schuh, W. Wegscheider, M.W. Wu, C. Schüller, cond-mat/0707.4111v (2007)

[42] A.N. Titkov, V.I. Safarov, G. Lampel, in *Proc. ICPS 14*, Edinburgh (1978), p. 1031; see also Ref. [1]

[43] A.E. Ruckenstein, S. Schmitt-Rink, R.C. Miller, Phys. Rev. Lett. **56**, 504 (1986)

[44] Ph. Rhoussignol, R. Ferreira, C. Delalande, G. Bastard, A. Vinattieri, J. Martinez-Pastor, L. Carraresi, M. Colocci, J.F. Palmier, B. Etienne, Surf. Sci. **305**, 263 (1994)

[45] T. Damen, L. Viña, J.E. Cunningham, J. Shah, L.J. Sham, Phys. Rev. Lett. **67**, 3432 (1991)

[46] T. Uenoyama, L.J. Sham, Phys. Rev. Lett. **64**, 3070 (1990)

[47] R. Ferreira, G. Bastard, Phys. Rev. B **43**, 9687 (1991)

[48] B. Baylac, T. Amand, X. Marie, B. Dareys, M. Brousseau, G. Bacquet, V. Thierry-Meg, Solid State Commun. **93**, 57 (1995)

[49] J. Kainz, P. Schneider, S.D. Ganichev, U. Rössler, W. Wegscheider, D. Weiss, W. Prettl, V.V. Bel'kov, L.E. Golub, D. Schuh, Physica E **22**, 418 (2004)

[50] Y. Ohno, R. Terauchi, T. Adachi, F. Matsukura, H. Ohno, Phys. Rev. Lett. **83**, 4196 (1999)

[51] W.H. Lau, M.E. Flatté, J. Appl. Phys. **91**, 8682 (2002)

[52] O.Z. Karimov, G.H. John, R.T. Harley, W.H. Lau, M.E. Flatté, M. Henini, R. Airey, Phys. Rev. Lett. **91**, 246601 (2003)

[53] S. Döhrmann, D. Hägele, J. Rudolph, M. Bichler, D. Schuh, M. Oestreich, Phys. Rev. Lett. **93**, 147405 (2004)

[54] A.A. Kiselev, K.W. Kim, Phys. Rev. B **61**, 13115 (2000)

[55] A.W. Holleitner, V. Sih, R.C. Myers, A.C. Gossard, D.D. Awschalom, Phys. Rev. Lett. **97**, 036805 (2006)

[56] A.R. Cameron, P. Riblet, A. Miller, Phys. Rev. Lett. **76**, 4793 (1996)

[57] C.P. Weber, J. Orenstein, B.A. Bernevig, S.-C. Zhang, J. Stephens, D.D. Awschalom, Phys. Rev. Lett. **98**, 076604 (2007)

3

Exciton Spin Dynamics in Semiconductor Quantum Wells

T. Amand and X. Marie

3.1 Two-dimensional Exciton Fine Structure

The spin properties of excitons in nanostructures are determined by their fine structure. Before analyzing the exciton spin dynamics, we give first a brief description of the exciton spin states in quantum wells. We will mainly focus in this chapter on GaAs or InGaAs quantum wells which are model systems. For more details, the reader is referred to the reviews in [1, 2]. As in bulk material, exciton states in II–VI and III–V quantum wells correspond to bound states between valence band holes and conduction band electrons. As will be seen later, exciton states are shallow two-particle states rather close to the nanostructure gap, i.e., their spatial extension is relatively large with respect to the crystal lattice, so that the envelope function approximation can be used to describe these states. A description of the exciton fine structure in bulk semiconductors can be found in [3].

In quantum well structures, as in bulk material, a conduction electron and a valence hole can bind into an exciton, due to the Coulomb attraction. However, the exciton states are strongly modified due to confinement of the carriers in one direction. As we have seen, this confinement leads to the quantization the single electron and hole states into sub-bands (cf. Chap. 2), and to the splitting of the heavy- and light-hole band states. The description of the excitons is obtained, through the envelope function approach, and the fine exciton structure is then deduced by a perturbation calculation performed on the bound electron–hole states without electron–hole exchange. However, this approach becomes then more complex in the context of two dimensional structures [4]. The full electron–hole wave function is usually approximated by

$$\Psi_\alpha(\boldsymbol{r}_\mathrm{e}, \boldsymbol{r}_\mathrm{h}) = \chi_{c, v_\mathrm{e}}(z_\mathrm{e}) \chi_{j, v_\mathrm{h}}(z_\mathrm{h}) \frac{e^{i\boldsymbol{K}_\perp . \boldsymbol{R}_\perp}}{\sqrt{A}} \phi_{jnl}(\boldsymbol{r}_\perp) u_s(\boldsymbol{r}_\mathrm{e}) u_{m_\mathrm{h}}(\boldsymbol{r}_\mathrm{h}), \qquad (3.1)$$

where, α represents the full set of quantum indexes characterizing the exciton quantum state, e.g., explicitly: $|\alpha\rangle = |s, m_h; \nu_e, \nu_h, K_\perp, j, n, l\rangle$. Here $\chi_{c,\nu_e}(z)$ and $\chi_{j,\nu_h}(z)$ are the single particle envelope functions describing the electron, heavy-hole ($j = h$) or light-hole ($j = l$) motion along the z-growth axis, $R_\perp$ and $K_\perp$ are the exciton center of mass position and wave vector, respectively, A is the quantum well quantization area, and $\phi_{jnl}(r_\perp)$ characterizes the electron–hole relative motion in the quantum well plane. This is in fact the function basis we shall take to formulate the electron–hole exchange in a quantum well exciton.

The electron–hole exchange is determined through the evaluation of the direct and exchange integrals:

$$D_{\beta,\alpha} = \int_{\text{structure}} \Psi_\beta^*(r_e, r_h) \frac{e^2}{\epsilon_b |r_e - r_h|} \Psi_\alpha(r_e, r_h) \, dr_e \, dr_h, \qquad (3.2a)$$

$$-E_{\beta,\alpha} = -\int_{\text{structure}} \Psi_\beta^*(r_e, r_h) \frac{e^2}{\epsilon_b |r_e - r_h|} \Psi_\alpha(r_h, r_e) \, dr_e \, dr_h. \qquad (3.2b)$$

In the calculations of integrals (3.2), two contributions appear: a short-range one, which corresponds to the case where the electron and the hole are in the same Wigner cell in the structure, and a long-range one, which corresponds to the case where they are not.[1] Such integrals have been computed in [5]. It turns out that, in narrow quantum wells, they are much smaller than the heavy-/light-hole splitting Δ_{hl}, as well as the one between the different single particle sub-band states $\nu_{e(h)}$, and finally the $1s/2s$ exciton splitting. Then the first order perturbation theory, applied to the degenerated exciton states associated with a given sub-band pair, allows us to evaluate the corrections brought by (3.2) perturbation.

3.1.1 Short-Range Electron–Hole Exchange

For the ground state of the heavy-hole exciton (XH), the short-range perturbation matrix is

$$H^{(\text{SR})} = D^{(\text{SR})} - E^{(\text{SR})}. \qquad (3.3)$$

In two-dimensional (2D) systems, due to the splitting Δ_{lh} between heavy-hole and light-hole excitons (labeled XH and XL, respectively), it is possible to use the restriction of $H^{(\text{SR})}$ to the XH subspace. The XH basis states are labeled according to their projection to the quantization axis z (the structure growth axis), according to $|M\rangle = |s_e + j_h\rangle$ ($s_e = \pm 1/2$, $j_h = \pm 3/2$), so that $\mathcal{B}_{\text{XH}} = \{|+2\rangle, |+1\rangle, |-1\rangle, |-2\rangle\}$. It turns out that $H^{(\text{SR})}$ is proportional to $|\phi_{hh,1s}(r = 0)|^2 I_{hh}$, where $I_{hh} = \int_{-\infty}^{+\infty} |\chi_{c,1}(z)|^2 \times |\chi_{h,1}(z)|^2 \, dz$, a measure of the probability for the electron and the hole to be at the same position in the quantum well. For XH excitons, the short-range exchange splits

---

[1] In the latter contribution, only the exchange integral has to be taken into account, since the direct long-range Coulomb interaction has already been considered in the equations of the 2D mechanical exciton (i.e., without exchange).

the $J = 1$ optically active exciton states with the non-optically active $J = 2$ ones. It is convenient to evaluate (3.3) with respect to the 3D case. Then the short-range 2D splitting takes the form:

$$\Delta_0 = \frac{3}{4} \Delta_0^{3D} \frac{|\phi_{h,1s}(0)|^2}{|\phi_{h,1s}^{3D}(0)|^2} I_{hh}, \tag{3.4}$$

where ϕ_{1s}^{3D} and ϕ_{1s} are the 3D and 2D exciton hydrogenic 1s function, respectively.

The short-range exchange correction Δ_0 is independent of the exciton wave vector $K_\perp$. In an infinite quantum well, the overlap integral is $I_{hh} = 3/(2L_W)$ where L_W is the quantum well width, so that, letting a_B^{3D} be the usual three-dimensional exciton Bohr radius, and E_B, E_B^{3D} the exciton binding energies in the 2D and 3D semiconductor, respectively, we obtain $\Delta_0 \approx (9/16)\Delta_0^{3D}(E_B/E_B^{3D})^2 a_B^{3D}/L_W$, showing that when the quantum well width decreases, the 2D short-range exchange first increases since the exciton binding energy increases [4]. This corresponds to the trend observed experimentally (see Fig. 3.15 later).

3.1.2 Long-Range Electron–Hole Exchange

An exciton can be viewed as a polarization wave propagating inside the semiconductor structure. For a given $|\alpha\rangle$ exciton eigenstate state, the associated polarization field $\boldsymbol{P}(r, t)$ is oriented along the dipole $er_{\alpha,\emptyset} = \langle \alpha | er | \emptyset \rangle$. The exciton state propagating with $\boldsymbol{K}_\perp$ parallel to $\boldsymbol{r}_{\alpha,\emptyset}$, called the longitudinal exciton, and the one propagating with $\boldsymbol{K}_\perp$ perpendicular to $\boldsymbol{r}_{\alpha,\emptyset}$, called the transverse exciton, are slightly split in energy, by the longitudinal-transverse splitting. In 3D semiconductor material, this splitting Δ_{LT}^{3D} does not depend on the exciton wave vector. For GaAs, $\Delta_{LT}^{3D} \approx 0.1$ meV [7].

In quantum wells structure, the calculation can be found in [5]. For the lowest heavy-hole excitons, it leads to

$$\Delta_{LT}(\boldsymbol{K}_\perp) = \frac{3}{8} \Delta_{LT}^{3D} \frac{|\phi_{h,1s}(0)|^2}{|\phi_{1s}^{3D}(0)|^2} K_\perp I_0(K_\perp), \tag{3.5}$$

where I_0 is a form factor which, for $K_\perp \ll \pi/L_W$, takes the simple form $I_0 \approx |\langle \chi_{e,1} | \chi_{h,1} \rangle|^2$, which corresponds to the overlap of the electron and hole single particle envelope functions (in the infinite barrier model, $I_0 = 1$). Finally, we obtain the approximation: $\Delta_{LT}(K_\perp) \approx (3/16)\Delta_{LT}^{3D}|\langle \chi_{e,1}|\chi_{h,1}\rangle|^2(E_B/E_B^{3D})^2 a_B^{3D} K_\perp$. Contrary to the 3D case, the 2D longitudinal transverse splitting is zero for $K_\perp = 0$, and increases linearly with $K_\perp$. For instance, if $a_B^{3D} K_\perp \approx 0.1$, one can estimate $\Delta_{LT} \approx 40\,\mu$eV typically for GaAs/AlGaAs quantum wells of 2D character. Similarly to the short-range exchange Δ_0, the long-range splitting Δ_{LT} increases when the confinement increases (i.e., when the well width decreases). As we shall see in Sect. 3.3.4, the long-range exchange interaction is at the origin of an important spin relaxation channel for excitons in type I quantum wells.

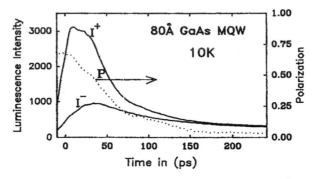

Fig. 3.1. Intensities I^+ and I^- and polarization $P = (I^+ - I^-)/(I^+ + I^-)$ of exciton luminescence as a function of time following picosecond excitation at the exciton energy. I^+ (I^-) corresponds to the intensity of σ^+ emission for σ^+ (σ^-) excitation [10]

3.2 Optical Orientation of Exciton Spin in Quantum Wells

Thanks to the development of stable ultrafast laser sources at the end of the 1980s, it has been possible to monitor directly in the time domain the carrier spin dynamics in semiconductors [8, 9]. Time-resolved polarization absorption measurements based on pump–probe techniques or time-resolved polarized photoluminescence experiments were extensively used to measure the spin relaxation of excitons in semiconductor quantum wells [10–12]. These time-resolved techniques are very complementary tools to the well-established measurements methods based on cw photoluminescence spectroscopy or Hanle type experiments [3, 13]. The measurement of the circular polarization dynamics of the exciton luminescence after a circularly polarized (σ^+) pulsed laser excitation allows one to measure both (i) the spin polarization of the exciton just after the δ-like optical pump and compare it with the theoretical value given by the optical selection rules (see Chap. 1) [4] and the band structure and (ii) the decay time of the exciton spin polarization, which allows one to deduce the dominant spin relaxation mechanism.

Figure 3.1 displays for a 8 nm GaAs/AlGaAs multiple quantum well the time evolution of I^+ and I^- for resonant excitation of the heavy-hole exciton XH (I^+ and I^- correspond to the right circularly polarized (σ^+) luminescence component following a right (σ^+) or left (σ^-) circularly polarized picosecond laser excitation, respectively). Since only the heavy-hole exciton is excited,[2] the initial polarization $P_L(t = 0)$ of the exciton luminescence is very large: $P_L(0) \approx 70\%$ (the time resolution of the streak camera used here as a detector is about 10 ps). The circular polarization in Fig. 3.1 decays with a time constant of ≈ 50 ps. The link between this decay time and the spin relaxation of exciton is not straightforward since several spin relaxation channels can occur simultaneously [14, 15]. This will be discussed in detail in Sect. 3.3.

---

[2] Here the spectral width of the laser pulse is much smaller than the heavy-light hole splitting.

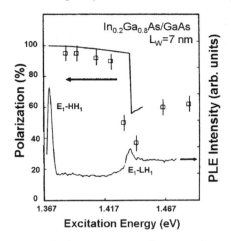

Fig. 3.2. Photoluminescence excitation spectrum under cw stationary excitation. The initial polarization degree $P_L(t = 0)$ in dynamical experiments of the heavy-hole exciton lumines-cence as a function of the laser excitation energy is also displayed; ($\square$): experimental data; solid line: calculated values. $T = 1.7\,K$ [14]

Figure 3.2 presents the variation of the initial polarization $P_L(0)$ of the exciton lu-minescence as a function of the picosecond laser excitation energy in a compressively strained InGaAs/GaAs multiple quantum well ($L_W = 7$ nm) [14]. When the incident photon energy is larger than the quantum well gap but smaller than the light-hole ex-citon transition (involving the E_1 and LH_1 sub-bands), the initial polarization $P_L(0)$ is as high as 95% (the time resolution of the up-conversion time-resolved photolumi-nescence spectroscopy technique used here is about 1 ps). This very high $P_L(0)$ value proves that the initial carrier thermalization process which occurs on a few hundred of femtosecond time scale, leads to very minor carrier depolarization (at least for the conduction electrons under non resonant excitation). The calculated initial polariza-tion for valence to conduction band transitions using the envelop function formalism and the effective Luttinger Hamiltonian is also plotted in Fig. 3.2 (full line) [16]. The variation of $P_L(0)$ versus the excitation energy is the result of valence band mixing. The mismatch around the $(E_1 - LH_1)$ excitation energy between the experiment and the calculation is just due to the fact that the latter does not take into account the absorption increase due to bound light-hole exciton state (XL) [17, 18]. As a fact, the XL oscillator strength may become stronger than the one of unbound $E_1 - HH_1$ electron–hole pair states at the same energy [4]. For a strictly resonant excitation of the light-hole exciton photoluminescence, the heavy hole exciton luminescence polarization can indeed be negative (opposite to the helicity of the excitation laser polarization), see the curve (3) in Fig. 3.3 for a $L_W = 4$ nm GaAs/AlGaAs multiple quantum well structure [17, 19].

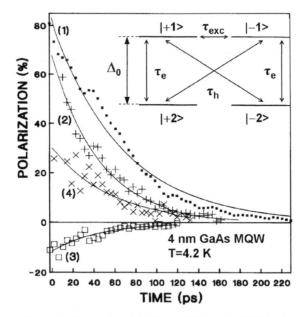

Fig. 3.3. Circular polarization dynamics of the heavy-hole exciton XH luminescence following a (σ^+)-polarized picosecond laser pulse. Four excitation energies: (1) $h\nu = \mathrm{XH} + 10\,\mathrm{meV}$; (2) $h\nu = \mathrm{XH} + 22\,\mathrm{meV}$; (3) $h\nu = \mathrm{XH} + 32\,\mathrm{meV}$, resonant with the light-hole exciton energy XL; (4) $h\nu = \mathrm{XH} + 74\,\mathrm{meV}$ [19]. Inset: Schematic diagram of the different exciton spin relaxation processes; τ_{exc}, τ_{e}, and τ_{h} represent the exciton, electron, and hole spin relaxation time, respectively (see text)

3.3 Exciton Spin Dynamics in Quantum Wells

Exciton luminescence polarization studies in semiconductor quantum wells have revealed the coexistence of two main mechanisms of exciton spin relaxation: the direct relaxation with simultaneous electron and hole spin flip due to the electron–hole exchange interaction [5] and an indirect one with sequential spin flips of the single particles (electron or hole), see the inset of Fig. 3.3. The rate of exciton spin relaxation in this indirect channel is limited by the slower single particle spin–flip rate, which is typically the electron one [20]. The relative efficiency of these mechanisms depends on the excitation conditions which can be *resonant* (the energy of the polarized excitation photons is equal to the exciton energy) or *non-resonant* (the photon excitation energy is typically above the quantum well gap energy $E_1 - \mathrm{HH}_1$). In the latter, the exciton spin dynamics is influenced by the exciton formation process [21].

3.3.1 Exciton Formation in Quantum Wells

In bulk semiconductors, two exciton formation processes are usually considered: straight hot exciton photogeneration, with the simultaneous emission of an LO

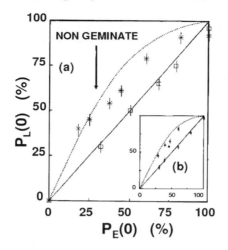

Fig. 3.4. (a) Initial circular polarization degree of the exciton photoluminescence $P_L(t = 0)$ versus the excitation light polarization degree P_E of the picosecond laser pulse in a $L_W = 7$ nm $In_{0.2}Ga_{0.8}As/GaAs$ multiple quantum well. The symbols represent the measured values for ($\square$) quasiresonant excitation: $h\nu = XH + 4$ meV; ($*$) non-resonant excitation: $h\nu = XH + 34$ meV. The continuous lines are, respectively, the calculated $P_L(0)$ values for a geminate and non-geminate (bimolecular) formation process. **(b)** Similar analysis on a $GaAs/Al_{0.3}Ga_{0.7}As$ multiple quantum well ($L_W = 4$ nm); ($\square$): resonant excitation, $h\nu = XH$; ($*$): non-resonant excitation, $h\nu = XH + 15$ meV [21]

phonon, in which the constitutive electron–hole pair is geminate; or bimolecular exciton formation which consists of the random binding of electrons and holes under the Coulomb interaction.

The analysis of the initial polarization $P_L(t = 0)$ of the exciton luminescence in time-resolved optical orientation experiments performed in GaAs/AlGaAs or In-GaAs/GaAs quantum wells reveals precious information about this exciton formation process [21]. The idea is to measure the initial photoluminescence polarization $P_L(0)$ using an elliptically polarized laser beam, characterized by its degree of circular polarization defined as $P_E = (\Sigma^+ - \Sigma^-)/(\Sigma^+ + \Sigma^-)$ where Σ^+ and Σ^- represents the intensities of the right and left circularly-polarized optical excitation components. Figure 3.4 presents the experimental photoluminescence circular polarization degree $P_L(0)$ versus P_E for non-resonant and resonant excitation conditions. The measurements are performed in a 7 nm InGaAs/GaAs multiple quantum well structure. The striking feature is that for non-resonant excitation ($h\nu = XH + 34$ meV $< XL$, where XH is the heavy-hole exciton energy and XL the light-hole one), the initial photoluminescence polarization degree is higher than the excitation light polarization. In contrast, in resonant excitation ($h\nu \approx XH$), below the quantum well gap, the behavior is completely different: within the experimental accuracy the initial photoluminescence polarization is equal to the excitation light polarization, whatever the P_E value is [21].

In resonant excitation conditions, the excitons are formed from geminate pairs which keep their initial spin orientation; the initial photoluminescence polarization is thus

$$P_L(0) = P_E \tag{3.6}$$

in agreement with the experimental results in Fig. 3.4.

In non-resonant excitation conditions (above the quantum well gap), the polarized excitation pulse creates electron–hole pairs with a total spin $M = +1$ and $M = -1$. The proportions are $(1 + P_E)/2$ and $(1 - P_E)/2$, respectively. If the excitons are formed from spin-unrelaxed non-geminate pairs by a bimolecular formation process the initial excitonic populations on optically active and inactive spin states, $|\pm 1\rangle$ and $|\pm 2\rangle$, are respectively: $N_{\pm 1} \propto (1 \pm P_E)^2/4$ and $N_{\pm 2} \propto (1 - P_E^2)/4$.

The coherence effects are neglected here since the electron and hole angular momenta are now uncorrelated. The initial polarization is then given by

$$P_L(0) = \frac{2P_E}{1 + P_E^2} \geq P_E; \tag{3.7}$$

an expression which shows that $P_L(0)$ is strictly higher than the polarization of the excitation light when $0 < P_E < 1$. This is due to the fact that the electron $|-1/2\rangle$ states are more populated than the $|+1/2\rangle$ ones, and have a higher probability to bind to a $|+3/2\rangle$ hole than to a $|-3/2\rangle$ one. Expression (3.7) is strictly independent both of the initially created electron–hole pair density and the value of the bimolecular formation coefficient [22–25]. The initial circular polarization $P_L(0)$ of the luminescence is plotted in Fig. 3.4 as a function of the excitation circular polarization P_E, according to (3.6) and (3.7); the full and dotted lines correspond, respectively, to the geminate and non-geminate exciton formation process. The comparison of the calculated and experimental polarization leads to the conclusion that following non-resonant excitation most of the excitons are formed by the bimolecular process.

3.3.2 Spin Relaxation of Exciton-Bound Hole

In contrast to bulk materials in which the hole spin relaxation time is very fast ($\lesssim 1$ ps, characteristic time of the momentum relaxation time) [3, 26], the lifting of the degeneracy in $k = 0$ between the heavy hole and light hole sub-bands in quantum wells yields a decrease of the valence band mixing and hence an increase of the hole spin relaxation time (see Chap. 2) [27–31]. The exciton spin dynamics can thus be strongly affected by the hole single particle spin relaxation time, which occurs on the same time-scale as the direct exciton spin relaxation which connects the two optically active $|+1\rangle$ and $|-1\rangle$ exciton states (see Sect. 3.3.4) [5]. However the exciton-bound hole spin relaxation time is usually shorter than the free hole spin relaxation in quantum wells. As a fact, the exciton is composed of holes states with wave-vectors ranging up to $(a_B)^{-1}$ typically (a_B is the ground state 2D exciton Bohr radius [4], and thus characterized by a significant valence band mixing).

Two experimental techniques have been used to measure directly the hole spin relaxation time (τ_h) within the 2D exciton [21, 32].

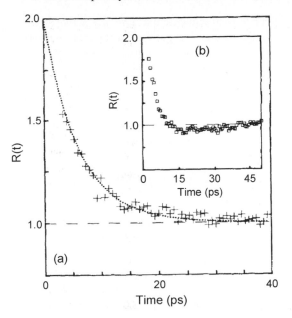

Fig. 3.5. (a) Time evolution of the ratio $R(t) = I_{\sigma^+}(t)/I_{\sigma^x}(t)$ where $I_{\sigma^+}(t)$ and $I_{\sigma^x}(t)$ are the total luminescence intensity following a (σ^+) circularly-polarized or (σ^x) linearly-polarized excitation pulse in a $L_W = 7$ nm $In_{0.2}Ga_{0.8}As$/GaAs multiple quantum well. *The solid line is an exponential fit of* $R(t)$ *according to* $R(t) = 1 + e^{-t/\tau_h}$, with $\tau_h = 5.5$ ps. (b) Same measurements for a GaAs/$Al_{0.3}Ga_{0.7}As$ multiple quantum well ($L_W = 4$ nm) [31]. The experiment is performed at 1.7 K

Measurement of the Hole Spin Relaxation Time by Monitoring the Total Luminescence Intensity Dynamics

This technique exploits the exciton bimolecular formation process in the non-resonant excitation conditions. As shown in Sect. 3.3.1, the exciton bimolecular formation process yields an initial population of the exciton in the optically inactive states $|\pm2\rangle$ with a proportion $(1 - P_E^2)/2$.

Let us consider two different excitation conditions: first, a 100% circularly σ^+ light excitation; second a linearly polarized σ^x light excitation. In each case, the total luminescence intensity I_{σ^+} and I_{σ^x} are recorded. These two measurements are performed at the same excitation energy (above the quantum well gap) and for the same excitation intensity. The ratio $R(t) = I_{\sigma^+}/I_{\sigma^x}$ is presented in Fig. 3.5 for the quantum well structures already presented in Fig. 3.4. When the excitation is linearly polarized ($P_E = 0$), the excitonic population is initially equidistributed over the four states. Consequently, only half of the excitons are initially active and this does not change with time, since this distribution corresponds to the thermal equilibrium of the electronic excitations.

When the excitation is 100% circular ($P_E = 1$), only $|+1\rangle$ states are initially populated so that all the excitons are optically active at $t = 0$. Consequently $R(0) = 2$.

The system will then tend to equalize the optically active and optically inactive excitonic population, due to the electron and hole single particle spin relaxation, so one expects a rapid decrease of $R(t)$ towards 1. The exciton spin-flip, governed by the exchange interaction between the electron and the hole (see Sect. 3.3.4), which changes the $|+1\rangle$ excitons into $|-1\rangle$ and vice versa, is strictly inoperative in the time evolution of $R(t)$, which decays according to $R(t) = 1 + e^{-t(1/\tau_h + 1/\tau_e)}$. As the single particle exciton bound electron spin relaxation time is longer than the exciton bound hole spin relaxation time (as can be inferred later from Sect. 3.3.3), the evolution of $R(t)$ in Fig. 3.5 reflects directly the hole spin relaxation time. The fit of the experimental curves yields a hole spin relaxation time of $\tau_h = 5.5$ ps and $\tau_h = 2.5$ ps respectively in the InGaAs/GaAs and GaAs/AlGaAs quantum well structures presented. The advantage of this method is the direct measurement of τ_h in the exciton, independently of the determination of the exciton spin relaxation time. Moreover, it does not require the modeling of the exciton energy relaxation and the effective radiative recombination processes as they are identical for the two I_{σ^+} and I_{σ^x} recordings. As expected in resonant excitation conditions (geminate formation of excitons), $R(t)$ does not depend on time and equals 1, as expected [21].

The energy dependence of the hole spin relaxation time has been studied by Baylac et al. with this technique [31]. These authors found an hole spin relaxation time of $\tau_h \approx 15$ ps for an excitation energy near the InGaAs/GaAs quantum well gap E_g, dropping down to $\tau_h \approx 6$ ps for $h\nu > E_g + 8$ meV as a consequence of the valence band mixing and the increasing of the electron–hole temperature with the increase of the excitation energy.

Measurement of the Hole Spin Relaxation with a Two-photon Excitation Process

A different experiment allows direct measurement of the conversion rate of $J = 2$ to $J = 1$ excitons in GaAs quantum wells due to hole single particle spin relaxation [32]. The experiment is basically as follows. First, $J = 2$ excitons are created via two-photon infrared excitation, using an optical parametric oscillator. Following the generation of the excitons, the single-photon recombination luminescence ($\approx$ visible or near infrared) from the $J = 1$ excitons is detected with a streak camera. The $J = 2$ excitons can by created by resonant excitation and observed immediately thereafter (after the conversion to $J = 1$ states), without unwanted background from the laser light (since the streak camera does not respond to the infrared exciting light).

This experiment relies on the fact that just as single photon emission from $J = 2$ states is forbidden, two-photon absorption by $J = 1$ excitons is forbidden by dipolar selection rules but two-photon absorption by $J = 2$ excitons is allowed.

The lower curve of Fig. 3.6 shows as a function of time the XH ($J = 1$) exciton luminescence at 730 nm, from a 3 nm quantum well at 2 K, excited by circularly polarized optical parametric oscillator light, i.e., following two-photon excitation of the $1s$ heavy-hole resonance. The rise time of the luminescence intensity after the two-photon excitation is mainly governed by the hole spin relaxation time τ_h (which is shorter than the single particle electron spin relaxation time [20], see Sect. 3.3.3).

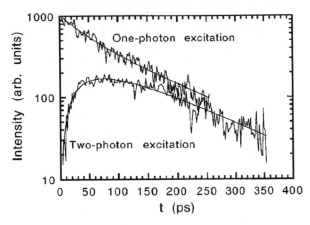

Fig. 3.6. *Lower curve*: Luminescence intensity dynamics of the heavy-hole exciton $J = 1$ in a $L_W = 3$ nm GaAs/AlGaAs quantum well, following the generation in the $J = 2$ spin state by a 100 fs circularly polarized laser pulse at 1 471 nm. *Upper curve*: Luminescence intensity dynamics of the heavy-hole exciton $J = 1$ in the same quantum well, following the generation in the $J = 1$ spin state by a 730 nm laser pulse (the relative intensity scales of the two curves are arbitrary) [32]

On the basis of simple rate equations for the $J = 1$ and $J = 2$ exciton states, Snoke et al. concluded that the time scale for the hole spin-flip process in a narrow ($L_W = 3$ nm) GaAs quantum well is of the order of 60 ps [32], which corresponds here to resonantly created XH excitons.

3.3.3 Spin Relaxation of Exciton-Bound Electron

The exciton-bound electron spin relaxation has been calculated by de Andrada e Silva and La Rocca taking into account the conduction band splitting due to the spin–orbit interaction [20]. They have shown that the off-diagonal matrix element between optically active and inactive exciton states (e.g., $|+1\rangle$ and $|+2\rangle$, see insert in Fig. 3.3) can be represented by an effective magnetic field with two contributions. The first one, induced by spin–orbit interaction (cf. Chap. 2), changes randomly as the exciton is elastically scattered and is responsible for the electron spin relaxation. The second one arises due to the electron–hole short-range exchange splitting Δ_0 between the optical active and inactive states (cf. (3.4)). It acts as a static external magnetic field, which reduces the electron spin relaxation rate due to energy mismatch. The estimated rate of the bound electron spin flip agrees well with values obtained from fitting the experimental data (see Sect. 3.3.4) [14, 15].

This spin relaxation rate $W_e = 1/2\tau_e$ is [20]

$$W_e = \frac{4\alpha_{so}^2 K^2}{\hbar} \frac{\tau^*}{1 + (\Delta_0\tau^*/\hbar)^2}, \qquad (3.8)$$

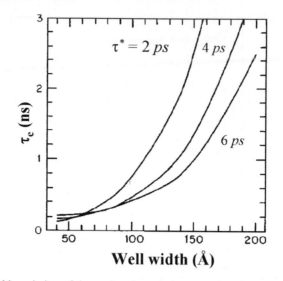

Fig. 3.7. Well width variation of the exciton-bound electron spin relaxation time τ_e for different values of the elastic momentum scattering time τ^* in GaAs/AlGaAs quantum wells [20]

where τ^* is the exciton elastic momentum scattering time, Δ_0 is the exchange splitting between the optical active and inactive exciton states, K is the exciton wave vector and α_{so} a constant depending on spin–orbit interaction in the conduction band.

This means that, when $(\Delta_0\tau^*/h) > 1$, the exciton-bound electron spin dynamics presents a motional narrowing type of relaxation analogous to the Dyakonov–Perel free-electron spin relaxation [33, 34]. Figure 3.7 displays the well-width dependence of this exciton-bound electron spin relaxation τ_e for different values of the elastic momentum scattering τ^* in GaAs/AlGaAs quantum well. The spin relaxation time increases with the well width due to the corresponding decrease in the average spin–orbit splitting in the conduction band that the bound electron feels.

Except in the narrow well limit, we observe the usual motional narrowing behavior with the exciton-bound spin-relaxation time roughly inversely proportional to the momentum scattering time. Experimental investigations of the exciton-spin dynamics in high-quality GaAs/Al$_x$Ga$_{1-x}$As multiple quantum wells ($x = 0.3$ and $L_W = 15$ nm) have determined through detailed fitting procedures that the exciton-bound electron-spin relaxation rate lays in the range 3×10^8 s^{-1} < W_e < 3×10^9 s^{-1} [14, 15], in agreement with the calculated values plotted in Fig. 3.7.

3.3.4 Exciton Spin Relaxation Mechanism

Exciton Spin Relaxation Due to Electron–Hole Exchange: The Maille, Andrada e Silva, and Sham Mechanism

The main exciton spin depolarization mechanism in quantum wells occurs via the exchange Coulomb interaction between the electron and the hole. The theory of this

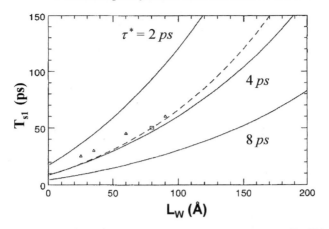

Fig. 3.8. Calculated exciton spin relaxation time ($T_{s1} \equiv \tau_{exc}$) versus well width for different values of the momentum scattering time in GaAs/AlGaAs quantum wells [5]. The experimental points are from [29] ($\triangle$) and [10] ($\square$)

mechanism has been developed by Maille, de Andrada e Silva, and Sham [5]. The process may be viewed as due to a fluctuating effective-magnetic field in the well interface plane, which originates in the electron–hole exchange (cf. Sects. 3.1.1, 3.1.2). The magnitude and direction of this field depends on the exciton center of mass momentum K, vanishing for $K = 0$ states (cf. (3.5)). The fluctuations, due to the scattering of the exciton center of mass momentum, are responsible for the exciton spin relaxation, in the same manner as any other motional narrowing spin–flip processes, with the characteristic dependence of the spin-relaxation time on the inverse momentum scattering time. In this process, the long range electron–hole exchange contribution dominates, its strength being characterized by $\Delta_{LT}(K) = \hbar\Omega_{LT}(K)$. The inverse exciton spin relaxation time (τ_{exc}, often labeled T_{s1}) is given, in the motional narrowing regime ($\Omega_{LT}(K)\tau^* \ll 1$), by

$$\frac{1}{T_{s1}} \approx \langle \Omega_{LT}^2 \rangle \tau^*, \tag{3.9}$$

where the square of the precession angular frequency $\Omega_{LT}^2(K)$ is averaged on the whole exciton population. The time T_{s1} is called *longitudinal spin relaxation time*. It corresponds to the relaxation between the $|+1\rangle$ and $|-1\rangle$ exciton states (i.e., circular depolarization time of exciton luminescence).

Maille, de Andrada e Silva, and Sham [5] have also calculated the *transverse spin relaxation time* T_{s2} which corresponds to the relaxation time of the coherence between $|+1\rangle$ and $|-1\rangle$ states. In the motional narrowing regime, and at low exciton density, $T_{s2} \approx 2T_{s1}$. The transverse exciton spin relaxation can be measured by recording the decay time of the exciton linear depolarization in optical alignment experiments (see Sect. 3.4.2) [35].

Figure 3.8 presents the calculated exciton spin relaxation time T_{s1} as a function of the well width for different values of the exciton momentum scattering time τ^*. The

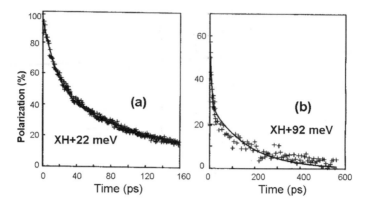

Fig. 3.9. Time evolution of the measured luminescence circular polarization for two excitation energies in a InGaAs/GaAs quantum well structure: (**a**) $\hbar\omega = $ XH + 22 meV and (**b**) $\hbar\omega = $ XH + 92 meV, i.e., larger than the light-hole exciton energy. The solid line corresponds to the fit with the rate equation model (see text) [14]

quantum well confinement enhances the exchange interaction compared to its value in bulk, as shown in Sect. 3.1.2. The long-range exchange interaction is found to be the dominant contribution to the spin-relaxation process. The short-range contribution is rendered less important, since the coupling between heavy- and light-holes is reduced due to sub-band formation in low-dimensional systems [5].

Measurement of the Maille, Andrada e Silva, and Sham Spin Relaxation Time

The measurement of the exciton spin relaxation time τ_{exc} (or T_{s1}) requires a fitting procedure of the experimental data, taking into account the single particle spin relaxation time of electrons (τ_e) and holes (τ_h) within the exciton and the direct exciton spin relaxation time (τ_{exc}), see the inset in Fig. 3.3. If the experiments are performed in non-resonant excitation conditions, the model must also take into account the bimolecular formation process (see Sect. 3.3.1).

The balance equations describing the evolution of different exciton spin state populations $n_M(t)$ ($M = \pm 1, \pm 2$), once they have been created, where solved in [5, 14, 15] as a function of the electron, hole and exciton spin transition rates, $W_e = 1/2\tau_e$, $W_h = 1/2\tau_h$ and $W_{exc} = 1/2\tau_{exc}$, respectively (taking into account the recombination time τ_r).

The calculated photoluminescence polarization is simply given by $P_{cal}(t) = (N_1 - N_{-1})/(N_1 + N_{-1})$. The comparison between calculated and experimental evolution of the exciton luminescence polarization is presented in Fig. 3.9. The straight lines in Figs. 3.9(a) and (b) correspond to least square fits of the experimental curves of the exciton photoluminescence circular polarization dynamics using the rate equations for a $L_W = 7$ nm InGaAs/GaAs quantum well structure. The depolarization dynamics are well described by an exciton spin relaxation ($\tau_{exc} \approx 58$ ps and 79 ps, respectively, for the two excitation conditions), and a shorter time (17 ps and 7 ps,

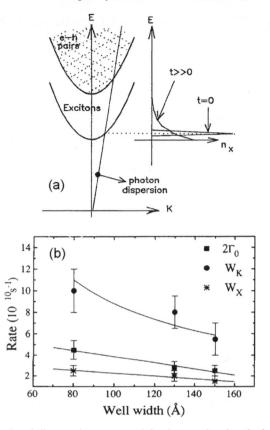

Fig. 3.10. (a) Energy band diagram in a two-particle picture, showing the initial created exciton distribution and a thermalized exciton distribution [10]. (b) Radiative recombination rate $(2\Gamma_0)$ of the exciton population at $K_\parallel = 0$, exciton effective scattering rate with phonons (W_K) and exciton spin relaxation time $(W_x = 1/2\tau_{exc})$. These rates are obtained from fits to the measured exciton polarization photoluminescence dynamics at $T = 12\,\text{K}$ in GaAs/AlGaAs quantum well structures[15]

respectively) which is identified as τ_h. The fit gives a much longer third time (τ_e), greater than 1 ns: as a matter of fact, the fit is not very sensitive to this third time. It is impossible to fit the data with only τ_e and τ_h: a finite excitonic spin relaxation τ_{exc} is compulsory to get a good agreement. But the excitonic spin relaxation time alone cannot explain the polarization decay as it leads to a calculated curve which is mono-exponential whereas the experimental ones are not.

In contrast to what could be expected, the modeling of the exciton spin depolarization dynamics measured by luminescence spectroscopy in strictly resonant excitation is not straightforward [10, 15]. The measured temporal dynamics of resonantly-excited luminescence is determined by the relaxation, thermalization and recombination dynamics of this initial non-thermal distribution of exciton. Figure 3.10(a) schematically displays the relevant energy diagram in the two particle or exciton representation [10]. The absorption of photons takes place only within the homogeneous width of the exciton. The homogeneous exciton line width of high quality multiple quantum well samples is usually less than the thermal energy kT (even at 10 K). As the photoexcited cold excitons thermalize, their distribution becomes wider than the initial distribution so that fewer excitons remain within the homogeneous line width of the exciton with increasing time. Since only excitons within the homogeneous line width couple to light as a consequence of the wave vector conservation, this leads to a decrease in the luminescence intensity, even though the total number of excitons has not decreased [15, 36]. Thus this process has to be taken into account in addition to the spin relaxation mechanisms previously described. Vinattieri and co-workers performed a comprehensive investigation of the dynamics of resonantly excited excitons in GaAs/AlGaAs quantum wells on picosecond time-scales [15]. With systematic multiparameter fits, they managed to extract the different relaxation rates, as shown in Fig. 3.10(b). They found, e.g., $W_x = 1.5 \times 10^{10}\,\mathrm{s}^{-1}$, $W_h = 0.7 \times 10^{10}\,\mathrm{s}^{-1}$, and $3 \times 10^8\,\mathrm{s}^{-1} < W_e < 3 \times 10^9\,\mathrm{s}^{-1}$ for a 15 nm GaAs/AlGaAs quantum well structure.

Non-degenerate, spectrally, and spin-resolved differential transmission experiments allow us also to determine different spin-relaxation times within the exciton [37, 38]. In these pump–probe experiments, the picosecond σ^+ pump pulse is resonant with the $|+1\rangle$ quantum well excitons formed with $+3/2$ heavy holes (hh) and $-1/2$ electrons; the non-degenerate probe pulse measures the absorption at the light hole (lh) transition. The transmission change of this probe pulse as a function of time with polarization σ^- is not sensitive to the population at the (hh) states with the angular momentum $+3/2$ but it is sensitive to the population of electrons with spins $-1/2$. In the same way, the σ^- probe transmission change at the heavy-hole excitonic transition is only sensitive to the population of $+1/2$ electrons and $-3/2$ holes. The last two bands are not initially populated by the pump pulse and, therefore, the population of these states results from electron or hole spin-flip processes. Using this experimental technique, it is possible to extract unambiguously the time constants corresponding to the spin relaxation of one of the three types of quasiparticles. Ostatnicky et al. measured for instance $\tau_e = 250\,\mathrm{ps}$ and $\tau_h = 30\,\mathrm{ps}$ in a 10 nm thick GaAs multiple quantum well structure [37].

The spin dynamics of neutral (X) and positively charged excitons (X^+ made of a hole singlet and one electron) have been measured and compared in modulation p-doped CdTe/CdMgZnTe quantum wells [39]. Thanks to the larger binding energy of the charged exciton (X^+) in II–VI quantum wells compared to the one in GaAs quantum wells [40–42], it is possible to study the neutral exciton X photoluminescence dynamics after a resonant excitation of X [noted $X(X)$ in the following], the X^+ photoluminescence dynamics after a resonant excitation of X [noted $X^+(X)$], and the X^+ photoluminescence dynamics after a resonant excitation of X^+ [noted $X^+(X^+)$]. Figure 3.11 illustrates the corresponding decay of the photoluminescence circular polarization for the three configurations.

The neutral excitonic polarization [$X(X)$ spectrum] decreases with a time constant of 12 ps, four times shorter than in typical III–V quantum wells of comparable

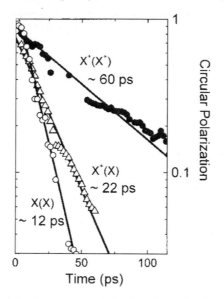

Fig. 3.11. Circular photoluminescence polarization dynamics in a modulation p-doped CdTe/CdMgZnTe quantum well ($L_W = 7.7$ nm), $T = 10$ K. The neutral exciton X photoluminescence dynamics is recorded after a resonant excitation of X [noted $X(X)$], the X^+ photoluminescence dynamics is recorded after a resonant excitation of X [noted $X^+(X)$], and the X^+ photoluminescence dynamics after a resonant excitation of X^+ is also studied [noted $X^+(X^+)$] [39]

sizes because of the larger exchange interaction (see Sect. 3.1) [15]. As the neural excitons X are created resonantly, i.e., without kinetic energy, this time reflects mainly the excitonic spin-flip time τ_{exc}, i.e., the simultaneous spin flip of the electron and the hole within the neutral exciton due to electron–hole exchange [5].

The $X^+(X^+)$ circular polarization decreases with a significantly longer time (≈ 60 ps). As the X^+ is formed with two heavy holes of opposite spin (i.e., $m_h = +3/2$ and $m_h = -3/2$, respectively), the electron–hole exchange cancels in this charged exciton complex, so that the polarization of the charged excitons reflects the spin relaxation τ_e of the electron only, due to spin–orbit interaction.

The polarization decay time of the X^+ generated via X states [$X^+(X)$ spectrum] is intermediate, with an average time constant ≈ 22 ps. This intermediate behavior originates directly from the continuous creation of X^+ by the neutral X: the X^+ created at short times $t < \tau_{exc}$ result from highly polarized X and those retain their polarization for quite a long time (τ_e), while nonpolarized X^+ are generated at slightly longer delays from excitons that have already lost their spin orientation. The fact that the $X^+(X)$ exhibits a strong initial polarization shows that the creation of X^+ via X states does not affect the spin orientation.

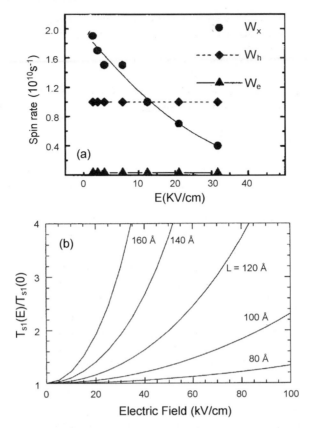

Fig. 3.12. (a) Measured dependence of the different spin relaxation rates W_x, W_h, and W_e on the applied electric field for a 15 nm GaAs/AlGaAs quantum well, $T = 20\,K$ [15]. (b) Calculated dependence of the exciton-spin relaxation time for various well widths [5]

Electric Field Dependence of the Exciton Spin Relaxation Time

An electric field applied along the quantum well growth axis will increase the separation between the electron and the hole within the exciton. The reduction of the overlap between the electron and the hole wave function will yield a decrease of the long-range part of the exchange interaction (see Sect. 3.1.2). As a result, the exciton spin relaxation rate decreases when the applied electric field increases (Fig. 3.12(a)). The measured variation of $W_x = 1/2\tau_{exc}$ is in rather good agreement with the calculated one (Fig. 3.12(b)) [5, 15].

3.4 Exciton Exchange Energy and g-Factor in Quantum Wells

The exciton exchange energy and g-factor are strongly modified compared to bulk values because of the confinement of the electron and hole wave functions along the quantum well growth direction. Both cw and time-resolved optical spectroscopy techniques have been used to measure these parameters in various quantum well structures [44, 46–48].

3.4.1 Exchange Interaction of Excitons and g-Factor Measured with cw Magneto-Photoluminescence Spectroscopy

Exciton Exchange Energy

The value of the short-range exchange interaction in the GaAs quantum well was first deduced from the measurements of the degree of circular polarization versus magnetic field of photoexcited luminescence [49]. The results presented below show evidence of exciton level crossings, which have been analyzed to give the short-range exciton exchange energy, which is about $\Delta_0 \approx 150\,\mu\text{eV}$ for a narrow GaAs/AlGaAs quantum well ($L_\text{W} \lesssim 5\,\text{nm}$).

The elegant technique used by Blackwood et al. relies on the measurement of the degree of circular polarization of the luminescence as a function of the applied magnetic field B_z (applied along the growth axis), under non-resonant linearly-polarized cw laser excitation. Figure 3.13 presents the variation of the circular polarization degrees P as a function of B_z for three quantum well structures with different well widths [49]. There is a general monotonic increase of $|P|$ with applied field, the sign depending on the direction of the field, with a superimposed peak at a field which varies with quantum well width.

This peak is due to magnetic field induced exciton level crossing (see Fig. 3.14). As the excitation is non-resonant (photogeneration of electron–hole pairs in the quantum well continuum), the bimolecular formation process of exciton will yield heavy-hole excitons in each of the four spin states ($|M\rangle = |+1\rangle, |-1\rangle, |+2\rangle, |-2\rangle$) with equal probability and the relative populations of the states under cw excitation will be determined by the balance of recombination processes and phonon-assisted relaxation between the levels. Thus populations of the two optically allowed levels will tend towards the Boltzmann thermal distribution, with the degree of thermalization depending on the relaxation rates between the levels. If these rates vary smoothly the population difference of the optically allowed levels will increase steadily with applied field. However, the transition rate between a pair of levels will increase sharply if their energies become equal, because the transition can then occur without the intervention of a phonon. This will be reflected in an anomaly (presence of a peak) in the population difference of the optically allowed levels and therefore in the degree of circular polarization P of the integrated luminescence.

Referring to Fig. 3.14, there are in general two fields at which levels cross. The calculation of the position of these level crossing allows one to estimate the zero-field exchange energy Δ_0. The effective Hamiltonian representing the interaction of a $1s$ exciton with a longitudinal magnetic field B_z can be written generally, according to

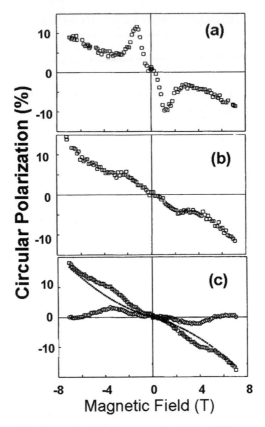

Fig. 3.13. Circular polarized cw photoluminescence for GaAs/AlGaAs multiple quantum well samples with well widths (**a**) 2.5 nm, (**b**) 5.6 nm, and (**c**) 7.3 nm [49]

[1], as

$$
\begin{aligned}
H_{ex} &= H_{hh}^{(SR)} + \mathcal{H}_{B\parallel} \\
&= 2\Delta_0 S_{ez} S_{hz} + \Delta_1 (S_{ex} S_{hy} + S_{ey} S_{hx}) + \Delta_2 (S_{ex} S_{hx} + S_{ey} S_{hy}) \\
&\quad + \mu_B B_z (g_{e\parallel} S_{ez} + g_{h\parallel} S_{hz}),
\end{aligned} \tag{3.10}
$$

where S_e is the electron spin operator and S_h is an effective spin operator representing the two heavy-hole states $|\pm 3/2\rangle \equiv |\mp 1/2\rangle_h$. The parameters $g_{e\parallel}$ and $g_{h\parallel}$, which are the electron and effective heavy-hole magnetic g-factors, and Δ_i ($i = 0, 1, 2$), which represent the short-range electron–hole exchange interaction, are functions of the quantum well width [50, 52] (see Sect. 3.1.1). Note that the expression (3.10) is valid down to C_{2v} symmetry.

The energies of the four heavy-hole exciton states for applied field parallel to z-axis are

$$
\delta E_{\pm 1} = \frac{\Delta_0}{2} \mp \frac{1}{2}\sqrt{\mu_B^2 B_z^2 (g_{h\parallel} + g_{e\parallel})^2 + \Delta_1^2}, \tag{3.11a}
$$

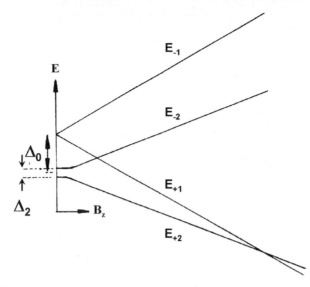

Fig. 3.14. Exciton energy levels as a function of the longitudinal magnetic field B_z. E_{+1} and $E_{\pm 2}$ correspond to the $|\pm 1\rangle$ optically active exciton states, and to the $|\pm 2\rangle$ non-optically active states, respectively [49]

$$\delta E_{\pm 2} = -\frac{\Delta_0}{2} \mp \frac{1}{2}\sqrt{\mu_B^2 B_z^2 (g_{h\parallel} - g_{e\parallel})^2 + \Delta_2^2}. \qquad (3.11b)$$

The levels are plotted in Fig. 3.14 for the ideal D_{2d} symmetry and for $\Delta_1 \ll \Delta_0$ [49]. The z component of exchange (Δ_0) causes a zero-field splitting between the optically active and non optically active states and the Δ_1 and Δ_2 components cause small additional zero-field splittings. D_{2d} has a fourfold rotation–reflection axis along the growth direction (z) which dictates $\Delta_1 = 0$, so that E_{+1} and E_{-1} are degenerate in zero field. If this symmetry is broken a zero-field splitting Δ_1 appears (see Sect. 3.5 on type II quantum wells or Chap. 4 on quantum dots).

The two fields, at which the exciton levels cross, are given by

$$B_z^{(h)} \approx \frac{\Delta_0}{g_{h\parallel}\mu_B} \quad \text{and} \quad B_z^{(e)} \approx \frac{\Delta_0}{g_{e\parallel}\mu_B}. \qquad (3.12)$$

From the measurements of the electron and hole g factors [50, 51], it turns out that $|g_{h\parallel}| > |g_{e\parallel}|$. The peaks observed in Fig. 3.13 are thus associated to $B_z^{(h)}$ ($B_z^{(e)}$ is beyond the range of measurement). So, the measurement of $B_z^{(h)}$ in Fig. 3.13 leads to the value of the exciton exchange energy Δ_0 plotted in Fig. 3.15. The exchange energy increases rapidly as the quantum well width decreases and as the barrier height increases. The values are in satisfactory agreement with calculations of the enhancement of the exchange relative to the bulk value ($\approx 10 \pm 5\,\mu eV$) due to enhanced electron–hole overlap [49], as expected from Sect. 3.1.1.

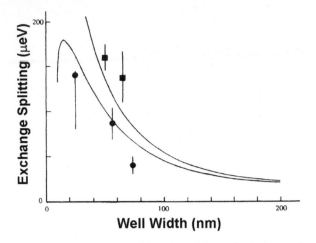

Fig. 3.15. Exciton exchange energy Δ_0 as a function of the well widths in a GaAs/AlGaAs (•) and GaAs/AlAs (■) quantum well. *The full lines* are the calculated values [49]

Exciton g-Factor

The effective Landé g-factor for the heavy-hole exciton in GaAs/AlGaAs quantum well has been determined as a function of well width from the Zeeman splitting of the cw luminescence spectra for moderate longitudinal magnetic fields (to avoid level crossings presented above) [50]. Figure 3.16(a) shows the measured Zeeman splittings up to $B_z = 2\,\mathrm{T}$ for different well widths. The variations as a function of the magnetic field are linear within the experimental uncertainties and the slopes give the values of $g_{\mathrm{exc\parallel}}(J = 1) = g_{\mathrm{e\parallel}} + g_{\mathrm{h\parallel}}$ which are plotted in Fig. 3.16(b), showing the change of sign for L_W between 7 and 11 nm.

3.4.2 Exciton Spin Quantum Beats Spectroscopy

Thanks to the development of ultrafast lasers and sensitive detectors, it has been possible to measure in the time domain the interaction of the exciton states with the external magnetic field [45–47, 53, 54]. This leads to measurements with a great accuracy of the exciton g-factor and exciton exchange energy.

The principle is the following. When two energetically closely spaced transitions are excited with a short optical pulse (with a spectral width larger than the splitting between the transitions), the two-induced polarizations in the medium oscillate with their slightly different frequencies. Their interference manifests itself in a modulation of the net polarization, the so-called Quantum Beats [55]. This allows energy splittings to be determined with higher resolution than in the spectral domain, provided that the beats period is shorter than their damping time.

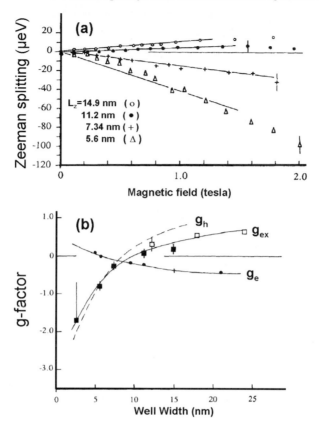

Fig. 3.16. (a) Low-field Zeeman splitting of the XH exciton luminescence lines in GaAs/AlGaAs quantum well at $T = 1.8$ K. (b) Electron ($g_e \equiv g_{e\parallel}$), heavy-hole ($g_h \equiv g_{h\parallel}$) and exciton [$g_{exc} \equiv g_{exc\parallel}(J = 1)$] g-factors in a GaAs/AlGaAs quantum well [50]

Exciton Spin Quantum Beats in Longitudinal Magnetic Fields

The exciton spin dynamics in longitudinal magnetic field (applied along the quantum well growth axis, Faraday configuration) has been measured with different experimental techniques, including time-resolved pump–probe transmission [53], time-resolved Faraday rotation [45, 54, 56] and time-resolved photoluminescence [46].

In a longitudinal magnetic field, the optically active exciton states are the $|+1\rangle$ and $|-1\rangle$ states split by the Zeeman energy $\hbar\Omega_\parallel = g_{exc}\mu_B B_z$, where $g_{exc} = g_{e\parallel} + g_{h\parallel}$. A linearly-polarized optical excitation pulse, resonant with the exciton energy, will thus create a coherent superposition of $|+1\rangle$ and $|-1\rangle$ states, making the observation of quantum beats as a function of time possible. Figure 3.17 shows the transient birefringence from the heavy-hole exciton in a 2.75 nm GaAs/AlGaAs multiple quantum well structure for various longitudinal magnetic fields [45]. The pump pulse (linearly-polarized) is resonant with the exciton absorption and the probe pulse has a linear polarization tilted by an angle of 45° with respect to the pump polarization.

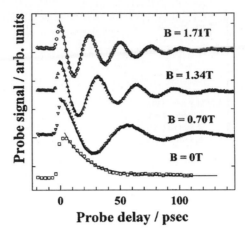

Fig. 3.17. Quantum beats observed in the transient birefringence from the XH exciton in a 2.75 nm GaAs/AlGaAs quantum well at $T = 1.8$ K for various applied longitudinal magnetic fields [45]

In this time-resolved Kerr rotation experiment, the transient pump-induced birefringence plotted in Fig. 3.17 corresponds to the degree of induced elliptization of the probe pulse reflected from the sample.

At zero field, there is an exponential decay, which corresponds to the coherent decay of the exciton linear polarization; the corresponding decay time T_{s2}^* is given by $1/T_{s2}^* = 1/T_{s2} + 1/\tau_{rad}$ where T_{s2} is the so-called exciton transverse spin relaxation time presented in Sect. 3.3.4 and τ_{rad} is the radiative lifetime [5]. As the magnetic field increases, the quantum beats observed in Fig. 3.17 correspond to the coherent oscillation between the Zeeman-split exciton levels ($M = \pm 1$) at the pulsation $\Omega_\parallel = g_{exc}\mu_B B_z/\hbar$. The fit of the data in Fig. 3.17 gives the Zeeman splitting from which the exciton g factor $|g_{exc}| = 1.52 \pm 0.01$ is obtained for a $L_W = 2.75$ nm multiple quantum well. This measurement is much more accurate than the ones performed previously in the spectral domain presented in Fig. 3.16(b) [50].

Exciton Spin Quantum Beats in Transverse Magnetic Fields

The spin Hamiltonian of the heavy-hole exciton in a transverse magnetic field (applied in the quantum well plane, $\boldsymbol{B}//x$) can be approximated by [47]

$$H = \hbar\omega S_x - \frac{2\Delta_0}{3}J_z S_z, \tag{3.13}$$

where $\hbar\omega = g_{e,x}\mu_B B_x$ and Δ_0 is the zero-field exciton exchange splitting between the optically active states $|\pm 1\rangle$ and the two dark states $|\pm 2\rangle$ (the much smaller splitting between the $|+2\rangle$ and $|-2\rangle$ states is neglected, as well as Δ_1) [1, 49]. We assume here that the transverse g-factor of the $j_{h,z} = \pm 3/2$ heavy hole is zero (spin quantum beats experiments performed in n-doped GaAs quantum well show that $g_{h,x} \approx 0.04$ [57], so that $g_{h,x} \ll g_{e,x} = g_{e,y}$).

The exciton quasistationary states $|\Psi_+\rangle$ in the transverse magnetic field are two linear combinations $E_\pm$ of optically active and inactive states split by the energy $\hbar\Omega_{exc}$ and are written as [46, 48, 59]

$$|\Psi_+\rangle \approx \hbar\omega|1\rangle + (\hbar\Omega_{exc} - \Delta_0)|2\rangle, \tag{3.14}$$

$$|\Psi_-\rangle \approx -(\hbar\Omega_{exc} - \Delta_0)|1\rangle + \hbar\omega|2\rangle, \tag{3.15}$$

where $\hbar\Omega_{exc} = (\Delta_0^2 + (\hbar\omega)^2)^{1/2}$. A (σ^+)-polarized pulsed excitation resonant with the exciton energy will thus create a coherent superposition of $|\Psi_+\rangle$ and $|\Psi_-\rangle$ states. Ignoring any spin relaxation processes (as well as recombination), the right (I^+) and left (I^-) circularly-polarized luminescence components are proportional to $|\langle\pm 1|\Psi(t)\rangle|^2$:

$$I^+(t) = 1 - \left(\frac{\omega}{\Omega_{exc}}\right)^2\left(\frac{1 - \cos(\Omega_{exc}t)}{2}\right), \tag{3.16}$$

$$I^-(t) = 0. \tag{3.17}$$

As a consequence, we expect to observe, in time-resolved photoluminescence, oscillations of the polarized emission $I^+(t)$ which should occur with a pulsation Ω_{exc}, i.e., the pulsation should not depend linearly on the applied transverse field. The co-polarized luminescence intensity I^+, modulated at the pulsation Ω_{exc} has an amplitude reduced by a factor $(\omega/\Omega_{exc})^2$, while the counter-polarized component is unmodulated in this simplified approach.

These exciton-like spin quantum beats were indeed observed in narrow multiple quantum well samples. Figure 3.18(b) presents the luminescence intensity dynamics co-polarized (I^+) and counter-polarized (I^-) with the resonant (σ^+)-polarized picosecond laser in a $L_W = 3$ nm multiple quantum well GaAs/AlGaAs sample. Quantum beats are observed only at strong magnetic field values; they appear as a weak amplitude modulation on the I^+ component but are not observable on I^- [18, 60]. If the excitation energy is higher than the quantum well band gap $(E_1 - HH_1)$, all the multiple quantum well samples exhibit quantum beats on I^+ and I^- with an oscillation frequency proportional to the magnetic field, see Fig. 3.18(a) [61, 62]. Oscillations on I^+ and I^- are phase shifted by π. These oscillations are attributed to the Larmor precession of the free electron with pulsation ω. This yields the accurate measurement of $g_{e,x} = 0.50 \pm 0.01$ in Fig. 3.18(a). When the laser excitation is resonant, we see clearly in Fig. 3.18(c) that the beat period is very different than in the non-resonant case. It is attributed to the exciton quantum beats and can be used to measure the exciton exchange energy $\Delta_0 = \hbar(\Omega_{exc}^2 - \omega^2)^{1/2}$; $\Delta_0 = 130 \pm 15$ μeV and $\Delta_0 = 105 \pm 10$ μeV are measured in a $L_W = 3$ nm and $L_W = 4.8$ nm GaAs/AlGaAs multiple quantum well structure respectively [46].

The following question arises now: why in most of the experiments performed in transverse magnetic fields do the authors observe electron quantum beats (with a pulsation $\omega = g_{e,\perp}\mu_B B/\hbar$) and not the exciton quantum beats [with a pulsation $\Omega_{exc} = ((\Delta_0/\hbar)^2 + \omega^2)^{1/2}$] though the recorded signal corresponds to exciton transitions [54, 61]. This enigma has been explained by Dyakonov et al. [47]. It turns

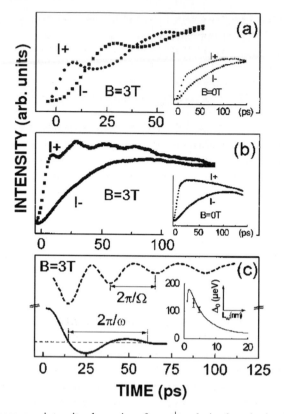

Fig. 3.18. Luminescence intensity dynamics after σ^+ polarized excitation in a $L_{\mathrm{w}} = 3\,\mathrm{nm}$ GaAs multiple quantum well, at $T = 1.7\,\mathrm{K}$. (**a**) The excitation energy is non-resonant ($E_1 - \mathrm{HH}_1 < h\nu < \mathrm{XL}$, where XL is the light-hole exciton energy) and $B = 3\,\mathrm{T}$ (inset, $B = 0$). (**b**) The excitation energy is resonant with XH and $B = 3\,\mathrm{T}$ (inset, $B = 0$). (**c**) The oscillations of the luminescence intensity component I^+ in resonant excitation (*dashed line*) and of the luminescence polarization photoluminescence in non-resonant excitation $E_1 - \mathrm{HH}_1 < h\nu < \mathrm{XL}$ (*full line*), under the same magnetic field $B = 3\,\mathrm{T}$. For the sake of clarity, the monotonous component has been subtracted from I^+. Inset: well-width dependence of the exciton exchange energy δ_0, from this experiment (dots with error bars) and theory [49, 46] (*full line*)

out that the observation of quantum beats on the excitonic luminescence at the electronic or excitonic pulsation (ω or Ω_{exc}, respectively) is related to the stability of the hole-spin orientation within the exciton. The argument is the following. Within the exciton, the correlation between electron and hole spins is held by the electron–hole exchange interaction. However, if this correlation is not strong enough to reduce the single-particle hole spin flip at a rate lower than $\Delta_0/\hbar$, the exchange interaction splitting Δ_0 no longer plays a role in the quantum beats. Then the quantum beats appear at the pulsation ω. Finally an electron bound into an exciton precesses like a free electron in the transverse magnetic field provided that $\tau_h \ll \hbar/\Delta_0$ where τ_h is the single-particle hole spin–flip time.

This condition can be fulfilled in large and narrow quantum wells but for different reasons. In large quantum wells (a fortiori in bulk material) such a hole spin flip occurs as a consequence of the mixing of states in the valence band due to spin–orbit interaction and small exchange interaction; the observation of the electron precession in a quantum well of 25 nm well width *under the resonant or non-resonant excitation conditions* reported initially by Heberle et al. is understood on this ground [61]. In narrow quantum wells the hole spin flip, which results in the observation of quantum beats at the pulsation ω in non-resonant excitation, is related to the formation-dissociation process of excitons and the related long cooling of the excited system [22, 23]. Quantum beats of the excitonic kind have been observed only in narrow quantum wells ($L_W < 10$ nm) *under resonant excitation*. This indicates again that the hole-spin orientation is rather stable in cold two-dimensional excitons ($\tau_h > \hbar/\Delta_0$), see Sect. 3.3.2. A model based on a density matrix approach has been developed in [47], which can reproduce the characteristic experimental features described above.

3.5 Exciton Spin Dynamics in Type II Quantum Wells

In the previous sections, we discussed the exciton spin properties in the so-called type I quantum well structures, i.e., where the conduction electron and the valence holes are confined in the same material and the same region in space. In GaAs/AlGaAs quantum wells, depending on the well width and Al percentage, it turns out that two types of lowest energy transitions are possible. For low Al content the conduction band-confined state in the well has the lowest energy (type I quantum well). For large Al content and small well width, the lowest conduction band-confined state in the well has a higher energy than the lowest X-confined state in the barrier [63]. As the holes are still confined in the GaAs well, the recombination takes therefore place between the hole in the well and the electrons in the barrier (type II quantum well). This accounts for the long exciton photoluminescence lifetime, of the order of a few microseconds [9].

The spin dynamics in these type II quantum well systems has been extensively studied by optical orientation experiments in stationary or time-resolved regime [1, 52, 58, 59, 64]. Because of the very small overlap between the electron and hole wave function, the spin relaxation mechanism induced by the exchange interaction between the electron and the hole does not play a significant role in contrast to type I quantum wells (see Sects. 3.1.2, 3.3.4). However the strong localization of the carrier wave function at the quantum well interface will (i) modify drastically the exciton fine structure and (ii) yield very long electron spin relaxation times ($\approx$a few tens of ns) compared to type I quantum wells [65]. It has been shown that the symmetry of the system is reduced form D_{2d} to C_{2v} and the two optically active exciton eigenstates are linearly polarized, split by an energy of a few μeV, and aligned along the $X' \equiv [1, 1, 0]$ and $Y' \equiv [1, -1, 0]$ crystal directions [52, 58, 64]. The symmetry

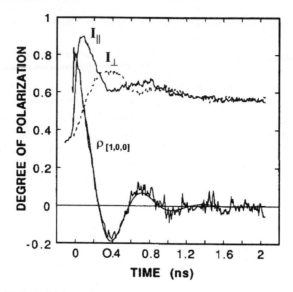

Fig. 3.19. Type II GaAs/AlAs 2.2/1.15 nm superlattice; $T = 4.2$ K. Degree of photoluminescence linear polarization $\rho_{[1,0,0]}$ as a function of time. The excitation picosecond laser pulse is linearly-polarized along the [1, 0, 0] axis. The intensities $I_{//}$ and $I_{\perp}$ are detected with polarization parallel and perpendicular to the excitation, respectively [58]

reduction in these type II GaAs superlattices was first explained by the presence of a random local deformation (due to the presence of bonds of different nature (Al–As or Ga–As) on each side of the interface taken as an As plane) which mixes the heavy and light hole states [58]. It was then shown that the splitting, called anisotropic exchange splitting, arises from an intrinsic effect: the mixing of heavy- and light-hole states at the interface due to the low (C_{2v}) symmetry of the interface.

Since the splitting of the X' and Y' excitonic sublevels (which is much smaller than $k_B T$), is much smaller than the laser spectral width, the two sublevels can be coherently excited at time $t = 0$ by a short pulse polarized along one of the [1, 0, 0] axes (i.e., 45° angle with respect to the exciton eigenstates orientations) [64]. As a consequence, the time-resolved luminescence signal, detected with polarization either parallel or perpendicular to the excitation, decays and oscillates with a period T' inversely proportional to the splitting Δ_1 between the two sublevels ($T' = h/\Delta_1$, see Sect. 3.1.1). The time dependence of the photoluminescence linear polarization is shown in Fig. 3.19 for a 2.2/1.5 nm type II GaAs/AlAs superlattice [58]. The period T' is about 640 ps, which corresponds to an energy splitting of $\approx$6.3 µeV. The application of a transverse magnetic field leads to a complex oscillation pattern with several frequencies, since, besides the coupling between $|+1\rangle$ and $|+2\rangle$ and $|-1\rangle$ and $|-2\rangle$ exciton states induced by the external magnetic field, the anisotropic exchange also couples the $|+1\rangle$ and $|-1\rangle$ states [59].

3.6 Spin Dynamics in Dense Excitonic Systems

When the exciton density becomes non-negligible with respect to the critical density defined by $n_c = [32\pi a_B^2]^{-1}$, the exciton mutual interactions start to modify significantly the single exciton picture we used up to now.[3] The above-mentioned critical density n_c corresponds to the one where the exciton binding energy is zero, due to phase space filling and screening of the Coulomb interaction. If the areal exciton density n_{ex} approaches n_c ($n_{ex} \lesssim n_c$), the exciton energies E_K must be corrected by a complex self-energy term, of which the real part corresponds to the energy shift and the imaginary part to the broadening of the single exciton states due to the mutual Coulomb interactions [66]. The experimental manifestations of these two complementary aspects in the case of two-dimensional structures, namely the spin dependent exciton energy shift and the spin dependent exciton–exciton collisions at high exciton densities, as revealed in elliptically polarized exciton populations, have been investigated in [72, 76].

In the case of excitation by elliptical light, excitons are created in the elliptical states:

$$|E_\theta\rangle = \sin(\theta + \pi/4)|+1\rangle + \cos(\theta + \pi/4)|-1\rangle \qquad (3.18)$$

so that linear excitons are given by $|X\rangle = |E_0\rangle$ and $i|Y\rangle = |E_{\pi/2}\rangle$. Excitons $|+1\rangle = |E_{\pi/4}\rangle$ and $|-1\rangle = |E_{-\pi/4}\rangle$, excited by σ^+ or σ^- light, respectively, are called circular excitons. The circular polarization of the state $|E_\theta\rangle$ is simply $P_c(\theta) = \sin(2\theta)$, while the linear one is $P_{\text{lin}}(\theta) = \cos(2\theta)$.

We are interested here in describing experiences performed at low temperature and under resonant (or quasiresonant) excitation with the heavy-hole excitons. The latter are thus created in the $1s$ state with very small, or even zero wave vector, i.e., with very small kinetic energy. As a consequence, the scattering probability to $2s$ or $2p$ states, which are close to the quantum well gap [74], is low. We restrain thus to the heavy-hole exciton subspace. The final result is that the interaction Hamiltonian can be approximated for an exciton pair (i, j), in a cold exciton population with low density, as [4]

$$H_{\text{exch}}^{ij} \approx \frac{6e^2 a_B}{\epsilon_0 A} \left(\sigma_e^{(i)} . \sigma_e^{(j)} + \sigma_h^{(i)} . \sigma_h^{(j)} + 2 \right), \qquad (3.19)$$

where $\sigma_{e(h)}^{(i)}$ represent Pauli matrices vector operators for electron spins and heavy-hole effective spins, and A is the quantum well quantization area. This expression results from the fact that the electron–electron or the hole–hole exchange dominate over direct Coulomb interaction as well as exciton exchange as a whole, provided that the initial wave vector K and K' of the excitons which interact are small with respect to $(a_B)^{-1}$ [77]. In such conditions, it neither depends on K, K', nor on the wave vector Q transferred during the collision.

---

[3] Note that the Bohr radius in two dimensions is 4 times smaller than the conventional Bohr radius.

The first evidence of exciton state energy shift at high density was obtained by Hulin et al., who observed, in femtosecond pump–probe experiments performed on GaAs/AlGaAs multiquantum well structures under a linearly polarized pump, a transient blue-shift of the exciton absorption line [67, 68]. The latter was shown to be tied to the reduced dimensionality of excitons, being well apparent in GaAs wells of thickness of the order of 5 nm, but disappearing rapidly for larger well sizes. The authors interpreted this effect in terms of a strong reduction of long-range many-body interactions in a 2D system, in agreement with the theory of Schmitt-Rink et al. [69].

It is well documented that in 3D systems, the exciton absolute energy remains unchanged, even at high densities [70]. This energy constancy is attributed to the almost exact compensation between two many-body effects acting in opposite directions: an interparticle attraction which, for bound electron–hole pairs at $T \approx 0$, is similar to a van der Waals interaction, and a repulsive contribution having its origin in the Pauli exclusion principle acting on the Fermi particles (electron and holes) forming the excitons. The argument of Schmitt-Rink et al. is that the long-range attractive component is strongly reduced in a 2D system, so that the short-range repulsive part becomes now unbalanced.

Using circularly polarized excitation in time resolved polarized luminescence, it has been shown that this blue shift is spin dependent, and that in a dense and circularly polarized exciton gas, a splitting occurs between the line co-polarized with the excitation, and the counter-polarized one, the former experiencing a blue shift, while the latter is red shifted [19, 30, 71]. Figure 3.20 presents the result of an experiment performed under resonant excitation on a high quality GaAs/AlGaAs multiquantum well structure (i.e., with the Stokes shift less than 0.1 meV and cw photoluminescence line width $\Gamma \approx 0.9$ meV at $T \approx 1.7$ K). Two-color time resolved up-conversion photoluminescence spectroscopy was used to perform such experiments, the excitation laser pulse duration being $\delta t \approx 1.5$ ps, so that only the $1s$ heavy-hole exciton state is excited [72]. Just after a σ^+ circularly polarized excitation resonant with the XH exciton ($P_E \approx 1$, $h\nu_E = $ XH), the strongly polarized emission ($P_L \approx 0.9$) displays a splitting between the co-polarized emission line I^+ and the counter-polarized one I^-.

In addition, the I^+ emission is strongly blue shifted, while the I^- is slightly red shifted. When increasing the excitation power, the energetic positions first vary linearly. A saturation then occurs at $P_{\text{sat}} \approx 3$ mW, when the I^+ line shifts to energies higher than the XH energy by the laser line width ($\delta_E \approx 2$ meV). The absorption then drops, due to energy mismatch between the laser and the renormalized XH exciton energy. This situation corresponds to the exciton density n_{sat} estimated at $n_{\text{sat}} \sim 2 \times 10^{10}$ cm^{-2}. The splitting amounts then to about $\delta E_{+1} \approx 1.9$ meV. For $P > P_{\text{sat}}$, a self-regulation effect appears for the exciton density. When the excitation energy is increased at XH $+ 3$ meV, the saturation effect occurs at higher excitation power. The blue shift can thus become a significant fraction of the $1s$ exciton binding energy, here estimated at $E_B \approx 8$ meV. The saturation exciton density is below the critical density n_c, here estimated to about 3×10^{11} cm^{-2}. A good phenomenological description of the line positions in the linear density regime ($n_{\text{ex}} \ll n_c$ and $n_{\text{ex}} <$

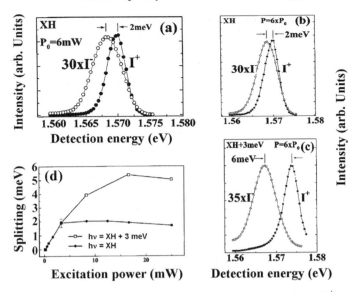

Fig. 3.20. Spectra of the exciton luminescence components co-polarized (I^+) and counter-polarized (I^-) with the circularly polarized excitation ($P_E \approx 1$) at time delay $t = 4$ ps. (a) The excitation energy is set to $h\nu_E = $ XH. The average excitation power P is $P_0 = 6$ mW. (b) $h\nu = $ XH and $P = 6P_0$. (c) $h\nu = $ XH + 3 meV and $P = 6P_0$. (d) Splitting energy between the two luminescence components I^+ and I^- as a function of P: (●) $h\nu = $ XH; (□) $h\nu = $ XH + 3 meV [72]

n_{sat}) can be obtained for a cold exciton gas, according to [71]:

$$\delta E_{\pm 1} = K_1 n_{\pm 1} + \frac{1}{2} K_1 (n_{+2} + n_{-2}) - K_2 n_{ex}, \qquad (3.20a)$$

$$\Delta E \equiv E_{+1} - E_{-1} = K_1 (n_{+1} - n_{-1}), \qquad (3.20b)$$

where the n_M are the exciton population densities corresponding to the $|M\rangle$ states ($n_M = N_M/A$), n_{ex} is the total exciton density, and K_1 and K_2 are positive constants. The first one, K_1, represents the strength of the repulsive part of the interaction between $J = 1$ excitons of the *same* angular momentum projection M, which takes its origin in the Pauli repulsion principle. Since the $|+1\rangle$ exciton shares its electron spin states with the $|-2\rangle$ exciton, but not its hole state, the contribution to the exchange energy shift of $|+1\rangle$ excitons by the $|-2\rangle$ excitons is taken as $(K_1/2)n_{-2}$. A similar reason holds for the contribution of n_{+2} excitons to δE_{+1}. The constant K_2, also positive, represents the weak attractive part of the interaction between excitons; for the sake of simplicity, it is assumed spin independent [69, 73].

The experimental values of K_1 and K_2 can be determined from the initial splitting ΔE and energy shift E_{-1} just after the resonant excitation by σ^+ light, so that $n_M(0) = n_{ex}(0)\delta_{M,+1}$. It is found that $10^{-10} \lesssim K_1 \lesssim 1.6 \times 10^{-10}$ meV cm^2, depending on the quantum well, and $K_2/K_1 \sim 0.15$ typically. From the theoretical calculation of Schmitt-Rink et al. for a non-polarized exciton gas [69], we can infer that $K_1 \approx 2 \times 3.86\pi (a_B)^2 E_B \approx 4 \times 6.06 e^2/(2\epsilon_0 a_B)$. Considering now a_B as a variational parameter a_{eff}, which is a good approximation for narrow quantum wells with marked 2D character [74],[4] the theoretical estimation of K_1 is in reasonable agreement with the experimental values.

Besides its contribution to the energy shift, the exchange interaction between excitons at high density is at the origin of specific exciton spin relaxation processes. In a dense gas of elliptical excitons, the exchange interaction between excitons becomes stronger than the internal electron–hole exchange within single excitons, and destroys the intra-exciton spin coherence. These effects have been investigated in [73, 76]. Spin-dependent mutual exciton scattering, as well as exchange assisted transfer between dark and bright states were thus revealed. An important consequence is the fast decay of both circular and linear exciton polarizations, in a dense and strictly elliptically polarized exciton system, while, for a strictly circularly polarized exciton system, the circular polarization decay does not depend on the exciton density, provided the latter remains below the critical density.

To conclude this section, let us mention that experimental studies of spin dynamics in the context of strong coupling of excitons with the electromagnetic field in semiconductor microcavities have been also performed and analyzed. The quasiparticle resulting from this coupling is called 2D excitonic-polariton, which is the 2D analog of Hopfield 3D polaritons [78]. The exciton-polaritons present a more marked bosonic character than bare excitons, due to their photon component. Specific aspects of 2D polaritons spin dynamics which rely on their exchange driven spin dependent scattering, such as spin dependent blue shift or parametric conversion of $|X\rangle$ to $|Y\rangle$ linearly polarized states can be found, e.g., in [79–84]. Finally, let us mention that optical Spin-Hall effect has been recently observed in such microcavities [85].

References

[1] E.L. Ivchenko, G.E. Pikus, *Superlattices and Other Heterostructures*. Springer Series in Solid States Sciences, vol. 110 (Springer, Berlin, 1997)
[2] L.C. Andreani, Optical transitions, excitons, and polaritons in bulk and low-dimensional semiconductor structures, in *Confined Electrons and Photons*, ed. by E. Burstein, C. Weisbuch (Plenum Press, New York, 1995)
[3] F. Meier, B.P. Zakharchenya (eds.), *Optical Orientation* (North-Holland, Amsterdam, 1984)
[4] T. Amand, X. Marie, cond-mat/0711.2030
[5] M.Z. Maialle, E.A. de Andrada e Silva, L.J. Sham, Phys. Rev. B **47**, 15776 (1993)
[6] G.F. Koster, J.O. Dimmock, R.G. Wheeler, H. Statz, *Properties of the Thirty-two Point Groups* (MIT, Cambridge, 1963)
[7] C. Weisbuch, R.G. Ulbrich, Resonant light scattering mediated by excitonic polaritons, in *Semiconductors*, ed. by M. Cardona, G. Güntherrodt. Light Scattering in Solids, vol. III (Springer, Berlin, 1982)

---

[4] The relation $E_B a_{\text{eff}} = e^2/(2\epsilon_0)$ interpolates correctly between the 2D and the 3D case [69].

[8] H.S. Chao, K.S. Wong, R.R. Alfano, H. Unlu, H. Morkoc, *Ultrafast Laser Probe Phenomena in Bulk and Microstructure. II.* SPIE, vol. 942 (1988), p. 215

[9] W.A.J.A. van der Poel, A.L.G.J. Severens, H.W. van Kesteren, C.T. Foxon, Superlattices Microstruct. **5**, 115 (1989)

[10] T.C. Damen, K. Leo, J. Shah, J.E. Cunningham, Appl. Phys. Lett. **58**, 1902 (1991)

[11] M.R. Freeman, D.D. Awschalom, J.M. Hong, Appl. Phys. Lett. **57**, 704 (1990)

[12] A. Tackeuchi, S. Muto, T. Inata, T. Fuji, Appl. Phys. Lett. **56**, 2213 (1990)

[13] K. Zerrouati, F. Fabre, G. Bacquet, J. Bandet, J. Frandon, G. Lampel, D. Paget, Phys. Rev. B **37**, 1334 (1988)

[14] B. Dareys, T. Amand, X. Marie, B. Baylac, J. Barrau, M. Brousseau, I. Razdobreev, D.J. Dunstan, J. Phys. IV, C5 **3**, 351 (1993)

[15] A. Vinattieri, J. Shah, T.C. Damen, D.S. Kim, L.N. Pfeiffer, M.Z. Maialle, L.J. Sham, Phys. Rev. B **50**, 10868 (1994)

[16] J. Barrau, G. Bacquet, F. Hassen, N. Lauret, T. Amand, M. Brousseau, Superlattices Microstruct. **14**, 27 (1993)

[17] S. Pfalz, R. Winkler, T. Nowitzki, D. Reuter, A.D. Wieck, D. Hägele, M. Oestreich, Phys. Rev. B **71**, 165305 (2005)

[18] M. Oestreich, D. Hägele, J. Hubner, W.W. Rühle, Phys. Stat. Sol. (a) **178**, 1 (2000)

[19] B. Dareys, X. Marie, T. Amand, J. Barrau, Y. Shekun, I. Razdobreev, R. Planel, Superlattices Microstruct. **13**, 353 (1993)

[20] E.A. de Andrada e Silva, G.C. La Rocca, Phys. Rev. B **56**, 9259 (1997)

[21] T. Amand, B. Dareys, B. Baylac, X. Marie, J. Barrau, M. Brousseau, D.J. Dunstan, R. Planel, Phys. Rev. B **50**, 11624 (1994)

[22] J. Szctytko, L. Kappei, J. Berney, F. Morier-Genoud, M.T. Portella-Oberli, B. Deveaud, Phys. Rev. Lett. **93**, 137401 (2004)

[23] D. Robart, X. Marie, B. Baylac, T. Amand, M. Brousseau, G. Baquet, G. Debart, R. Planel, J.M. Gérard, Solid State Commun. **95**, 287 (1995)

[24] K. Siantidis, V.M. Axt, T. Kuhn, Phys. Rev. B **65**, 35303 (2001)

[25] C. Piermarocchi, F. Tasone, V. Savona, A. Quattropani, P. Schwendimann, Phys. Rev. B **55**, 1333 (1997)

[26] P. Le Jeune, X. Marie, T. Amand, E. Vanelle, J. Barrau, M. Brousseau, R. Planel, *Proceedings of ICPS 24* (World Scientific, Singapore, 1998)

[27] G. Bastard, R. Ferreira, Surf. Sci. **267**, 335 (1992)

[28] T. Uenoyama, L.J. Sham, Phys. Rev. B **42**, 7114 (1990)

[29] Ph. Roussignol, R. Ferreira, C. Delalande, G. Bastard, A. Vinattieri, J. Martinez-Pastor, L. Carraresi, M. Colocci, J.F. Palmier, B. Etienne, Surf. Sci. **305**, 263 (1995)

[30] T.C. Damen, L. Vina, J.E. Cunningham, J. Shah, L.J. Sham, Phys. Rev. Lett. **67**, 3432 (1991)

[31] B. Baylac, X. Marie, T. Amand, M. Brousseau, J. Barrau, Y. Shekun, Surf. Sci. **326**, 161 (1995)

[32] D.W. Snoke, W.W. Rühle, K. Köhler, K. Ploog, Phys. Rev. B **55**, 13789 (1997)

[33] M.I. Dyakonov, V.I. Perel, Sov. Phys. Solid State **13**, 3023 (1972)

[34] M.I. Dyakonov, V.Yu. Kachorovskii, Sov. Phys. Semicond. **20**, 110 (1986)

[35] X. Marie, P. Le Jeune, T. Amand, M. Brousseau, J. Barrau, M. Paillard, Phys. Rev. Lett. **78**, 3222 (1997)

[36] B. Deveaud, F. Clérot, N. Roy, K. Satzke, B. Sermage, D.S. Katzer, Phys. Rev. Lett. **67**, 2355 (1991)

[37] T. Ostatnicky, O. Crégut, M. Gallart, P. Gilliot, B. Hönerlage, J.-P. Likformann, Phys. Rev. B **75**, 165311 (2007)

[38] H. Rahimpour Soleimani, S. Cronenberger, M. Gallard, P. Gilliot, J. Cibert, O. Crégut, B. Hönerlage, J.P. Likforman, Appl. Phys. Lett. **87**, 192104 (2005)

[39] E. Vanelle, M. Paillard, X. Marie, T. Amand, P. Gilliot, D. Brinkmann, R. Levy, J. Cibert, S. Tatarenko, Phys. Rev. B **62**, 2696 (2000)

[40] K. Kheng, R.T. Cox, Y. Merle d'Aubigné, F. Bassani, K. Saminadayar, S. Tatarenko, Phys. Rev. Lett. **71**, 1752 (1993)

[41] Z. Chen, R. Bratschitsch, S.G. Carter, S.T. Cundiff, D.R. Yakovlev, G. Karczewski, T. Wojtowicz, J. Kossut, Phys. Rev. B **75**, 115320 (2007)

[42] R.I. Dzhioev, V.L. Korenev, M.V. Lazarev, V.F. Sapega, D. Gammon, A.S. Bracker, Phys. Rev. B **75**, 33317 (2007)

[43] E. Tsitsishvili, R. von Baltz, H. Kalt, Phys. Rev. B **71**, 155320 (2005)

[44] R.T. Harley, M.J. Snelling, Phys. Rev. B **53**, 9561 (1996)

[45] R.E. Worsley, N.J. Traynor, T. Grevatt, R.T. Harley, Phys. Rev. Lett. **76**, 3224 (1996)

[46] T. Amand, X. Marie, P. Le Jeune, M. Brousseau, D. Robart, J. Barrau, R. Planel, Phys. Rev. Lett. **78**, 1355 (1997)

[47] M. Dyakonov, X. Marie, T. Amand, P. Le Jeune, D. Robart, M. Brousseau, J. Barrau, Phys. Rev. B **56**, 10412 (1997)

[48] J. Puls, F. Henneberger, Phys. Stat. Sol. (a) **164**, 499 (1997)

[49] E. Blackwood, M.J. Snelling, R.T. Harley, S.R. Andrews, C.T.B. Foxon, Phys. Rev. B **50**, 14246 (1994)

[50] M.J. Snelling, E. Blackwood, C.J. McDonagh, R.T. Harley, Phys. Rev. B **45**, 3922 (1992)

[51] M.J. Snelling, G.P. Flinn, A.S. Plaut, R.T. Harley, A.C. Tropper, R. Eccleston, C.C. Phillips, Phys. Rev. B **44**, 11345 (1991)

[52] H.W. van Kesteren, E.C. Cosman, W.A.J.A. van der Poel, C.T. Foxon, Phys. Rev. B **41**, 5283 (1990)

[53] S. Bar-Ad, I. Bar-Joseph, Phys. Rev. Lett. **66**, 2491 (1991)

[54] D.D. Awschalom, D. Loss, N. Samarth, *Semiconductor Spintronics and Quantum Computation, NanoScience and Technology* (Springer, Berlin, 2002)

[55] S. Haroche, in *Topics in Applied Physics*, vol. 13, ed. by K. Shimoda (Springer, Berlin, 1976), p. 253

[56] J. Baumberg, S.A. Crooker, D.D. Awschalom, N. Samarth, H. Luo, J.K. Furdyna, Phys. Rev. Lett. **72**, 712 (1994)

[57] X. Marie, T. Amand, P. Le Jeune, M. Paillard, P. Renucci, L.E. Golub, V.M. Dymnikov, E.L. Ivchenko, Phys. Rev. B **60**, 5811 (1999)

[58] C. Gourdon, P. Lavallard, Phys. Rev. B **46**, 4644 (1992)

[59] I.V. Mashkov, C. Gourdon, P. Lavallard, D.Y. Roditchev, Phys. Rev. B **55**, 13761 (1997)

[60] Ya. Gerlovin, Yu.K. Dolgikh, S.A. Eliseev, V.V. Ovsyankin, Yu.P. Efimov, V.V. Petrov, I.V. Ignatiev, I.E. Kozin, Y. Masumoto, Phys. Rev. B **65**, 35317 (2001)

[61] A.P. Heberle, W.W. Rühle, K. Ploog, Phys. Rev. Lett. **72**, 3887 (1994)

[62] R.M. Hannak, M. Oestreich, A.P. Heberle, W.W. Rühle, K. Köhler, Solid State Commun. **93**, 313 (1995)

[63] P. Dawson, B.A. Wilson, C.W. Tu, R.C. Miller, Appl. Phys. Lett. **48**, 541 (1986)

[64] W.A.J.A. van der Poel, A.L.G.J. Severens, Opt. Commun. **76**, 116 (1990)

[65] E.A. de Andrada e Silva, G.C. La Rocca, Physica E **2**, 839 (1998)

[66] G.D. Mahan, *Many Particle Physics* (Plenum, New York, 1981)

[67] N. Peyghambarian, H.M. Gibbs, J.L. Jewell, A. Antonetti, A. Migus, D. Hulin, A. Mysyrowicz, Phys. Rev. Lett. **53**, 2433 (1984)

[68] D. Hulin, A. Mysyrowicz, A. Antonetti, A. Migus, W.T. Masselink, H. Morkoc, H.M. Gibbs, N. Peyghambarian, Phys. Rev. B **33**, 4389 (1986)

[69] S. Schmitt-Rink, D.S. Chemla, D.A.B. Miller, Phys. Rev. B **32**, 6601 (1985)

[70] H. Haug, S. Schmitt-Rink, Prog. Quantum Electron. **9**, 3 (1984)

[71] T. Amand, X. Marie, B. Baylac, B. Dareys, J. Barrau, M. Brousseau, R. Planel, D.J. Dunstan, Phys. Lett. A **193**, 105 (1994)

[72] P. Le Jeune, X. Marie, T. Amand, F. Romstad, F. Perez, J. Barrau, M. Brousseau, Phys. Rev. B **58**, 4853 (1998)

[73] J. Fernandez-Rossier, C. Tejedor, L. Muoz, L. Via, Phys. Rev. B **54**, 11582 (1996)

[74] G. Bastard, *Wave Mechanics Applied to Semiconductor Heterostructures*. Les Éditions de Physique, Paris (1989)

[75] C. Ciuti, P. Swendimann, B. Deveaud, A. Quattropani, Phys. Rev. B **62**, R4825 (2000)

[76] T. Amand, D. Robart, X. Marie, M. Brousseau, P. Le Jeune, J. Barrau, Phys. Rev. B **55**, 9880 (1997)

[77] C. Ciuti, V. Savona, C. Piermarocchi, A. Quattropani, P. Swendimann, Phys. Rev. B **58**, 7926 (1998)

[78] J.J. Hopfield, Phys. Rev. **112**, 1555 (1955)

[79] X. Marie, P. Renucci, S. Dubourg, T. Amand, P. Le Jeune, J. Barrau, J. Bloch, R. Planel, Phys. Rev. B **59**, R2494 (1999). Rapid. Com.

[80] I. Shelykh, G. Malpuech, K.V. Kavokin, A.V. Kavokin, P. Bigenwald, Phys. Rev. B **70**, 115301 (2004)

[81] A. Kavokin, P.G. Lagoudakis, G. Malpuech, J.J. Baumberg, Phys. Rev. B **67**, 195321 (2003)

[82] P. Renucci, T. Amand, X. Marie, P. Senellart, J. Bloch, B. Sermage, K.V. Kavokin, Phys. Rev. B **72**, 075317 (2005)

[83] D.N. Krizhanovskii, D. Sanvitto, I.A. Shelykh, M.M. Glazov, G. Malpuech, D.D. Solnyshkov, A. Kavokin, S. Ceccarelli, M.S. Skolnick, J.S. Roberts, Phys. Rev. B **73**, 073303 (2006)

[84] D.D. Solnyshkov, I.A. Shelykh, M.M. Glazov, G. Malpuech, T. Amand, P. Renucci, X. Marie, A.V. Kavokin, *Physique et Technique des Semiconductors*, vol. 41 (Springer, Berlin, 2007), p. 1099

[85] C. Leyder, M. Romanelli, J.Ph. Karr, E. Giacobino, T.C.H. Liew, M.M. Glazov, A.V. Kavokin, G. Malpuech, A. Bramati, Nat. Phys. Lett. **3**, 628 (2007)

4

Exciton Spin Dynamics in Semiconductor Quantum Dots

X. Marie, B. Urbaszek, O. Krebs, and T. Amand

4.1 Introduction

Semiconductor quantum dots are nanometer sized objects that contain typically several thousand atoms of a semiconducting compound resulting in a confinement of the carriers in the three spatial directions. They can be synthesized by a large variety of methods based on colloidal chemistry [1, 2], molecular beam epitaxy or metalorganic chemical vapor deposition. Quantum dots can be formed at interface steps of thin quantum wells [3, 4] or by self-assembly in the Stransky–Krastanov growth mode during molecular beam epitaxy. This process is driven by the strain resulting from the difference in lattice parameter between the matrix (barrier) and the dots, for example 7% for InAs dots in GaAs. The quantum dots obtained in this well-studied system are typically 20 nm in diameter and 5 nm in height and are formed on a thin quantum well referred to as a wetting layer, see Fig. 4.1 for a transmission electron microscope image [5]. Samples used for optical spectroscopy are then covered again by the barrier material. The Stransky–Krastanov growth mode is applied to a large variety of III–V and II–VI compounds [6–8]. An interesting alternative for fabricating GaAs or InAs dots is provided by a technique which is not strain driven, called molecular droplet epitaxy [9]. Quantum dots defined by electrostatic potentials have also shown very interesting effects at very low temperature [10, 11].

Quantum dots are often referred to as artificial atoms because of the discrete nature of their valence and conduction states. The discrete nature of the electronic levels involved in the optical transitions lies at the origin of fascinating experiments that use dots as emitters of single, indistinguishable, and entangled photons [12–14]. Long optical coherence times [15] have made coherent control experiments possible [7, 16]. In this chapter we describe the carrier spin dynamics in quantum dots with a direct band gap. The absence of translational motion prolongs the carrier spin lifetimes as compared to bulk (3D) and quantum well (2D) structures, because spin relaxation mechanisms based on the spin–orbit interaction are strongly inhibited—see Chap. 1.

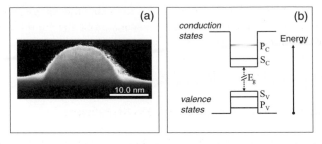

Fig. 4.1. (**a**) Electron microscopy image of an InAs dot in GaAs [5]. (**b**) Sketch of the lowest lying, discrete energy states in a quantum dot, where the energy separation E_g is determined by the band gap energy and strain in the semiconducting quantum dot material

However, two spin interactions are enhanced by the strong quantum confinement, namely the Coulomb exchange interaction between carriers and the hyperfine interaction between electron and nuclear spins [17]. The latter is a fascinating subject in its own right and can give rise, for example, to dynamical nuclear polarization [18, 19], electron spin dephasing [10, 11, 20, 21] and bistable nuclear spin configurations [22, 23]. The hyperfine effects will be evoked in this chapter where appropriate, but for a more detailed discussion the reader is referred to Chap. 11. In the following sections we will focus on the Coulomb exchange interaction which controls the spin phenomena in dots containing an electron–hole pair (neutral exciton) and plays a key role for spin dynamics of dots containing charged excitons, i.e., an electron–hole pair and an additional hole or electron.

The crystal growth axis in quantum dot samples is the z axis. Along this spatial direction the quantum confinement is the strongest. In optical spectroscopy, the light propagation direction is often chosen to be along the z axis. An optically active exciton in a neutral dot has a total angular momentum projection onto the z axis of either $|J_z = +1\rangle$ or $|J_z = -1\rangle$, as will be detailed in Sect. 4.2. The optically active exciton states have been found to be very robust against spin–flip mechanisms, although they exhibit spin quantum beats already at zero external magnetic field due to the anisotropic exchange interaction—see Sect. 4.3. In Sect. 4.4, the effect of an external magnetic field on the neutral exciton spin dynamics is investigated. The spin dynamics of charged excitons without applied fields is discussed in Sect. 4.5, and interesting effects, such as exchange driven negative polarization and gate voltage controlled hole spin relaxation, are highlighted. Electron spin coherence and polarization in charged excitons in magnetic fields parallel or perpendicular to the z axis are examined in Sect. 4.6.

4.2 Electron–Hole Complexes in Quantum Dots

The electronic structure of quantum dots can be analyzed through techniques such as capacitance–voltage measurements, electron spin resonance and a large variety of optical spectroscopy experiments. The latter allows a detailed analysis of the optically active eigenstates and their symmetry by analyzing the energy and polarization of absorbed or emitted photons. These experiments probe the interplay between carrier confinement, direct and exchange Coulomb terms and eventually the hyperfine interaction. In the following section the orders of magnitude of the different effects that determine the exciton fine structure are outlined.

4.2.1 Coulomb Corrections to the Single Particle Picture

Quantum dots can be populated by valence holes and conduction electrons through optical excitation and/or through controlled tunneling in charge tuneable structures [6, 24]. The single particle energies may be reasonably well determined by assuming a parabolic confinement potential. For self-assembled as well as interface fluctuation dots the vertical confinement energies are almost an order of magnitude larger than the lateral confinement energies. The quantization energies of both electrons and holes are larger than the Coulomb energies. The Coulomb effects can therefore be treated as perturbations to the single particle energy spectrum [24, 25]. At zero magnetic field the lowest lying conduction (valence) level S_c (S_v) is twofold degenerate and the adjacent P_c (P_v) level is fourfold degenerate, see Fig. 4.1(b) for the energy level diagram. For brevity a Coulomb correlated electron–hole pair trapped inside a dot by the confinement potential will be referred to as an *exciton*.

4.2.2 Fine Structure of Neutral Excitons

In early measurements on single quantum dots, fine structure splittings due to electron–hole exchange interaction of several tens of μeV were observed [3]. Although small in energy, a thorough understanding of the exciton fine structure is necessary to determine the eigenstates of the electron–hole complexes and the resulting polarization direction and amplitude of the photons emitted or absorbed during optical or electrical excitation. In its general form the exchange energy is proportional to the integral:

$$E_{eh}^X \propto \int\!\!\int \frac{\Psi_{eh}^*(r_1, r_2)\Psi_{eh}(r_2, r_1)}{|r_1 - r_2|} \, dr_1 \, dr_2, \tag{4.1}$$

where $\Psi_{eh}(r_e, r_h)$ is the exciton wave function and $r_{e(h)}$ is the electron (hole) coordinate vector. The electron–hole pair fine structure is strongly affected by the interplay between the exchange interaction and the reduced dot symmetry, usually C_{2v}. In the latter symmetry, the projection on the quantization axis z of the total angular momentum is not a good quantum number. Yet, the conduction (valence) single particle states can still be described, to zeroth order, as the S_c (S_v) and P_c (P_v) 2D atomic like orbitals. Taking into account the spin–orbit interaction, it is convenient to start with the same basis as in (001)-grown type I quantum wells (for which the relevant symmetry is D_{2d}), see Chap. 3. The lowest conduction state is then S-like, with two spin states $s_{e,z} = \pm 1/2$; the upper valence band is split into a heavy-hole band with the angular momentum projection $j_{h,z} = \pm 3/2$ and a light-hole band with $j_{h,z} = \pm 1/2$ at the center of the Brillouin zone. To simplify the discussion we focus on quantum dots with built-in strain such as self assembled dots. In this system the light-hole states lie at much lower energy in the valence band than the heavy-hole ones. Therefore, the optical transitions with the lowest energy will be between heavy holes and conduction band electrons. Only these so called heavy-hole exciton states will be considered in the following.

The heavy-hole exciton states can be described using the basis set $|J_z\rangle = |j_{h,z}, s_{e,z}\rangle$, i.e., $|J_z = +1\rangle = |3/2, -1/2\rangle$, $|J_z = -1\rangle = |-3/2, 1/2\rangle$, $|J_z = +2\rangle = |3/2, 1/2\rangle$, $|J_z = -2\rangle = |-3/2, -1/2\rangle$. The $|\pm 1\rangle$ states are accessible through one photon absorption or emission and are called radiative (optically active) states, in contrast to the non-radiative $|\pm 2\rangle$ states. At this point, it is convenient to describe the heavy-hole states in the presence of spin–orbit interaction by a pseudospin $1/2$, the $|\pm 1/2\rangle_h$ states corresponding to the $|\mp 3/2\rangle$ states [26, 27]. The heavy-hole exciton exchange Hamiltonian can then be written, in the C_{2v} symmetry, in the general form [28]:

$$H_{e,h}^{ex} = 2\delta_0 j_z s_z + \delta_1 (j_x s_x - j_y s_y) + \delta_2 (j_x s_x + j_y s_y), \qquad (4.2)$$

where the $\{s_\alpha\}$ and $\{j_\alpha\}$ ($\alpha = x, y, z$) are Pauli matrices representing the electron spin and the heavy-hole pseudospin operators. δ_0, δ_1, δ_2 are phenomenological parameters corresponding to the energy distances between centers of the radiative and non-radiative doublets, and between levels within the radiative and nonradiative doublets, respectively. The splittings δ_0 and δ_2 originate from the short range electron–hole Coulomb exchange contribution and usually obey the relation $\delta_2 \ll \delta_0$ [26]. The anisotropic part of the exchange interaction splits the $|\pm 1\rangle$ radiative exciton doublets into two eigenstates labeled $|X\rangle = (|+1\rangle + |-1\rangle)/\sqrt{2}$ and $|Y\rangle = (|+1\rangle - |-1\rangle)/i\sqrt{2}$, linearly polarized along the [110] and [1$\bar{1}$0] crystallographic directions, respectively [29, 30]. The corresponding energy splitting δ_1, which includes short as well as long range contributions, is comparable to δ_0 [31]. The two corresponding linearly polarized lines have been clearly identified in photoluminescence (PL) experiments on individual interface fluctuation [3, 4] and self-organized dots [29, 32, 33].

The anisotropic exchange splitting δ_1 originates from quantum dot elongation and/or interface optical anisotropy [3, 27] and varies from a few μeV in InAs dots, see Figs. 4.2(a) and (c), up to several hundreds of μeV in CdSe dots [34]. A nice demonstration of Kramers theorem is presented in Fig. 4.2(a) and (b): the neutral exciton X^0 shows a split doublet, while the charged exciton X^- (two electrons in a spin singlet state and one hole) consists of a single line [33]. This theorem states indeed that particle complexes with an integer total spin (X^0 case) may show a splitting at zero applied magnetic field for low symmetry systems,[1] whereas half-integer spin complexes do not split irrespective of the quantum dot symmetry [26, 28, 35].

---

[1] By adjusting the growth method or applying a magnetic or electric field perpendicular to the growth direction [36] the fine structure splitting δ_1 of neutral excitons can be tuned close to zero, i.e., smaller than the homogeneous broadening of the optical transition. This is an important feature for application of quantum dots as a source of entangled photon pairs [14].

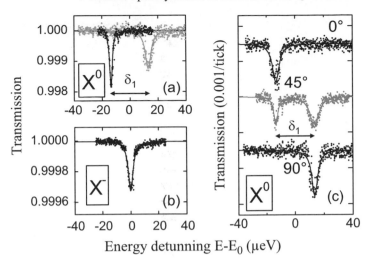

Fig. 4.2. Differential transmission of an individual InAs dot embedded in a charge tuneable structure measured with an ultra-narrow line width laser [33]. (**a**) The two curves for the neutral exciton were recorded with orthogonal linear polarizations, the splitting corresponding to the anisotropic exchange energy is δ_1. (**b**) The singly charged exciton and (**c**) dependence of the transmission on the angle of the excitation laser field with respect to the [110] direction

4.3 Exciton Spin Dynamics in Neutral Quantum Dots without Applied Magnetic Fields

Optical pumping experiments in Wurtzite-type nanocrystals have been among the first to show a slowing down of the carrier spin relaxation processes in dot compared to bulk or quantum well structures [37, 38]. These investigations have revealed the importance of both the quantum dot size and the carrier trapping at surface sites for the spin polarization time τ_s. Early experiments by Gotoh et al. [39] show exciton spin relaxation times of about 900 ps in InGaAs quantum disks, which is almost twice as long as the radiative recombination lifetime. These first results have been obtained mostly in non-resonant excitation conditions (i.e., photogenerating the carriers in the barrier). In this case the observed spin dynamics of the ground state reflects the spin relaxation occurring in the bulk barrier, in the excited dot states, and finally in the dot ground state, including any spin-flip scattering due to the energy relaxation process itself. Hence a strictly resonant excitation of the ground state is desirable to study the *intrinsic* quantum dot spin dynamics.

4.3.1 Exciton Spin Dynamics under Resonant Excitation

Figure 4.3(a) displays the time-resolved PL intensity for self-assembled InAs quantum dots in GaAs. The detection polarization is parallel (I^X) or perpendicular (I^Y) to the linearly polarized σ^X excitation laser. The latter emits picosecond pulses every 12.5 ns [40]. The detection energy is strictly the same as the excitation one, so that

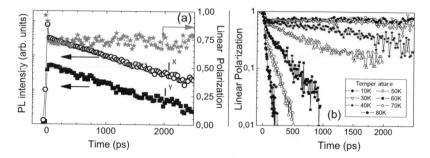

Fig. 4.3. Time resolved PL measurements on self-assembled InAs dots [40]. (**a**) PL intensity with polarization parallel I^X (*hollow circles*) and perpendicular I^Y (*solid squares*) to the linearly polarized (σ^X) excitation laser ($T = 10$ K); the temporal evolution of the corresponding linear polarization $\mathcal{P}_{\text{lin}}$ (*stars*) is also displayed. The laser excitation and detection energies are set to 1.137 eV (resonant excitation). (**b**) Temperature dependence of the linear polarization dynamics of neutral excitons

only quantum dots with coinciding exciton ground state and laser energy contribute to the PL signal. The corresponding kinetics of the linear polarization defined as $\mathcal{P}_{\text{lin}} = (I^X - I^Y)/(I^X + I^Y)$ is also plotted. The experiment is performed at low laser excitation power (which results in an estimated average density of photo-excited carriers less than one electron–hole pair per quantum dot).

The PL intensity decays with a characteristic time $\tau_{\text{rad}} \sim 800$ ps. After the pulsed excitation, the quantum dot emission exhibits a strong linear polarization ($\mathcal{P}_{\text{lin}} \sim 0.75$) which remains constant during the exciton emission time (i.e., over ~ 2.5 ns). This behavior differs strongly from the exciton linear polarization dynamics in bulk or type I quantum well structures, characterized by a linear polarization decay time of a few tens of picoseconds [41]. The experimental observation of a linear polarization of an exciton which does not decay with time is the proof that neither the electron, nor the hole spin relax on the exciton lifetime scale. It also shows that the exciton spin eigenstate is maintained during the whole exciton lifetime. From this observation, we can infer that the exciton spin relaxation time is longer than 20 ns, i.e., at least *25 times larger* than the radiative lifetime. Much shorter linear polarization decay times in the order of 100 ps have been measured in large, individual GaAs interface fluctuation dots in transient differential transmission experiments [42]. Since the carriers are more weakly confined in this kind of structures, spin relaxation mechanisms related to the spin–orbit interaction for the valence states are not entirely suppressed.

Let us now examine the robustness of the exciton spin when raising the temperature. Figure 4.3(b) displays the linear polarization of the quantum dot ground state emission as a function of the temperature after a strictly resonant linearly polarized excitation. Up to 30 K, the decay time is longer than 20 ns. At 80 K, the linear decay time drops down to 20 ps with an activation energy $E_{\text{a}} \sim 30$ meV. The dependence of the linear polarization on the temperature can be well explained in terms of the second order quasielastic interaction between LO phonons and carriers as proposed by Tsitsishvili et al. [43]. The scattering events occur via the virtual excited states

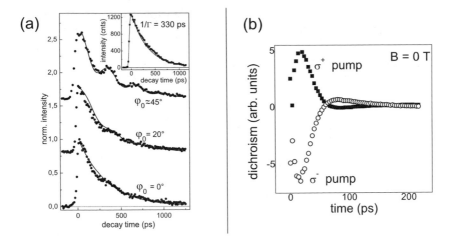

Fig. 4.4. (**a**) PL transients of a *single* CdSe/ZnSe quantum dot for excitation at the 1 LO-phonon resonance for three polarization configurations. *Dotted curves*: experiment; For experimental details and fitting procedure see [34]. (**b**) Resonant transient dichroism in an *ensemble* of self-assembled InAs/GaAs dots from pump–probe measurements for σ^- (*open circles*) and σ^+ (*solid squares*) pump polarization [44]

of the exciton built with an S_c and an $P_v^{x(y)}$ state, which are coupled to both $|X\rangle$ and $|Y\rangle$ exciton states via quasiresonant absorption and emission of LO phonons. This process leads to exciton transitions between the ground $|X\rangle$ and $|Y\rangle$ sublevels and determines the temporal evolution of the linear polarization. The transition rate is proportional to $N_{LO}(1 + N_{LO}) \approx \exp(-\hbar\omega_{LO}/k_B T)$, where $\hbar\omega_{LO} = 32\,\text{meV}$ is the InAs LO phonon energy and N_{LO} their occupation factor. The calculated decay times $\tau_s \sim 3.5\,\text{ns}$ (40 K) and $\tau_s \sim 44\,\text{ps}$ (80 K) closely agree with the experimental results.

4.3.2 Exciton Spin Quantum Beats: The Role of Anisotropic Exchange

A linearly polarized laser excitation along the [100] or [010] directions (45° tilt with respect to the [110] and [1$\bar{1}$0] directions) should lead to the observation of beats of $\mathcal{P}_{\text{lin}}$ at an angular frequency corresponding to the anisotropic exchange interaction splitting δ_1 (see the discussion in type II quantum wells in Chap. 3). This has been clearly observed in an experiment performed on individual CdSe dots by Flissikowski et al. presented in Fig. 4.4(a) showing that the exciton spin coherence time T_2 is longer than the radiative lifetime [34].

Similarly a circularly polarized (σ^+) excitation should lead to the observation of circular polarization $\mathcal{P}_c$ quantum beats at the angular frequency $\delta_1/\hbar$ [44–46]. Following σ^+ excitation, a coherent superposition of the two linearly polarized eigenstates $|X\rangle$ and $|Y\rangle$ is generated as the line width of the pulsed excitation laser is larger than the anisotropic exchange interaction splitting δ_1. This results in oscillations of the circular polarization (dichroism) with a period corresponding to the energy split-ting between $|X\rangle$ and $|Y\rangle$. An example of these oscillations in an ensemble of InAs quantum dots in resonant transient dichroism measurements is shown in Fig. 4.4(b) [44]. An oscillation period of 130 ps, corresponding to an average energy splitting $\delta_1 = 30\,\mu\text{eV}$, and an inhomogeneous decay time of $T_2^* = 30\,\text{ps}$ are found. This fast damping of the oscillations is due to the fluctuations of δ_1 from dot to dot. Similar values for δ_1 have been reported in different III–V samples [29, 45, 47].

To clarify the physical origin of the anisotropic exchange splitting, InAs quantum dots of different sizes and grown under different conditions have been compared. The splitting δ_1 is usually found to decrease when going from dots with high confinement energies (low emission energies) to dots with low confinement energies (high emis-sion energies). This trend has been reported by several groups [47–50] and in general smaller values of δ_1 are found for samples annealed after growth. It is difficult to dis-tinguish between the effects of Ga or In diffusion into the dot and a change in shape or size, as all these parameters have an influence on the confinement potential and hence on the measured emission energy. The observed fine structure is a fingerprint of the symmetry of the dot investigated. Pseudopotential calculations [51] predict the existence of an anisotropic exchange splitting even for dots with a cylindrical or square base.[2] This suggests that small, near zero values of δ_1 found in annealed sam-ples reflect the cancellation of different contributions to the fine structure splitting, rather than the fact that all contributions to the fine structure splitting are negligible.

4.4 Exciton Spin Dynamics in Neutral Quantum Dots in External Magnetic Fields

In the absence of an applied external field the energy splittings measured in single dot spectroscopy and quantum beats in time resolved experiments have revealed two linearly polarized eigenstates $|X\rangle$ and $|Y\rangle$ along the [110] and [1$\bar{1}$0] crystallographic directions. The linear polarization degree $\mathcal{P}_{\text{lin}}$ measured in resonant PL experiments is close to 100%, whereas the circular polarization degree $\mathcal{P}_c$ is close to zero.

In the following paragraphs the exciton fine structure revealed in continuous wave and time resolved experiments will be discussed for magnetic fields parallel to the growth axis z (Faraday configuration) and perpendicular to z, i.e., in the quan-tum well plane (Voigt geometry). In particular, it will be shown that for a Zeeman splitting that is much larger than the anisotropic exchange splitting δ_1 the optically active eigenstates are circularly polarized.

4.4.1 Zeeman Effect Versus Anisotropic Exchange Splittings in Single Dot Spectroscopy

Faraday Configuration

Strong exchange effects are observed in II–VI interface fluctuation dots such as CdTe dots in CdMgTe [4]. Figure 4.5 shows micro-photoluminescence spectra for an indi-

---

[2] The quantum dot symmetry is reduced to C_{2v} as soon as the symmetry along z is broken.

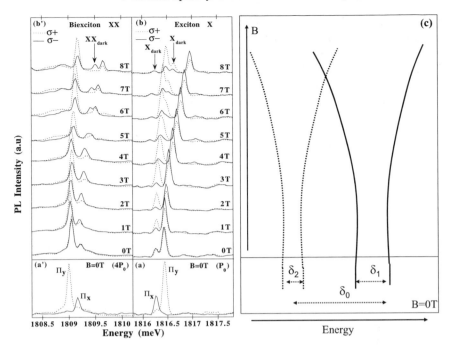

Fig. 4.5. Micro-PL of an individual CdTe quantum dot in CdMgTe after reference [4]. (a) Linearly polarized PL of an exciton in a single quantum dot with reduced symmetry at zero magnetic field for a laser excitation power P_0 of about $1\,\mathrm{W\,cm^{-2}}$. (b) Circularly polarized PL spectra (σ^- and σ^+ for *solid* and *dotted lines*, respectively) for various magnetic fields increasing from *bottom* to *top*. Optically forbidden states (*dark exciton*) are marked with arrows. (a$'$) and (b$'$) show the recombination of a biexciton state generated with a laser excitation power of $4P_0$. (c) Sketch of the crossing of the bright and dark states

vidual dot at different magnetic fields. The anisotropic exchange splitting observed for the neutral exciton at zero field in Fig. 4.5(a) is reversed for the biexciton peak in Fig. 4.5(a$'$). The biexciton is a four particle complex made up of a ground state valence hole singlet and a ground state conduction electron singlet. In this configuration the anisotropic exchange effects are canceled out. For the spectra in Fig. 4.5(a$'$) the exchange split exciton is the final state after photon emission, whereas in Fig. 4.5(a) the exciton is the initial state before photon emission. This explains the reversal of the exchange splitting for the two situations.

The magnetic field dependence in Figs. 4.5(b) and (b$'$) shows two important effects: (i) an increase of the observed splitting with the applied field due to the Zeeman effect and (ii) a conversion of the polarization of the detected photons from linear to circular, as reported in [29, 40, 52, 53]. The energy separation ΔE between the two bright exciton states depends on the longitudinal electron and hole g factor, the anisotropic exchange interaction splitting, and the applied magnetic field B:

$$\Delta E = \sqrt{\delta_1^2 + \Delta_Z^2},\qquad(4.3)$$

where $\Delta_Z = (g_{e,z} + g_{h,z})\mu_B B$ and $\mu_B = 57.9\,\mu eV/T$ is the Bohr magneton. Taking the example of an InAs dot with $\delta_1 \simeq 50\,\mu eV$, $g_{e,z} = -0.8$ and $g_{h,z} = -2.2$ [29] gives $\Delta_Z \simeq \delta_1$ for an applied field of $B = 0.3\,T$. In this case for fields $B \gg 0.3\,T$ the splitting δ_1 can be neglected in (4.3).

The possible crossing between bright and dark states is sketched in Fig. 4.5(c). In this magnetic field range an anticrossing is observed experimentally. In the case of a dot with a fully broken symmetry a distinction between bright and dark states is no longer possible as the $|\pm1\rangle$ states and the $|\pm2\rangle$ states are coupled [29]. The appearance of additional transitions at high applied fields in Figs. 4.5(b) and (b') is due to this coupling and is a fingerprint of the particularly low symmetry of the investigated dot.

Voigt Configuration

The application of an in-plane magnetic field destroys the remaining rotational symmetry of the system. This causes a mixing of the bright and dark states, so that dark states become observable in the spectra. Typically, four mixed transitions are observed at high in-plane magnetic fields [29], while in zero magnetic field two lines are observed, which are split due to the anisotropic electron–hole exchange interaction.

4.4.2 Exciton Spin Quantum Beats in Applied Magnetic Fields

Faraday Configuration

As explained above, a magnetic field applied along the growth axis of the sample can be used to control the exciton spin eigenstates. A necessary but not sufficient condition for the observation of spin quantum beats is a long carrier spin lifetime, as discussed in Sect. 4.3. The effect of the anisotropic exchange interaction can be visualized as an effective magnetic field acting in the quantum dot plane for the two exciton bright states $|\pm1\rangle$ considered as a pseudospin $1/2$ [27]. When applying an external field B_z along z the total field felt by the carriers will be the sum of the effective and the applied field ($\Delta E = \hbar|\boldsymbol{\Omega}|$), as sketched in Fig. 4.6(b). The carrier spin precess around the total field $\boldsymbol{\Omega}$ which results in a periodic oscillation of the projection of the spin polarization onto the z axis.

In analogy to the quantum beats observed at $B = 0$ for $t < 300\,ps$ (Fig. 4.4(a)), one expects to measure quantum beats in magnetic fields with an oscillation period $T = h/\Delta E$. The values of ΔE deduced from the measured beat periods for an InAs dot ensemble are plotted in Fig. 4.6(c) for different fields and are fitted with (4.3). In Fig. 4.6(a) oscillations of the PL intensity co- and cross-polarized with the linearly polarized excitation laser are observed because quasiresonant laser excitation creates a superposition of $|+1\rangle$ and $|-1\rangle$ states. Note that the damping of the oscillations observed in Fig. 4.6(a) is not due to an intrinsic decoherence process. It simply comes from the inhomogeneous character of the experiment performed on an ensemble of dots. Similarly to CdSe dots [34], the intrinsic exciton spin coherence time T_2 is of the order of the radiative lifetime (a few hundred picoseconds).

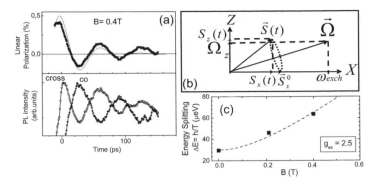

Fig. 4.6. Ensemble of InAs/GaAs quantum dots [44]. (**a**) *Lower panel*: Time resolved PL signal co-(stars) and cross polarized (*open circles*) with the linearly polarized excitation laser; *Upper panel*: A beat period of the linear polarization of $T = 65$ ps is extracted. (**b**) Pseudospin formalism with $\hbar\omega_{\mathrm{exch}} = \delta_1$ and $\hbar\Omega_z = g\mu_B B_z$. (**c**) Evolution of the exciton level splitting in an applied magnetic field along z fitted with (4.3)

Voigt Configuration

Yugova and co-workers have studied quantum beat phenomenon as a function of the angle between the direction of light propagation parallel to z and the applied magnetic field in InP/InGaP dots [8]. Their work revealed beats between the dark and bright states that are mixed by the transverse component of the magnetic field, similar to the results obtained in quantum wells, see Chap. 3. In ensemble measurements that probe not only neutral but also unintentionally doped dots, quantum beats due to the Larmor precession of the electron or hole in the transverse field are observed at different frequencies [54–57], as will be discussed in detail in the following sections.

4.5 Charged Exciton Complexes: Spin Dynamics without Applied Magnetic Fields

Singly charged excitons X^- or X^+ are 3-particle complexes (named therefore "trions") : in an X^- trion two electrons occupy the conduction ground state in a singlet configuration, whereas a single hole lies in the valence ground state. Similarly the X^+ trion is formed by a pair of holes in a singlet configuration and a conduction band electron.

Trions offer the interesting possibility to investigate the spin physics of a single particle (electron or hole) because the exchange interaction vanishes in the trion ground state. For example, the spin of an X^+ trion reduces to that of the single electron (both paired holes being then spectators) which makes this complex particularly suitable for studying the electron–nuclei hyperfine interaction in a 3D confined system (see Chap. 11). Like excitons, they are usually formed by an optical excitation and recombine radiatively over a comparable timescale of a few hundred picoseconds. This time limitation can be however overcome in the case of trions. Indeed,

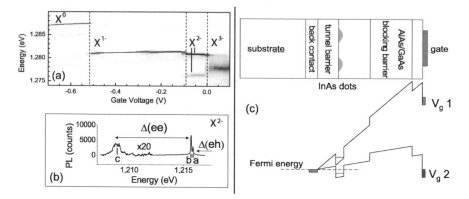

Fig. 4.7. (a) Contour plot of the PL transition energy as a function of applied bias for a single InAs quantum dot embedded in a charge tuneable structure, after reference [31, 60]. Black corresponds to strong PL signals (several thousand counts) for the neutral exciton X^0, the charged exciton X^{1-} (trion, one hole and two electrons), the X^{2-} (one hole and three electrons) and the X^{3-} (one hole and four electrons). **(b)** X^{2-} PL for a single InAs dot (different from **(a)**). The separation marked as $\Delta(eh)$ between the peaks a and b is given by the electron–hole Coulomb exchange, whereas $\Delta(ee)$ is given by the electron–electron Coulomb exchange [31]. **(c)** Schematic drawing of the conduction band line up of a n-type charge tuneable structure for two different applied gate voltages V_g [60]

since they are generally formed from singly charged dots, optical excitation and recombination offer a unique means to investigate and possibly manipulate the spin state of a single carrier trapped in a quantum dot over a longer timescale.

4.5.1 Formation of Trions: Doped and Charge Tuneable Structures

Controlling the charge state of excitons relies on the possibility of doping semiconductor materials with n-type or p-type impurities as reviewed in Chap. 1. The quantum dots behave like traps for the free carriers released by the ionized impurities and thus acquire one or several resident charge(s). In some cases the residual doping of the materials is sufficient to obtain singly charged dots [58], but usually a delta-doping layer is grown a few nanometers below the dot layer with a density adjusted to reach the desired average charge [56, 59]. This modulation doping technique can be significantly improved by controlling the position of the quantum dot levels with respect to the Fermi level of an heavily doped region (n- or p-type). This is achieved by applying an electric voltage between a top gate and the doped region grown 20 nm below the layer of quantum dots [60, 61]. In such a charge-tuneable structure a given quantum dot is coupled to a reservoir of free carriers through a tunneling barrier, see Fig. 4.7(c), which enables to control the excess charge with the precision of a single electron. This effect is characterized in micro-photoluminescence spectra by an abrupt shift of the recombination energy from the neutral to the charged exciton line (see Fig. 4.7(a)).[3]

4.5.2 Fine Structure and Polarization of X^+ and X^- Excitons

As confirmed by high resolution transmission spectroscopy [33] (see Fig. 4.2(b)), the trion ground state is perfectly degenerate in contrast to the neutral exciton. The zero field quantum beats are thus suppressed [58] and optical orientation of trions with circularly polarized light becomes possible providing a straightforward insight into the spin relaxation of a single carrier in a quantum dot. A hole spin lifetime of $\tau_s \sim$ 20 ns (remarkably long compared to bulk or quantum wells) was found for X^- trions both in CdSe and in InAs quantum dots [61, 62]. However, even though Kramers degeneracy holds for trions, the exchange interaction within each pair of particles still plays an important role for the spin dynamics, when excitons are created under *non-resonant* laser excitation. Indeed, hot trions ($X^{\pm*}$) with one carrier in a P-like state (an electron for X^-, a hole for X^+) form an intermediate state for the relaxation towards the trion ground state with a non-vanishing fine structure as shown in Fig. 4.8. The latter is dominated by the exchange between identical carriers, namely between S_c and P_c electrons in X^{-*} with $\Delta_{ee} \approx 5$ meV, while the total electron–hole exchange Δ_{eh} (with both S_c and P_c electrons) is a relatively weak perturbation (a few hundred μeV) which splits the triplet configuration.[4]

An important difference between the singlet and triplet configuration concerns their relaxation towards the trion ground state. The electron triplet requires a spin-dependent interaction in order to relax towards a singlet configuration, whereas the hot singlet can relax much faster (in a few picoseconds) thanks to the interaction with phonons.[5] Yet, the projection of the total spin of trion states indicated in the right-hand side of Fig. 4.8 is a good quantum number only when the dot in-plane anisotropy is neglected. In actual quantum dots, the anisotropic electron–hole exchange gives rise to a coupling $\tilde{\delta}_1^-$ between the $\pm 1/2$ triplet states and the $\mp 3/2$ singlet states as represented in Fig. 4.8 by dashed arrows. Experiments and theoretical estimates show that this coupling is particularly strong for the P_c level [64–66] and must therefore be taken into account when analyzing the spin dynamics during the hot trion relaxation.

---

[3] This shift is due to small change of the direct Coulomb interaction between the carriers confined in the quantum dot, which in turn modifies the binding energy of an electron hole pair.

[4] This fine structure can be observed in the micro-photoluminescence spectrum of charged biexcitons $XX^\pm$ or doubly charged excitons $X^{2\pm}$ (see, e.g., Fig. 4.7(b)) [31, 58], and a similar fine structure is predicted for X^{+*} [35].

[5] The difference in lifetimes manifests in the line width of the X^{2-} transitions shown in Fig. 4.7(b) [31, 60].

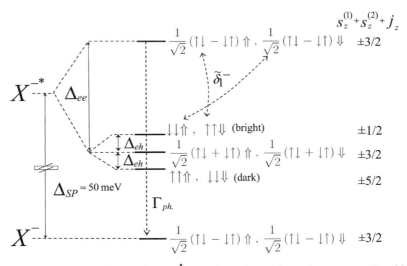

Fig. 4.8. Fine structure of a hot trion X^{-*}, namely a trion with an electron on a P_c orbital of the quantum dot. Only the spin degree of freedom is represented, with simple arrows ($\uparrow$, $\downarrow$) for the electrons and double arrows ($\Uparrow$, $\Downarrow$) for the hole pseudospin j_z. The hot trion singlet relaxes in a few picoseconds by a phonon-mediated process (Γ_{ph}), while triplet states require a spin dependent interaction which is provided by the anisotropic electron–hole exchange $\tilde{\delta}_1^-$, yet in the "bright" configuration only

4.5.3 Spin Dynamics in Negatively Charged Exciton Complexes X^{n-}

Optical orientation of charged excitons in a quantum dot follows a very specific mechanism that is succinctly described in this section [59, 63, 67]. As shown in Fig. 4.9 for different samples of self-assembled InGaAs/GaAs quantum dots with different n-doping levels, the photoluminescence circular polarization is negative, i.e., opposite to the excitation polarization, although the optical selection rules for both excitation and radiative recombination imply only heavy-hole states. Besides, time-resolved measurements reveal a quite unusual increase (in absolute value) of this negative circular polarization in the timescale of the radiative recombination (~ 1 ns).

To understand this effect, let us consider the case of quantum dots doped with a single resident electron ($\langle n \rangle = 1$). When a spin polarized electron, say with spin $\downarrow$ is captured after optical generation in the barrier material, it undergoes a strong exchange interaction with the resident electron. It forms either a triplet pair $\downarrow\downarrow$ when both spins are parallel, or a spin configuration (singlet or triplet) with zero projection along z, namely $\frac{1}{\sqrt{2}}(\uparrow\downarrow \pm \downarrow\uparrow)$, when both spins are antiparallel. On the other hand, the captured hole is unpolarized, because its spin orientation relaxes in the energy continuum where it is created (see Chap. 3).[6] Four types of hot trion states are thus

---

[6] Under quasiresonant excitation, such that the spin of the photo-created hole is conserved, it is possible to select transitions giving rise to a strong (above 90%) positive optical orien-

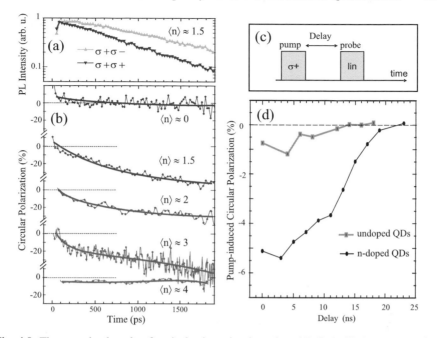

Fig. 4.9. Time-resolved study of optical orientation in n-doped InGaAs/GaAs quantum dots under non-resonant excitation at low temperature and zero magnetic field [59, 67]. (**a**) Traces of the polarization resolved PL intensity of a sample with average charge $\langle n \rangle \approx 1.5$ and (**b**) PL polarization dynamics for different samples with increasing doping. (**c**) Schematics of the pump–probe excitation sequence employed for measuring the spin lifetime of resident electrons. (**d**) Evolution of the circular polarization of the probe PL following a circularly polarized excitation

populated, among which only the trions $\downarrow\downarrow\Uparrow$ and $\downarrow\downarrow\Downarrow$ still carry information about the initial spin orientation. The former can optically recombine in contrast to the latter which is dipole inactive. They are respectively called "bright" and "dark" trions in analogy to the neutral excitons (see Sect. 4.2.2). The bright trion initially emits a positively polarized photon (i.e., with the same helicity as the excitation laser) when recombining, but it relaxes in a few tens of picoseconds down to the singlet ground state with a flipped hole spin $\Downarrow$ due to the anisotropic coupling $\tilde{\delta}_1^-$ (see Fig. 4.8). This specific "flip–flop"-induced relaxation which transfers the electron spin orientation to that of the hole changes the sign of the trion polarization. This explains the increase of negative polarization during the radiative lifetime. Similar mechanisms relying on the anisotropic exchange interaction take place for quantum dots with more than

tation [61, 69]. The reversal of polarization is however clearly observed in the PL excitation spectroscopy of an individual charged quantum dot [63].

one resident electron and lead also to a negative circular polarization, with yet a somewhat different dynamics [67, 68].[7]

4.5.4 Spin Memory of Trapped Electrons

One of the motivations for studies of the optical orientation of trions is to initialize the spin state of a resident electron in a quantum dot by a laser pulse and then to measure the spin memory on a timescale longer than the radiative lifetime. This effect can be observed in a pump–probe experiment as schematically represented in Fig. 4.9(c). A circularly polarized excitation pulse (the pump) is used to polarize the spin of resident electrons. After a certain delay, the spin memory is read out by analyzing the circular polarization of the PL emitted from a linearly polarized excitation pulse (the probe).

This technique was first employed by Cortez et al. [59] who reported a spin memory effect of ∼15 ns in an ensemble of n-doped InGaAs/GaAs quantum dots (see Fig. 4.9(d)). This result was rather surprising because a longer living memory (∼100 μs) is expected when assuming that electron spin relaxation is governed by the hyperfine interaction with the lattice nuclei [20]. This observation stimulated several studies to further clarify the role of the hyperfine interaction which actually manifests itself in a variety of new effects (see Chap. 11). Pump–probe experiments similar to those in [59] but under different excitation conditions (which probably enhance dynamic nuclear polarization) have been recently carried out by Oulton et al. [72]. They actually reveal a very long spin lifetime (up to ∼0.1 s) of the coupled electron to nuclear spin system in n-doped InAs quantum dots.

4.6 Charged Exciton Complexes: Spin Dynamics in Applied Magnetic Fields

The X^+ exciton is formed in quantum dots which contain a resident hole. When the laser excitation energy is tuned to the heavy-hole to electron transition energy of the wetting layer, the optically created carriers are subsequently captured by a dot and relax in energy to the quantum dot ground state. The probability of the hole spin losing its initial spin orientation during the capture and energy relaxation process is very high [59]. This yields the formation of a hole spin singlet in the S_v valence state, i.e., the total spin of the two holes is zero. It has been shown that the electron created in the wetting layer can keep its initial spin orientation when reaching the S_c state to form the X^+ exciton with the hole spin singlet. The average electron spin is related to the circular polarization degree $\mathcal{P}_c$ of the X^+ emission simply by: $\langle \hat{S}_z^e \rangle = -\mathcal{P}_c/2$. The photoluminescence polarization is therefore a sensitive probe of the electron spin orientation on the radiative lifetime scale, of typically 1 ns.

---

[7] Negative circular polarization under non-resonant excitation is also observed in charged GaAs/Ga$_x$Al$_{1-x}$As quantum dots, but is most likely due to another mechanism based on the accumulation of "dark trions" [70, 71].

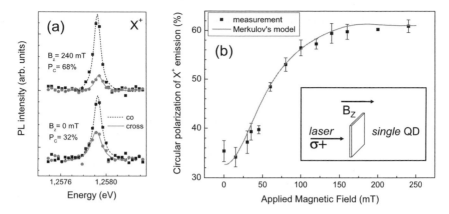

Fig. 4.10. The excitation polarization is provided by a 50 kHz photo-elastic modulator. (**a**) Polarization resolved PL spectra of the X^+ transition of a single InAs quantum dot without (*lower part*) and with (*upper*) an applied magnetic field after [76]. (**b**) Circular polarization of the X^+ line as a function of magnetic field (*squares*). Fit (*grey line*) using model in reference [20]. Inset: experimental geometry

After the radiative recombination of the X^+ exciton, a hole with a certain spin orientation is left behind in the dot. How long this resident hole keeps its spin orientation is still an open question, that could be addressed, for example, in pump probe type measurements—see Sect. 4.5.4. Preliminary results showed a hole spin lifetime $\tau_s > 20$ ns for zero external fields [73] and $\tau_s = 270 \pm 180$ μs in an external field of 1.5 T applied along the z axis [74]. The hole spin relaxation time for dots with strong confinement potentials should in principle be prolonged compared to the equivalent measurements for electrons as the hyperfine interaction plays a less important role. The spin coherence of resident *electrons* left behind after the radiative recombination of the X^- [56] is discussed in detail in Chap. 6.

4.6.1 Electron Spin Polarization in Positively Charged Excitons in Longitudinal Magnetic Fields

Since the electron–hole Coulomb exchange interaction vanishes during the radiative lifetime of the X^+ exciton, it is the hyperfine interaction between electron and nuclear spins that determines the X^+ polarization dynamics observed experimentally. In the following the effect of the nuclear spin orientation on the electron is considered. The reciprocal effect of the electron spin on the nuclear spin polarization can lead to dynamical polarization of nuclear spins [18, 19, 23, 69] and is discussed in detail in Chap. 11. Yet, since the nuclear polarization builds up on a timescale of seconds, it is negligible in the experiments discussed below (e.g., see Fig. 4.10) where the excitation laser polarization was changed from σ^+ to σ^- at a frequency $f = 50$ kHz [18, 75].

At zero applied magnetic field the PL measurements on a single InAs dot show a circular polarization degree of only 32% of the X^+ emission, Fig. 4.10(a). By applying a relatively small magnetic field of only 150 mT this polarization degree can be doubled, as can be seen in Fig. 4.10(b) [76]. This behavior is very unusual for a non-magnetic semiconductor. It should be noted that the thermal energy is much larger than the Zeeman splitting in these experiments, $k_B T \gg g\mu_B B$. It will be argued in the following that the initially low circular polarization is a consequence of the interaction of the electron spins with the nuclear spins and that the effects of this interaction can be inhibited by applying a magnetic field along the z-axis [21].

Theoretical studies have predicted that the dominant mechanism of electron spin relaxation in quantum dots at low temperature is due to the hyperfine interaction with nuclear spins [20, 77, 78]. An electron spin in a quantum dot interacts with a large but finite number of nuclei $N_L \approx 10^3$–10^5. In the frozen fluctuation model, the sum over the interacting nuclear spins gives rise to a finite magnetic like field $\boldsymbol{B}_N$, about which the electron spin coherently precesses [20, 77]. However, the amplitude and direction of $\boldsymbol{B}_N$ vary randomly from dot to dot, so that the average electron spin $S_z^e(t)$ of a dot ensemble decays rather fast ($\sim$100 ps) as shown experimentally [21].[8] For the sake of simplicity, this spin dephasing mechanism affecting a quantum dot ensemble is termed here *spin relaxation*.

This spin relaxation mechanism manifests also in single dot measurements because the orientation of the nuclear field $\boldsymbol{B}_N$ in a given quantum dot evolves randomly on a time scale of 100 μs. The latter is much longer than the radiative lifetime ($\sim$1 ns), but much shorter than the typical integration time of several seconds used for single dot measurements. Therefore, the electron spin dephasing due to averaging over different nuclear field orientations still holds.

The effect of the fluctuating hyperfine field $\boldsymbol{B}_N$ on the electron spin during the radiative lifetime of the X^+ exciton is revealed by applying a small magnetic field $\boldsymbol{B}$ along the z axis—see Fig. 4.10. In a classical picture the electron spin precesses about the total field $\boldsymbol{B} + \boldsymbol{B}_N$. Applying a field $|\boldsymbol{B}|$ which largely exceeds the rms value of $\boldsymbol{B}_N$ ($\sim$30 mT in InAs quantum dots) results in a precession vector almost parallel to the z axis. In this case the electron spin keeps its initial spin orientation along z if no spin relaxation due to another mechanism takes place. The gradual suppression of precession about the in-plane component of $\boldsymbol{B}_N$ can be fitted by the model of Merkulov et al. [20] applied to the case of X^+ trions [76]—see Fig. 4.10(b). This confirms the role of the hyperfine interaction as the main cause of electron spin relaxation in III–V quantum dots.[9] Electron paramagnetic resonance spectroscopy suggests that room temperature electron spin dynamics in free standing ZnO quantum dots is also governed by the hyperfine interaction with nuclear spins [81]. Interactions between carrier and nuclear spins have also to be taken into account in diodes that emit polarized light following the electrical injection of spin polarized carriers from a magnetic barrier into quantum dots [82].

---

[8] The spin polarization decay is limited in amplitude because $\boldsymbol{B}_N$ is almost collinear to the spin direction for about 1/3 of the quantum dots. See Chap. 11 for details.

[9] In a magnetic field of several Tesla, spin relaxation times in the ms range has been reported for resident electrons in charge tuneable self-assembled InAs dots [79] and in electrostatically defined dots [80].

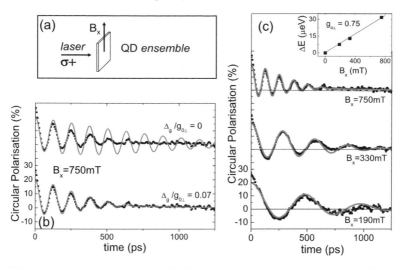

Fig. 4.11. (a) Experimental geometry. **(b)** Electron spin quantum beats in ensembles of X^+ excitons in p-doped InAs quantum dots measured via the circular PL polarization. Data fitted with and without fluctuations of the transverse electron g factor, after [55]. **(c)** Measurement as in **(b)** for different fields and fits performed with the complete model developed in [55]

4.6.2 Electron Spin Coherence in Positively Charged Excitons in Transverse Magnetic Fields

The eigenstates of the X^+ exciton levels in a transverse magnetic field $\boldsymbol{B}$ parallel to the x axis are the $|X^+, \uparrow\rangle_x$ and the $|X^+, \downarrow\rangle_x$ states (see Fig. 4.11(a) for the experimental geometry).[10] Since the hole spin singlet is not split in external fields, the Zeeman splitting during the radiative lifetime of the X^+ exciton is given by the conduction electron spin splitting $g_{e,\perp}\mu_B B$.

With a σ^+ polarized excitation pulse the $|X^+, \downarrow\rangle_z$ state is created as a linear superposition of $|X^+, \uparrow\rangle_x$ and $|X^+, \downarrow\rangle_x$. The system evolves in time with beats between these states which can be evidenced by monitoring the evolution of the degree of circular polarization during the radiative lifetime—see Fig. 4.11.[11] The damping of the oscillations gives the effective coherence time of the system. Here dephasing effects due to averaging over dot ensembles and intrinsic sources of decoherence have to be distinguished. The model by Merkulov et al. that includes only spin dephasing due to hyperfine effects fails to reproduce the fast damping of the oscillations. The main contribution to the observed dephasing time of 400 ps comes from the inhomogeneous deviation of electron transverse g factor ($\sim 7\%$). Obviously, this

---

[10] Dynamical nuclear polarization does not play any role in this geometry because the average electron spin is always perpendicular to the applied field [18].

[11] The transverse electron g factor can be extracted from the dependence of the beat frequency on the magnetic field strength. The value of $g_{e,\perp} = 0.75$ obtained in this way is in agreement with the values measured in single dot spectroscopy [29].

dephasing process vanishes for an individual quantum dot, a situation where the hyperfine interaction would remain the dominant source of decoherence at low temperature. From the measurements presented in Fig. 4.11 a dephasing time of 1 ns can be extracted for this contribution, which should be observable in time-resolved experiments on a single dot.

4.7 Conclusions

In this chapter dedicated to exciton spin dynamics in quantum dots, we have reviewed some of the most striking properties recently uncovered in this system. The three-dimensional confinement of carriers in quantum dots strongly enhances the spin lifetime of excitons as compared to bulk or quantum wells. This is evidenced in a large number of experiments where optical orientation or exciton dipole alignment were investigated. However, the strong quantization of carriers on a few nanometer length scale strongly alters their spin dependent properties by strengthening the electron–hole exchange interaction. This effect gives rise to the splitting δ_1 of the exciton eigenstates into linearly polarized states at zero magnetic field, as a result of the actual anisotropy of quantum dots. A neutral exciton can thus be seen as a pseudospin $1/2$ submitted to an internal magnetic-like field, which in general cannot be suppressed, even though a longitudinal external field of a few Tesla can restore circularly-polarized eigenstates. The exchange interaction influences also the spin dynamics of singly charged excitons by introducing coupling between singlet and triplet spin configuration in the fine structure of excited states. This gives rise to a singular negative circular polarization of negatively charged excitons X^- under non-resonant circularly polarized excitation. Eventually, we introduced the role of the hyperfine interaction which couples the spin of conduction electron to that of the lattice nuclei. It manifests itself as a slowly fluctuating magnetic field of a few 10 mT in III–V quantum dots leading to spin dephasing of unpaired electrons. In principle this is a weak perturbation which can be almost canceled out by applying a small magnetic field (typically 150 mT for InAs dots), but requires yet great attention whenever dynamical nuclear polarization takes place (see Chap. 11).

Acknowledgements. We thank J.M. Gérard, V. Ustinov and A. Lemaître for the sample growth. We are grateful to M. Paillard, M. Sénès, P-.F. Braun, L. Lombez, D. Lagarde, S. Laurent, B. Eble, P. Renucci, H. Carrère, P. Voisin, K.V. Kavokin, and V.K. Kalevich for their contributions to this work. We thank A. Högele, F. Henneberger and L. Besombes for providing the original figures.

References

[1] V.I. Klimov, A.A. Mikhailovsky, S. Xu, A. Malko, J.A. Hollingsworth, C.A. Leatherdale, H.-J. Eisler, M.G. Bawendi, Science **290**, 314 (2000)
[2] N.P. Stern, M. Poggio, M.H. Bartl, E.L. Hu, G.D. Stucky, D.D. Awschalom, Phys. Rev. B **72**, 161303(R) (2005)

[3] D. Gammon, E.S. Snow, B.V. Shanabrook, D.S. Katzer, D. Park, Phys. Rev. Lett. **76**, 3005 (1996)

[4] L. Besombes, K. Kheng, D. Martrou, Phys. Rev. Lett. **85**, 425 (2000)

[5] J.P. McCaffrey, M.D. Robertson, S. Fafard, Z.R. Wasilewski, E.M. Griswold, L.D. Madsen, J. Appl. Phys. **88**, 2272 (2000)

[6] J.-Y. Marzin, J.-M. Gerard, A. Izraël, D. Barrier, G. Bastard, Phys. Rev. Lett. **73**, 716 (1994)

[7] T. Flissikowski, A. Betke, I. Akimov, F. Henneberger, Phys. Rev. Lett. **92**, 227401 (2004)

[8] I.A. Yugova, I.Ya. Gerlovin, V.G. Davydov, I.V. Ignatiev, I.E. Kozin, H.W. Ren, M. Sugisaki, S. Sugou, Y. Masumoto, Phys. Rev. B **66**, 235312 (2002)

[9] N. Koguchi, S. Takahashi, T. Chikyow, J. Cryst. Growth **111**, 688 (1991)

[10] J.R. Petta, A.C. Johnson, J.M. Taylor, E.A. Laird, A. Yacoby, M.D. Lukin, C.M. Marcus, M.P. Hanson, A.C. Gossard, Science **309**, 2180 (2005)

[11] F.H.L. Koppens, C. Buizert, K.J. Tielrooij, I.T. Vink, K.C. Nowack, T. Meunier, L.P. Kouwenhoven, L.M.K. Vandersypen, Nature **442**, 766 (2006)

[12] E. Moreau, I. Robert, L. Manin, V. Thierry-Mieg, J.M. Gerard, I. Abram, Phys. Rev. Lett. **87**, 183601 (2001)

[13] C. Santori, D. Fattal, J. Vucovic, G.S. Solomon, Y. Yamamoto, Nature **419**, 594 (2002)

[14] R.M. Stevenson, R.J. Young, P. Atkinson, K. Cooper, D.A. Ritchie, A.J. Shields, Nature **439**, 179 (2006)

[15] P. Borri, W. Langbein, S. Schneider, U. Woggon, R.L. Sellin, D. Ouyang, D. Bimberg, Phys. Rev. Lett. **87**, 157401 (2001)

[16] X. Li, Y. Wu, D. Steel, D. Gammon, T.H. Stievater, D.S. Katzer, D. Park, C. Piermarocchi, L.J. Sham, Science **301**, 809 (2003)

[17] B. Eble, O. Krebs, A. Lemaître, K. Kowalik, A. Kudelski, P. Voisin, B. Urbaszek, X. Marie, T. Amand, Phys. Rev. B **74**, 081306(R) (2006)

[18] F. Meier, B. Zakharchenya, *Optical Orientation, Modern Problem in Condensed Matter Sciences*, vol. 8 (North-Holland, Amsterdam, 1984)

[19] D. Gammon, A.L. Efros, T.A. Kennedy, M. Rosen, D.S. Katzer, D. Park, S.W. Brown, V.L. Korenev, I.A. Merkulov, Phys. Rev. Lett. **86**, 5176 (2001)

[20] I.A. Merkulov, Al.L. Efros, M. Rosen, Phys. Rev. B **65**, 205309 (2002)

[21] P.-F. Braun, X. Marie, L. Lombez, B. Urbaszek, T. Amand, P. Renucci, V.K. Kalevich, K.V. Kavokin, O. Krebs, P. Voisin, Y. Masumoto, Phys. Rev. Lett. **94**, 116601 (2005)

[22] V.K. Kalevich, V.L. Korenev, JETP Lett. **56**, 253 (1992)

[23] P.-F. Braun, B. Urbaszek, T. Amand, X. Marie, O. Krebs, B. Eble, A. Lemaître, P. Voisin, Phys. Rev. B **74**, 245603 (2006)

[24] R.J. Warburton, B.T. Miller, C.S. Dürr, C. Bödefeld, K. Karrai, J.P. Kotthaus, G. Medeiros-Ribeiros, P.M. Petroff, S. Huant, Phys. Rev. B **58**, 16221 (1998)

[25] M. Grundmann, O. Stier, D. Bimberg, Phys. Rev. B **52**, 11969 (1995)

[26] E.L. Ivchenko, *Springer Series in Solid-State Science*, vol. 110 (Springer, Berlin, 1997)

[27] R.I. Dzhioev, B.P. Zakharchenya, E.L. Ivchenko, V.L. Korenev, Y.G. Kusraev, N.N. Ledentsov, V.M. Ustinov, A.E. Zhukov, A.F. Tsatsulnikov, Phys. Solid State **40**, 790 (1998)

[28] I.E. Kozin, V.G. Davydov, I.V. Ignatiev, A.V. Kavokin, K.V. Kavokin, G. Malpuech, H.-W. Ren, M. Sugisaki, S. Sugou, Y. Masumoto, Phys. Rev. B **65**, 241312 (2002)

[29] M. Bayer, G. Ortner, O. Stern, A. Kuther, A.A. Gorbunov, A. Forchel, P. Hawrylak, S. Fafard, K. Hinzer, T.L. Reinecke, S.N. Walck, J.P. Reithmaier, F. Klopf, F. Schafer, Phys. Rev. B **65**, 195315 (2002), and references therein

[30] E.L. Ivchenko, A.Y. Kaminski, I.L. Aleiner, JETP **77**, 609 (1993)

[31] B. Urbaszek, R.J. Warburton, K. Karrai, B.D. Gerardot, P.M. Petroff, J.M. Garcia, Phys. Rev. Lett. **90**, 247403 (2003)
[32] V.D. Kulakovskii, G. Bacher, R. Weigand, T. Kummell, A. Forchel, E. Borovitskaya, K. Leonardi, D. Hommel, Phys. Rev. Lett. **82**, 1780 (1999)
[33] A. Högele, S. Seidl, M. Kroner, K. Karrai, R.J. Warburton, B.D. Gerardot, P.M. Petroff, Phys. Rev. Lett. **93**, 217401 (2004)
[34] T. Flissikowski, A. Hundt, M. Lowisch, M. Rabe, F. Henneberger, Phys. Rev. Lett. **86**, 3172 (2001)
[35] K.V. Kavokin, Phys. Stat. Sol. (a) **195**, 592 (2003)
[36] B.D. Gerardot, S. Seidl, P.A. Dalgarno, R.J. Warburton, D. Granados, J.M. Garcia, K. Kowalik, O. Krebs, K. Karrai, A. Badolato, P.M. Petroff, Appl. Phys. Lett. **90**, 041101 (2007)
[37] M. Chamaro, C. Gourdon, P. Lavallard, J. Lumin. **70**, 222 (1996)
[38] J.A. Gupta, D.D. Awschalom, X. Peng, A.P. Alivisatos, Phys. Rev. B **59**, 10421(R) (1999)
[39] H. Gotoh, H. Ando, H. Kamada, A. Chavez-Pirson, J. Temmyo, Appl. Phys. Lett. **72**, 1341 (1998)
[40] M. Paillard, X. Marie, P. Renucci, T. Amand, A. Jbeli, J.M. Gérard, Phys. Rev. Lett. **86**, 1634 (2001)
[41] X. Marie, P. LeJeune, T. Amand, M. Brousseau, J. Barrau, M. Paillard, R. Planel, Phys. Rev. Lett. **79**, 3222 (1997)
[42] T.H. Stievater, X. Li, T. Cubel, D.G. Steel, D. Gammon, D.S. Katzer, D. Park, Appl. Phys. Lett. **81**, 4251 (2002)
[43] E. Tsitsishvili, R.V. Baltz, H. Kalt, Phys. Rev. B **66**, 161405 (2002)
[44] M. Senes, B. Urbaszek, X. Marie, T. Amand, J. Tribollet, F. Bernardot, C. Testelin, M. Chamarro, J.-M. Gérard, Phys. Rev. B **71**, 115334 (2005)
[45] A.S. Lenihan, M.V. Gurudev Dutt, D.G. Steel, S. Ghosh, P.K. Bhattacharya, Phys. Rev. Lett. **88**, 223601 (2002)
[46] F. Bernardot, E. Aubry, J. Tribollet, C. Testelin, M. Chamarro, L. Lombez, P.-F. Braun, X. Marie, T. Amand, J.-M. Gérard, Phys. Rev. B **73**, 085301 (2006)
[47] W. Langbein, P. Borri, U. Woggon, V. Stavarache, D. Reuter, A.D. Wieck, Phys. Rev. B **69**, R161301 (2004)
[48] A. Greilich, M. Schwab, T. Berstermann, T. Auer, R. Oulton, D.R. Yakovlev, M. Bayer, V. Stavarache, D. Reuter, A. Wieck, Phys. Rev. B **73**, 045323 (2006)
[49] C. Testelin, E. Aubry, M. Chaouache, M. Maaref, F. Bernardot, M. Chamarro, J.-M. Gérard, Phys. Stat. Sol. (c) **3**, 3900 (2006)
[50] C. Testelin, E. Aubry, M. Chaouache, M. Maaref, F. Bernardot, M. Chamarro, J.-M. Gérard, Phys. Stat. Sol. (c) **4**, 1385 (2007)
[51] G. Bester, S. Nair, A. Zunger, Phys. Rev. B **67**, 161306(R) (2003)
[52] R.I. Dzhioev, B.P. Zakharchenya, E.L. Ivchenko, V.L. Korenev, Y.G. Kusraev, N.N. Ledentsov, V.M. Ustinov, A.E. Zhukov, A.F. Tsatsulnikov, JETP **65**, 804 (1997)
[53] K. Kowalik, PhD thesis, Université Pierre et Marie Curie (Paris 6), http://tel.archives-ouvertes.fr
[54] K. Nishibayashi, T. Okuno, Y. Masumoto, H.-W. Ren, Phys. Rev. B **68**, 035333 (2003)
[55] L. Lombez, P.-F. Braun, X. Marie, P. Renucci, B. Urbaszek, T. Amand, O. Krebs, P. Voisin, Phys. Rev. B **75**, 195314 (2007)
[56] A. Greilich, D.R. Yakovlev, A. Shabaev, Al.L. Efros, I.A. Yugova, R. Oulton, V. Stavarache, D. Reuter, A. Wieck, M. Bayer, Science **313**, 341 (2006)
[57] I.A. Yugova, A. Greilich, E.A. Zhukov, D.R. Yakovlev, M. Bayer, D. Reuter, A.D. Wieck, Phys. Rev. B **75**, 195325 (2007)

[58] I.A. Akimov, A. Hundt, T. Flissikowski, F. Henneberger, Appl. Phys. Lett. **81**, 4730 (2002)

[59] S. Cortez, O. Krebs, S. Laurent, M. Senes, X. Marie, P. Voisin, R. Ferreira, G. Bastard, J.-M. Gérard, T. Amand, Phys. Rev. Lett. **89**, 207 (2002)

[60] R.J. Warburton, C. Schaflein, D. Haft, F. Bickel, A. Lorke, K. Karrai, J.M. Garcia, W. Schoenfeld, P.M. Petroff, Nature **405**, 926 (2000)

[61] S. Laurent, B. Eble, O. Krebs, A. Lemaître, B. Urbaszek, X. Marie, T. Amand, P. Voisin, Phys. Rev. Lett. **94**, 147401 (2005)

[62] T. Flissikowski, I.A. Akimov, A. Hundt, F. Henneberger, Phys. Rev. B **68**, 161309(R) (2003)

[63] M.E. Ware, E.A. Stinaff, D. Gammon, M.F. Doty, A.S. Bracker, D. Gershoni, V.L. Korenev, S.C. Badescu, Y. Lyanda-Geller, T.L. Reinecke, Phys. Rev. Lett. **95**, 177403 (2005)

[64] I.A. Akimov, K.V. Kavokin, A. Hundt, F. Henneberger, Phys. Rev. B **71**, 75326 (2005)

[65] M.M. Glazov, E.L. Ivchenko, R.V. Baltz, E.G. Tsitsishvili, cond-mat/0501635 (2005)

[66] M. Ediger, G. Bester, B.D. Gerardot, A. Badolato, P.M. Petroff, K. Karrai, A. Zunger, R.J. Warburton, Phys. Rev. Lett. **98**, 036808 (2007)

[67] S. Laurent, M. Senes, O. Krebs, V.K. Kalevich, B. Urbaszek, X. Marie, T. Amand, P. Voisin, Phys. Rev. B **73**, 235302 (2006)

[68] V.K. Kalevich, I.A. Merkulov, A.Y. Shiryaev, K.V. Kavokin, M. Ikezawa, T. Okuno, P.N. Brunkov, A.E. Zhukov, V.M. Ustinov, Y. Masumoto, Phys. Rev. B **72**, 045325 (2005)

[69] C.W. Lai, P. Maletinsky, A. Badolato, A. Imamoglu, Phys. Rev. Lett. **96**, 167403 (2006)

[70] R.I. Dzhioev, B.P. Zakharchenya, V.L. Korenev, P.E. Pak, D.A. Vinokurov, O.V. Kovalenkov, I.S. Tarasov, Phys. Solid State **40**, 1745 (1998)

[71] A.S. Bracker, E.A. Stinaff, D. Gammon, M.E. Ware, J.G. Tischler, A. Shabaev, Al.L. Efros, D. Park, D. Gershoni, V.L. Korenev, I.A. Merkulov, Phys. Rev. Lett. **94**, 047402 (2005)

[72] R. Oulton, A. Greilich, S.Yu. Verbin, R.V. Cherbunin, T. Auer, D.R. Yakovlev, M. Bayer, I.A. Merkulov, V. Stavarache, D. Reuter, A.D. Wieck, Phys. Rev. Lett. **98**, 107401 (2007)

[73] P.-F. Braun, L. Lombez, X. Marie, B. Urbaszek, M. Senes, T. Amand, V. Kalevich, K. Kavokin, O. Krebs, P. Voisin, V. Ustinov, Phys. Stat. Sol. (b) **242**, 1233 (2005)

[74] D. Heiss, S. Schaeck, H. Huebl, M. Bichler, G. Abstreiter, J.J. Finley, D.V. Bulaev, D. Loss, arXiv:0705.1466

[75] D. Gammon, S.W. Brown, E.S. Snow, T.A. Kennedy, D.S. Katzer, D. Park, Science **277**, 85 (1997)

[76] O. Krebs, B. Eble, A. Lemaître, B. Urbaszek, K. Kowalik, A. Kudelski, X. Marie X, T. Amand, P. Voisin, Phys. Stat. Sol. (a) **204**, 202 (2007)

[77] A.V. Khaetskii, D. Loss, L. Glazman, Phys. Rev. Lett. **88**, 186802 (2002)

[78] Y.G. Semenov, K.W. Kim, Phys. Rev. B **67**, 73301 (2003)

[79] M. Kroutvar, Y. Ducommun, D. Heiss, M. Bichler, D. Schuh, G. Abstreiter, J.J. Finley, Nature **432**, 81 (2004)

[80] J.M. Elzerman, R. Hanson, L.H.W. van Beveren, B. Witkamp, L.M.K. Vandersypen, L.P. Kouwenhoven, Nature **430**, 431 (2004)

[81] W.K. Liu, K.M. Whitaker, A.L. Smith, K.R. Kittilstved, B.H. Robinson, D.R. Gamelin, Phys. Rev. Lett. **98**, 186804 (2007)

[82] L. Lombez, P. Renucci, P.F. Braun, H. Carrère, X. Marie, T. Amand, B. Urbaszek, J.L. Gauffier, P. Gallo, T. Camps, A. Arnoult, C. Fontaine, C. Deranlot, R. Mattana, H. Jaffrès, J.-M. George, P.H. Binh, Appl. Phys. Lett. **90**, 081111 (2007)

5

Time-Resolved Spin Dynamics and Spin Noise Spectroscopy

J. Hübner and M. Oestreich

This chapter gives an overview of the major experimental techniques employing optical spectroscopy to unravel, exploit, and control the dynamics of the entity spin in semiconductors. We describe the areas of application as well as the advantages and drawbacks of the different techniques by means of selected examples from up-to-date research in semiconductor spin physics.

5.1 Introduction

Already since 1969, [1] time integrated optical Hanle experiments have uncovered interesting spin physics in semiconductors. Time-resolved Hanle experiments and time-resolved Faraday rotation spectroscopy are the powerful successors which directly measure the precession frequency of the carrier spins in an external or internal effective magnetic field. Most time-resolved optical experiments today employ short optical laser pulses on the picosecond or femtosecond timescale to excite electrons from the valence to the conduction band of the semiconductor. Excitation by circular polarized light leads in most III–V and II–VI semiconductors to a preferential spin orientation of the excited electrons in the conduction band and the remaining hole in the valence band. The reasons for the preferential spin orientation of electrons and holes are the optical selection rules which depend on the sample structure, the semiconductor material, and the direction of excitation. The relaxation of the hole spin towards thermal equilibrium occurs in many cases on the timescale of the momentum relaxation time, due to the strong spin–orbit interaction and mixing of light- and heavy-hole spin states. The spin relaxation time of free holes in bulk GaAs is for example at moderate temperatures faster than 100 fs [2] and a significant hole spin polarization can thus be neglected in many experiments. The electron spin relaxation is usually much slower so that most time-resolved optical experiments only observe the electron spin dynamics.

The excitation of a sample by a polarized laser pulse initializes the spin orientation and defines the time zero. Subsequently, the time-resolved experiment measures the temporal evolution of the spin orientation. The dynamics can be dominated by internal magnetic fields, which can be in the focus of the investigations themselves, or external magnetic fields, which yield further fundamental parameters, such as, for example, the electron g-factor [3]. In time-resolved photoluminescence experiments the spin polarization is detected by measuring the degree of circular polarization of the photoluminescence sent out from the sample after excitation. In time-resolved Faraday rotation experiments, the spin polarization is obtained by measuring the rotation of the linear polarization of a probe laser pulse. Usually, a single laser pulse excites not a single electron but an ensemble of electrons and the experiments impart the dynamics of the average spin polarization of the electron ensemble. This average spin polarization is defined as $P_s = (n_+ - n_-)/(n_+ + n_-)$, where n_+ and n_- are the number of electrons with spin $+1/2$ and spin $-1/2$, respectively. The exponential decay of the initial spin polarization to thermal equilibrium is described by the spin relaxation time τ_s. Though in general, the spin relaxation depends also on temporally changing parameters like, for example, carrier density and carrier temperature. Therefore the spin relaxation can not be exactly described by a mono-exponential decay.

Time-integrated Hanle experiments, time-resolved Hanle experiments, and time-resolved Faraday rotation spectroscopy employ optical excitation to create a preferential spin orientation. A very new kind of Faraday rotation experiment circumvents this perturbation of the studied systems and uses the fluctuating natural spin polarization of a system in thermodynamical equilibrium. This technique was first introduced in quantum optics in rubidium gas in the year 2000 [4] and a short time later was successfully demonstrated in semiconductors [5]. The technique is called spin noise spectroscopy and is based on the Faraday rotation of laser light in the non-absorptive regime (i.e., of laser light with an energy below the band–gap of the semiconductor) without a preceding excitation. Astonishingly, the technique is rather sensitive and promises eminent advantages for distinctive types of spin measurements in semiconductors.

In the following, Sect. 5.2 gives an overview about the time- and polarization-resolved photoluminescence techniques and shows exemplarily photoluminescence results concerning a new electron spin relaxation mechanism in (110) GaAs quantum wells and on the coherent electron hole spin dynamics in natural GaAs quantum dots. Afterwards, Sect. 5.3 describes the technique of time-resolved Faraday rotation spectroscopy and introduces spin amplification as an experimental example. At last, Sect. 5.4 introduces spin noise spectroscopy as a non-perturbative spin detection technique and presents results on the spin dynamics in bulk GaAs.

5.2 Time- and Polarization-Resolved Photoluminescence

Optically excited semiconductors are not in thermal equilibrium and the photoexcited electrons in the conduction band recombine with holes in the valence band to

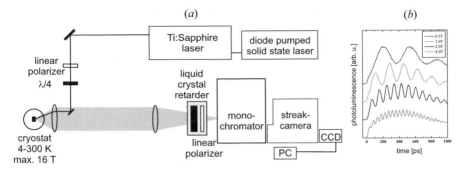

Fig. 5.1. (**a**) Typical setup for time-resolved photoluminescence spectroscopy. (**b**) One circularly polarized component of the time-resolved photoluminescence after excitation of the sample by a circularly polarized laser pulse. The decrease in oscillation amplitude with time results from spin relaxation. The oscillation depth is a measure of the spin polarization, although, the decrease of the beat modulation depth at 4 T results from lack of temporal resolution in this specific experiment

restore the state of the lowest free energy of the system again. The optical selection rules for recombination are identical to the selection rules for absorption and thereby the degree of circular polarization P_σ is a measure of the spin polarization. The maximum degree of circular polarization of the photoluminescence from bulk semiconductors like GaAs for 100% electron spin polarization and unpolarized holes is $P_\sigma = 50\%$ (see Fig. 1.2 from Sect. 1.3.9). This is due to the fact that the energy of heavy and light holes are degenerate at the Γ-point, where it is most likely that an excitation takes place ($p = 0$ for direct gap semiconductors). Additionally the heavy hole transition is three times stronger than the light hole transition, which yields $P_\sigma = (3 - 1)/(3 + 1) = 0.5 = 50\%$. But since photo-created electrons possess in most bulk semiconductors a maximum spin polarization of 50%,[1] the maximum P_σ is equal to $0.5 \times 0.5 = 0.25 = 25\%$ in photoluminescence experiments. In lower-dimensional semiconductors the degeneracy of heavy and light holes is lifted and the degree of circular polarization can become nearly 100% for 100% electron spin polarization.[2] Knowing the optical selection rules, time- and polarization-resolved photoluminescence becomes a powerful tool to study the spin dynamics in semiconductors.

5.2.1 Experimental Technique

Figure 5.1(a) schematically depicts a typical setup for time- and polarization-resolved photoluminescence measurements. A 10 W diode pumped, continuous wave solid

---

[1] In some crystals symmetries, e.g., with hexagonal wurtzite structures, the valence band degeneracy is lifted, enhancing the degree of polarization.

[2] Please note that the depicted selection rules in Fig. 1.2 in Sect. 1.3.9 are strongly simplified and a more rigorous theoretical treatment by $k \cdot p$ theory including excitonic effects is necessary to obtain more realistic predictions (see [6]).

state laser pumps a pulsed picosecond or femtosecond Ti:Sapphire laser having a repetition rate of 80 MHz and a maximum average output power of 2 W. The laser light first passes an optional linear polarizer to enhance the degree of linear polarization of the laser. Next, the linear polarization of the laser light is converted to circular polarization by, e.g., a Soleil Babinet or Berek compensator, a liquid crystal retarder, or a photoelastic or electro-optic modulator acting as a $\lambda/4$-plate. All these retarders are adjusted to get circularly polarized light at the sample, i.e., changes of light polarization at successive optics-like mirrors—especially dielectric or coated mirrors show a strong birefringence—have to be compensated.[3] After the retarder, the circularly polarized laser pulse is focused on the sample that is mounted in a variable temperature magneto-cryostat in Voigt geometry, i.e., the magnetic field is perpendicular to the direction of excitation. The photoluminescence from the sample is collected in backwards direction and passed through a $\lambda/4$-retarder followed by a linear polarizer to select one circular component. The energy resolution is performed by a monochromator and time resolution by a photomultiplier tube or a streak camera system. A streak camera system has the significant advantage over a photomultiplier tube in that it can detect simultaneously a two-dimensional data set due to a spatially mapped photocathode. The data set can contain such information as, e.g., the photoluminescence versus time and light frequency, polarization, or position of the collected photoluminescence on the sample. Figure 5.1(b) shows, as an example for time-resolved photoluminescence, the temporal dynamics of one circular component of the photoluminescence of a 25 nm GaAs/(AlGaAs) quantum well after excitation with a 2 ps, circularly polarized laser pulse. A magnetic field B perpendicular to the excitation direction leads to a Larmor precession of the optically excited electron spins around the axis of the magnetic field with a Larmor frequency $\Omega_L = g\mu_B B/\hbar$, where g is the electron g-factor and μ_B is the Bohr magneton. The precession of the electron spin is directly visible as modulation (beats) in the polarization resolved photoluminescence intensity. The measured beat period is—as expected from the equation for the Larmor frequency—proportional to the magnetic field. High precision measurements reveal a slight deviation from this linear proportionality since g depends on the electron energy in the conduction band and thereby on the applied magnetic field B (see [7] for further details). The photoluminescence intensity modulation (beats) start at time $t = 0$ with a minimum since the circular polarization of excitation and detection are opposite. The overall increase of the photoluminescence in the first 400 ps does not depend on the spin dynamics but results from the cooling of initially hot electrons and holes to lattice temperature. The overall decrease of the photoluminescence at later times results from the reduction of carrier density due to recombination. In general the modulation of the normalized average photoluminescence intensity can be modeled in first approximation with the following formula:

---

[3] Liquid crystal retarders have the advantage that the polarization can be controlled electrically with a variable frequency of up to some 10 Hertz. Photoelastic modulators have the advantage of higher but fixed frequencies of around 50 kHz while electro-optic modulators are used up to MHz, but require a fast and accurate high voltage source. The best choice concerning the retarder depends on the photoluminescence detection system.

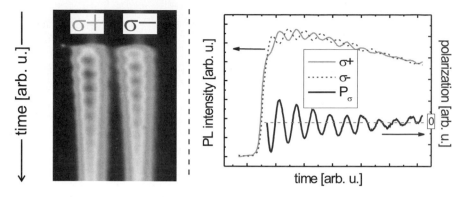

Fig. 5.2. Circular polarization resolved photoluminescence acquired with a streak camera system. *Left*: The raw data depicted is photoluminescence from a GaAs quantum well sample at low temperatures. The energy resolution follows the horizontal axis (*right to left*). In this special case, the two orthogonal polarizations are sent through the monochromator at the same time, shifted by a fixed distance horizontally with the help of a Wollaston prism. *Right*: Depicted are the time transients of the intensity and of the extracted polarization

$$I(t) = I_0\big(1 - e^{-t/\tau_c}\big)e^{-t/\tau_r} \times M(t)$$

with

$$M(t) = 1 \pm a_M \times e^{-t/\tau_s} \cos(\Omega_L t),$$

where τ_c, τ_r, τ_s denote the cooling, recombination, and transverse spin relaxation time. The sign of the modulation depth $\pm a_M$ reflects the detection with the same or opposite circular polarization, respectively, and Ω_L is the modulation (Larmor) frequency. The effects of the carrier dynamics of the spin independent photoluminescence dynamics might seem to obscure the interpretation of the data yielding the parameters of $M(t)$ but subsequent or simultaneous measurement of the right (σ^-) and left (σ^+) circular photoluminescence components allows for the direct extraction of the time-resolved optical polarization $P_\sigma = (\sigma^- - \sigma^+)/(\sigma^- + \sigma^+)$, i.e., unobscured extraction of the spin polarization $P_s = a_M \times e^{-t/\tau_s} \cos(\Omega_L t)$. A corresponding data set is depicted in Fig. 5.2, where the two orthogonal circular polarizations have been acquired simultaneously, allowing the distinctive and exact extraction of the degree of polarization, the spin quantum beat frequency, and the spin relaxation time.

5.2.2 Experimental Example I: Spin Relaxation in (110) Oriented Quantum Wells

This first experimental example shows results from temperature dependent photoluminescence measurements on n-doped (110) GaAs quantum wells with a thickness of 20 nm. The experiment yields a strong anisotropy of the electron spin relaxation time and a new spin relaxation mechanism due to inter-sub-band scattering [8]. At

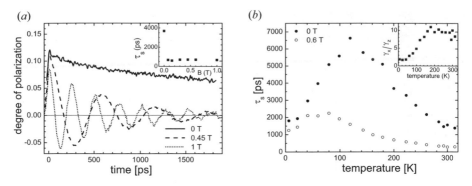

Fig. 5.3. *Left*: Degree of polarization P_σ of the time-resolved photoluminescence for magnetic fields of 0 T (*solid line*), 0.45 T (*dashed line*), and 1 T (*dotted line*) measured at 200 K. *Inset*: Dependence of measured τ_s on magnetic field. *Right*: Temperature dependence of spin lifetime τ_s for $B = 0$ T (*closed circles*) and $B = 0.6$ T (*open circles*). *Inset*: Corresponding temperature dependence of spin relaxation anisotropy τ_x/τ_z. (From [8])

the same time, the experiment demonstrates nicely a major drawback of photoluminescence measurements to study the electron spin dynamics, i.e., the inevitable presence and influence of holes.

We have seen in Sect. 1.4.2 that the random magnetic field depending on the momentum p—that leads to spin relaxation—always points in (110) GaAs quantum wells toward a growth direction. Thereby, spin relaxation due to the Dyakonov–Perel mechanism [9] is suppressed for electron spins aligned in the z-direction. On the other hand, electron spins in the quantum well plane precess around the p-dependent magnetic field and thereby experience spin relaxation since every single electron changes its momentum p stochastically. The resulting anisotropy of the spin relaxation time is directly observed by time- and polarization-resolved photoluminescence measurements: Fig. 5.3(a) shows the transients of P_σ after excitation in the z-direction for three different external magnetic fields in y-direction. At zero magnetic field ($B = 0$ T) (solid line), the temporal decrease of P_σ is slow, i.e., the spin relaxation is slow for spins in the z-direction. At finite magnetic field B (dashed and dotted lines), the spins precess around the external magnetic field and the envelope of P_σ (oscillation amplitude) decreases more rapidly since the precessing electron spins are in average half the time oriented within the quantum well plane where they experience fast Dyakonov–Perel spin relaxation. The inset of Fig. 5.3(a) shows τ_s versus B. The almost constant spin lifetime for $B > 0$ T excludes a direct influence of the external magnetic field on the spin relaxation, i.e., τ_s depends strongly on the spin orientation and only weakly on the magnitude of the external magnetic field.

Next, we study the temperature dependence of τ_s for $B = 0$ T and $B = 0.6$ T. The data acquired at $B = 0$ T yield τ_s in the z-direction and thereby the spin relaxation rate in the z-direction $\gamma_z = 1/\tau_s$. Measuring at $B = 0.6$ T yields the composed spin relaxation time in x- and z-directions and thereby the spin relaxation rate $(\gamma_x + \gamma_z)/2$. Since γ_z can be extracted from the $B = 0\,\text{T}$ measurement, the $B = 0.6\,\text{T}$ measurement yields γ_x.[4]

Figure 5.3(b) shows an increase of the spin relaxation time with increasing temperature for temperatures between 2 K and 80 K and $B = 0.6\,\text{T}$ (open circles). This increase is a typical artifact of photoluminescence measurements: Photoluminescence depends on the presence of holes and these holes lead to electron spin relaxation by the Bir–Aronov–Pikus mechanism [13]. The spin relaxation via hole spins decreases with increasing temperature since the electron hole exchange interaction becomes weaker at higher temperatures. Above 80 K, τ_s decreases for $B = 0.6\,\text{T}$ again since the Dyakonov–Perel mechanism becomes more effective at higher temperatures due to the occupation of higher momentum states. In contrast, the spin relaxation time increases for spins in the z-direction (closed circles) up to 6.5 ns at 120 K since the Dyakonov–Perel mechanism is not effective for spins in the z-direction and the Bir–Aronov–Pikus mechanism decreases even further with higher temperatures. However, above 120 K, τ_s decreases again and that is astonishingly since the Dyakonov–Perel mechanism is suppressed, the hole spin influence becomes weaker at higher temperatures, and as well the Elliott–Yafet mechanism as the hyperfine interaction with nuclear spins are also too weak to explain the decrease. The explanation of this puzzle is a new spin relaxation mechanism that depends on inter-sub-band scattering, the so-called inter-sub-band spin relaxation mechanism. This mechanism can be understood in the following way: Without spin–orbit interaction, scattering between sub-bands, e.g., the first (1) and second (2) sub-band would be spin-conserving. If spin–orbit interaction present, which in parallel leads to Dyakonov–Perel spin dephasing, transitions in the form of $|\psi_{p_i,\uparrow}^{(1)}\rangle \leftrightarrow |\psi_{p_j,\downarrow}^{(2)}\rangle$ become possible with a scattering rate given by the sub-band splitting and the strength of the spin–orbit interaction, i.e., the Dresselhaus spin splitting [8]. This example shows that even in the presence of holes time-resolved photoluminescence experiments yield be spin physics.

---

[4] Direct measurement of γ_x by excitation and photoluminescence detection in an x-direction is not possible due to the optical selection rules in quantum wells. The interband matrix elements of the heavy hole transition in quantum wells read $u_{hh1} = -1/\sqrt{2}|X + iY\uparrow\rangle$ and $u_{hh2} = 1/\sqrt{2}|X - iY\downarrow\rangle$, i.e., the heavy hole transition has no z-component and quantum wells emit linear polarized photoluminescence in x-direction irrespectively from the electron or hole spin polarization. Note that this fact has been ignored in several publications (see for example [10, 11]). The interband matrix element of the light hole transition reads $u_{lh1} = -1/\sqrt{6}|X + iY\downarrow\rangle + \sqrt{2/3}|Z\uparrow\rangle$ and $u_{lh2} = 1/\sqrt{6}|X - iY\uparrow\rangle + \sqrt{2/3}|Z\downarrow\rangle$, i.e., the light hole transition has a z-component and spin polarization and circular polarization are interconnected. This interconnection can be seen at resonant excitation of the light hole [12] and in photoluminescence in very wide quantum wells with strong heavy-light hole mixing and specially strained quantum wells, where the light hole has a higher energy than the heavy hole.

5.2.3 Experimental Example II: Coherent Dynamics of Coupled Electron and Hole Spins in Semiconductors

This second photoluminescence experiment demonstrates that the interaction of electrons and holes also yields interesting spin physics [14]. The spin relaxation of holes is not always fast. The large energy splitting of the heavy and light hole states in thin quantum wells and quantum dots leads for example to very long spin relaxation times of free holes [15–18]. Also, localized holes exhibit long spin relaxation times. In the following, we show, as a nice example for slow hole spin relaxation, results from photoluminescence experiments on undoped 3 nm GaAs/AlGaAs quantum wells. Surface fluctuations lead in thin quantum wells to strongly localized states which are called natural quantum dots. These natural quantum dots are excited resonantly by ps laser pulses at low temperatures. The excited electrons and holes form excitons with a large electron hole exchange energy in the direction of strongest confinement, i.e., in growth direction. The spin dynamics of an electron and a hole within an exciton is described for an arbitrary magnetic field $\boldsymbol{B} = (B_x, B_y, B_z)$ by the Hamiltonian

$$\hat{H} = \frac{1}{2} \sum_{i,j=x,y,z} \left\{ \mu_B B_j \left(g_{ij}^{(e)} \hat{\sigma}_i^{(e)} + g_{ij}^{(h)} \hat{\sigma}_i^{(h)} \right) - c_{ij} \hat{\sigma}_i^{(h)} \hat{\sigma}_j^{(e)} \right\}, \tag{5.1}$$

where $g_{ij}^{(e)}$ and $g_{ij}^{(h)}$ are the g-factor tensors, $\hat{\sigma}_i^{(e)}$ and $\hat{\sigma}_i^{(h)}$ are the Pauli-spin operators of the electron and hole, respectively, c_{ij} is the electron–hole exchange tensor, μ_B is Bohr's magneton, and z is the growth and quantization direction. The light hole is neglected in the Hamiltonian since the heavy and light hole energy splitting is large in thin quantum wells. The temporal evolution of the corresponding exciton spin states $\psi(\sigma_i^{e,h}, t)$ can be easily solved with standard methods and the time dependent Schrödinger's equation.

Figure 5.4 shows the complex dynamics of one *linear* polarization component of the photoluminescence after *linearly* polarized optical excitation of the quantum dots. The weak modulation depth of the experimentally observed oscillations results from the limited temporal resolution of the setup. The observed decay of the fast oscillations within the observation time of 50 ps is not due to decoherence but due to destructive interference of spin quantum beats from individual excitons with varying parameters for the exchange energy and g-factors. The photoluminescence dynamics shows such a rich structure since the linear polarized light coherently excites the four eigenstates of the Hamiltonian. The four eigenstates are composed by the two electronic spin states $|+1/2\rangle$ and $|-1/2\rangle$ and the two hole spin states $|+3/2\rangle$ and $|-3/2\rangle$ yielding the so-called dark and bright states which are mixed by the electron–hole exchange interaction and the external magnetic field. The complete dynamics is destroyed as soon as the electron or the hole spin relaxes, i.e., the observation of the linearly polarized photoluminescence beatings manifests the long hole spin relaxation times. Most interestingly, the dynamics yields directly a *controlled not* gate operation for the electron–hole spin states if the orientation and strength of the magnetic field are chosen properly. Thereby, the electron–hole spin interaction is in principle a candidate for new concepts of solid state quantum devices.

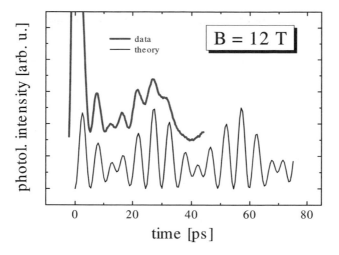

Fig. 5.4. Measured (*thick line*) and calculated (*thin line*) temporal evolution of the photoluminescence for linear polarized excitation and detection and a magnetic field tilted by 45° from the growth direction, i.e., from the excitation and detection direction. From [14]

5.2.4 Photoluminescence and Spin-Optoelectronic Devices

The preceding two examples have shown advantages and disadvantages of photoluminescence measurements concerning the spin dynamics in semiconductors. Besides many other experiments, polarization resolved photoluminescence has probably the most attractive success story in the field of spin-optoelectronic device research. Photoluminescence measurements have demonstrated spin transport in high electric field [19], the efficient electrical spin injection [20, 21], polarized emission from electrically pumped spin light emitting diodes [22–24], the GHz modulation of the stimulated emission of a vertical cavity surface emission laser by modulation of the spin orientation [25], the threshold reduction of such a laser by injection of spin polarized electrons [26–28], and many more results on spin devices, i.e., photoluminescence and spin-optoelectronic devices are strongly linked with each other and the reader is referred to the literature for further information.

5.3 Time-Resolved Faraday/Kerr Rotation

Time-resolved Faraday or Kerr rotation are two established pump–probe techniques to measure the spin dynamics in doped semiconductors. Assume for example n-doped GaAs where the electrons in the conduction band are unpolarized in thermal equilibrium, i.e., we neglect any stochastic spin polarization which is insignificant in this experiment but plays the dominant role in spin noise spectroscopy (see Sect. 5.4). A short, circularly polarized laser pulse pumps similarly to the photoluminescence experiment spin polarized electrons from the valence band to the conduction band

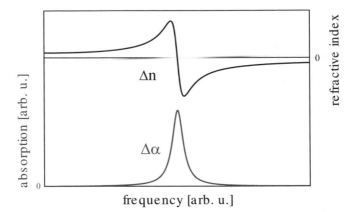

Fig. 5.5. A change in absorption $\Delta\alpha$ is always connected with a change of the refractive index Δn (shown here for a single, homogeneously broadened transition which is described by a Lorentzian absorption). The change of the refractive index becomes zero slightly below the maximum of $\Delta\alpha$. At optical frequencies, Δn is in very good approximation zero at the maximum of $\Delta\alpha$

and partially polarizes the electron ensemble. The degree of polarization is weak for small excitation densities compared to the doping density and large for high excitation densities. After excitation, the excited holes recombine with the electron ensemble composed of doping and optically created electrons and a partially polarized electron ensemble remains. The degree of polarization of the partially polarized electron ensemble is detected by a second laser pulse that is linearly polarized and temporally delayed to the pump pulse. This second laser pulse is called the probe pulse. The probe pulse is either transmitted through the sample (Faraday) [29] or reflected (Kerr) rotation geometry [30, 31]. The degree of rotation of the linearly polarization of the probe pulse is for a normal incidence of the probe pulse proportional to the carrier spin orientation in the sample. This rotation is again the result of optical selection rules: The spin polarized electrons block optical transitions around the Fermi energy for one circular polarization stronger than for the other one. The difference in absorption $\Delta\alpha$ leads, via the Kramers–Kronig relation, to a difference in the refractive index Δn for σ^+ and σ^- polarized light (see Fig. 5.5). Linear polarized light is a coherent superposition of σ^+ and σ^- polarized light. The different refractive indexes $n_{\sigma+}$ and $n_{\sigma-}$ induce a phase shift between the two components, leading to a rotation of the linear polarization direction. As a consequence the time-resolved amplitude of the rotation of the linear polarization of the probe pulse directly reflects the degree and dynamics of the electron spin polarization.[5]

---

[5] In addition to the rotation, the slight absorption will cause a reduction in the amplitude of one circular component. This leads to an effective rotated elliptical polarization. The ellipticity is usually small compared to the phase shift, especially in non-magnetic semiconductors , i.e., the polarization bridge set-up in the described configuration is not sensitive to this so-called circular dichroism [32].

One advantage of the Faraday rotation technique over the described time-resolved photoluminescence technique in Sect. 5.2.1 is the better temporal resolution which is limited in the Faraday rotation experiments mainly by the temporal length of the pump and probe pulse. Another advantage of the Faraday rotation technique is the smaller influence of optically created holes on the electron spin relaxation in n-doped systems.[6] The influence of holes is smaller since time-resolved photoluminescence measurements usually require much higher degrees of electron spin polarization for a reliable signal. Additionally the Faraday rotation signal does not depend on the presence of holes at the time of measurement, i.e., at large delay times where the optically injected holes by the pump pulse have already recombined and only the polarized electron ensemble remains.

5.3.1 Experimental Set-Up

Figure 5.6 schematically depicts a typical setup for a Faraday rotation experiment. The linear polarized pico- or femtosecond laser pulses from a Ti:Sapphire laser are split by a beam splitter in a strong pump pulse and a significantly weaker probe pulse (e.g., in a ratio of 10:1 in intensities). The pump pulse is time-delayed versus the probe pulse by a mechanical delay stage to scan the temporal dynamics.[7] Subsequently, the intensities of pump and probe beams are modulated in the kHz regime with two different frequencies by an optical chopper. The pump beam is circularly polarized and focused on the sample which can be mounted in a magneto-optical cryostat. The probe beam is focused on the same spot on the sample,[8] passes after transmission through the sample and through a Wollaston prism (linear polarization sensitive beam splitter), and is focused on the two photodiodes of a balanced receiver. A $\lambda/2$-retarder plate before or after the sample is used to rotate the linear polarization of the probe beam in order to balance the intensities on the two photodiodes. A small change of the linear polarization direction due to the spin polarization

---

[6] In n-doped systems, another effect can arise from the increased influence of nuclear spins since especially localized electrons have a strong interaction with nuclear spins due to their s-type wave function, i.e., a local non-vanishing probability distribution at its center. Nuclear spin effects arise on very different timescales: from femtoseconds on the microscopic sample scale, to micro- and milliseconds on the mesoscopic scale, to hours on the macroscopic (diffusion) scale and are basically present in every optical excitation involving either circular polarized light or magnetic fields: Polarized electron spins can relax via flip–flop processes with nuclear spins and thereby pump the nuclear spin system. The same holds even for an excitation with linear polarized light if an external magnetic field is applied, since the magnetic field induces a macroscopic electronic spin-polarization in thermal equilibrium. In this case, the optically excited unpolarized electrons make a spin flip–flop with the nuclear spins to reach the thermal equilibrium condition of the macroscopic spin polarization (see Chap. 11).

[7] It is the pump pulse, not the probe pulse that is delayed by the mechanical delay stage, since even small unintentional changes of the probe pulse direction usually change the intensity on the photodiodes of the balanced receiver.

[8] The probe beam is often focused to a smaller diameter than the pump beam to detect an area of nearly constant excitation density of the pump beam.

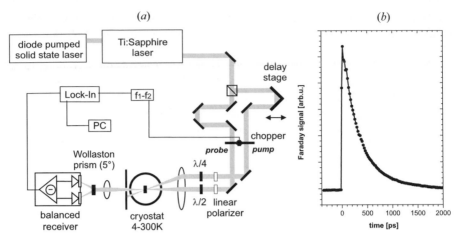

Fig. 5.6. (a) Set-up for time-resolved Faraday rotation spectroscopy. (b) Typical transient of the difference signal of the balanced receiver, i.e., the time-resolved dynamics of the pump-induced magnetization of a sample. In this case, the sample is a 10 nm n-doped GaAs/(Al, Ga)As multiple quantum well at 10 K. The pump and probe intensities are 2 mW and 80 μW, respectively

of the electron ensemble increases the intensity of one photodiode and decreases the intensity on the other photodiode. This difference in intensity is measured by the balanced receiver[9] and sent to a lock-in amplifier that is locked to the difference or sum frequency of the pump and probe frequencies, i.e., the lock-in signal directly reflects the Faraday rotation signal which is proportional to the electron spin polarization. A typical Faraday rotation signal is shown in Fig. 5.6(b).

Most Faraday rotation experiments are frequency degenerate measurements where the pump and probe pulses have the same light frequency. One should keep in mind that this degeneracy in the frequency can lead to artifacts especially at nearly resonant excitation of the sample. Figure 5.5 shows that Δn is zero at the maximum of $\Delta \alpha$. A short, resonant probe pulse has a finite width in energy and integrates Δn

---

[9] A balanced receiver is used for several reasons. First, the signal of an balanced receiver is nearly independent on intensity fluctuations of the laser. Second, one might be tempted not to use a Wollaston prism and a balanced receiver but to measure the change of intensity through a polarizer, that crosses out the probe beam after the sample, i.e., the rotation of the linear polarization in the sample results in a linear polarization that is partially transmitted by the succeeding polarizer. This does usually not work out since in most cases the Faraday rotation angle φ is very small. The transmitted intensity through a polarizer is $I_T = I_0 \cos^2(\phi)$, where I_0 is the incoming intensity and ϕ is the angle between the light polarization and the axes of the polarizer. A Taylor expansion for crossed polarization directly yields that the transmitted intensity is not proportional to φ but to φ^2, i.e., negligibly small for small φ. For the described Wollaston prism the light polarization is 45° to the polarizing axis and the Taylor expansion is linear to φ, i.e., the balanced receiver signal is directly proportional to the Faraday rotation angle.

symmetrically around the center frequency, where Δn changes the sign. The consequence is that the average signal is vanishing since Δn is antisymmetric, i.e., the Faraday rotation signal is zero although the electrons are spin polarized. For non-resonant excitation, the photoexcited electrons and thereby $\Delta \alpha$ and Δn relax in time to lower energy and the Faraday rotation signal changes solely due to carrier cooling. This problem can be solved by non-degenerate Faraday rotation if the Faraday rotation signal is also measured in dependence on the probe beam frequency.[10]

5.3.2 Experimental Example: Spin Amplification

This experimental example shows results on the spin relaxation in n-doped GaAs at low temperatures. The experiment demonstrates nicely the strength of the Faraday rotation spectroscopy in n-doped samples with very long spin relaxation times.

The spin relaxation at low temperatures in bulk GaAs strongly depends on the doping level. At low doping levels, electrons are localized to the donor, electron spin interaction with nuclear spins is large, and electron spin relaxation is fast. At high doping levels, the electrons occupy high momentum states due to the Pauli exclusion principle, and spin relaxation due to the Dyakonov–Perel mechanism is fast. At intermediate doping levels close the Mott insulator transition, the electron spin relaxation becomes long. Actually, at doping levels around $1 \times 10^{16}\,\mathrm{cm}^{-3}$ the spin relaxation becomes much longer than the 12.5 ns repetition rate of a standard Ti:Sapphire laser and resonant spin amplification can be observed, as demonstrated by Kikkawa et al. [34]. This experiment is discussed in the following.

The left side of Fig. 5.7 shows the time-resolved Faraday rotation measurements on undoped and various Si-doped bulk GaAs samples at a temperature of 5 K and a magnetic field in Voigt geometry of 4 T. The 50 μm thick samples are excited by 100 fs pulses from an 80 MHz Ti:Sapphire laser. All samples nicely show spin quantum beats, i.e., the precession of the electron spin around the magnetic field, as observed in the time-resolved photoluminescence measurements in Sect. 5.2.1. The highly doped $(1 \times 10^{18}\,\mathrm{cm}^{-3})$ sample has a slower precession frequency than the lower doped samples since electrons occupy higher energies and the g-factor is energy dependent.

But first, we will focus on the envelope of the Faraday rotation signal. The undoped sample exhibits a moderately fast decay of the envelope of the Faraday rotation signal. However, this decay is not necessarily dominated by the spin relaxation but might be dominated by electron–hole recombination. Faraday rotation measurements cannot distinguish between spin relaxation and electron–hole recombination which is a clear disadvantage to time-resolved photoluminescence measurements.[11]

---

[10] Non-degenerate Faraday rotation can for example reveal spin dependent many-body effects [33].

[11] A similar problem in Faraday rotation arises from holes. Holes can have extremely long lifetimes at low hole densities and influence the electron spin relaxation by the Bir–Aronov–Pikus mechanism. These long hole lifetimes are not directly accessible to Faraday rotation measurements and are sometimes underestimated. Pump intensity dependent measurements

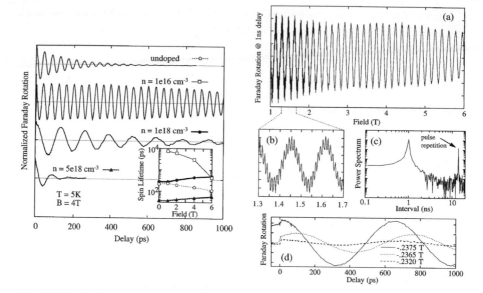

Fig. 5.7. *Left*: Time-resolved Faraday rotation signal for undoped and doped GaAs. *Right*: (a) Time-resolved Faraday rotation signal for $n = 10^{16}\,cm^{-3}$ taken at $\Delta t = 1$ ns. (c) Power spectrum of (a) taken over the interval $B = 1$ to 2.5 T. (d) Time-resolved Faraday rotation signal showing resonant and off-resonant behavior. From [34]

If the spin relaxation time becomes much longer than the time between the laser pulses, i.e., $\tau_s^{-1} \ll 80\,\text{MHz}$, another interesting effect can be observed: Fig. 5.7(a) shows the Faraday rotation signal at a fixed pump probe delay of 1 ns in dependence on the magnetic field. The Faraday rotation signal exposes periodic changes since the Larmor frequency changes with the magnetic field. Consequently the Faraday rotation signal has a maximum if $n \times T_p = 1$ ns and a minimum if $(n + \frac{1}{2}) \times T_p = 1$ ns, where n is an integer and $T_p = 2\pi\hbar/(g^*\mu_B B)$ is the precession period of the electrons. This behavior is also expected for relatively fast spin relaxation (≤ 1 ns). However, Fig. 5.7(b) shows that a second period appears at low magnetic fields. The power spectrum of Fig. 5.7(a) taken over the interval from $B = 1$ to 2.5 T shows that the frequency of the fast oscillations corresponds to the pulse repetition frequency of the laser (see Fig. 5.7(c)). This fast oscillation signal is unique to long spin relaxation samples and arises from an increased polarization when the spin precession is commensurable with the pulse repetition rate, meaning polarized electrons are injected at exactly that moment, when the processing ensemble—injected by the previous pulse—has the same spin orientation. This "hitting the nail right on the head" effect piles up and leads to a bigger than usual spin polarization. The width of the peak is inverse proportional to the transverse spin relaxation time and a T_2^* time of more than

and extrapolation to pump density zero are often used to account for the influence of holes. In contrast to Faraday rotation, the Kerr rotation also measures the total reflectivity of the sample which gives a good indication on the carrier density.

100 ns can be extracted (see also Sect. 5.4). Figure 5.7(d) shows this "resonant spin amplification" even more clearly for different magnetic fields corresponding to different resonant conditions. An almost identical signature would also be measurable in time-resolved photoluminescence measurements employing a pulsed laser source.

5.4 Spin Noise Spectroscopy

Time-resolved Faraday and Kerr rotation and photoluminescence measurements are very powerful tools to directly probe the dynamics on ultrashort timescales but they bear two intrinsic disadvantages. First of all, the optical creation of holes which is connected with every optical interband excitation leads to electron spin relaxation due to the Bir–Aronov–Pikus mechanism. Secondly, at low lattice temperatures even the resonant excitation of carriers leads in doped samples to a heating of the carrier system.[12] These two disadvantages are not present in all optical spin noise spectroscopy.

In general noise spectroscopy is an elegant method to unravel properties of a system in thermal equilibrium without exposing the system to unnecessary excitations. While electrical noise measurements yield information about the physical processes of the underlying electrical transport mechanisms, all optical spin noise spectroscopy yields information about the intrinsic spin dynamics and the g-factor g^*. For simplicity let us assume at first a weakly n-doped bulk GaAs sample where the electrons located at the donors are independent from each other and do not interact. The spins of these non-interacting electrons are Poisson distributed at thermal equilibrium, i.e., if the number of electrons is N, the average spin polarization in thermal equilibrium at arbitrary times is according to statistical physics $\sqrt{N}$. The projection of the spin polarization on the direction of light propagation leads again to a Faraday rotation which can be sensed by linearly polarized, below band gap probe light.

5.4.1 Experimental Realization

Figure 5.8 schematically depicts a typical experimental set-up for spin noise spectroscopy. The laser is an easy to use, tunable, continuous wave laser diode in either Littmann or Littrow configuration, which is temperature and current stabilized. The laser light passes a Faraday isolator to avoid backscattering of laser light into the diode. A linear polarizer and a spatial filter ensure a high linear polarization ratio and a good quality of the spatial mode. The below band-gap laser light is focused on the antireflection coated sample, recollimated after transmission through the sample, and the Faraday rotation of the linear polarization is measured by a polarizing beam

---

[12] Assume for example an n-doped bulk GaAs sample: The excitation of cold electrons is synonymous to the excitation of electrons at the Fermi energy and thereby of electrons with a finite quasimomentum p. The excited holes have the same quasimomentum due to momentum conservation and thereby a finite excess energy. The relaxation of the photoexcited holes to $p = 0$ thus heats the electron system due to the strong Coulomb interaction.

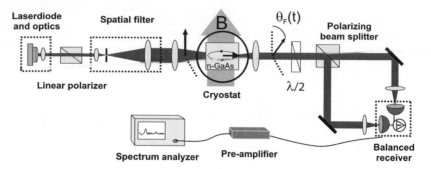

Fig. 5.8. Experimental set-up for spin noise spectroscopy in semiconductors. The spectrum analyzer can be either read out by a computer or for better averaging completely replaced by a fast PC, which acquires the data with a high speed time domain digitizer. The power spectrum is then computed by a FFT routine. From [35]

splitter or a Wollaston prism and a balanced receiver. A small external magnetic field B perpendicular to the beam propagation direction is used to increase the sensitivity of the method: the magnetic field causes the stochastic spin polarization to precess around the magnetic field axis with the Larmor frequency Ω_L and shifts the signal to frequencies where electrical and optical noise is mainly shot noise limited and thereby frequency independent. The transients recorded at $B = 0\,\mathrm{mT}$ then serves as background spectra.[13] Additionally the magnetic field enables the measurement of the electron g-factor. In contrast to the time-resolved Faraday rotation experiments, spin noise spectroscopy measures the radio frequency power spectrum of the Faraday rotation by a spectrum analyzation technique as shown in Fig. 5.8. The amplitude, center frequency, and width contain the information on the spin dynamics in the sample (see [35] for experimental details).

Figure 5.9 shows a typical spin noise spectrum for an external magnetic field of $30\,\mathrm{mT}$ in Voigt geometry at low temperatures. The maximum of the spin noise signal is centered at the Larmor precession frequency $\Omega_L = g\mu_B B/\hbar$ of the stochastic electron spin polarization. The depicted spin noise spectrum can be excellently fitted by a single Lorentz fit which is a clear indication for a single spin relaxation time.[14] The full width at half maximum of the Lorentzian lineshape w_f directly yields the spin relaxation time $\tau_s = (\pi w_f)^{-1}$. For a Poisson distribution, the area A below the

---

[13] Instead of switching the magnetic field on and off to obtain signal and background spectra, a liquid crystal retarder can be placed after the sample to transmit or suppress the Faraday rotation signal: If the liquid crystal retarder is set to $\lambda/4$-retardance the rotation of the linear probe light is effectively projected equally onto the two diodes of the balanced receiver, i.e., a background spectrum is obtained, containing only the electrical and photon shot noise. If the liquid crystal retarder is set to $\lambda/2$-retardance, it is basically ineffective and the Faraday rotation signal can be measured as usual. Care has to be taken for the orientation of the optical axis with respect to the other polarizing components.

[14] The Fourier transformation of a mono-exponential decay yields a Lorentzian lineshape in frequency.

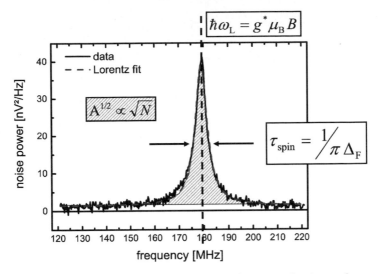

Fig. 5.9. Typical spin noise spectrum. Shown is the noise power density vs. frequency. The information that can be extracted from the Lorentz curve fit is described in the text

lineshape is proportional to the total number of electron spins contributing to the spin noise signal. For interacting electrons, A can be reduced due to Pauli blocking or enhanced due to positive (ferromagnetic) spin–spin correlations, i.e., spin noise spectroscopy can be employed as an elegant method to study usually frail spin–spin interactions in the many body regime.

5.5 Spin Noise Measurements in *n*-GaAs

The spin noise signal is generally very weak compared to the white photon shot noise and the electrical noise of the amplifiers. However, spin noise can be most easily extracted by subtracting two noise spectra at different magnetic fields, since only the spin noise is magnetic field dependent. The sensitivity of spin noise spectroscopy is therefore astonishingly high and calculations show that single spin measurements should feasibly be relatively easy.[15] In Fig. 5.10(left) the dependence of the spin relaxation time as well as the transmission as a measure for the light absorption in the sample is presented as a function of the probe wavelength. The larger transmission is due to less light absorption in the sample and the spin relaxation time increases. This directly reflects the described suppressed perturbation due to holes and carrier heating. Figure 5.10(right) shows as an example the measurement of an ultra-long

---

[15] The sensitivity of spin noise spectroscopy is physically limited by the photon shot noise. Electrical noise and laser noise play, even in the simple setup of Fig. 5.8, a secondary role. Since photon shot noise is white noise, it is the art of averaging which brings this technique forward.

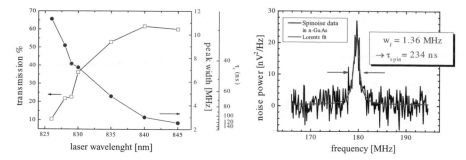

Fig. 5.10. *Left*: *The open squares* (*left axis*) denote the overall transmission through (in this case uncoated, $n_d = 1.8 \times 10^{16}$ cm^{-3}) the GaAs sample in dependence on wavelength at low temperatures. *The filled circles* (*right axis*) give the corresponding peak width of the spin noise spectrum and accordingly the spin relaxation time ($B = 30$ mT, $T = 18$ K). *Right*: Spin noise spectrum recorded in n-GaAs ($B = 30$ mT, $T = 1.4$ K, and $\lambda_{\text{Laser}} = 1.485$ eV) yielding very long spin relaxation times

spin relaxation time in n-doped bulk GaAs at low temperatures for a very large energy distance to the electronic transition. In this case, the width at half maximum of the frequency power spectrum is only 1.36 MHz corresponding to a spin relaxation time of $\tau_s = 234$ ns, much longer compared to time-resolved Faraday rotation measurements in the same material system with comparable doping densities.

This experiment is only a convincing first step, and the very young technique of spin noise spectroscopy will surely demonstrate its fascinating capabilities in further experiments which are inaccessible to the long established time-integrated and time-resolved Hanle and photoluminescence experiments.

5.6 Conclusions

This chapter presented the three major optical techniques: picosecond spin-resolved photoluminescence, femtosecond Faraday rotation, and frequency resolved spin noise spectroscopy. All three techniques rest upon the optical selection rules, win overall electrical measurements by their ease of use, and possess the quality to unveil the spin dynamics in direct semiconductors from different starting points: Time and polarization resolved photoluminescence yield the most direct results concerning the complete electron and hole dynamics following an optical spin-selective excitation. Compared to the photoluminescence measurements, the pump–probe Faraday rotation technique yields a higher time resolution and has the advantage to be sensitive even after the optically excited holes have recombined. Spin noise spectroscopy is the most juvenescent technique and reveals intrinsic properties of the unperturbed spin system. The optical techniques have been substantiated in this chapter with experimental data from works on anomalous spin relaxation in (110) quantum wells, electron hole spin interaction, spin amplification, and ultra long spin relaxation. The examples given above are a modest selection of experiments, being only the tip of an iceberg of a vast variety of exciting contributions to the field of spin dynamics in semiconductors.

References

[1] R.R. Parsons, Phys. Rev. Lett. **23**, 1152 (1969)
[2] D.J. Hilton, C.L. Tang, Phys. Rev. Lett. **89**, 146601 (2002)
[3] M. Oestreich, W.W. Rühle, Phys. Rev. Lett. **74**, 2315 (1995)
[4] S.A. Crooker, D.G. Rickel, A.V. Balatsky, D.L. Smith, Nature **431**, 49 (2004)
[5] M. Oestreich, M. Römer, R. Haug, D. Hägele, Phys. Rev. Lett. **95**, 216603 (2005)
[6] S. Pfalz, R. Winkler, T. Nowitzki, D. Reuter, A.D. Wieck, D. Hägele, M. Oestreich, Phys. Rev. B **71**, 165305 (2005)
[7] M. Oestreich, S. Hallstein, W. Rühle, J. Sel. Top. Quantum Electron. **2**, 747 (1996)
[8] S. Döhrmann, D. Hägele, J. Rudolph, M. Bichler, D. Schuh, M. Oestreich, Phys. Rev. Lett. **93**, 147405 (2004)
[9] M.I. Dyakonov, V.I. Perel, Sov. Phys. Solid State **13**, 3023 (1971)
[10] Y. Ohno, D.K. Young, B. Beschoten, F. Matsukura, H. Ohno, D.D. Awschalom, Nature **402**, 790 (1999)
[11] M. Oestreich, Nature **402**, 735 (1999)
[12] M. Oestreich, D. Hägele, H.C. Schneider, A. Knorr, A. Hansch, S. Hallstein, K.H. Schmidt, K. Köhler, W.W. Rühle, Solid State Commun. **108**, 753 (1998)
[13] G.L. Bir, A.G. Aronov, G.E. Pikus, Sov. Phys. JETP **42**, 705 (1976)
[14] D. Hägele, J. Hübner, W.W. Rühle, M. Oestreich, Solid State Commun. **120**, 73 (2001)
[15] S.D. Ganichev, S.N. Danilov, V.V. Bel'kov, E.L. Ivchenko, M. Bichler, W. Wegscheider, D. Weiss, W. Prettl, Phys. Rev. Lett. **88**, 057401 (2002)
[16] P. Schneider, J. Kainz, S.D. Ganichev, S.N. Danilov, U. Rössler, W. Wegscheider, D. Weiss, W. Prettl, V.V. Bel'kov, M.M. Glazov, L.E. Golub, D. Schuh, J. Appl. Phys. **96**, 420 (2004)
[17] M. Syperek, D.R. Yakovlev, A. Greilich, M. Bayer, J. Misiewicz, D. Reuter, A. Wieck, AIP Conf. Proc. **893**, 1303 (2007)
[18] X. Marie, T. Amand, P. Le Jeune, M. Paillard, P. Renucci, L.E. Golub, V.D. Dymnikov, E.L. Ivchenko, Phys. Rev. B **60**, 5811 (1999)
[19] D. Hägele, M. Oestreich, W.W. Rühle, N. Nestle, K. Eberl, Appl. Phys. Lett. **73**, 1580 (1998)
[20] M. Oestreich, J. Hübner, D. Hägele, P.J. Klar, W. Heimbrodt, W.W. Rühle, D.E. Ashenford, B. Lunn, Appl. Phys. Lett. **74**, 1251 (1999)
[21] B.T. Jonker, G. Kioseoglou, A.T. Hanbicki, C.H. Li, P.E. Thompson, Nat. Phys. **3**, 542 (2007)
[22] R. Fiederling, M. Keim, G. Reuscher, W. Ossau, G. Schmidt, A. Waag, L.W. Molenkamp, Nature **89**, 787 (1999)
[23] H.J. Zhu, M. Ramsteiner, H. Kostial, M. Wassermeier, H.P. Schönherr, K.H. Ploog, Phys. Rev. Lett. **87**, 016601 (2001)
[24] M. Holub, P. Bhattacharya, J. Phys. D: Appl. Phys. **40**, R179 (2007)
[25] S. Hallstein, J.D. Berger, M. Hilpert, H.C. Schneider, W.W. Rühle, F. Jahnke, S.W. Koch, H.M. Gibbs, G. Khitrova, M. Oestreich, Phys. Rev. B **56**, R7076 (1997)
[26] J. Rudolph, D. Hägele, H.M. Gibbs, G. Khitrova, M. Oestreich, Appl. Phys. Lett. **82**, 4516 (2003)

[27] J. Rudolph, S. Döhrmann, D. Hägele, M. Oestreich, W. Stolz, Appl. Phys. Lett. **87**, 241117 (2005)
[28] M. Oestreich, J. Rudolph, R. Winkler, D. Hägele, Supperlattices Microstruct. **37**, 306 (2005)
[29] J.J. Baumberg, S.A. Crooker, D.D. Awschalom, N. Samarth, H. Luo, J.K. Furdyna, Phys. Rev. B **50**, 7689 (1994)
[30] R.T. Harley, O.Z. Karimov, M. Henini, J. Phys. D: Appl. Phys. **36**, 2198 (2003)
[31] N.I. Zheludev, M.A. Brummell, R.T. Harley, A. Malinowski, S.V. Popov, D.E. Ashenford, B. Lunn, Solid State Commun. **89**, 823 (1994)
[32] A.K. Zvezdin, V.A. Kotov, *Modern Magnetooptics and Magnetooptical Materials* (IoP Publishing, Bristol, 1997)
[33] P. Nemec, Y. Kerachian, H.M. van Driel, A.L. Smirl, Phys. Rev. B **72**, 245202 (2005)
[34] J.M. Kikkawa, D.D. Awschalom, Phys. Rev. Lett. **80**, 4313 (1998)
[35] M. Römer, J. Hübner, M. Oestreich, Rev. Sci. Instrum. **78**, 103903 (2007)

6

Coherent Spin Dynamics of Carriers

D.R. Yakovlev and M. Bayer

6.1 Introduction

In this chapter we discuss the coherent spin dynamics of carriers confined in semi-conductor nanostructures. To monitor this dynamics, the spins are oriented optically by polarized pulsed laser excitation and their subsequent coherent precession about an external magnetic field is detected. The precession frequency is determined by the splitting between the spin levels. Any zero field splitting induced by hyperfine or spin–orbit interaction can be tailored through the magnetic field. Destruction of spin coherence by scattering leads to a change of the phase of the precession, so that through a phase measurement detailed insight in the spin dynamics can be taken. In addition to its generation the spin coherence can be monitored also optically using ultra-short pulses of polarized light. Aside from these fundamental studies, recently the control and manipulation of spin coherence also attract more and more attention. Initial steps along these lines have been taken and further progress can be expected in the near future.

The experimental studies in this chapter are based on quantum well and quantum dot structures containing resident carriers (electrons or holes). An ensemble of carrier spins is addressed, which complicates the task due to variations of the precession frequencies, which are particularly prominent for the quantum dot case. However, a remarkable tool to overcome this limitation has been discovered in our experiments. The spin ensemble can be synchronized by and with a periodic train of laser pulses, thus providing a locking of several electron spin precession modes. On a time scale of a microsecond the excitation protocol selects only a fraction of dots, namely those which satisfy the phase synchronization conditions. However, on a longer time scale of seconds up to minutes the nuclear fields in the quantum dots are rearranged such that all dots contribute to the coherent signal.

6.1.1 Spin Coherence and Spin Dephasing Times

In the presence of an external magnetic field B the spin states are Zeeman split. If oriented normal to the field the spins undergo a precession with the Larmor frequency $\Omega = g\mu_B B/\hbar$. Here g is carrier g-factor along the field and μ_B is the Bohr magneton. The two relaxation times T_1 and T_2 describe the relaxation of the longitudinal and transverse spin components (with respect to B), respectively [1]. During the T_1 time the relaxation from the upper to the lower Zeeman state occurs, for example. This process requires energy dissipation, which in most cases is matched by acoustic phonon emission or by energy exchange with the carrier kinetic energy. The phase of the precession of the transverse spin component is lost during the T_2 time, which we will refer to as *spin coherence time*. The phase relaxation does not cost any energy and an arbitrary scattering event can lead to spin decoherence. In quantum mechanics terms the Larmor spin precession is described as the time evolution of the coherent superposition state of the two spin states split by the magnetic field. The populations of these states oscillate with period $2\pi/\Omega$ for which the term *spin beats* is commonly used.

An ensemble of carrier spins with a dispersion of precession frequencies loses its macroscopic spin coherence considerably faster than the individual spin coherence time T_2. This happens because a phase shift between spins with different precession frequencies develops fast in time, while the spin coherence of each individual spin is still preserved. The ensemble coherence is lost during the *spin dephasing time T_2^**. We can introduce the inhomogeneous spin relaxation time T_2^{inh} in the following way:

$$\frac{1}{T_2^*} = \frac{1}{T_2} + \frac{1}{T_2^{inh}}. \tag{6.1}$$

In strong magnetic fields, the dispersion of the precession frequency is due to variations of the carrier g-factors. This results in a $1/B$ dependence of T_2^{inh}. At zero field the T_2^{inh} time is limited by variations of the nuclear fluctuation fields about which each electron precesses.

It is obvious that T_2^* cannot exceed T_2, and indeed in semiconductors the difference between these times can reach a few orders of magnitude. Usually it is the T_2^* that is measured experimentally, and it is a challenging task requiring sophisticated techniques to measure the T_2 time of an individual spin, which theoretically can be as long as twice T_1. One such technique is the well-established spin-echo technique [2] and another one is based on the novel phenomenon of mode-locking of the electron spin coherence, as presented in Sect. 6.3.3.

The spin coherence of electronic states has been studied in semiconductor structures of different dimensionality, including bulk-like thin films, quantum wells, and quantum dots [3, 4]. The spin coherence time of an electron has been found to vary over a wide range from a few picoseconds up to a few microseconds. A T_2^* time of 300 ns has been measured in bulk GaAs at liquid helium temperature [5]. For quantum wells the longest spin dephasing times reported so far are 10 ns for GaAs/(Al, Ga)As [6] and 30 ns for CdTe/(Cd, Mg)Te [7, 8]. They have been found for

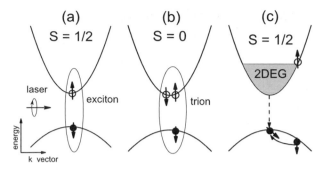

Fig. 6.1. Schematic presentation of a generation of carrier spin coherence by circular polarized laser pulses. The three cases differ with respect to the density of the 2DEG in the quantum well: (**a**) empty well, only photogenerated carriers are present, which become bound to form excitons; (**b**) low density 2DEG, trions with a singlet ground state are formed by a photogenerated exciton and a background electron. The interaction of the trion with the 2DEG is negligible; (**c**) dense 2DEG with a Fermi energy exceeding the exciton binding energy. Bound complexes such as excitons and trions are suppressed

structures with a very diluted two-dimensional electron gas (2DEG). For (In, Ga)As/GaAs quantum dots $T_2 = 3\,\mu s$ and $T_2^* = 0.4\,ns$ have been measured at $B = 6\,T$ [9, 10].

6.1.2 Optical Generation of Spin Coherent Carriers

Time-resolved optical spectroscopy using ultra-short laser pulses as short as $\sim 100\,fs$ allow one to explore carrier spin coherence. The corresponding techniques were described in Chap. 5.

Figure 6.1 illustrates three different situations occurring in experiments on the coherent spin dynamics in quantum wells under resonant optical excitation. These cases differ in the density of resident 2D electrons, n_e. In undoped samples (panel (a), $n_e = 0$) spin oriented excitons are photogenerated. Depending on the experimental details, the coherent spin dynamics of either an exciton or an electron in the exciton can be measured in this case (Chap. 3). However, this dynamics can be monitored only during the exciton lifetime, which is typically in the range from 30 ps to 1 ns.

For high density 2DEGs (panel (c), $n_e a_B^2 > 1$, where a_B is the exciton Bohr radius), exciton formation is suppressed because of state-filling and screening effects. After photogeneration the hole loses its spin and energy and recombines with an electron from the Fermi see. However, the spin oriented electron photogenerated at the Fermi level has infinite lifetime, which allows one to study its long-lived spin coherence. As a result, a circularly polarized photon can change the spin polarization of the 2DEG by $S = \pm 1/2$.

In the case of diluted 2DEGs (panel (b), $n_e a_B^2 \ll 1$) the mechanism for generation of electron spin coherence is not as obvious. The lowest in energy optical transition corresponds to a negatively charged exciton (trion), which consists of two electrons and one hole [11]. The ground state is a singlet trion with antiparallel orientation of the electron spins. When excited resonantly, this state does not contribute directly to the spin polarization because the hole undergoes fast decoherence and the total spin of the two electrons is $S = 0$.

However, a generation of electron spin coherence has been observed experimentally under resonant excitation of trions, both for quantum wells [7, 12] and quantum dots [9]. There are two equivalent approaches to explain this generation. The first one suggests that a coherent superposition of electron and trion states is excited by a circular polarized light pulse when the system is subject to an external magnetic field [9, 12, 13]. The second one is based on considering the 2D electrons which are involved in the trion formation: under circular polarized excitation electrons with a specific spin orientation will be taken from the 2DEG and, consequently, a spin polarization with opposite sign is induced [7, 16]. More details will be given in this chapter.

6.1.3 Experimental Technique

The experimental results described in this chapter have been obtained by applying the techniques of time-resolved pump–probe Faraday rotation or Kerr rotation, see Chap. 5 and [14, 15]. In order to have sufficient spectral resolution all experiments have been performed with laser pulses with 1.5 ps duration, corresponding to about 1.5 meV spectral width. 100 fs pulses, on the other hand, have a spectral width of 20 meV, potentially also leading to the excitation of excited states. The pulses were taken from a mode-locked Ti:Sapphire oscillator generating pulses at a repetition rate of 75.6 MHz, i.e., with a period of 13.2 ns between the pulses. A scheme of the Faraday rotation experiment is given in Fig. 6.2(a).

The basic features of the experiment can be summarized as follows. The quantum well sample with the 2DEG is excited along the structure growth axis (z-axis) by an intense circular-polarized pump pulse, which induces resonant interband transitions and generates spin oriented electrons and holes. Then a much weaker, linearly-polarized probe pulse with the frequency either identical to or different from the pump frequency arrives at the sample. The rotation of the polarization plane of the transmitted probe pulse is analyzed as a function of the delay between the pump and probe pulses. An external magnetic field $\boldsymbol{B}$ is applied in the quantum well plane, say, along the x-axis and leads to precession of the z and y electron spin components with the Larmor frequency $\Omega \equiv \Omega_x = g_e \mu_B B / \hbar$, where g_e is the electron g-factor along the magnetic field. For 2D heavy-holes bound in excitons or trions, the in-plane g-factor is very small and can be ignored. The Faraday rotation angle of the probe beam gives direct access to the induced spin coherence.

A typical Faraday rotation signal is shown in Fig. 6.2(b). It contains information about the (static) spin splitting and the (dynamic) spin coherence. The Zeeman splitting and the underlying g-factor values are evaluated from the oscillation period: $T_L = 2\pi / \Omega = h / g \mu_B B$. The exponential decay of the signal amplitude gives the spin dephasing time T_2^*. Experimentally observed signals may look more complicated, for example two types of carriers or localized and free carriers of the same type, which differ in g-factor and dephasing time, may contribute.

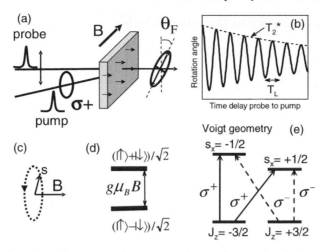

Fig. 6.2. (a) Experimental geometry for pump–probe Faraday rotation. The pump is circular polarized and the probe is linearly polarized. (b) Scheme of the Faraday rotation signal due to precession of the electron spin. (c) Classical approach to Larmor electron spin precession about the magnetic field direction. (d) Two electron Zeeman states split by a magnetic field. The quantum mechanical approach to spin precession is based on a coherent excitation of the two states. (e) Energy level diagram for heavy-hole exciton transitions in quantum wells in a magnetic field along the x-axis (Voigt geometry)

The Faraday/Kerr rotation technique can be used in two regimes. For the degenerate case, the pump and probe beams have the same photon energy, denoted as one-color experiment in the following. In a non-degenerate two-color experiment the pump and probe energies can be tuned independently. This can be realized either by two synchronized Ti:Sapphire lasers [16] or by pump and probe pulses shaped by spectral filtering of 100 fs laser pulses [17].

The scheme in Fig. 6.2(e) explains how the coherent superposition of two electron spin states is excited in a quantum well subject to an external magnetic field. The optical transitions involve the heavy-holes at the top of the valence band and the electrons at the bottom of the conduction band. The heavy-hole state has no spin projection on the field applied along the x-axis, and therefore it does not split. However the electron state is split. σ^+ polarized light, for example, couples the $J_z = -3/2$ hole state with both electron states $s_x = -1/2$ and $+1/2$. To generate a coherent superposition of these states the laser spectral width must exceed the Zeeman splitting.

In this chapter we describe the current status of the coherent optical manipulation of carrier spins in semiconductor nanostructures. In Sect. 6.2 electron and hole spin coherence in CdTe and GaAs based quantum wells will be examined. We concentrate on low free carrier densities, i.e., on a regime in which the optical spectra in the vicinity of the band edge are dominated by the neutral and charged exciton resonances. Section 6.3 describes experimental results for (In, Ga)As/GaAs quantum dots charged by a single electron.

6.2 Spin Coherence in Quantum Wells

Carriers in ideal quantum wells are strongly confined along the structure growth direction (along the z-axis) but are free to move in the well plane. This modifies the spin relaxation, as discussed in Chap. 1. The main localization mechanism for the in-plane carrier motion is provided by well width fluctuations, which result from monolayer steps at the heterointerfaces formed during molecular-beam epitaxy growth. At low temperatures and in structures with relatively low carrier density, where the Fermi energy does not exceed the localization potential, carriers are localized. As a result, most of spin relaxation mechanisms which require carrier motion become inefficient. In this regime a long coherence time is expected for the carriers, as confirmed by the experimental data given below. This is valid for both electrons and holes.

In quantum wells based on semiconductors with a zinc-blende lattice structure, like GaAs or CdTe, the band gap is given by the bottom of the isotropic conduction band and the top of the strongly anisotropic heavy-hole band. The heavy-hole has a spin projection $J_z = \pm 3/2$ along the structure growth axis and its in-plane components are equal to zero: $J_{x,y} = 0$. Therefore, a transverse magnetic field applied in Voigt geometry does not induce a hole spin precession, but acts only on the electrons. Note that in real quantum wells this statement based on symmetry arguments may not hold strictly, so that a finite in-plane component of the heavy-hole spin may appear due to mixing with the light-hole states.

The carrier confinement in the quantum wells enhances the electron–hole Coulomb interaction, which results in an increase of the exciton (X) binding energy. The confinement also stabilizes the charged exciton complexes (trions) consisting of either two electrons and one hole (negatively charged trion, T^-) or two holes and one electron (positively charged trion, T^+). Both types of trions have been well demonstrated for GaAs, CdTe, and ZnSe based quantum wells [18–20]. They are formed in structures with low carrier density, when the interaction between the resident carriers can be neglected. The trion line appears in absorption and emission spectra below the exciton resonance. It is shifted from the exciton by the trion binding energy, which is about 10% of the exciton binding energy. The ground trion state has a singlet spin structure, i.e., for T^- the two electrons have an antiparallel spin configuration. Therefore the probability of trion generation by a resonant light depends on the spin orientation of the resident electron. This is of key importance for optical generation, control, and manipulation of the electron spin coherence in quantum wells and quantum dots.

Samples. In this chapter we present selected results of two quantum well samples to illustrate carrier spin coherence phenomena. Both samples have been fabricated by molecular beam epitaxy. The first one is a CdTe/Cd$_{0.78}$Mg$_{0.22}$Te structure with five 20-nm-thick CdTe wells, each containing a 2DEG of low density, $n_e = 1.1 \times 10^{10}$ cm^{-2} [16]. The second structure contains a two-dimensional hole gas, $n_h = 1.51 \times 10^{11}$ cm^{-2}, which is confined in a single 15-nm-thick GaAs quantum well sandwiched between Al$_{0.34}$Ga$_{0.66}$As barriers [21]. Both samples have been grown on GaAs substrates which are opaque at the energies of the exciton and trion resonances in the quantum wells. Therefore, the Faraday rotation technique, where the probe

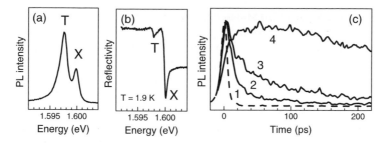

Fig. 6.3. (a) Photoluminescence spectrum of a 20-nm-thick CdTe/(Cd, Mg)Te quantum well measured under nonresonant continuous-wave excitation with a photon energy of 2.33 eV. The exciton (X) and trion (T) resonances are separated by the trion binding energy of 2 meV. **(b)** Reflectivity spectrum of the same structure. **(c)** Kinetics of the photoluminescence measured by a streak camera under 1.5 ps pulsed excitation resonant at the trion energy (*curve 2*) and detuned from it by 0.8 meV (*curve 3*) and 27 meV (*curve 4*) to higher energies. The laser pulse is shown by the *dashed line* [16]

beam is detected after transmission through the sample, cannot be applied and Kerr rotation was used in which the reflected probe beam is analyzed.

6.2.1 Electron Spin Coherence

n-type doped structures are best suited to study electron spin coherence because the lifetime of the resident electrons is not limited by radiative recombination with holes. Further, in such structures the resident electrons are not perturbed by the presence of holes at times exceeding the exciton lifetime.

Optical Spectra of the CdTe/CdMgTe Quantum Well

A photoluminescence (PL) spectrum of the n-type CdTe/Cd$_{0.78}$Mg$_{0.22}$Te quantum well measured at temperature $T = 1.9$ K is given in Fig. 6.3(a). The spectrum shows the exciton and trion recombination lines separated by 2 meV due to the trion binding energy. The exciton binding energy is 12 meV. The full width at half maximum of the exciton line is about 0.5 meV and arises mainly from exciton localization in well width fluctuations.

The reflectivity spectrum of the same quantum well is given in Fig. 6.3(b). Following the procedure described in [22] we have found that the exciton oscillator strength is ten times larger than that of the trion. This fact should be taken into account when the intensities of the Kerr rotation signals at the exciton and trion energies are compared. Both the probe response and the number of photogenerated carriers are proportional to the oscillator strength.

The interpretation of the observed spin dynamics requires information on the recombination times of excitons and trions under resonant excitation. We performed corresponding measurements under linearly polarized excitation using a streak camera. The results for pumping into the exciton and trion resonances are very similar to each other. The typical recombination kinetics for the trion is given in Fig. 6.3(c). For resonant excitation at the trion energy (curve 2, detuning $\Delta E = 0$) about 80% of the luminescence intensity decays on a time scale of 30 ps and the rest decays with a time of 100 ps. When the excitation energy is detuned 0.8 meV above the trion resonance a redistribution of the two exponential decays with 30 and 100 ps time occurs in favor of the longer decay component. Such a behavior is typical for quantum well emission [23, 24]. The shorter decay time, 30 ps, can be attributed to radiative recombination of trions and excitons generated in the radiative cone where their wave vectors match those of the photons. For excitons scattered out of the radiative cone, a longer time of about 100 ps is required to be returned to the cone via emission of acoustical phonons. The exciton luminescence lifetime is slowed down to 100 ps in this case.

The recombination of trions is not restricted to the radiative cone since the remaining electron can take the required momentum for momentum conservation. Therefore, we may expect a fast decay of the trion emission of about 30 ps even for nonresonant photoexcitation. In most cases, however, the trions are formed from photogenerated excitons under these experimental conditions, and the excitons dominate the absorption due to their larger oscillator strength. As a result, the decay of the trion luminescence is determined not by trion recombination but by trion formation and is contributed by the trion formation time and exciton lifetime. For small detuning (curve 3), the fast 30 ps process coexists with the longer 100 ps one. When the excitation energy is tuned to the band-to-band absorption (curve 4, $\Delta E = 27$ meV), the emission decay is prolongated to 250 ps as additional time is required for the free carriers to be bound to excitons.

Long-Lived Electron Spin Coherence

In this part we examine the electron spin coherence with the goal to identify the experimental conditions under which the longest relaxation times are achieved and to collect information about spin dephasing mechanisms.

A typical time evolution of Kerr signals measured for resonant excitation in the trion state is given in Fig. 6.4(a). Several characteristic features are to be noted here: (i) The spin beat oscillation frequency increases linearly with the magnetic field (panel (b)) and the slope of this dependence gives the value of the electron g-factor $|g_e| = 1.64$. In CdTe based quantum wells of comparable thickness, the g-factor is negative [25]. (ii) The spin beats at $B = 0.25$ T show a very weak damping at positive delays traced up to 4 ns and are clearly seen also at negative delays. This means that they do not fully decay over the time interval of 13.2 ns between the pump pulses. With exciton and trion lifetimes below 100 ps the long-living spin beats can be therefore assigned to the coherent spin dynamics of the resident electrons. (iii) The dephasing accelerates with increasing magnetic field so that no beats at negative delays are seen at 0.65 T and higher.

The *magnetic field dependence* of the spin dephasing time T_2^* is shown in Fig. 6.4(c). The dephasing increases with increasing field strengths. In the field range from 0.5 to 7 T, T_2^* follows a $1/B$ dependence (solid line) and tends to saturate for

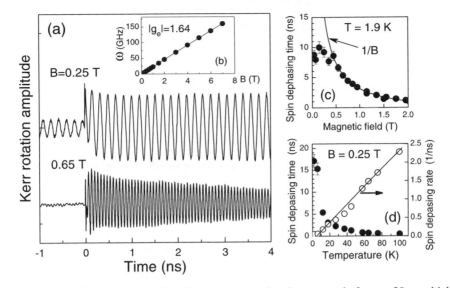

Fig. 6.4. (a) Dynamics of the Kerr rotation signal measured for a 20-nm-thick CdTe/(Cd, Mg)Te quantum well for resonant trion excitation. $T = 1.9$ K. Inset (b) displays the beat frequency vs. magnetic field. (c) Spin dephasing time of electrons as function of magnetic field. Line is a $1/B$ interpolation. (d) Spin dephasing time T_2^* (*closed circles*) and dephasing rate $1/T_2^*$ of electrons (*open circles*) as functions of bath temperature. The line is a linear interpolation [7, 24]

weaker fields. As an ensemble of electron spins is addressed in this experiment, an inhomogeneous dephasing is caused by a spread Δg_e of the electron g-factors which is given by

$$T_2^{inh}(\Delta g_e) = \frac{\hbar}{\Delta g_e \mu_B B}. \tag{6.2}$$

From the $1/B$ interpolation of experimental data we evaluate $\Delta g_e = 0.001$, which is only 0.6% of the mean g_e value. The saturation value of 10 ns gives a lower limit for T_2, see (6.1).

The *temperature dependencies* of the dephasing time T_2^* and dephasing rate $1/T_2^*$ are plotted in Fig. 6.4(d). The dephasing rate is about constant in the range from 1.9 to 7 K and it increases linearly up to 100 K. The characteristic depth of the electron localization potential evaluated from the exciton line width is about 0.5 meV, which corresponds to an activation temperature of 6 K. For higher temperatures a $T_2^* \propto 1/T$ dependence is observed. Such behavior is characteristic for the Dyakonov–Perel mechanism of spin relaxation. Theoretical consideration of this mechanism gives the following expression for the electron spin relaxation rate [26, 27]:

$$\frac{1}{\tau_s} = \alpha_c^2 \tau_p(T) \frac{E_{1e}^2 k_B T}{\hbar^2 E_g}. \tag{6.3}$$

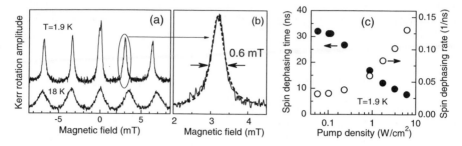

Fig. 6.5. (**a**), (**b**) Resonant spin amplification signal for a 20-nm-thick CdTe/(Cd, Mg)Te quantum well measured at a weak excitation density of 0.05 W/cm^2. (**c**) Spin dephasing time T_2^* (*closed circles*) and dephasing rate $1/T_2^*$ (*open circles*) of electrons as functions of pump power [7]

Here τ_p is the momentum relaxation time, α_c is a parameter related to the spin splitting of the conduction band, E_g is the band gap energy, and E_{1e} is the electron confinement energy. Equation (6.3) is valid for $E_{1e} \gg k_B T$ as well as for relaxation of the spin components along the growth direction in a (100)-oriented quantum well. The linear T-dependence measured experimentally allows us to suggest that in the studied sample τ_p is independent of temperature for $T < 100$ K, which seems reasonable. The electron mobility in CdTe based quantum wells is not high (usually does not exceed a few tens of thousands V cm^2/s) and τ_p is rather short falling in the picosecond range. Therefore, the Dyakonov–Perel spin relaxation mechanism controls the electron spin coherence in the temperature range from 7 to 100 K.

The technique of resonant spin amplification is ideally suited for studying long-lived spin beats, when the signal has a considerable amplitude at negative delays and interferes with the signal at positive delays [28] (Chap. 5). It allows one to measure T_2^* times also at zero magnetic field. In this method the external magnetic field is scanned from small negative to small positive field strengths and a small negative time delay for the probe pulse is chosen. This delay was $\Delta t = -100$ ps for the data shown in Fig. 6.5. The T_2^* time can be directly evaluated from the width of the peaks. At $T = 1.9$ K the dephasing time $T_2^* = 30$ ns. To achieve such a long time, the pump density was reduced as much as possible to 0.05 W/cm^2 where the signal is still detectable with reasonable intensity. Increasing the temperature causes a broadening of the peaks, which reflects a shortening of the dephasing time. A similar behavior is observed when the pump density is increased at a fixed temperature (Fig. 6.5(c)). This effect is related to heating and delocalization of the electrons localized in well width fluctuations. Free electrons have more relaxation channels so that their spin coherence decays faster.

Generation Mechanism: Model Considerations

We now turn to the problem of the generation of carrier spin coherence in quantum wells with low electron density (Fig. 6.1(b)). The formation of trions plays a key role in this process. We first discuss the main points of the model consideration developed by Ivchenko and Glazov [16] and then present experimental data supporting this model.

Resonant Excitation of Trions

We start the analysis from the case of resonant trion excitation. According to the selection rules, the absorption of a circular polarized photon leads to the formation of an electron–hole pair with fixed spin projections: $(e, -1/2; hh, +3/2)$ and $(e, +1/2; hh, -3/2)$ for right (σ^+) and left (σ^-) circular polarized photons, respectively. At weak and moderate magnetic fields the ground state of the negatively charged trion has a singlet electron configuration with antiparallel orientation of the electron spins. Therefore for resonant excitation only resident electrons with orientation opposite to the photogenerated electrons can contribute to trion formation. This results in a depletion of electrons with a z-spin component $S_z = +1/2$ under σ^+ pumping and of $S_z = -1/2$ electrons for σ^- pumping. As a result the remaining resident electrons become spin polarized. An external magnetic field applied in the QW plane leads to precession of the spin polarization of the resident electrons and, therefore, to oscillations of the Kerr signal.

The rate equations describing the spin dynamics of the electrons and trions after resonant, pulsed excitation of trions are given by [16]

$$\frac{dS_z}{dt} = S_y\Omega - \frac{S_z}{\tau_s} + \frac{S_T}{\tau_0^T}, \qquad \frac{dS_y}{dt} = -S_z\Omega - \frac{S_y}{\tau_s}, \qquad \frac{dS_T}{dt} = -\frac{S_T}{\tau^T}. \quad (6.4)$$

Here $S_T = (T_+ - T_-)/2$ is the effective trion spin density with $T_\pm$ being the densities of negatively-charged trions with heavy-hole spin $\pm3/2$, S_y and S_z are the corresponding components of the electron–gas spin density, τ^T is the lifetime of the trion spin including the trion lifetime τ_0^T and the spin relaxation time τ_s^T, i.e., $\tau^T = \tau_0^T\tau_s^T/(\tau_0^T + \tau_s^T)$, and τ_s is the electron spin relaxation time. This time can be identified as the ensemble transverse spin relaxation time T_2^*. For normal incidence of the σ^+ polarized pump pulse the initial conditions for solving the rate equations are $S_y(0) = 0$, $S_T(0) = -S_z(0) = n_0^T/2$ with n_0^T being the initial density of photogenerated trions. The x component of the electron spin density is conserved.

At *zero magnetic field* one finds

$$S_z(t) = -\frac{n_0^T}{2}\left[\eta_0\exp(-t/\tau^T) + (1 - \eta_0)\exp(-t/\tau_s)\right], \quad (6.5)$$

where $\eta_0 = (\tau_0^T)^{-1}/[(\tau^T)^{-1} - \tau_s^{-1}]$. This result can be understood as follows. Right after photoexcitation by σ^- polarized light, the system contains n_0^T singlet trions with $-3/2$ polarized holes, and n_0^T electrons with uncompensated spin $+1/2$, because the same number of electrons with spin $-1/2$ were taken from the 2DEG for trion formation, as illustrated by Fig. 6.6(b). In the absence of spin relaxation, trions decay by emitting σ^- photons and leave behind electrons with spin $-1/2$. As a result, the initially generated electron spin polarization is compensated by these

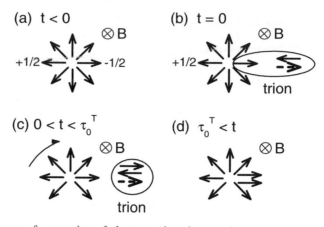

Fig. 6.6. Scheme of generation of electron spin coherence in an external magnetic field by resonant photogeneration of trions. (a) Initial state of a 2DEG whose polarization in the plane perpendicular to the magnetic field is zero. The spins precess around $\boldsymbol{B}$. (b) A σ^- polarized photon generates an (e, +1/2; hh, −3/2) electron–hole pair, which captures a −1/2 resident electron to form a trion. The 2DEG becomes polarized due to the uncompensated +1/2 electron spin in the 2DEG. (c) During the trion lifetime, τ_0^T, the 2DEG polarization precesses around the magnetic field. The trion state does not precess in a magnetic field because of the electron singlet configuration and the zero in-plane hole g-factor. (d) After trion recombination the −1/2 electron contributes again to the 2DEG spin polarization (we neglect here the spin relaxation of the hole in the trion). Shown is the final state of the 2DEG with the induced spin polarization [16]

'trion' electrons and goes to zero during the trion decay. Spin relaxation of electrons and/or trions leads, however, to an imbalance. Namely, the spins of the unbound and the 'trion' electrons do not completely compensate each other, and therefore, a spin polarization is induced.

For an *in-plane magnetic field* the imbalance appears even if spin relaxation is absent. Indeed, the heavy-hole and the singlet electron pair bound in a trion are not affected by the magnetic field, whereas the spins of the resident electrons precess (Fig. 6.6(c)). The trion recombines by emission of a σ^- photon, leaving behind a −1/2 electron. As a result, even after the trions have decayed, the electron spin polarization is non-zero and oscillates with frequency Ω.

For resonant excitation, the initial number of photogenerated trions, n_0^T, cannot exceed $n_e/2$. Thus, n_0^T increases linearly with pump intensity for small excitation density and then saturates at $n_e/2$. In the simplest model

$$n_0^T = \frac{n_e}{2} G\tau_0^T / (1 + G\tau_0^T), \qquad (6.6)$$

where G is the generation rate proportional to the pump power. As a result, the initial electron spin polarization also shows a saturation behavior, caused by the saturation of trion absorption. This conclusion is confirmed by the experiments (Fig. 6.8).

Resonant Excitation of Excitons in a Diluted 2DEG

If the pump photon energy is tuned to the exciton transition, the photogenerated excitons tend to bind to trions at low temperatures as long as they find resident electrons with proper spin orientation.

At *low excitation intensities* satisfying the condition $n_0^X \ll n_e/2$ (here n_0^X is the number of photogenerated excitons) the total spin of the electron gas after decay of all excitons can be estimated as

$$|S_e| = \frac{\tau^X}{2\tau_b}n_0^X,$$ (6.7)

where τ^X is the total lifetime of the exciton spin, including the radiative decay time, the time of exciton binding into a trion, $\tau_b \sim (\gamma n_e)^{-1}$, and the spin relaxation time $(\tau^X)^{-1} = (\tau_0^X)^{-1} + \tau_b^{-1} + (\tau_s^X)^{-1}$. In order to simplify the analysis we assume that the time of exciton binding into a trion, $\tau_b \sim (\gamma n_e)^{-1}$, is shorter than the exciton radiative lifetime τ_0^X and also shorter than the spin relaxation time of the electron in an exciton τ_s^X. In this case, shortly after pulsed excitation all excitons are bound to trions and n_0^X trions are formed. As a result, n_0^X spins of the resident electrons contribute to the Kerr rotation signal.

At *higher excitation intensities*, $n_0^X \geq n_e/2$. A fraction of the spin-polarized excitons determined by the number of electrons $n_e/2$ immediately form trions. Therefore, in absence of electron-in-exciton spin relaxation processes the trion density cannot exceed $n_e/2$, and the total number of spin coherent electrons will be $n_e/2$. Note, that this is the maximum value, which can be achieved for resonant trion excitation. However, in presence of electron-in-exciton spin relaxation also the remaining $n_0^X - (n_e/2)$ excitons can be converted to trions. Obviously, the maximum number of formed trions cannot exceed the concentration of resident electrons, n_e. This process decreases the number of coherent resident electrons and at high excitation densities can bring it to zero. The reason is that all resident electrons will be bound to trions. Their release from the trions by radiative recombination is spread in time and can not provide any spin synchronization.

Detection Aspects

Turning to detection, we find that selective addressing of the exciton resonance results in temporal oscillations of the probe–pulse Kerr rotation signal. The modulation comes from the photoinduced difference in the resonance frequencies $\omega_{0,\pm}$ and/or in the non-radiative damping rates $\Gamma_\pm$. Both differences, $\omega_{0,+} - \omega_{0,-}$ and $\Gamma_+ - \Gamma_-$, appear because of the exchange interaction between an electron in an exciton and the resident electrons: the first difference is related to the Hartree–Fock renormalization of the electron energy in the spin-polarized electron gas, and the second one is related to the spin dependence of the electron–exciton scattering [22]. As a result, the rotation of the total electron spin leads to a modulation of the exciton resonance frequency and non-radiative broadening, and thus to oscillations of the Kerr rotation angle. Note, that an in-plane magnetic field results also in spin precession of the electron in an exciton and the total Kerr signal will be a superposition of 2DEG and exciton signals.

The situation for the probe tuned to the trion resonance is qualitatively the same. The Kerr rotation signal contains components arising from the spin precession of the 2DEG and the electron-in-exciton spin precession. However, one can expect that detection at the trion resonance is less sensitive to the exciton spin dynamics as compared to detection at the exciton resonance.

Two-Color Pump–Probe Experiments

A two-color Kerr rotation technique allows independent tuning of the energies of the pump and probe beams. Consequently either the excitation or detection conditions can be kept constant, which simplifies the identification of relaxation processes.

Figure 6.7 shows Kerr rotation signals detected at the trion and exciton resonances for three different pump energies: (a) resonant with the trion, (b) resonant with the exciton, and (c) non-resonant 72 meV above the exciton energy. A common feature of all signals is the appearance of long-living spin beats related to coherent spin precession of resident electrons. This coherence is excited efficiently for all pump energies and can be detected by probing the trion or exciton resonance.

Some of the signals shown in Fig. 6.7 contain a short-living contribution right after the pump pulse with a decay time of 50–70 ps. This part is especially pronounced for the "pump X/probe X" situation (i.e., pump and probe degenerate with the exciton resonance), see panel (b). The fast component is related to the exciton contribution to the Kerr rotation signal. To extract the times and relative amplitudes of the short and long-living components in the spin beat signals each trace has been fitted with a biexponential decay function:

$$y(t) = \left(A \exp(-t/\tau_1) + B \exp(-t/\tau_2) \right) \cos(g_e \mu_B B t/\hbar), \qquad (6.8)$$

where A and B are constants describing the amplitudes of the fast (τ_1) and slow (τ_2) components, respectively. The parameters extracted from these fits are collected in Table 6.1.

All signals, except the one for "pump T/probe T", are symmetric with respect to the abscissa. The "pump T/probe T" signal shows an initial relaxation of the center-of-gravity of the electron beats with a time of 75 ps, which can be attributed to hole spin relaxation in the trions, see Sect. 6.2.2.

The decay times and relative amplitudes in Table 6.1 are given for $B = 1$ T and $T = 1.9$ K. In general, these parameters depend strongly on pump intensity, magnetic field strength and lattice temperature. Here we focus on the pump energy dependence and address first the amplitudes because they are the key to understanding the generation of electron spin coherence and the role of the trions in this process.

In the "pump T/probe T" experiment only the long-living 2DEG signal is observed (Fig. 6.7(a)). This is in line with the model expectations since in this case only trions are photoexcited. Moving the pump energy to resonance with the exciton and further to the interband transition leads to appearance of the fast decaying

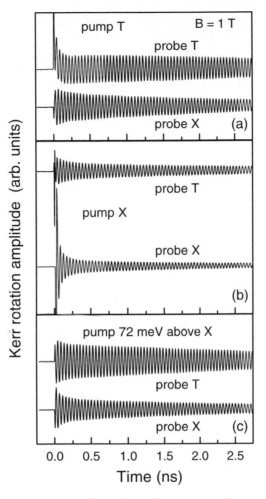

Fig. 6.7. Kerr rotation signals of a CdTe/(Cd, Mg)Te quantum well measured by a two-color technique at $T = 1.9$ K using various pump excitation energies: (**a**) resonant with the trion at 1.5982 eV, (**b**) resonant with the exciton at 1.6005 eV, and (**c**) nonresonant at 1.6718 eV, which is 72 meV above the exciton resonance. Pump density 56 W/cm^2 and probe density 8 W/cm^2 [16]

Table 6.1. Decay times τ_1/τ_2 and amplitude ratios A/B extracted from bi-exponential fits (6.8) to the experimental data in Fig. 6.7

	Pump T	Pump X	Non-resonant
Probe T	–/5.7 ns	40 ps/3.5 ns	56 ps/3.6 ns
	0/1	0.5/0.5	0.2/0.8
Probe X	–/2.6 ns	50 ps/2.0 ns	70 ps/2.8 ns
	0/1	0.9/0.1	0.5/0.5

component when the signal is probed at the trion energy. This appearance can be attributed to the spin dynamics of the exciton, which is excited either resonantly or non-resonantly.

The Kerr rotation signal probed at the exciton energy has two contributions given (i) by the coherent precession of electrons in excitons and (ii) by the spin precession of free electrons. The former decays with the exciton recombination time. This fast exciton component is clearly seen in panels (b) and (c) but is absent for pumping at the trion resonance. For non-resonant excitation its relative amplitude does not exceed 50%. In this case electrons and holes are photogenerated 72 meV above the exciton resonance, and therefore they have a high probability to scatter and relax independently to the bottoms of their bands, where they become bound to trions and excitons. The relative amplitudes of the fast and long-living signals reflect the probability of trion and exciton formation. We expect that trion formation is preferable as the 2DEG density exceeds the concentration of photocarriers by at least an order of magnitude.

A very different ratio of the relative amplitudes, 90% for the fast decay and 10% for the long-living dynamics, is observed for resonant pumping of the exciton and detecting at the exciton (Fig. 6.7(b)). There are at least two factors which favor the exciton population in comparison with that of the trion under resonant pumping of the exciton. First, the photogeneration leads to the formation of excitons with very low kinetic energy, and therefore, they remain within the radiative cone and quickly recombine (during 30–50 ps) before becoming bound to trions [23]. Second, an exciton fraction is localized, so that they are not mobile and cannot reach the sites in the quantum well where the background electrons are localized. Consequently, the formation of trions out of this exciton reservoir is suppressed. Moreover, the ratio of the contributions of the excitons and the 2DEG to the Kerr rotation signal is spectrally dependent: detection at the exciton resonance is more sensitive to the spin precession of the electron in the exciton.

The results of the two-color experiments in Fig. 6.7 are in good agreement with our model; namely, (i) the signal oscillating with the electron Larmor frequency is contributed by precessing electrons which are resident or exciton constituents, and (ii) the trion formation resulting either from resonant photoexcitation or from capture of excitons and resident electrons is a very efficient mechanism for spin coherence generation in a diluted 2DEG.

Pump Power Dependence of the Kerr Rotation Amplitude

The Kerr rotation amplitudes of the 2DEG measured at the trion and exciton resonances as function of pump density are shown in Fig. 6.8. The signal was measured at a delay of 0.5 ns, where the contribution of the fast decaying component is vanishingly small. At low excitation density both dependencies show a linear behavior. At higher density the amplitudes demonstrate pronounced non-linear behavior: for the "pump T/probe T" configuration the signal saturates, while for the "pump X/probe X" configuration it decreases with increasing excitation density. Both results are in agreement with the model calculations [16]. The dashed curve corresponds to trion

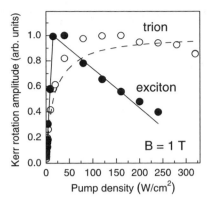

Fig. 6.8. Normalized long-lived amplitude of the 2DEG Kerr rotation signal from a CdTe/(Cd, Mg)Te quantum well measured at a delay of 0.5 ns under resonant pumping of the excitons (*closed circles*) and the trions (*open circles*). Model calculations are shown by the lines [16]

excitation, while the solid line is for exciton pumping. For the dashed curve the only fit parameter was the saturation level. For the solid line the only fit parameter was the ratio between the electron-in-exciton spin relaxation time and the exciton radiative lifetime, $\tau_s^X/\tau_0^X = 10$.

6.2.2 Hole Spin Coherence

The spin coherence of holes can also be studied by a pump–probe Kerr rotation. However, its experimental observation is more challenging. First, the heavy holes have zero in-plane spin components, and cannot precess about an in-plane magnetic field. Tilting the field out of plane allows one to observe the hole spin beats, which however have a long period (proportional to the small out-of-plane field component). Usually only a few oscillation periods can be observed during the hole dephasing time [21, 29]. Second, the spin–orbit interaction in the valence band is considerably stronger than in the conduction band, and consequently spin relaxation is very efficient for free holes. The reported relaxation times vary from 4 ps [30] up to ~1 ns [29, 31] demonstrating a strong dependence on the doping level, doping density and excitation energy. For localized holes most of the spin relaxation mechanisms should be suppressed, including the one due to hyperfine interaction with the nuclear spins, which is inherent for electrons [32]. Therefore, a long spin coherence time for holes is expected.

In *n-type quantum wells* the lifetime of the photogenerated holes is limited by recombination. For non-resonant excitation of, e.g., GaAs/(Al, Ga)As quantum wells it can last up to a few hundred picoseconds [29], but for resonant trion excitation it does not exceed a few tens of picoseconds. In the latter case, which is typical for pump–probe Kerr rotation experiments, the spin dynamics of holes bound to trions can be seen as an asymmetric shift of the center-of-gravity of the electron oscillations. An

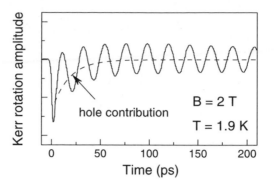

Fig. 6.9. Time-resolved Kerr rotation signal measured on a *n*-type CdTe/(Cd, Mg)Te quantum well. The laser is resonant with the trion energy. The experimental data are shown by the *solid line*. The *dashed line* gives the exponentially decaying part of the center-of-gravity of the electron spin beats which is assigned to the hole spins in the trions [24]

example of such behavior in an *n*-type CdTe/Cd$_{0.78}$Mg$_{0.22}$Te quantum well is shown in Fig. 6.9. The decay of this shift on a time scale of about 20 ps (dashed line) is contributed by hole spin dephasing and by trion recombination. It is not easy to separate these contributions and to extract the spin dephasing time.

In *p-type quantum wells* the lifetime of the resident holes is infinitely long, which makes these structures attractive for studies of the hole spin coherence. An example considered here is a 15-nm-thick GaAs/Al$_{0.34}$Ga$_{0.66}$As quantum well with $n_h = 1.51 \times 10^{11}$ cm^{-2}.

The Kerr rotation signal for resonant excitation of the positively charged trion is shown by the upper curve in Fig. 6.10(a). It consists of two coherent signals. After decomposition by a fit, the two contributions are shown in panel (a). The one with a faster precession frequency corresponds to $|g_e| = 0.285$, which is typical for electrons in GaAs-based quantum wells. It is observed only during ~200 ps after pump arrival and decays with a time of 50 ps, which coincides with the lifetime of the resonantly excited T^+. The other contribution with a small precession frequency is assigned to the hole spin beats. In a magnetic field of 7 T they decay with a time of about 100 ps. The hole beats can be followed up to 500 ps delay. At these long times the Kerr rotation signal is solely due to the coherent hole precession.

Experimentally it is difficult to observe the hole spin quantum beats due to the very small in-plane hole *g*-factor. To enhance the visibility, the magnetic field was tilted slightly out of plane by an angle $\vartheta = 4°$ to increase the hole *g*-factor by mixing the in-plane component ($g_{h,\perp}$) with the one parallel to the quantum well growth axis ($g_{h,\parallel}$), which typically is much larger: $g_h(\vartheta) = (g_{h,\parallel}^2 \sin^2 \vartheta + g_{h,\perp}^2 \cos^2 \vartheta)^{1/2}$. For the studied structure $|g_{h,\perp}| = 0.012 \pm 0.005$ and $|g_{h,\parallel}| = 0.60 \pm 0.01$ [21].

Figure 6.10(b) shows the hole contribution to the Kerr rotation signal for different magnetic fields. The hole spin dephasing time T_2^* is plotted versus B in the inset. A long-living hole spin coherence with $T_2^* = 650$ ps is found at $B = 1$ T. With increasing B up to 10 T it shortens to 70 ps. The field dependence is well described

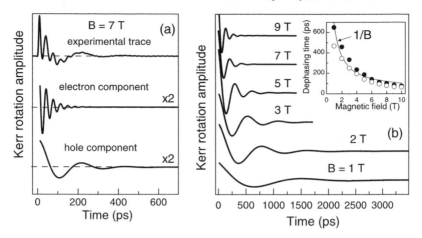

Fig. 6.10. Kerr rotation traces for a p-type 15-nm-thick GaAs/(Al, Ga)As quantum well. The magnetic field was tilted by $\vartheta = 4°$ out of the quantum well plane. The laser energy of 1.5365 eV is resonant with T^+. The powers were set to 5 and 1 W/cm^2 for pump and probe, respectively. (**a**) *Top trace* is the measured signal. *Bottom traces* are obtained by separating electron and hole contributions. (**b**) Hole component extracted from a fit to the Kerr rotation signals at different magnetic fields. *The inset* shows the magnetic field dependence of the hole spin dephasing time T_2^*. *The solid line* is a $1/B$ fit to the data. *The closed* and *open circles* are data measured for pump to probe powers of 1 to 5 W/cm^2 and 5 to 1 W/cm^2, respectively. $T = 1.6$ K [21]

by a $1/B$-form, from which we conclude that the dephasing shortening arises from the inhomogeneity of the hole g-factor $\Delta g_h = 0.0007$. The hole spin dephasing time decreases drastically by elevating the lattice temperature to 5–10 K [21]. The main reason for that is hole delocalization, which activates spin relaxation caused by the spin–orbit interaction.

6.3 Spin Coherence in Singly Charged Quantum Dots

Semiconductor quantum dots have attracted considerable interest due to the three-dimensional confinement of carriers. For confined electrons, most of the spin relaxation mechanisms related to the spin–orbit interaction are inefficient. For a singly charged dot containing only one electron the carrier-carrier interaction is also absent. However, the hyperfine interaction of the electron with the dot nuclei is enhanced because of the localization (Chap. 11).

In order to avoid the inhomogeneity of the quantum dot ensemble, which leads to a considerable broadening of the optical spectra, techniques for single dot spectroscopy have been developed. These techniques have also been applied to study the energy and spin structure of neutral and charged dots and the recombination and spin relaxation dynamics [33] (Chap. 4). However, these techniques require the growth of diluted ensembles of dots and/or post-growth sample structuring, i.e., making opaque

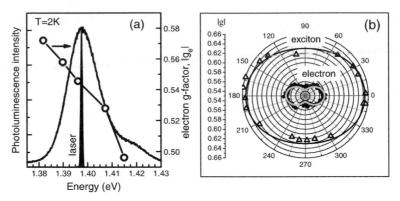

Fig. 6.11. (a) Photoluminescence spectrum of an (In, Ga)As/GaAs quantum dot sample. The filled trace gives the spectrum of the excitation laser used in the Faraday rotation experiments, which could be tuned across the inhomogeneously broadened emission band. The symbols give the electron in-plane g-factor along the [1$\bar{1}$0] direction across this band. (b) In-plane angular dependence of the electron (*circles*) and exciton (*triangles*) g-factors obtained from circular dichroism experiments. Lines are fits to the data using (6.10). $B = 5\,T$. Angle zero corresponds to a field orientation along the x-direction which is defined by the [110] crystal axis [34]

masks or etching mesas in order to select one or only a few dots. Also the optical signals from a single dot are pretty weak and long accumulation times of minutes or hours are often required. In this section we show some examples of studies of an ensemble of (In, Ga)As/GaAs dots by time-resolved techniques. They allow us to obtain detailed information on the coherent spin dynamics of excitons and resident electrons.

Samples. The (In, Ga)As/GaAs quantum dot sample contained 20 layers of dots separated from each other by 60 nm GaAs barriers [9]. The density of dots in each layer was about $10^{10}\,cm^{-2}$. 20 nm below each dot layer an n-doping δ-sheet with a Si-dopant density roughly equal to the dot density was positioned. The as-grown InAs/GaAs sample shows a ground state luminescence at a wavelength around 1.2 μm at cryogenic temperatures. After thermal annealing for 30 s at 960°C, which causes intermixing of the In and Ga atoms, the ground state emission was shifted to 0.89 μm which lies in the sensitivity range of a Si-detector. From Faraday rotation studies [9] we estimate that about 75% of the dots are occupied by a single electron, while 25% contain no residual charge. Transmission electron microscopy studies have shown that after overgrowth the as-grown dots are still about dome-shaped and are rather large with a diameter of about 25 nm and a height of about 5 nm. Thermal annealing increases these parameters.

A photoluminescence spectrum of this sample measured under excitation into the wetting layer at 1.46 eV is shown in Fig. 6.11(a). The emission band has a full width at a half maximum of 15 meV. For pump–probe Faraday rotation experiments spectrally narrow laser pulses and resonant excitation were used.

6.3.1 Exciton and Electron Spin Beats Probed by Faraday Rotation

The energy splittings of the exciton states in quantum dots due to the electron–hole exchange interaction (the so-called fine structure) can be changed by a magnetic field (Zeeman splitting). The size of the splittings is in the range from 0.01 to 1 meV (Chap. 4), which can be typically studied by high-resolution spectroscopy in the spectral domain using luminescence or absorption [35]. To reach the required spectral resolution in such experiments, single dots have to be isolated. An alternative possibility is spectroscopy in the temporal domain.

For measuring fine structure splittings between two levels, the quantum beat spectroscopy can be used. The levels are excited coherently by a pulsed laser into a superposition state. As a result, the probability of the excited superposition shows oscillations in time with a period corresponding to the level splitting. The technique is also suited for ensemble measurements, but one should keep in mind that the results correspond to an average over a large number of addressed dots. A detailed study of the exciton fine structure in (In, Ga)As/GaAs dots is presented in [34]. We present here only a short overview of these results to give comprehensive information on carrier g-factors and exciton fine structure in the dots for which the spin coherence has been investigated.

Experiment. Two modifications of the time-resolved pump–probe Faraday rotation technique similar to those developed in [36, 37] have been used to address the (In, Ga)As/GaAs dots. The first technique, optically induced linear dichroism, uses a linearly polarized pump beam, which leads to optical alignment of excitons in the dots. The second technique uses an intense circular polarized pump pulse for inducing circular dichroism by optical orientation of the carrier spins. In both cases, the optical anisotropy induced by the pump pulse is analyzed by measuring the rotation angle of the polarization plane of a linearly polarized probe pulse. The circular dichroism results from orientation of either the electron or the hole spin and, therefore, persists until the spin orientation of both carriers is destroyed. In case of linear dichroism, a coherent superposition of the $+1$ and -1 exciton states is created. This superposition is destroyed by any spin relaxation process and its lifetime is limited by the fastest process.

Linear dichroism signals in longitudinal magnetic field $(B \parallel z)$ are shown in Fig. 6.12(a). The strongly damped oscillation at zero field is caused by the anisotropic exchange splitting of the bright exciton states with $\delta_1 = 4 \pm 4 \,\mu\text{eV}$. From the precession frequency in high fields plotted in Fig. 6.12(b) the exciton g-factor, $g_{X\parallel} = g_{h\parallel} - g_{e\parallel}$, can be extracted: $|g_{X\parallel}| = |g_{h\parallel} - g_{e\parallel}| = 0.16 \pm 0.11$. Experiments in tilted magnetic fields allow to measure the longitudinal g-factors for electrons and holes: $|g_{e\parallel}| = 0.61$ and $|g_{h\parallel}| = 0.45$.

Circular dichroism signals in transverse magnetic field are presented in Fig. 6.12(c). Quantum beats with at least two different frequencies are clearly observed, resulting in a modulation of the signal at short delay times. After 300 ps, which is close to the exciton lifetime, the modulation vanishes, and a monotonic decay of the beats amplitude is seen. The decay time strongly depends on magnetic field, decreasing from 3 ns at 1 T to 0.5 ns at 6 T.

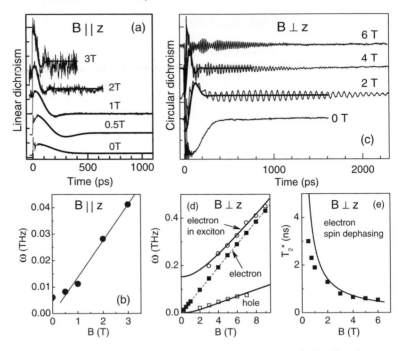

Fig. 6.12. (**a**) Linear dichroism signals in longitudinal magnetic fields (Faraday geometry). *Thick lines* are fits to the data by exponentially damped harmonics. (**b**) Field dependence of the precession frequency extracted from the fits. The line is a *B*-linear fit. (**c**) Circular dichroism traces in transverse magnetic fields (Voigt geometry). Lines at 2 and 4 T give fits to the initial part of the traces by exponentially damped harmonic functions. (**d**) Spin beat frequencies vs. magnetic field for the long-lasting oscillations (*solid squares*) as well as for the initial part oscillations (*open circles* and *open squares*). *Dashed line* is a *B*-linear fit. *Solid lines* are fits of the exciton fine structure. (**e**) Spin dephasing time T_2^* of the long-lasting oscillations as function of magnetic field. The line gives a $1/B$ fit. $T = 2\,\mathrm{K}$ [34]

The long-lived component of the circular dichroism signal is caused by a spin precession of the resident electrons in the singly charged quantum dots [9]. Its frequency is $\omega_e = g_{e\perp}\mu_B B/\hbar$, where $g_{e\perp}$ is the electron g-factor in the dot plane. We find $|g_{e\perp}| = 0.54$ (dashed line in Fig. 6.12(d)). The dephasing time of the long-lived beats shows a $1/B$ dependence (Fig. 6.12(e)) from which we obtain $|\Delta g_{e\perp}| = 0.005$ using (6.1) and (6.2).

A Fourier analysis of the circular dichroism signals during the first 0.5 ns shows three distinct frequencies. One of them coincides with the electron frequency seen at longer delays. The two others (shown by the open symbols in Fig. 6.12(d)) are related to the exciton fine structure. By fitting the experimental data (the solid lines) with the non-approximate forms given in [34] one obtains an isotropic exchange splitting between the bright and dark excitons $\delta_0 = 0.10 \pm 0.01\,\mathrm{meV}$ and the hole g-factor $|g_{h\perp}| = 0.15$.

Spectral Dependence of the Electron g-Factor

The energy dispersion of the electron g-factor within the dot ensemble has been measured by varying the excitation energy across the emission band. Figure 6.11(a) shows that from the low to the high energy side the g-factor decreases from 0.57 to 0.49. This variation can be understood if one makes the assumption that the main effect of the confinement is an increase of the band gap E_g. The deviation of g_e from the free electron g-factor $g_0 = 2$, determined from $\mathbf{k} \cdot \mathbf{p}$ calculations, is given by [38, 39]

$$g_e = g_0 - \frac{4m_0 P^2}{3\hbar^2} \frac{\Delta}{E_g(E_g + \Delta)}. \tag{6.9}$$

Here m_0 is the free electron mass, P is the matrix element describing the coupling between valence and conduction band, and Δ is the spin–orbit splitting of the valence band. The decrease of the g-factor modulus with increasing emission energy could be then only explained if the g-factor has a negative sign. This argument is supported by measurements of the dynamic nuclear polarization [34], similar to those described in [40]. This allows us to determine also signs for the exciton and hole g-factors.

Anisotropy of Electron g-Factor in Quantum Dot Plane

By varying the field orientation in the quantum dot plane one can determine the in-plane anisotropy of the electron g-factor. For an arbitrary direction, characterized by the angle α relative to the x-axis, the electron g-factor is

$$\left|g_{e\perp}(\alpha)\right| = \sqrt{g_{e,x}^2 \cos^2\alpha + g_{e,y}^2 \sin^2\alpha}, \tag{6.10}$$

where $g_{e,x}$ and $g_{e,y}$ are the g-factors along the x and y axes, [110] and [1$\bar{1}$0], respectively. The circular dichroism signal has been measured as a function of α. Figure 6.11(b) shows the resulting angular dependence of the electron (circles) and the exciton (triangles) g-factors. One finds from fits by (6.10) that $|g_{e,x}| = 0.57$ and $|g_{e,y}| = 0.54$. The origin of the in-plane anisotropy is not understood. It could possibly be explained by piezoelectric effects, modifying the band structure.

6.3.2 Generation of Electron Spin Coherence

We turn now to the generation of electron spin coherence in singly charged quantum dots under resonant excitation into a trion. The problem is similar to the one in quantum wells with a diluted carrier gas (Sect. 6.2.1). Instead of the classical approach used for the quantum wells we present here the quantum mechanical formulation of this problem. The detailed consideration can be found in [9, 10, 13].

It is well documented by the experimental results in Fig. 6.12(c) that spin coherence of the resident electron in a singly charged quantum dot can be generated optically. Deeper insight into the underlying mechanism is provided by the excitation density dependence of the generation efficiency. Figure 6.13(a) shows Faraday

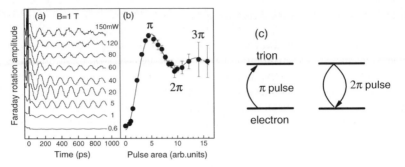

Fig. 6.13. Pump power dependence of the spin coherence generation in (In, Ga)As/GaAs quantum dots. (**a**) Faraday rotation signals for different pump powers. (**b**) Faraday rotation amplitude versus laser pulse area Θ. The line is a guide to the eye [9]. (**c**) The scheme illustrates the generation process for π and 2π-pulses

rotation signals for different pump powers. The corresponding amplitudes are plotted in Fig. 6.13(b) versus the laser pulse area Θ defined as $\Theta = 2 \int [d E(t)] dt / \hbar$ in dimensionless units. d is the dipole matrix element for the transition from the valence to the conduction band. $E(t)$ is the electric field amplitude of the laser pulse. For pulses of constant duration, Θ is proportional to the square root of excitation power.

The Faraday rotation amplitude shows a non-monotonic behavior with increasing pulse area. First, it rises to reach a maximum, then drops to about 60%. Thereafter it shows another strongly damped oscillation. This behavior is similar to the one known from Rabi oscillations of a Bloch vector, whose z-component describes the electron–hole population [41, 42]. The laser pulse drives coherently this population, leading to coherent oscillations as function of the pulse area Θ. In this case the ground state is an empty dot, and the excited state represents a dot with a photogenerated electron–hole pair.

For the case of a singly charged dot, the ground state is a dot with an electron and the excited state is a dot with a trion, i.e., a dot containing two electrons and one hole (see Fig. 6.13(c)). A laser pulse with $\Theta = \pi$ drives the system from the ground to the excited state, which corresponds to maximal generation efficiency. Further, an increase of the laser power does not increase the generated spin polarization. The reason is that the very same pulse with $\Theta > \pi$ starts to drive the system back to the ground state. For $\Theta = 2\pi$ the system returns to the ground state and no spin polarization is generated. Thus the Faraday rotation amplitude achieves a maximum for a π-pulse, and a minimum for a 2π-pulse. The damping of the oscillations most likely is due to ensemble inhomogeneities of quantum dot properties such as the dipole moment d [43].

With these observations at hand one can understand the origin of the observed spin coherence. We discuss first neutral dots. Resonant optical pulses with σ^- polarization create a superposition state of vacuum and exciton:

$$\cos\left(\frac{\Theta}{2}\right)|0\rangle - i \sin\left(\frac{\Theta}{2}\right)|\uparrow\Downarrow\rangle, \tag{6.11}$$

where $|0\rangle$ describes the deexcited semiconductor. The hole spin orientations $J_z = \pm 3/2$ are symbolized by the arrows $\Uparrow$ and $\Downarrow$, respectively. The electron and hole spins become reversed for σ^+ excitation. The exciton component precesses in magnetic field for a time, which cannot last longer than the exciton lifetime. In an ensemble, the precession can be seen until the coherence is destroyed by spin scattering of either electron or hole. The strength of the contribution to the ensemble Faraday rotation signal is given by the square of the exciton amplitude $\sin^2(\Theta/2)$.

Let us turn now to singly charged quantum dots, for which the resonant excitation can lead to the excitation of trions. We assume that the deexcited quantum dot state is given by an electron with arbitrary spin orientation:

$$\alpha|\uparrow\rangle + \beta|\downarrow\rangle, \tag{6.12}$$

with $|\alpha|^2 + |\beta|^2 = 1$. A σ^--polarized laser pulse would create an exciton with spin configuration $|\uparrow\Downarrow\rangle$. This is, however, restricted by the Pauli-principle, due to which the optically excited electron must have a spin orientation opposite to the resident one in order to form a trion singlet state $|\uparrow\downarrow\Downarrow\rangle$. Therefore, the pulse excites only the second component of the initial electron state, i.e., $\beta|\downarrow\rangle$. As a consequence, a coherent superposition state of an electron and a trion is created:

$$\alpha|\uparrow\rangle + \beta\cos\left(\frac{\Theta}{2}\right)|\downarrow\rangle - i\beta\sin\left(\frac{\Theta}{2}\right)|\downarrow\uparrow\Downarrow\rangle, \tag{6.13}$$

which consists of two spin singlet electrons and a hole in state $|\Downarrow\rangle$. We assume that decoherence does not occur during the excitation process, i.e., the pulse length is much shorter than the radiative decay and the carrier spin relaxation times. One sees that the electron–hole population oscillates with pulse area Θ. The excitation is most efficient for $\Theta = \pi$, which gives the superposition state:

$$\alpha|\uparrow\rangle - i\beta|\downarrow\uparrow\Downarrow\rangle. \tag{6.14}$$

After some time the electron–hole pair will relax, leaving the resident electron in the quantum dot. This occurs on the time scale of the trion radiative lifetime. At zero magnetic field and in the absence of hole spin relaxation within the trion, the system will return to its initial state described by (6.12). Therefore, no electron spin coherence will be generated. However, spin coherence will be generated, if hole spin relaxation takes place before the trion recombination.

In an external magnetic field in the Voigt geometry, the hole relaxation is no more the crucial factor for spin coherence generation. The reason is that the electron part of (6.14), i.e., $\alpha|\uparrow\rangle$ will precess around the magnetic field, while the singlet state of trion does not precess. Therefore, full compensation of the induced spin polarization after trion recombination is impossible and spin coherence is induced.

Modeling the electron and trion spin dynamics under pulsed resonant excitation [9] gives the following equation for the amplitude of the long-lived electron spin polarization after trion recombination:

$$S_z(t) = \mathrm{Re}\left\{\left(S_z(0) + \frac{0.5J_z(0)/\tau_0^{\mathrm{T}}}{\gamma_{\mathrm{T}} + i(\omega_e + \Omega_{\mathrm{h}})} + \frac{0.5J_z(0)/\tau_0^{\mathrm{T}}}{\gamma_{\mathrm{T}} + i(\omega_e - \Omega_{\mathrm{h}})}\right)\exp(i\omega_e t)\right\}, \tag{6.15}$$

where $S_z(0)$ and $J_z(0)$ are the electron and trion spin polarizations created by the pulse, $\omega_e = \Omega_e + \Omega_{N,x}$ is the electron precession frequency in the magnetic field resulting from the external field and the effective nuclear field. $\gamma_T = 1/\tau_0^T + 1/\tau_s^T$ is the total trion decoherence rate contributed by the trion radiative recombination time τ_0^T and the hole spin relaxation in the trion τ_s^T. If the radiative relaxation is fast $\tau_0^T \ll \tau_s^T$, $\Omega_{e,h}^{-1}$, the induced spin polarization $S_z(t)$ is nullified on average by trion relaxation, as $S_z(0) = -J_z(0)$. This corresponds to the situation at zero magnetic field. In contrast, if the spin precession is fast, $\Omega_{e,h} \gg (\tau_0^T)^{-1}$, the electron spin polarization is maintained after trion decay [13, 44]. It is the case for the studied (In, Ga)As/GaAs dots.

6.3.3 Mode Locking of Spin Coherence in an Ensemble of Quantum Dots

The ensemble dephasing does not lead to a destruction of the individual spin coherence, but masks it due to the rapid accumulation of phase differences among different spins. The T_2 time may be obtained by sophisticated spin-echo techniques [2], which typically are quite laborious. A less complicated and robust measurement scheme would be therefore highly desirable. Such a scheme may be also useful for the processing of quantum information, including initialization, manipulation, and read-out of a coherent spin state.

To address this point, we look again at Faraday rotation traces, especially for negative delays, shown in Fig. 6.14(a). Long-lived electron spin quantum beats are seen at positive delays, as discussed before. Surprisingly, for negative delays strong spin beats with frequencies identical to those of the electron precession are also observed. The amplitude of these beats increases when approaching zero delay $t = 0$. Note, that spin beats at negative delay have been reported for experimental situations in which the dephasing time exceeds the time interval between the pump pulses: $T_2^* \geq T_R$, see Fig. 6.4(a) and [3]. This is clearly not the case here because the Faraday rotation signal has fully vanished after 1.5 ns at $B = 6$ T, so that the dephasing is much faster than the pulse repetition period. The frequency and rise time of the signal at negative delays are the same as at positive delays, indicating that the negative delay signal is given by the electron spin precession.

Figure 6.14(b) shows the signal when scanning the delay over a larger time interval, in which three pump pulses, separated by 13.2 ns from each other, are located. At each pump pulse arrival electron spin coherence is created, which is dephased after a few ns. Before the next pump arrival coherent signal from the electrons reappears. This negative delay precession can occur only if the coherence of the electron spin in each single dot prevails for much longer times than T_R, i.e., if $T_2 \gg T_R$.

Spin Coherence Time of an Individual Electron

Independent of the origin of the coherent signal at negative delays, its observation opens a pathway towards measuring the spin coherence time T_2: When increasing the pump pulse separation until it becomes comparable with T_2, the amplitude of the

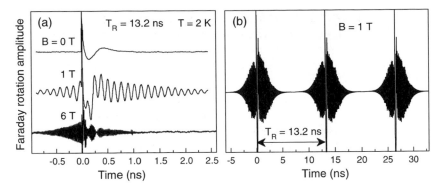

Fig. 6.14. (a) Pump–probe Faraday rotation signals at different magnetic fields in singly charged (In, Ga)As/GaAs quantum dots. The pump power density is $60\,\mathrm{W/cm^2}$, the probe density is $20\,\mathrm{W/cm^2}$ [10]. (b) Faraday rotation signal recorded for a longer delay range in which three pump pulses were located

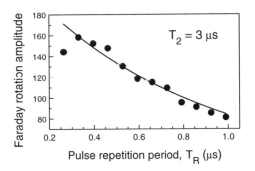

Fig. 6.15. Faraday rotation amplitude at negative delay as function of the time interval between subsequent pump pulses measured at $B = 6\,\mathrm{T}$ and $T = 6\,\mathrm{K}$. The line shows calculations with a single fit parameter $T_2 = 3\,\mu\mathrm{s}$ [10]

negative delay signal should decrease. Corresponding data are given in Fig. 6.15. The Faraday rotation amplitude detected at a fixed negative delay shortly before the next pump arrival is shown there as function of T_R. T_R is increased from 13.2 up to 990 ns. A decrease of the amplitude is seen, demonstrating that T_R becomes comparable to T_2. The result of model calculations shown by the line allow us to determine the coherence time of a single dot, $T_2 = 3.0 \pm 0.3\,\mu\mathrm{s}$, which is four orders of magnitude longer than the ensemble dephasing time $T_2^{\star} = 0.4\,\mathrm{ns}$ at $B = 6\,\mathrm{T}$.

Mechanism of Spin Synchronization

In order to understand the striking fact that the single quantum dot coherence time can be seen in an ensemble measurement, let us consider the excitation of a single quantum dot by a periodic train of circular polarized π-pulses. The first impact of the pulse train is a synchronization of the electron spin precession. We define the

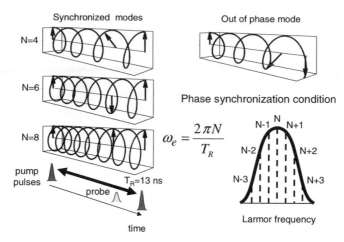

Fig. 6.16. Scheme of the phase synchronization condition for the electron spin coherence with a periodic train of laser pulses. *On the left* modes satisfying the PSC (6.16) are shown. *On the right* a non-PSC dot is given. *The arrows* show the spin orientation of a resident electron in a quantum dot. *Bottom right: The thick solid line* shows the distribution of electron precession frequencies in the dot ensemble caused by the dispersion of the electron g-factor. The PSC modes selected from this distribution are shown by *the dashed lines*

degree of spin synchronization by $P(\omega_e) = 2|S_z(\omega_e)|$. Here the z-component of the electron spin vector, $S_z(\omega_e)$, is taken at the moment of pulse arrival. If the pulse period, T_R, is equal to an integer number N times the electron spin precession period, $2\pi/\omega_e$, the action of such π-pulses leads to almost complete electron spin alignment along the light propagation direction z [13]. In general, the degree of synchronization for π-pulses is given by $P_\pi = \exp(-T_R/T_2)/[2 - \exp(-T_R/T_2)]$. In our case it reaches almost its largest value $P_\pi = 1$, corresponding to 100% synchronization, because for excitation with high repetition rate such as 75.6 MHz $T_R \ll T_2$ so that $\exp(-T_R/T_2) \approx 1$.

We remind the reader that in an ensemble of quantum dots, the electrons do not precess with the same frequency, but have a frequency distributions with a broadening γ. The latter is determined by the electron g-factor dispersion and the spectral width of the pump laser (Fig. 6.11(a)).

The ensemble contains quantum dots whose precession frequencies fulfil a synchronization relation with the laser, which we term phase synchronization condition (PSC) in the following:

$$\omega_e = 2\pi N/T_R \equiv N\omega_R. \tag{6.16}$$

Here ω_R is the repetition frequency of the pump pulses. The PSC modes selected from the continuous distribution of electron precession frequencies are shown by the dashed lines in Fig. 6.16. Since the electron spin precession frequency is typically much bigger than the laser repetition rate, we have $N \gg 1$ for not too small magnetic fields. Since in addition the spread of precession frequencies is also much larger than the laser repetition rate, multiple subsets within the optically excited quantum dot ensemble satisfy (6.16) for different N. This is illustrated schematically by Fig. 6.16. In the left side three precession modes satisfying the PSC (6.16) with $N = 4, 6, 8$ are given. The spin precession with a frequency different from the PSC is shown in the right side of the figure. Two important conclusions can be drawn from this scheme.

First, for the PSC dots spin synchronization will be accumulated until it reaches its maximum value, see discussion above for the single dot synchronization. The reason is that at the moment of the pump pulse arrival the spin coherence generated by the previous pulse has the same orientation as the one which the subsequent pulse induces. In other words the contributions of all pulses in the train are constructive. In contrast, the contributions have arbitrary orientations in the non-PSC dots (right panel of Fig. 6.16) and for these dots the degree of spin synchronization will be always far from saturation. Practically, this means that a PSC dot gives a stronger signal to the Faraday rotation signal than a non-PSC dot.

Second, in the PSC dots (left panels) the electron spins indicated by the arrows have the same phase at the moments of pump pulse arrival, but they have different orientation between the pulses. As a result, the signal from the dot subsets satisfying the PSC dephases shortly after each pump pulse. However it is revived before the next pulse leading to the characteristic signal shown in Fig. 6.14(b). To be more specific, the spins in each PSC subset precess between the pump pulses with frequency $N\omega_R$, starting with an initial phase which is the same for all subsets. Their contribution to the spin polarization of the ensemble at a time t after the pulse is given by $-0.5\cos(N\omega_R t)$. The sum of oscillating terms from all synchronized subsets leads to a constructive interference of their contributions to the Faraday rotation signal around the times of pump pulse arrival. The rest of the quantum dots does not contribute to the average electron spin polarization $\overline{S}_z(t)$ at times $t \gg T_2^*$, due to dephasing. The synchronized spins therefore move on a background of dephased electrons, which however still precess individually during the spin coherence time. The number of synchronized PSC subsets can be estimated by $\Delta N \sim \gamma/\omega_R$. It increases linearly with magnetic field and T_R.

The π-pulse excitation is not critical for the electron spin phase synchronization by the pulse train. Any resonant pulse train of arbitrary intensity creates a coherent superposition of trion and electron in a quantum dot, leading to a long-lived coherence of the resident electron spins, because the coherence is not affected by the radiative decay of the trion component. Each pulse of σ^+ polarized light changes the electron spin projection along the light propagation direction by $\Delta S_z = -(1 - 2|S_z(t \to t_n)|)W/2$, where $t_n = nT_R$ is the time of the nth pulse arrival, and $W = \sin^2(\Theta/2)$ [9, 12]. Consequently, a train of such pulses orients the electron spin opposite to the light propagation direction, and it also increases the degree of electron spin synchronization P. Application of $\Theta = \pi$-pulses (corresponding to $W = 1$) leads to a 99% degree of electron spin synchronization already after a dozen pulses. However, if the electron spin coherence time is long enough ($T_2 \gg T_R$), an extended train of pulses leads as well to a high degree of spin synchronization, also for $\Theta \ll 1$ ($W \approx \Theta^2/4$).

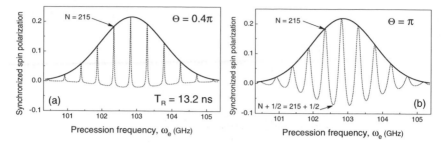

Fig. 6.17. Spectra of phase synchronized electron spin precession modes created by a train of circularly polarized pulses calculated for the pulse area $\Theta = 0.4\pi$ and π at the moment of pulse arrival. $T_R = 13.2$ ns. **(a)** At low pumping intensity the pulse train synchronizes the electron spin precession in a very narrow frequency range around the phase synchronization condition: $\omega_e = 2\pi N / T_R$. **(b)** π-pulses widen the range of synchronized precession frequencies. In addition, electron spins with opposite polarization at frequencies between the phase synchronization condition become significantly synchronized. Calculations have been done for $B = 2$ T, $|g_e| = 0.57$, $\Delta g_e = 0.005$ and $T_2 = 3$ µs [10]

The effect of the pump intensity (pump area) on the distribution of the spin polarization, which is synchronized with the pulse train for $\Theta = 0.4\pi$ and π, is shown in Fig. 6.17. The density of the electron spin precession modes is shown by the solid line, which gives the envelop of the spin polarization distribution. The quasidiscrete structure of the distribution created by the pulse train (the dashed lines) is the most important feature, which allows us to measure the long spin coherence time of a single quantum dot within an ensemble: A continuous density of spin precession modes would cause fast dephasing on a time inversely proportional to the total width of the frequency distribution: $T_2^* = \hbar / \gamma$. However the gaps in the density of precession modes facilitate the constructive interference at negative delay times. These gaps are created by mode locking of the electron spin precession with the periodic pulse sequence.

Control of Ensemble Spin Precession

In this part we turn to testing the degree of control over the spin coherence of an ensemble of singly charged dots that can be achieved by periodic laser excitation. For that purpose, a train of two pump pulses is used. Adjustments of the delay between these pulses and their polarization are used to control the shape and phase of the spin coherent signal. The robustness of the mode-locked spin coherence with respect to variations of lattice temperature and magnetic field strength is demonstrated.

Two Pump Pulse Excitation Protocol

Each pump pulse in a train is now split into two pulses with a fixed delay $T_D < T_R$ between them. The results for $T_D = 1.84$ ns are shown in Fig. 6.18(a). Both pumps are circular co-polarized and have the same intensities. When the quantum dots are

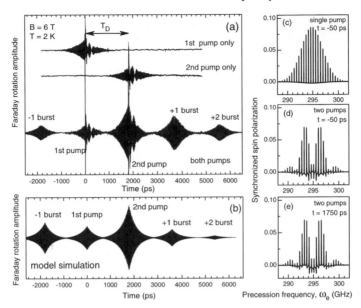

Fig. 6.18. Control of the electron spin synchronization in (In, Ga)As/GaAs quantum dots by two trains of pump pulses with $T_R = 13.2$ ns, shifted in time by $T_D = 1.84$ ns. (**a**) Experimental Faraday rotation signal measured for separate action of the first or the second pump (*the two upper curves*) and for joint action of both pumps (*the bottom curve*). The pumps were co-polarized (σ^+). (**b**) Modeling of the Faraday rotation signal in the two pump pulse experiment with the parameters $\Theta = \pi$ and $\gamma = 3.2$ GHz. (**c, d, e**) illustrate the modification of the PSC mode spectrum under two pump pulse action. Model parameters: $\Theta = 0.4\pi$, $\gamma = 3.2$ GHz, $|g_e| = 0.57$ and $\Delta g_e = 0.005$. Panel (**c**) is for the single pump protocol, measured just before the pump pulse, see upper curve in panel (**a**). Panels (**d**) and (**e**) are for two pump pulses shown by the lower curve in (**a**) measured before pump 1 and pump 2, respectively [10]

exposed to only one of the pump pulses (the two upper traces), the Faraday rotation signals are identical except for a shift by T_D. The signal changes drastically under excitation by the two pulse train (lower trace): Around the arrival of pump 1 the same Faraday rotation response is observed as before in the one-pump experiment. Also around pump 2 qualitatively the same signal is observed with a considerably larger amplitude. This means that the coherent response of the synchronized quantum dot ensemble can be amplified by the second laser pulse.

Even more remarkable are the echo-like responses showing up before the first and after the second pump pulse. They have a symmetric shape with the same decay and rise times T_2^*. The temporal separation between them is a multiple of T_D. Note that these Faraday rotation bursts show no additional modulation as seen at positive delays when a pump is applied. The reason is that excitons generated in charge neutral dots have already recombined and do not contribute to these signals.

Apparently, the electron spins in the quantum dot sub-ensemble, which is synchronized with the laser repetition rate, have been clocked by introducing a second

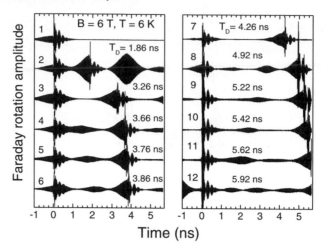

Fig. 6.19. Faraday rotation traces measured as function of delay between probe and first pump pulse at time zero. A second pump pulse was delayed relative to the first one by T_D, as indicated at each trace. The top left trace is measured without the second pump [45]

frequency, which is determined by the laser pulse separation T_D. The clocking results in multiple bursts in the Faraday rotation response. The conditions for this clocking are analyzed in further detail in the next paragraph. Here we demonstrate how the two pump pulses modify the spectrum of the spin precession modes. The changes become very clear when one compares the single pump spectrum in Fig. 6.18(c) with the two pump spectra in panels (d) and (e). The model calculations of the Faraday rotation signal shown in panel (b) reproduce the experimental burst signals.

Signal Shaping by Changing Delay between Pump Pulses

Figure 6.19 shows Faraday rotation traces excited by a two-pulse train with a repetition period $T_R = 13.2$ ns between the pump doublet. The two pulses have the same intensity and polarization. The delay between these pulses T_D was varied between $\sim T_R/7$ and $\sim T_R/2$. The signal varies strongly, depending on whether the delay time T_D is commensurate with the repetition period T_R, $T_D = T_R/i$ with $i = 2, 3, 4, \ldots$, or incommensurate, $T_D \neq T_R/i$. For commensurability the signal shows strong periodic bursts of quantum oscillations only at times equal to multiples of T_D, as seen for $T_D = 1.86$ ns $\approx T_R/7$. Commensurability is also achieved for $T_D = T_R/4 \approx 3.26$ ns and $T_D = T_R/3 \approx 4.26$ ns.

When T_D and T_R are incommensurate, the Faraday rotation signal shows bursts of spin beats between the two pulses of each pump doublet, in addition to the bursts outside of the doublet. One can see a single burst midway between the pumps for $T_D = 3.76$ and 5.22 ns. Two bursts, each equidistant from the closest pump and also equidistant from one another, appear at $T_D = 4.92$ and 5.62 ns. Three equidistant bursts occur at $T_D = 5.92$ ns.

Although the time dependencies of the Faraday rotation signals look very different for commensurate and incommensurate T_D and T_R, in both cases they result from constructive interference of synchronized spin precession modes [45]. For a train of pump pulse doublets the phase synchronization conditions involve the intervals T_D and $T_R - T_D$ in the laser excitation protocol

$$\omega_e = 2\pi N K / T_D = 2\pi N L / (T_R - T_D), \tag{6.17}$$

where K and L are integers. This condition imposes limitations on the T_D values, for which synchronization is obtained:

$$T_D = \left[K / (K + L) \right] T_R, \tag{6.18}$$

which for $T_D < T_R/2$ leads to $K < L$. This phase synchronization condition explains the position of all bursts in the signals from Fig. 6.19. For commensurability, one has $K = 1$ so that $T_D = T_R/(1 + L)$. In this case constructive interferences should occur with a period T_D as seen for $T_D = 1.86$ ns ($L = 6$).

For incommensurability the number of bursts between the pulses and the delays at which they appear can be tailored. There should be just one burst, when $K = 2$, because then the constructive interference must have a period $T_D/2$. In experiment, a single burst is indeed seen for $T_D = 3.76$ ns ($L = 5$) and 5.22 ns ($L = 3$), see Fig. 6.19. Two bursts are seen for $T_D = 4.92$ and 5.62 ns, corresponding to $K = 3$ and $L = 5$ and 4, respectively. Finally, the Faraday rotation signal with $T_D = 5.92$ ns shows three bursts between the pumps, which is described by $K = 4$ and $L = 5$. Thus a good agreement between experiment and theory is established, highlighting the high flexibility of the pump protocol.

Polarization Control of Signal Phase

To obtain further insight into the tailoring of electron spin coherence by a two-pulse train, we change from co- to counter-circular polarized pumps. T_D is fixed at $T_R/6 \approx 2.2$ ns. The appearances of the corresponding Faraday rotation signals are similar, as shown in Fig. 6.20. Besides the two bursts directly connected to the pump pulses one sees one further +1 burst. The insets show closeups of different bursts. The phase of the spin beats differs by π between the co- and counter-polarized configuration for the pump 1 and +1 bursts. On the other hand, there is no phase difference for the pump 2, −1 and +2 bursts (the last two signals are not shown here). The sign, κ, of the Faraday rotation amplitude in the counter-circular configuration undergoes T_D-periodic changes in time as compared to the constancy in the co-circular case, see Fig. 6.20(b). This demonstrates optical switching of the electron spin precession phase by π in an ensemble of quantum dots.

The observed effect of phase sign reversal is well described by the model [45]. A detailed consideration shows that the modulus of the spin synchronization exhibits constructive interference with a period $T_R/6$ and changes its sign with a period $2T_R/6$. The relative sign of the Faraday rotation amplitude for the counter- and co-circular cases, $\kappa = \mathrm{sign}\{\cos[\pi t/(T_R/6)]\}$, is in agreement with the experimental data in Fig. 6.20(b).

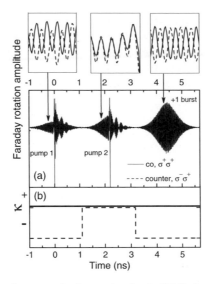

Fig. 6.20. (a) Faraday rotation traces in the co-circular (*solid line*) or counter-circular (*dashed line*) polarized two pump pulse experiments, measured for $T_D = 2.2$ ns. $B = 6$ T and $T = 6$ K. The signal amplitudes look very similar for the two configurations and can hardly be distinguished on the displayed time scale. Therefore the three additional panels give close-ups of the two traces showing the relative sign, κ, of the Faraday rotation amplitudes. κ is plotted in panel **(b)** vs. time [45]

Stability Against Temperature Increase and Magnetic Field Variation

The Faraday rotation bursts due to constructive interference of the electron spin contributions show remarkable stability against variations of the magnetic field in the range from 1 to 10 T. While the appearance of bursts changes with the field strength (the bursts are squeezed due to the decrease of T_2^* with increasing field), the delay times at which the bursts appear remain unchanged. Also the bursts amplitude does not vary strongly in this field range.

Further, both at positive and negative delays, the Faraday rotation signals remain almost unchanged for temperature changes from 2 up to 25 K [45]. The stability against variations of magnetic field and temperature is a consequence of the mode-locking generation mechanism, which is not fixed to the properties of specific dots, e.g., their spectral energy or spin precession frequency. The periodic pump laser train always selects the proper subsets of dots which satisfy the phase synchronization condition, even for strongly varying experimental conditions.

Requirements for Quantum Dot Ensemble

The mode locking mechanism in an ensemble of quantum dots with inhomogeneously broadened precession frequencies rises the question what properties a quantum dot ensemble should have for application in quantum coherent devices. In general, quantum dot ensembles whose spin states are homogeneously broadened would be optimal for quantum information processing. However, fabrication of such ensembles cannot be foreseen on the basis of current state-of-the-art technology, which always gives a sizable inhomogeneity. Under these circumstances, a distribution of the electron g-factor is favorable to enable realization of mode locking, because the phase synchronization condition is fulfilled by many quantum dot subsets, leading to a strong spectroscopic response. Further, it gives some flexibility when changing, for example, the laser protocol (e.g., wavelength, pulse duration and repetition rate) by which the quantum dots are addressed, and therefore changing the phase synchronization condition. In response to such a change, the ensemble involves other quantum dot subsets in the synchronization such that again the single dot coherence can be recovered. However, a very broad distribution of electron g-factors would lead to a very fast dephasing in the ensemble, making it difficult to observe the Faraday rotation both after and before pulse arrival. In this case the phase synchronization can be exploited only during a quite short range of time.

6.3.4 Nuclei Induced Frequency Focusing of Spin Coherence

In this section we describe an effect which originates from the hyperfine interaction of an electron spin with the nuclear spins. An in-depth consideration of hyperfine interaction effects can be found in Chap. 11. The spatial confinement in quantum dots protects the electron spins against most relaxation mechanisms (Chap. 1). However, the hyperfine interaction with the lattice nuclei is enhanced by confinement, leading to spin decoherence and dephasing [32, 46]. This problem may be overcome by polarizing the nuclear spins [47], but the high degree of polarization required, close to 100%, has not been achieved so far.

However, as we will show, the hyperfine interaction, rather than being detrimental, can be utilized as a precision tool. We demonstrate that it can modify the continuous mode spectrum of the electron spin precession in a dot ensemble into a few discrete modes. The information on this digital spectrum can be stored in the nuclear spin system for tens of minutes because of the long nuclear memory times [3, 48–50].

In a dot ensemble, fast electron spin dephasing arises not only from variations of the electron g-factor, but also from nuclear field fluctuations, leading to different spin precession frequencies. The dephasing due to the g-factor variations can be partly overcome by the described mode-locking [10], which synchronizes the precession of specific electron spin modes in the ensemble with the clocking rate of the periodic pulsed excitation laser. Still, it would leave a significant fraction of dephased electron spins, whose precession frequencies would not satisfy the phase synchronization condition. However, the nuclear spin polarization can adjust the electron spin precession frequency in each quantum dot such that the whole ensemble becomes locked on only a few frequencies.

The experiments were done on the same sample containing an ensemble of self-assembled (In, Ga)As/GaAs dots as used in the previous section. Further, the two pump pulse Faraday rotation technique was applied. Signals measured for single pump and two pump protocols are shown in Fig. 6.21(a). Surprisingly, the signal pattern created by the two pulse protocol is memorized over several minutes. One

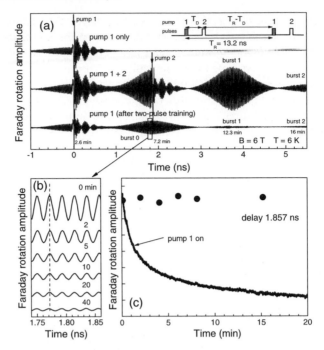

Fig. 6.21. (**a**) Faraday rotation traces measured on an ensemble of singly-charged (In, Ga)As/GaAs quantum dots. Details of the optical excitation protocol are given in the sketch. The top trace was measured using a train of single pump pulses. *The middle trace* was excited by a two pump pulse protocol with the second pump delayed by $T_D = 1.86$ ns relative to the first one. The measurement over the whole delay time range took about 20 min. *The lowest trace* was taken for a single pump pulse excitation protocol with pump 2 closed. Recording started *right* after measurement of *the middle trace*. Some times at which the different bursts were measured are indicated. The pump and probe power density were 50 and $10 \, \text{W/cm}^2$. (**b**) Faraday rotation signals measured over a small delay range at the maximum of 'burst 0' for different times after closing the second pump, while pump 1 and the probe were always on. (**c**) Relaxation kinetics of the Faraday rotation amplitude at a delay of 1.857 ns after switching off pump 2. Before this recording, the system was treated for 20 minutes by the two pump excitation. The curve was measured with pump 2 blocked at $t = 0$. The circles show the signal for different times in complete darkness (both pumps and probe were blocked). $B = 6 \, \text{T}$ and $T = 6 \, \text{K}$ [51]

would expect that blocking of the second pulse in a pump doublet would destroy the periodic burst pattern on a µs time scale according to the electron spin coherence time, T_2, in these dots [10]. Only the signal around the first pump should remain over the scanned range of pump–probe delays. The middle trace was recorded after the sample was illuminated for ∼20 minutes by the pump-doublet train. Immediately after this measurement, the second pump was blocked and a measurement using only the single pump train was started (bottom trace). Contrary to the expectations, the signal shows qualitatively all characteristic for a pump doublet protocol. A strong signal ('burst 0') appears around the delays where the second pump was located. Further signals, denoted 'burst 1' and 'burst 2', also appear. The system, therefore, remembers for minutes its previous exposure to a two pump protocol!

Additional Faraday rotation traces were recorded in a short delay range around 'burst 0' for different times after closing the second pump (Fig. 6.21(b)). The decay kinetics was measured at a fixed delay of 1.857 ns (corresponding to the maximum signal) vs. the time after switching off pump 2 (curve in Fig. 6.21(c)). A strong signal is seen even after 40 min. The observed dynamics is well described by a bi-exponential dependence on elapsed time t, $a_1 \exp(-t/\tau_1) + a_2 \exp(-t/\tau_2)$ with a memory time τ_1 of a minute and $\tau_2 = 10.4$ min. The decay, however, critically depends on the light illumination conditions. When the system is held in darkness (both pumps and probe are blocked), no relaxation occurs at all on an hour time scale. This is shown by circles in Fig. 6.21(c), which give the Faraday rotation amplitude when switching on pump 1 as well as the probe after a dark period t. Note, that the rise time of the effect also has a slow component on a minute time scale.

The observed long memory of the excitation protocol must be imprinted in the dot nuclei, for which long spin relaxation times up to hours or even days have been reported in high magnetic fields [48, 49]. The nuclei in a particular dot must have been aligned along the magnetic field through the hyperfine interaction with the electron during exposure to the pump train. This alignment, in turn, changes the electron spin precession frequency, $\omega_e = \Omega_e + \Omega_{N,x}$, where the nuclear contribution, $\Omega_{N,x}$, is proportional to the nuclear polarization. The slow rise and decay dynamics of the Faraday rotation signal indicate that the periodic optical pulse train stimulates the nuclei to increase the number of dots, for which the electron spin precession frequencies satisfy the phase synchronization condition for a particular excitation protocol. But what is driving the projection of the nuclear spin polarization on the magnetic field to a value that allows an electron spin to satisfy the PSC?

The nuclear polarization is changed by electron–nuclear spin flip–flop processes resulting from the Fermi-contact-type hyperfine interaction [52]. Such processes, however, are suppressed in a strong magnetic field due to the energy mismatch between the electron and nuclear Zeeman splittings by about three orders of magnitude. Flip–flop transitions, which are assisted by phonons compensating this mismatch, have a low probability due to the phonon-bottleneck [53, 54]. This explains the robustness of the nuclear spin polarization in darkness (Fig. 6.21(c)).

Consequently, the resonant optical excitation of the singlet trion becomes the most efficient mechanism in the nuclear spin polarization dynamics. The excitation process rapidly turns "off" the hyperfine field of a resident electron acting on the nuclei, and the field is subsequently turned "on" again by the trion radiative decay. Thereby it allows a flip–flop process during switching without energy conservation.

The nuclear spin–flip rate for this mechanism is proportional to the rate of optical excitation of the electron, $\Gamma_\uparrow(\omega_e)$. According to the selection rules, the probability of exciting the electron to a trion by σ^+ polarized light is proportional to $1/2 + S_z(\omega_e)$, where $S_z(\omega_e)$ is the component of the electron spin polarization along the light propagation direction taken at the moment of pump pulse arrival. Therefore, the excitation rate $\Gamma_\uparrow(\omega_e) \sim [1/2 + S_z(\omega_e)]/T_R$. For electrons satisfying the PSC,

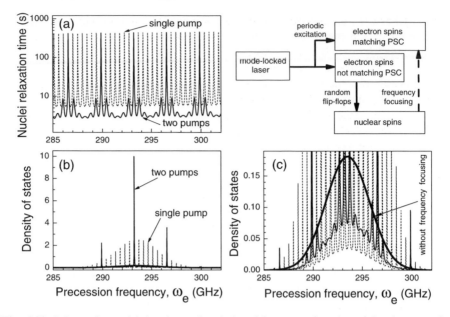

Fig. 6.22. Scheme for explaining the nuclear induced frequency focusing of the electron spin precession modes. The periodic resonant excitation by a mode-locked circular polarized laser synchronizes the precessions of electron spins whose frequency satisfy the phase synchronization conditions (PSC). At the same time this excitation leads to a fast nuclear relaxation time in quantum dots, which do not satisfy the PSCs, via optically stimulated electron–nuclei spin flip–flop processes. The random fluctuation of the nuclear spin modifies the electron spin precession frequency and becomes frozen when this frequency reaches a PSC. (**a**) Average spin relaxation time of the As nuclei vs. the electron spin precession frequency calculated for the single pump pulse (*dashed*) and the two pump pulse (*solid*) excitation protocols. (**b**) Density of electron spin precession modes in an ensemble of singly charged dots modified by the nuclei, calculated for the single pump (*dashed*) and the two pump (*solid*) excitation protocols. The *thick black line* shows the unmodified density of electron spin precession modes due to the dispersion of the electron g-factor in the dot ensemble and due to nuclear polarization fluctuations. This is better seen in panel (**c**), which is a close-up of panel (**b**). Calculations were done for $B = 6\,\mathrm{T}$, $|g_e| = 0.555$, $\Delta g_e = 0.0037$, $\gamma = 1\,\mathrm{GHz}$, $T_R = 13.2\,\mathrm{ns}$, $T_D = T_R/7\,\mathrm{ns}$, and $T_2 = 3\,\mathrm{\mu s}$ [51]

$S_z(\omega_e) \approx -1/2$, the excitation probability is very low due to Pauli blocking [10]. Due to the very long spin coherence time, T_2, the excitation rate for these electrons is reduced by two orders of magnitude to $1/T_2$ as compared to $1/T_R + 1/T_2$ for the rest of electrons (in the present experiments $T_2/T_R \approx 200$) [51].

Due to the factor $\Gamma_1(\omega_e)$, the nuclear relaxation rate has a strong and periodic dependence on ω_e, with the period determined by the PSC of a particular excitation protocol: $2\pi/T_R$ for the single pulse train and $2\pi/T_D$ for the double pulse train (Fig. 6.22(a)). The huge difference in the nuclear flip rate explains why $\Omega_{N,x}$ in each dot tends to reach a value allowing the electron spin to fulfil the PSC. In dots where the PSC is not fulfilled, the nuclear contribution to ω_e changes randomly due to the light stimulated nuclear flip–flop processes on a seconds time scale. The typical range $\Delta\Omega_{N,x}$ of this contribution to ω_e is limited by statistical fluctuations of the nuclear spin polarization. For the studied (In, Ga)As dots, $\Delta\Omega_{N,x}$ lies on a GHz scale [51] which is comparable with the separation between the phase synchronized modes $2\pi/T_R \sim 0.48\,\text{GHz}$. As a result, the nuclear contribution occasionally drives an electron to a PSC mode, where its precession frequency is virtually frozen on a minutes time scale. This leads to the frequency focusing in each dot and to accumulation of the dots, for which electron spins match the PSC.

The frequency focusing modifies the spin precession mode density of the dot ensemble (Fig. 6.22(b) and its close-up in panel (c)). Without focusing, the density of the electron spin precession modes is Gaussian with a width: $\Delta\omega_e = [(\Delta\Omega_{N,x})^2 + (\mu_B\Delta g_e B/\hbar)^2]^{1/2}$, where Δg_e is the g-factor dispersion. Frequency focusing modifies the original continuum density to a comb-like distribution. Eventually the whole ensemble participates in a coherent precession locked on only a few precession frequencies. This suggests that a laser protocol (defined by pulse sequence, width, and rate) can be designed such that it focuses the electron–spin precession frequencies in the dot ensemble to a single mode. It this case the spin coherence of the ensemble will dephase with the single electron coherence time T_2.

The focusing of electrons into PSC modes is directly manifested by the Faraday rotation signals in Fig. 6.21(a), as it causes comparable amplitudes before and after the pump pulses. The calculations demonstrate that, without frequency focusing, the amplitude at negative delays, A_{neg}, does not exceed 30% of the positive delay signal amplitude, A_{pos} (Fig. 6.23(a)). The strong optical pump pulses in the experiment address all quantum dots, and their total contribution should make the Faraday rotation signal much stronger after the pulse than before, when only mode-locked electrons are relevant. However, the nuclear adjustment increases the negative delay signal to more than 90% (Fig. 6.23(b)) of the positive delay signal. The large experimental value of A_{neg}/A_{pos} (panel (c)) confirms that in our experiment almost all electrons in the optically excited dot ensemble become involved in the coherent spin precession. Calculations of the pump intensity dependence of the ratio A_{neg}/A_{pos} show that the nuclear focusing increases the ratio of electrons involved in the coherent spin precession relative to their total number, n_{psc}/n, almost to unity, even at low excitation density (Figs. 6.23(e, d)).

Therefore, the nuclei in singly charged quantum dots exposed to a periodic pulsed laser excitation drive almost all the electrons in the ensemble into a coherent spin precession. The exciting laser acts as a metronome and establishes a robust macroscopic quantum bit in dephasing free subspaces. This may open new promising perspectives on the use of an ensemble of charged quantum dots during the single electron coherence time T_2.

The results presented in Sect. 6.3 show that the shortcomings which are typically attributed to quantum dot spin ensembles may be overcome by elaborated laser excitation protocols. The related advantages are due to the robustness of the phase synchronization of the quantum dot ensemble: (i) a strong detection signal with relatively small noise; (ii) changes of external parameters like repetition rate and mag-

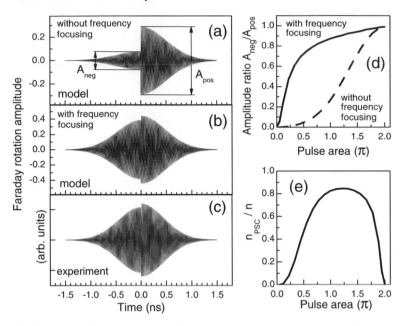

Fig. 6.23. Panels (**a**), (**b**) and (**c**) show Faraday rotation traces of an ensemble of charged quantum dots subject to a single pump pulse excitation protocol with pulse area $\Theta = \pi$ at a magnetic field $B = 6\,\text{T}$. (**a**) Faraday rotation traces calculated with the density of electron spin precession modes unchanged by the nuclei. (**b**) Faraday rotation traces calculated with the density of electron spin precession modes modified by the nuclei. (**c**) Experimental trace of the Faraday rotation signal obtained after extracting the contribution of neutral excitons. (**d**) Calculated ratio of the signal amplitudes, $A_{\text{neg}}/A_{\text{pos}}$, with (*solid*) and without (*dashed*) including the nuclear rearrangement, as function of the pump pulse area. (**e**) Dependence of the relative number of electrons, n_{psc}/n, in a quantum dot ensemble involved in the mode-locked precession as function of the pump pulse area. For the calculation parameters see Fig. 6.22 [51]

netic field strength can be accommodated for in the phase synchronization condition due to the broad distribution of electron spin precession frequencies in the ensemble and the large number of involved quantum dots.

6.4 Conclusions

We have shown that the spin coherence of carries in semiconductor nanostructures (quantum wells and quantum dots) can be addressed by time-resolved optical spectroscopy. Using short laser pulses spin coherence can be generated, detected, controlled, and manipulated. Still, there are plenty of open questions and challenging problems, especially concerning spin coherence manipulation. Here all-optical techniques can be combined with other established methods, e.g., the electron spin resonance or nuclei magnetic resonance.

Acknowledgements. This chapter is a result of our collaboration with A. Greilich, E.A. Zhukov, I.A. Yugova, R. Oulton, M. Syperek, A.L. Efros, A. Shabaev, I.A. Merkulov, M.M. Glazov, E.L. Ivchenko, G. Karczewski, T. Wojtowicz, J. Kossut, D. Reuter, and A.D. Wieck.

References

[1] A. Abragam, *The Principles of Nuclear Magnetism* (Oxford Science Publications, London, 1961)

[2] C.P. Slichter, *Principles of Magnetic Resonance* (Springer, Berlin, 1996)

[3] D.D. Awschalom, N. Samarth, in *Semiconductor Spintronics and Quantum Computation*, ed. by D.D. Awschalom, D. Loss, N. Samarth (Springer, Berlin, 2002), pp. 147–193

[4] I. Zutic, J. Fabian, S. Das Sarma, Rev. Mod. Phys. **78**, 323 (2004)

[5] R.I. Dzhioev, B.P. Zakharchenya, V.L. Korenev, D. Gammon, D.S. Katzer, JETP Lett. **74**, 182 (2001)

[6] R.I. Dzhioev, V.L. Korenev, B.P. Zakharchenya, D. Gammon, A.S. Bracker, J.G. Tischler, D.S. Katzer, Phys. Rev. B **66**, 153409 (2002)

[7] E.A. Zhukov, D.R. Yakovlev, M. Bayer, G. Karczewski, T. Wojtowicz, J. Kossut, Phys. Stat. Sol. (b) **243**, 878 (2006)

[8] H. Hoffmann, G.V. Astakhov, T. Kiessling, W. Ossau, G. Karczewski, T. Wojtowicz, J. Kossut, L.W. Molenkamp, Phys. Rev. B **74**, 073407 (2006)

[9] A. Greilich, R. Oulton, E.A. Zhukov, I.A. Yugova, D.R. Yakovlev, M. Bayer, A. Shabaev, Al.L. Efros, I.A. Merkulov, V. Stavarache, D. Reuter, A. Wieck, Phys. Rev. Lett. **96**, 227401 (2006)

[10] A. Greilich, D.R. Yakovlev, A. Shabaev, Al.L. Efros, I.A. Yugova, R. Oulton, V. Stavarache, D. Reuter, A. Wieck, M. Bayer, Science **313**, 341 (2006)

[11] G.V. Astakhov, V.P. Kochereshko, D.R. Yakovlev, W. Ossau, J. Nürnberger, W. Faschinger, G. Landwehr, T. Wojtowicz, G. Karczewski, J. Kossut, Phys. Rev. B **65**, 115310 (2002)

[12] T.A. Kennedy, A. Shabaev, M. Scheibner, Al.L. Efros, A.S. Bracker, D. Gammon, Phys. Rev. B **73**, 045307 (2006)

[13] A. Shabaev, Al.L. Efros, D. Gammon, I.A. Merkulov, Phys. Rev. B **68**, 201305(R) (2003)

[14] J.J. Baumberg, D.D. Awschalom, N. Samarth, H. Luo, J.K. Furdyna, Phys. Rev. Lett. **72**, 717 (1994)

[15] N.I. Zheludev, M.A. Brummell, A. Malinowski, S.V. Popov, R.T. Harley, D.E. Ashenford, B. Lunn, Solid State Commun. **89**, 823 (1994)

[16] E.A. Zhukov, D.R. Yakovlev, M. Bayer, M.M. Glazov, E.L. Ivchenko, G. Karczewski, T. Wojtowicz, J. Kossut, Phys. Rev. B **76**, 205310 (2007)

[17] Z. Chen, R. Bratschitsch, S.G. Carter, S.T. Cundiff, D.R. Yakovlev, G. Karczewski, T. Wojtowicz, J. Kossut, Phys. Rev. B **75**, 115320 (2007)

[18] G.V. Astakhov, D.R. Yakovlev, V.P. Kochereshko, W. Ossau, W. Faschinger, J. Puls, F. Henneberger, S.A. Crooker, Q. McCulloch, D. Wolverson, N.A. Gippius, A. Waag, Phys. Rev. B **65**, 165335 (2002)

[19] G. Finkelstein, H. Shtrikman, I. Bar-Joseph, Phys. Rev. B **53**, R1709 (1996)

[20] K. Kheng, R.T. Cox, Y. Merle d'Aubigne, F. Bassani, K. Saminadayar, S. Tatarenko, Phys. Rev. Lett. **71**, 1752 (1993)

[21] M. Syperek, D.R. Yakovlev, A. Greilich, J. Misiewicz, M. Bayer, D. Reuter, A.D. Wieck, Phys. Rev. Lett. **99**, 187401 (2007)

[22] G.V. Astakhov, V.P. Kochereshko, D.R. Yakovlev, W. Ossau, J. Nürnberger, W. Faschinger, G. Landwehr, Phys. Rev. B **62**, 10345 (2000)
[23] V. Ciulin, P. Kossacki, S. Haacke, J.-D. Ganiere, B. Deveaud, A. Esser, M. Kutrowski, T. Wojtowicz, Phys. Rev. B **62**, R16310 (2000)
[24] D.R. Yakovlev, E.A. Zhukov, M. Bayer, G. Karczewski, T. Wojtowicz, J. Kossut, Int. J. Mod. Phys. B **21**, 1336 (2007)
[25] A.A. Sirenko, T. Ruf, M. Cardona, D.R. Yakovlev, W. Ossau, A. Waag, G. Landwehr, Phys. Rev. B **56**, 2114 (1997)
[26] M.I. Dyakonov, V.Yu. Kachorovski, Sov. Phys. Semicond. **20**, 110 (1986)
[27] E.L. Ivchenko, *Optical Spectroscopy of Semiconductor Nanostructures* (Alpha Science, Harrow, 2005)
[28] J.M. Kikkawa, D.D. Awschalom, Phys. Rev. Lett. **80**, 4313 (1998)
[29] X. Marie, T. Amand, P. Le Jeune, M. Paillard, P. Renucci, L.E. Golub, V.D. Dymnikov, E.L. Ivchenko, Phys. Rev. B **60**, 5811 (1999)
[30] T.C. Damen, L. Viña, J.E. Cunningham, J.E. Shah, L.J. Sham, Phys. Rev. Lett. **67**, 3432 (1991)
[31] B. Baylac, T. Amand, X. Marie, B. Dareys, M. Brousseau, G. Bacquet, V. Thierry-Mieg, Sol. State Commun. **93**, 57 (1995)
[32] I.A. Merkulov, Al.L. Efros, M. Rosen, Phys. Rev. B **65**, 205309 (2002)
[33] P. Michler (ed.), *Single Quantum Dots*. Topics Appl. Phys., vol. 90 (Springer, Berlin, 2003)
[34] I.A. Yugova, A. Greilich, E.A. Zhukov, D.R. Yakovlev, M. Bayer, D. Reuter, A.D. Wieck, Phys. Rev. B **75**, 195325 (2007)
[35] M. Bayer, G. Ortner, O. Stern, A. Kuther, A.A. Gorbunov, A. Forchel, P. Hawrylak, S. Fafard, K. Hinzer, T.L. Reinecke, S.N. Walck, J.P. Reithmaier, F. Klopf, F. Schafer, Phys. Rev. B **65**, 195315 (2002), and references therein
[36] S.A. Crooker, D.D. Awschalom, J.J. Baumberg, F. Flack, N. Samarth, Phys. Rev. B **56**, 7574 (1997)
[37] R.E. Worsley, N.J. Traynor, T. Grevatt, R.T. Harley, Phys. Rev. Lett. **76**, 3224 (1996)
[38] P.Y. Yu, M. Cardona, *Fundamentals of Semiconductors* (Springer, Berlin, 1996)
[39] I.A. Yugova, A. Greilich, D.R. Yakovlev, A.A. Kiselev, M. Bayer, V.V. Petrov, Yu.K. Dolgikh, D. Reuter, A.D. Wieck, Phys. Rev. B **75**, 245302 (2007)
[40] D. Paget, G. Lampel, B. Sapoval, V.I. Safarov, Phys. Rev. B **15**, 5780 (1977)
[41] T.H. Stievater, X. Li, D.G. Steel, D.S. Katzer, D. Park, C. Piermarocchi, L. Sham, Phys. Rev. Lett. **87**, 133603 (2001)
[42] A. Zrenner, E. Beham, S. Stufler, F. Findeis, M. Bichler, G. Abstreiter, Nature **418**, 612 (2002)
[43] P. Borri, W. Langbein, S. Schneider, U. Woggon, R.L. Sellin, D. Ouyang, D. Bimberg, Phys. Rev. B **66**, 081306(R) (2002)
[44] S.E. Economou, R.-B. Liu, L.J. Sham, D.G. Steel, Phys. Rev. B **71**, 195327 (2005)
[45] A. Greilich, M. Wiemann, F.G.G. Hernandez, D.R. Yakovlev, I.A. Yugova, M. Bayer, A. Shabaev, Al.L. Efros, D. Reuter, A.D. Wieck, Phys. Rev. B **75**, 233301 (2007)
[46] A.V. Khaetskii, D. Loss, L. Glazman, Phys. Rev. Lett. **88**, 186802 (2002)
[47] W.A. Coish, D. Loss, Phys. Rev. B **70**, 195340 (2004)
[48] V.L. Berkovits, A.I. Ekimov, V.I. Safarov, Sov. Phys. JETP **38**, 169 (1974)
[49] D. Paget, Phys. Rev. B **25**, 4444 (1982)
[50] R. Oulton, A. Greilich, S.Yu. Verbin, R.V. Cherbunin, T. Auer, D.R. Yakovlev, M. Bayer, V. Stavarache, D. Reuter, A. Wieck, Phys. Rev. Lett. **98**, 107401 (2007)
[51] A. Greilich, A. Shabaev, D.R. Yakovlev, Al.L. Efros, I.A. Yugova, D. Reuter, A.D. Wieck, M. Bayer, Science **317**, 1896 (2007)

[52] M.I. Dyakonov, V.I. Perel, in *Optical Orientation*, ed. by F. Meier, B.P. Zakharchenja (North-Holland, Amsterdam, 1984)
[53] A. Khaetskii, Yu.V. Nazarov, Phys. Rev. B **61**, 12639 (2000)
[54] M. Kroutvar, Y. Ducommun, D. Heiss, M. Bichler, D. Schuh, G. Abstreiter, J.J. Finley, Nature **432**, 81 (2004)

7

Spin Properties of Confined Electrons in Si

W. Jantsch and Z. Wilamowski

7.1 Introduction

Silicon is a rather light element with its nuclear charge of $Z = 14$. Therefore spin–orbit interaction is weak in Si crystals as it is dominated also in solids by atomic effects (see Chap. 1). From a band structure point of view, the weak spin–orbit interaction of Si has been explained by the only possible interaction of the Δ_1 conduction band minimum with very deep core states [1, 2].

The spin–orbit interaction in solids arises from the effective magnetic field seen by electrons moving with respect to other charges in the solid (see Sect. 1.2.3 and [2]). The weakness of the spin–orbit interaction in Si manifests itself also by the small deviation of the g-factor of the conduction electrons which is very close to 2.000.[1] Apart from the (weak) influence of the spin–orbit interaction on the g-factor, it is also important for the spin lifetime and spin dephasing. Therefore we expect particularly long spin coherence in Si, and Si has been considered as an attractive material for spintronic applications right from the beginning [4–6].

In addition, Si has only one isotope, ^{29}Si, with nuclear spin ($I = 1/2$) with an abundance of less than 5%, in contrast to III–V compounds which have 100% nuclear spins, many of which have higher I values. The hyperfine interaction with nuclear spins is ruled by the Fermi contact term (see Sect. 1.2.4), and thus the probability of

---

[1] The precise value of the g-factor is not easy to determine in a three-dimensional semiconductor since at low temperatures, which are necessary to exclude effects due to thermal excitation, carriers are frozen-out and one can determine only the g-factor of electrons bound to donors. If they are shallow enough, then their g-factor should be close to that of the conduction band, but there are "chemical shifts" [3]. Therefore heavily doped samples have been investigated which, beyond the Mott transition, have free carriers also at low temperatures. The required high doping concentrations change, however, the band structure slightly and they cause heavy scattering. In two-dimensional structures, these effects can be avoided using modulation doping.

finding the electron at the site of the nucleus. For delocalized electrons, this probability is small. In contrast, when the electrons are confined to a small volume, like in a donor state or in a quantum dot, the role of coupling of electronic and nuclear spins increases. For Si this problem can be avoided using isotopically purified ^{28}Si and a 60 ms lifetime, and a 60 ms lifetime has been demonstrated for P-bound electrons [7]. The nuclear spins have still longer lifetimes (they can be hours). Therefore one concept for utilizing spin information is based on the nuclear spins of P-donors in Si [4].

Si also poses, however, a few problems when considered, e.g., for the realization of a quantum computer. Si has an indirect gap—its minimum of the conduction band is located in the [100] direction close to 85% of the diameter of the Brillouin zone and therefore the conduction band not only has the two-fold spin degeneracy but also the valley degeneracy. This may cause an additional spin dephasing and it would make control and detection of spin states difficult according to the present concepts. In the presence of strain—e.g., in heterostructures—the sixfold valley degeneracy can be lowered to two and, for still lower symmetry, in principle, it can be lifted [8] but nevertheless this poses additional complications as compared to the mostly direct gap III–V or II–VI semiconductors.

Interest in the spin properties of Si is much older though. It started in the 1950s stimulated by the possibility to learn a lot about defects in the most important material of electronics [9]. Electron spin resonance (ESR) was shown to yield a lot of detailed information on point defects, in particular on the highly localized, so-called deep levels in semiconductors. The corresponding deep states have a very small Bohr radius, comparable to the lattice constant, and, therefore, the hyperfine interaction of the bound electron with its nucleus is strong and can be easily resolved. Due to the $2I + 1$ different possible orientations of the nuclear spin, the bound electron may experience $2I + 1$ different magnetic field situations and therefore the ESR spectra exhibit the corresponding splitting. The ESR spectra thus reflect the natural abundance of the isotopes of that impurity which in many cases allows for an unequivocal chemical identification of the impurity species. In addition, the g-factor anisotropy and a possible "fine structure" splitting contain information on the spin magnitude and on the symmetry of the local environment.

A hyperfine interaction may also occur with nuclear spins of host atoms within the radius of the wave function of the bound state. This ligand hyperfine interaction provides information on the arrangement of the impurity or, more general, the atoms constituting the defect. This information can be obtained by investigating the angular dependencies of the hyperfine structure in the ESR spectra. The g-factor and the fine-structure of the ESR spectra contain additional information on the angular momentum of the electronic state. Altogether ESR yields very detailed information on the electronic state of the defect, but no direct information on its activation energy and thus on its effect on the electronic properties of the host. In order to get those, one has to combine ESR with other techniques, like photo-ESR [10], where the effect of illumination on the ESR amplitude is investigated, or optically detected magnetic resonance, where the effect of ESR is on the photoluminescence yield, or its polarization is used to identify the defects involved in some luminescence feature [11].

For free, delocalized carriers, ESR does not yield that much information at first glance. Usually there is only one fairly wide line and the g-factor can be evaluated from it. Nevertheless, in the case of Si quantum wells ESR proved to be a very valuable tool to investigate spin properties of the two-dimensional electron gas [12–16]. This was surprising since the sensitivity of ESR is characterized by a minimum number of spins in the microwave cavity of a traditional spectrometer. In such spectrometers, the magnetic field is modulated and lock-in detection is used to increase the sensitivity. Therefore the ESR linewidth enters the sensitivity as well, and the sensitivity can be characterized by the minimum number of detectable spins per Gauss linewidth. At liquid He sample temperature this figure is typically on the order of 10^{10} spins/G. For a sample with a sheet carrier density of 10^{11} cm^{-2} this would cause substantial sensitivity problems if the linewidth of the free carrier resonance would be comparable to that of shallow donors, namely a few Gauss. Fortunately, due to the long spin lifetimes in Si quantum wells, the linewidth of the conduction electron resonance in Si quantum wells is a hundred times smaller and therefore the signal is not huge but easily detectable. In addition, there is also an electric dipole contribution to the transition probability which becomes partially allowed in low symmetry samples [17] (see Sect. 7.5).

There are only very few other materials where conduction electron spin resonance has been reported for low-dimensional structures. Heterostructures of AlAs/GaAlAs and GaN/GaAlN are examples where the g-factor is also close to 2.000, indicative of small a spin–orbit interaction and accordingly the ESR linewidth is small enough to allow detection of a signal [18, 19]. ESR offers a few advantages in the investigation of spin properties:

- For sufficiently small linewidth, it allows a very accurate determination of the g-factor: in the case of Si quantum wells, a resolution of $1 : 10^5$ is easily possible and therefore small effects can be detected;
- The anisotropy of the g-factor and the linewidth can be easily measured since the sample can be rotated within the cavity;
- In many cases, other free carrier effects also appear in the spectra, like cyclotron resonance [14], or, more general, magnetoplasma effects [20], which can be very useful since they allow us to measure also the momentum relaxation time in situ [21]. Also in a few cases (but not in Si) Shubnikov–deHaas oscillations were seen which allow also a precise evaluation of the carrier density [20];
- The conduction electron resonance amplitude in principle allows us also to evaluate the contribution of the free carriers to the magnetic susceptibility separately, i.e., independent of other contributions, e.g., from defects in the substrate [22].
- There is also the possibility for time-resolved "spin–echo" experiments [23] which directly demonstrate the possibility to manipulate spins by microwave pulses [24]. In spin–echo experiments, dephasing due to spatial fluctuations of the magnetic field, which vary slowly on the timescale of the experiment (usually due to nuclear spins), can be recognized and repaired.

In this chapter we restrict the discussion to ESR results in Si with carrier confinement into two or "zero" dimensions, i.e., to quantum wells and to donors and "dots", respectively, as these results yield more detailed information on the effect of spin–orbit interaction in Si.

In Sect. 7.2 we introduce the spin–orbit effects in asymmetric quantum wells and their effect on the spin splitting which is seen as an additional anisotropy of the g-factor. In Sect. 7.3 we review the effects of spin–orbit interaction in asymmetric Si-quantum wells on the spin relaxation and dephasing. In Sect. 7.4 we discuss the recently found effect of a dc current on the ESR. Section 7.5 discusses the nature of the ESR excitation in asymmetric quantum wells and we show that there is a substantial contribution of the spin–orbit field which is induced by high frequency currents. Modeling of this effect indicates an increase in the efficiency of spin excitation—monitored by an increase in ESR sensitivity—by four orders of magnitude. In Sect. 7.6 we review work on shallow donors and we describe recent results for laterally confined electrons. Recently for the shallow donors in isotopically enriched ^{28}Si a spin coherence time of 60 ms was demonstrated. For quantum dots, the expected quenching of spin relaxation was so far not found—apparently it is limited by the interaction of simultaneously optically excited and closely located electron–hole pairs in these experiments. Nevertheless, the spin dephasing time of 0.3 ms demonstrates the superior potential of Si due to, among others, the small abundance of ^{29}Si.

7.2 Spin–Orbit Effects in Si Quantum Wells

7.2.1 The Bychkov–Rashba Field

In a non-centro-symmetrical quantum well the spin–orbit coupling leads to a zero field splitting. Within the picture of a one-electron band structure this splitting is described by the so-called Rashba [25], or Bychkov–Rashba [26] term in the Hamiltonian:

$$H_{BR} = \alpha_{BR}(\sigma \times k) \cdot n, \tag{7.1}$$

where σ_α are vectors of the Pauli matrices, or by the spin Hamiltonian:

$$H_{BR} = a_{BR}(s \times k) \cdot n, \tag{7.2}$$

using the spin operator, s. Here $\hbar k$ is the electron momentum, n is the unit vector perpendicular to the sample layer and the constant $a_{BR} = 2\alpha_{BR}$ reflects the strength of the spin–orbit coupling. Within a coordinate system related to the layer, where the z axis is along n, the Bychkov–Rashba term takes the form:

$$H_{BR} = a_{BR}(s_x k_y - s_y k_x). \tag{7.3}$$

In general, the lower symmetry can originate from a lack of inversion symmetry of the crystal or from an asymmetry of some layer structure grown by epitaxy. The first case, introduced by Dresselhaus [27] and Rashba [25], is sometimes also ascribed "bulk inversion asymmetry", while the second case, discussed by Bychkov and Rashba [26], to "structure inversion asymmetry". Because of the high symmetry of the diamond structure, Si based layers exhibit a structure inversion asymmetry only. The latter is dominated by the effect of a doping layer, accommodated at a distance of 10 to 20 nm on top of the quantum well. There is neither the linear bulk inversion asymmetry term nor its cubic "Dresselhaus" counterpart [27] which occur, e.g., in III–V compounds, but in principle, symmetry considerations would also allow for an additional third order term [28–30]. In this review, however, we will not treat that type of spin–orbit coupling because no experimental evidence has been found for it in Si/SiGe structures [28].

The occurrence of the Bychkov–Rashba coupling is equivalent to an effective, spin–orbit induced magnetic field, B_{BR}, acting on electron spins. This field can be defined by

$$g\mu_B B_{BR} = a_{BR}(k \times n). \tag{7.4}$$

The Bychkov–Rashba field is thus in-plane oriented. It is perpendicular to the k-vector and proportional to the magnitude of the k-vector. The total spin splitting, $\hbar\Omega_{BR}(k)$, i.e., the difference in energy between parallel and antiparallel spin orientation is given by $\hbar\Omega_{BR} = a_{BR}k$.

Thermal Distribution of the Bychkov–Rashba Field

A Si layer, grown pseudomorphically on the (100) surface plane of a relaxed $Si_{0.75}Ge_{0.25}$ buffer, will be under tensile strain. Therefore the six-fold valley degeneracy of the conduction band is partially lifted [31]: the four in-plane valleys (synonymous for valleys with their main axis in-plane) are shifted up in energy and only the two "perpendicular" valleys are occupied at low temperatures, and their degeneracy is thus $g_v = 2$. The in-plane motion is thus ruled by the small isotropic transverse effective mass, $m^* = 0.19m_0$, and all Fermi vectors are of the same magnitude:

$$k_F = \sqrt{\frac{4\pi n_s}{g_s g_v}} = \sqrt{\pi n_s}, \tag{7.5}$$

where $g_s = 2$ stands for the spin degeneracy. These two valleys are lower in energy than the conduction band edge of $Si_{0.75}Ge_{0.25}$ layers, which thus constitute barriers confining a two-dimensional electron gas in the embedded Si layer.

For all electrons at the Fermi circle, the Bychkov–Rashba field has the same magnitude. The directions of the k-vectors, and thus also the direction of $B_{BR}(k)$, are uniformly distributed in the two-dimensional sample plane. Therefore, at thermal equilibrium, the mean value of the Bychkov–Rashba field, averaged over the ensemble of all electrons vanishes, $\langle B_{BR} \rangle = 0$. The mean value of the Bychkov–Rashba field, averaged over all uncompensated spins, i.e., those participating in spin resonance, vanishes as well. On the other hand, each uncompensated spin in the vicinity of the Fermi k-vector experiences a Bychkov–Rashba field of $|B_{BR}(k)| = a_{BR}k_F/g\mu_B = \Omega_{BR}(k_F)/\gamma$, where γ is the gyromagnetic ratio. Therefore, the mean square value of the Bychkov–Rashba field, averaged over all uncompensated spins, is

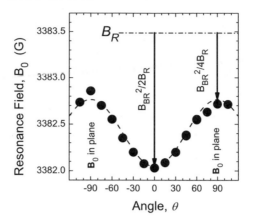

Fig. 7.1. Dependence of the resonance field, B_0, on the direction of the applied field for a modulation doped SiGe/Si/SiGe quantum well. The arrows indicate the shift of the resonance field caused by the Bychkov–Rashba field (see (7.10))

$$\langle B_{BR}^2 \rangle = \frac{\Omega_{BR}^2}{\gamma^2} = \left(\frac{a_{BR}k_F}{g\mu_B} \right)^2 = \left(\frac{a_{BR}}{g\mu_B} \right)^2 \pi n_s. \tag{7.6}$$

The distribution of the Bychkov–Rashba field affects the position of the resonance field and the magnitude of the linewidth. For the discussion of these phenomena it is convenient to consider mean values of squares of each component of the individual Bychkov–Rashba frequency. The in-plane components are

$$\Omega_{BRx}^2 = \Omega_{BRy}^2 = \gamma^2 \langle B_{BRx}^2(k) \rangle = \gamma^2 \langle B_{BRy}^2(k) \rangle = \frac{\gamma^2}{2} \langle B_{BR}^2 \rangle = \frac{a_{BR}^2 k_F^2}{2\hbar}, \tag{7.7}$$

and the out-of-plane component is

$$\Omega_{BRz}^2 = \gamma^2 \langle B_{BRz}^2(k) \rangle = 0. \tag{7.8}$$

Experimentally, one can evaluate $\langle B_{BR}^2 \rangle$ from the anisotropy of the resonance field [21].

g-Factor Anisotropy—Bychkov–Rashba Field in Si/SiGe Structures

The anisotropic distribution of the Bychkov–Rashba field also causes anisotropy of the resonance field. Resonance occurs when the sum of the applied and the Bychkov–Rashba field is equal to the resonance field $B_R = B_0 + B_{BR}$. The momentum relaxation is much faster than the spin relaxation: in high mobility Si/SiGe layers $1/\tau_p \approx 10^{11}$ to $10^{12}\,\mathrm{s}^{-1}$ while $1/T_2 \approx 10^6\,\mathrm{s}^{-1}$. Therefore the time fluctuations are well averaged and the resonance position reflects the mean resonance field only.

For perpendicular field, $B_0 \| \hat{n}$ (i.e., $\theta = 0$), the Bychkov–Rashba fields of all electrons are perpendicular to B_0. Therefore, in spite of the fact that the direction of

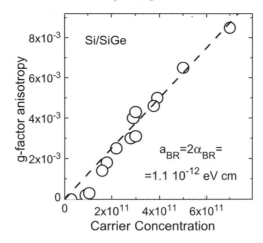

Fig. 7.2. Measured g-factor anisotropy (*circles*) vs. electron concentration. *Dashed line*: fit resulting in a Bychkov–Rashba parameter of $\alpha_{BR} = 1.1 \times 10^{12}\,\mathrm{eV\,cm}$. After [21]

B_{BR} fluctuates in time, the magnitude of B_R is constant and equal for all electrons. Resonance occurs when the external field is given by

$$B_0 = \sqrt{B_R^2 - B_{BR}^2} \approx B_R - \frac{\langle B_{BR}^2 \rangle}{2 B_R} \tag{7.9}$$

and the resulting resonance field is smaller than without Bychkov–Rashba field and the reduction is indicated in Fig. 7.1.

For the angular dependence of the resonance field the distribution of B_{BR}, caused by the distribution of the k-vectors has to be taken into account. For in-plane orientation of B_0 the shift is smaller by a factor 2. Here the longitudinal components of B_{BR} with respect to B_0 average to zero and only its perpendicular component affects the mean value of the resonance field. Consequently, the mean value of the square of the perpendicular component is equal to $B_{BR}^2/2$. The anisotropy of the resonance field, i.e., the difference between B_0 for in-plane and for perpendicular orientation of the applied field, is equal to

$$B_0(90°) - B_0(0°) = \frac{B_{BR}^2(E_F)}{4 B_R}. \tag{7.10}$$

When the Bychkov–Rashba field dominates the g-factor anisotropy, (7.10) allows us to evaluate the absolute value of the Bychkov–Rashba field for electrons at the Fermi energy.

This type of analysis of the experimental data for Si/SiGe structures shows that B_{BR} is of the order of $100\,\mathrm{G}$ and it increases with increasing Fermi k-vector. The observed g-factor anisotropy is shown in Fig. 7.2 as a function of electron concentration, n_s, and the evaluated dependence of B_{BR} on the electron concentration is shown in Fig. 7.3 [21].

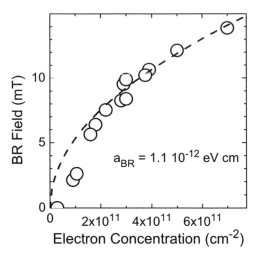

Fig. 7.3. Dependence of the Bychkov–Rashba field, as evaluated from g-factor anisotropy, on the electron concentration. *The dashed line* is plotted assuming the constant value $a_{BR} = 1.1 \times 10^{12}$ eV cm, independent of the carrier concentration. After [21]

7.3 Spin Relaxation of Conduction Electrons in Si/SiGe Quantum Wells

7.3.1 Mechanisms of Spin Relaxation of Conduction Electrons

At low temperatures three mechanisms dominate the spin relaxation of conduction electrons. Elliott and Yafet considered the probability for a spin flip which accompanies momentum scattering events [2, 32]. This probability arises from the fact that in the presence of spin–orbit interaction the pure spin states are not eigenstates anymore: each state has a finite admixture of the opposite spin state which depends on the k-vector. The Elliott–Yafet rate is thus proportional to the momentum scattering rate [2, 32]:

$$\Delta\omega = \frac{\alpha_{EY}}{\tau_p}. \tag{7.11}$$

Bir, Aronov, and Picus analyzed the electron hole scattering which can be very effective in semiconductors when both types of carriers occur [33]. The Dyakonov–Perel mechanism is the most common type of spin relaxation [34]. The classical Dyakonov–Perel relaxation originates from a dependence of the resonance frequency on electron k-vector, $\omega(k)$, which leads to a spread of the resonance frequency, and thus to a broadening of the resonance line. Since the momentum relaxation usually is much faster than the spin relaxation, the original spread of the resonance frequency is motionally averaged. As a result, the resonance line is of the Lorentzian shape with a width of

$$\Delta\omega = \Omega^2 \tau_c. \tag{7.12}$$

Here Ω^2 is the variance of the distribution of $\omega(\boldsymbol{k})$ and τ_c is the correlation time. In the simplest case, when the resonance frequencies before and after a scattering event are not correlated, the correlation time is just equal to the momentum scattering time, $\tau_c = \tau_p$. When small angle scattering occurs, then τ_c can be dependent on the peculiar distribution of the resonance frequencies. In general, the correlation time is defined by the correlator $\langle \omega(0)\omega(t) \rangle$ [35, 36]. The detailed analysis of the Dyakonov–Perel mechanism requires also a discussion of the cyclotron motion [28, 37, 38], electron–electron scattering and electron–electron exchange [29, 39].

The Dyakonov–Perel relaxation is sometimes generalized for a broad class of spin relaxation mechanisms which can be analyzed within the Dyakonov–Perel formula, including the motional narrowing due to electron hopping, where the spread of frequencies is caused by hyperfine coupling to nuclear spins or by dipolar spin–spin interaction, in a manner similar to the motional narrowing in nuclear magnetic resonance [36].

The spin relaxation of the two-dimensional electrons in high mobility Si/SiGe structures is dominated by the Dyakonov–Perel mechanism where the spread of the resonance frequency is caused by a spin–orbit field of the Bychkov–Rashba type. Since the momentum relaxation rate in the high mobility layers is of the order of $1/\tau_p \simeq 10^{11}\,\mathrm{s}^{-1}$, it is easy to fulfill the condition $\omega_c \tau_p > 1$, which implies that the modulation frequency is dominated by the cyclotron motion rather than by the momentum scattering [28, 38]. For the low temperature range and moderate electron concentrations no evidence of the effect of electron–electron scattering [29] has been found so far.

7.3.2 Linewidth and the Longitudinal Relaxation Time of the Two-dimensional Electron Gas in Si/SiGe

Normally, in an ESR apparatus, the first derivative of the microwave absorption is measured with respect to the applied magnetic field. Unless a special setup is used, the in-phase component of the absorption is detected and dispersion effects are eliminated using an automatic frequency control circuit. Nevertheless, the ESR absorption signal of a two-dimensional electron gas in Si/SiGe is characterized by a complex line shape which results from a superposition of absorptive, $\chi''(\omega)$, and dispersive, $\chi'(\omega)$, components of the dynamic magnetic susceptibility. Moreover, with increasing microwave power the absorption component changes its sign from the classical positive to negative sign (see line shapes in Fig. 7.4).

The explanation of the ESR signal shape requires detailed studies of the mechanisms of excitation and power absorption. In particular, one has to consider the resonance excitation by both the magnetic microwave field and the effective spin–orbit field [40–42] and various mechanisms of dissipation of the absorbed power. Some details are discussed below in Sect. 7.5 showing that because of the complex dependence of the electric conductivity on spin polarization, the discussion of the signal amplitude is difficult and does not allow for a simple analysis of the signal saturation in order to evaluate the longitudinal spin relaxation rate. But, without any

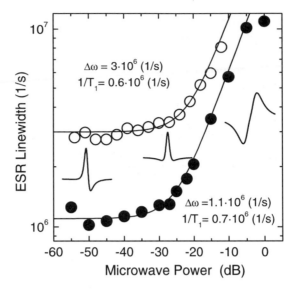

Fig. 7.4. Dependence of the ESR linewidth for perpendicular orientation of the applied field on microwave power for two different Si/SiGe samples (*solid* and *open dots*). The line broadening caused by saturation of the ESR signals allows us to evaluate the longitudinal relaxation rate $1/T_1$. The icons show the evolution of the line shape for different microwave powers

analysis of the power absorption mechanisms, one can evaluate the power dependent resonance linewidth, $\Delta\omega$, and from that the longitudinal spin relaxation rate $1/T_1$.

A phenomenological deconvolution of the experimental line shape as a combination of $\chi''(\omega)$ and $\chi'(\omega)$ allows us to evaluate $\Delta\omega$ and its dependence on the microwave power, P. An example of the line broadening at high microwave power is shown in Fig. 7.4. The fitting curves correspond to the known dependence $\Delta\omega(P) = \Delta\omega_0(1 + \gamma^2 B_1^2 T_1/\Delta\omega_0)^{1/2}$ [43, 44], where $\Delta\omega_0$ is the unsaturated, narrowed linewidth, and B_1^2 is the square of the microwave magnetic field, proportional to the microwave power, P.

The single ESR line of Si/SiGe quantum well structures is very narrow. Depending on the electric properties of the two-dimensional electrons and on the direction of the applied field the linewidth varies from 3 to 100 μT. The smallest width is observed for layers with the smallest electron concentration.

The dependence of $\Delta\omega$ on the momentum relaxation rate is shown in Fig. 7.5. The narrowing of the linewidth with increasing momentum scattering frequency is well visible in the intermediate scattering regime, indicating that the spin relaxation is dominated by the Dyakonov–Perel mechanism. The dashed line shows the upper limit of the Elliott–Yafet mechanism, which can play a role for low mobility only. For low mobility, the total spin relaxation rate (see Fig. 7.5) approaches a value half of the longitudinal relaxation rate, $\Delta\omega \simeq 1/2T_1$. This observation demonstrates, that the linewidth is dominated by the longitudinal spin relaxation, which is an attribute of the Elliott–Yafet relaxation.

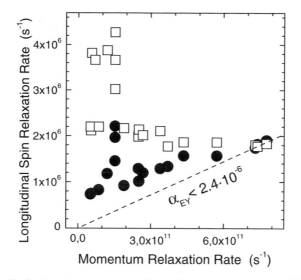

Fig. 7.5. Longitudinal spin relaxation rate, $1/T_1$ (*solid dots*) and the doubled linewidth, $2\Delta\omega$, (*open squares*) measured at $\theta = 0$ as a function of the momentum relaxation rate. *The dashed line* stands for the upper limit of the Elliott–Yafet rate: $\alpha_{EY} = 2.4 \times 10^{-6}$. After [28]

Another argument for the Elliott–Yafet spin relaxation can be also found in the temperature dependence of the resonance linewidth. For samples with a moderate electron mobility, $\Delta\omega$ slightly decreases first with increasing temperature, indicating enhanced motional narrowing—characteristic for the Dyakonov–Perel relaxation. At higher temperatures, when the mobility decreases, it saturates, however, and then it increases in the high temperature range. This leads to the conclusion that Elliott–Yafet becomes the dominant mechanism at high temperatures. In addition, it shows that the real value of α_{EY} is not much smaller than the upper limit evaluated in Fig. 7.5.

In the high mobility range a non-monotonic dependence of $\Delta\omega$ on the scattering rate is observed. Such a behavior cannot be explained solely by the dependence of τ_p on the carrier concentration [22]. Low concentration layers are characterized by a lower mobility. Therefore, the Dyakonov–Perel linewidth is expected to approach the highest values for samples with high electron concentration and high mobility corresponding to the longest τ_m. In addition, large fluctuations of the spin–orbit field occur due to the large Fermi k-vector. The experimentally observed decrease of $\Delta\omega$, for small momentum relaxation rates, indicates that τ_m is not directly ruled by τ_p but rather by an additional mechanism of field modulation. The detailed analysis brought us to the conclusion that the cyclotron motion, where the direction of the carrier velocity changes continuously, leads to an additional modulation rate [28, 37, 38].

Experimental evidence for the effect of cyclotron motion on the modulation frequency can be found in the angular dependence of the linewidth. For a two-dimensional electron gas, the cyclotron frequency $\omega_c = eB\cos\theta/m^*$ scales with

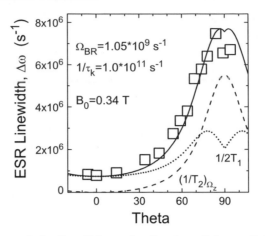

Fig. 7.6. Dependence of the linewidth on the direction of the applied field for a two-dimensional electron concentration of $n_s = 2 \times 10^{11} \, \mathrm{cm}^{-2}$ measured at a frequency of $\omega = 2\pi \cdot 9.4 \, \mathrm{GHz}$. At the ESR, the cyclotron frequency is $\omega_c = 3.1 \times 10^{11} \, \mathrm{s}^{-1}$. *The solid line* is described by (7.24), *the dotted one* corresponds to one half of the longitudinal relaxation rate and *the dashed one* to the contribution due to longitudinal fluctuations (7.25). Ω_{BR} and τ_p are fitting parameters listed in the figure. The corresponding Bychkov–Rashba field is $B_{BR} = 6 \, \mathrm{mT}$. After [28]

the transverse component of the applied field. Therefore, ω_c vanishes for in-plane orientation of the applied magnetic field ($\theta = 90°$) and comparison of the linewidth for different orientations of the magnetic field allows for an estimation of the role of the cyclotron motion on the modulation frequency.

An example of the angular dependence of the linewidth on the direction of the applied field is shown in Fig. 7.6. The linewidth is characterized by a pronounced anisotropy. For perpendicular orientation of the applied field, ($\theta = 0$), ω_c is big and $\Delta\omega$ is a few times smaller as compared to $\theta = 90°$, where no effect of the cyclotron motion occurs. In that sense the strong anisotropy of $\Delta\omega$ confirms the influence of cyclotron motion on τ_m. The quantitative description requires also a discussion of the anisotropy of the fluctuations of the Bychkov–Rashba field. The modeling is presented below in Sect. 7.3.3. When the cyclotron motion does not play any role, e.g., for strong momentum scattering, the linewidth for perpendicular orientation is 2/3 of the linewidth for in-plane orientation (see Fig. 7.7).

Further experimental evidence for the influence of the cyclotron motion on τ_c can be found in the dependence of the anisotropy of $\Delta\omega$ on electron mobility. For low mobility, where $\omega_c \tau_p \ll 1$, the k-vector does not change considerably between scattering events, and therefore τ_c is expected to approach τ_p. On the other hand, for $\omega_c \tau_p \gg 1$, an electron performs a few cyclotron orbits between successive scattering events, and then the modulation is expected to be defined by ω_c [37].

The dependence of the linewidth anisotropy, as a function of the momentum scattering rate, is shown in Fig. 7.7. The experimental values for the electron mobility were evaluated from the cyclotron resonance linewidth. These data give the proper

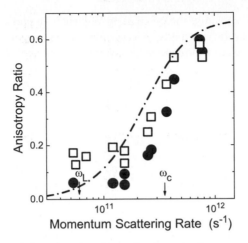

Fig. 7.7. Dependence of the spin relaxation anisotropy on the momentum relaxation rate. *Squares* stand for the transverse relaxation, $1/T_2(0°) = \Delta\omega(0°)$, and *circles* for $1/2T_1(0°)$, both normalized by the in-plane linewidth $1/T_2(90°)$ [28]. *The dash-dotted line is described by (7.24) and (7.25). For $1/\tau_p \gg \omega_c$, the anisotropy of the Dyakonov–Perel linewidth originating from Bychkov–Rashba splitting tends to a value of 2/3*

trend, but because measurements were done at low frequency and no correction due to the plasma shift has been taken into consideration [20] a moderate systematic experimental error may occur (the carrier density here is much smaller than in the case of the GaN quantum wells).

For high electron mobility (long τ_p) the anisotropy is most pronounced, confirming the role of the cyclotron modulation of the Bychkov–Rashba field in the Dyakonov–Perel spin relaxation.

7.3.3 Dephasing and Longitudinal Spin Relaxation

Transverse and Longitudinal Relaxation Originating from the Classical Dyakonov–Perel Relaxation

The discussion of the spin relaxation times usually is based on the Bloch equations, where the transverse relaxation rate, $1/T_2 = \Delta\omega$, defines the linewidth of the resonance line, at least in the case when the line shape can be described by a Lorentzian function [44]. The Lorentzian linewidth is related to the half-width by $\Delta\omega_{1/2} = 2\Delta\omega$ and with the peak-to-peak width of the differential line shape given by $\Delta\omega_{p-p} = 2\Delta\omega/\sqrt{3}$ [44].

The transverse relaxation rate is equivalent to the dephasing rate of an ensemble of precessing spins, and equal to the mean value of the relaxation rates of both transverse spin components, $1/T_2 = 1/T_\varphi = (1/T_{x'} + 1/T_{y'})/2$. The dashed coordinate system is related to the external magnetic field, $\boldsymbol{B}$, which is directed along the z'-axis, while the x' and y' axes are perpendicular to the applied field. The dephasing rate originates from the spread of the precession frequency. In particular, fluctuations of the z component of effective fields, as described by the variance $\Omega_{z'}^2$, directly result in the dephasing rate according to the Dyakonov–Perel model (see (7.12)):

$$(1/T_2)_{\Omega_{z'}} = \Omega_{z'}^2 \tau_c. \tag{7.13}$$

This is the result of motional narrowing of the initial line broadening caused by the variance of the z' component of the spin–orbit field and can be evaluated as the Fourier transform (at zero frequency) of the correlator $\langle \Omega(0)\Omega(t)\rangle$.

Fluctuations of the transverse effective field components contribute to the spin flip probability, $1/T_{sf}$, i.e., they enhance the longitudinal spin relaxation rate by

$$\frac{1}{T_1} = \frac{1}{T_{z'}} = \frac{2}{T_{sf}} = \left(\Omega_{x'}^2 + \Omega_{y'}^2\right)\tau_c'. \tag{7.14}$$

The dash at the modulation time indicates that the longitudinal relaxation rate is dominated by the fluctuations *at the Larmor frequency*, ω_L. Strictly speaking, $1/T_1$ corresponds to the Fourier transform of $\langle \Omega_x(0)\Omega_x(t)\rangle$ and $\langle \Omega_y(0)\Omega_y(t)\rangle$, both taken at the Larmor frequency. For the simplest case of an exponential decay of the correlation function, τ_c' is obtained as

$$\tau_c' = \frac{\tau_p}{1 + \omega_L^2 \tau_p^2}, \tag{7.15}$$

whereas $\tau_c = \tau_p$.

When $\omega_c \tau_p \geq 1$, the Bychkov–Rashba field, and the corresponding precession frequency of an individual spin, are modulated with the cyclotron frequency. Consequently, the correlation function oscillates as well, and this oscillation decays with the momentum relaxation time τ_p. Generally, the correlation times are described by

$$\tau_c' = \frac{\tau_p}{1 + (\omega_L - \omega_c)^2 \tau_p^2}, \tag{7.16}$$

and

$$\tau_c = \frac{\tau_p}{1 + \omega_c^2 \tau_p^2}. \tag{7.17}$$

The transverse components of the fluctuating fields contribute also to the resonance linewidth, i.e., to the transverse relaxation rate, $1/T_2$. There are two approaches. The classical one [2, 28] assumes that the total linewidth, i.e., the Bloch transverse relaxation rate, is the sum of the dephasing rate caused by longitudinal fluctuations, $\Omega_{z'}^2$, as described by (7.13), and the contribution originating from fluctuations of the transverse components, which is equal to half of the longitudinal relaxation rate (see (7.15)). The linewidth is equal to

$$\Delta\omega = \frac{1}{T_2} = \left(\frac{1}{T_2}\right)_{\Omega_{z'}} + \frac{1}{2T_1}. \tag{7.18}$$

The more systematic notation [29, 45] states that fluctuations of an effective field along an axis α result in dephasing of the two other components. In particular, the longitudinal relaxation rate is described by (7.14) and the transverse components are given by

$$\frac{1}{T_{x'}} = \Omega_{y'}^2 \tau_c' + \Omega_{z'}^2 \tau_c \tag{7.19}$$

and

$$\frac{1}{T_{y'}} = \Omega_{x'}^2 \tau_c' + \Omega_{z'}^2 \tau_c. \tag{7.20}$$

The total linewidth is thus given by

$$\Delta\omega = \frac{1}{T_2} = \frac{1}{2}\left(\frac{1}{T_{x'}} + \frac{1}{T_{y'}}\right) = \frac{1}{2}\left(\Omega_{x'}^2 + \Omega_{y'}^2\right)\tau_{c'} + \Omega_{z'}^2 \tau_c. \tag{7.21}$$

That expression is equivalent to (7.18) with contributions given by (7.13) and (7.14).

Angular Dependence of the Dyakonov–Perel Spin Relaxation

The variance of the Bychkov–Rashba frequency due to the distribution of k-vectors is described by (7.7) and (7.18). When the applied magnetic field is tilted by an angle θ from the normal to the layer, then in the primed coordinate system related to the applied field the components of the fluctuations in the Bychkov–Rashba frequency are

$$\Omega_{z'}^2 = \frac{\Omega_{BR}^2}{2}\sin^2\theta, \tag{7.22}$$

and

$$\Omega_{x'}^2 + \Omega_{y'}^2 = \frac{\Omega_{BR}^2}{2}\left(\cos^2\theta + 1\right). \tag{7.23}$$

The dependence of the transverse relaxation rate on the direction of the external magnetic field is [28, 38, 46][2]

$$\frac{1}{T_2} = \frac{\Omega_{BR}^2(\cos^2\theta + 1)}{4}\frac{\tau_p}{1 + (\omega_L - \omega_c)^2\tau_p^2} + \frac{\Omega_{BR}^2\sin^2\theta}{2}\frac{\tau_p}{1 + \omega_c^2\tau_p^2}. \tag{7.24}$$

---

[2] In [28] a factor 2 was missing in the denominator of the expression for $1/T_1$ as pointed out in [38]. This error came from the erroneous assumption that T_1 and the spin–flip time were equivalent. The expression in this paper has the same factors as in the paper by Glazov [38]. Additional discrepancies with [38] arose since there in Fig. 4 our data from Fig. 5 in [28] were used, but for an unknown reason they were considered as the double linewidth. Here we are analyzing the same data in Fig. 3.3. Our fitting value of τ_p is the same but Ω_{BR} in [38] is smaller (because the double linewidth is assumed). The second point concerns a detail in the denominator in the part describing the contribution to $1/T_2$ due to the transverse fluctuations. In our approach, we claim it is exactly half of $1/T_1$ (compare (7.24) and (7.25)). Consequently, for the modulation time we take the Fourier transform at the Larmor frequency, which gives a factor in the denominator. Glazov does the same to evaluate $1/T_1$, but he omits ω_L when discussing the same contribution to $1/T_2$.

The first term is half of the longitudinal relaxation rate:

$$\frac{1}{T_1} = \frac{\Omega_{BR}^2 (\cos^2 \theta + 1)}{2} \frac{\tau_p}{1 + (\omega_L - \omega_c)^2 \tau_p^2} \tag{7.25}$$

and the second one is the contribution originating from the longitudinal fluctuations: $(1/T_2)_{\Omega_{z'}}$.

7.3.4 Comparison with Experiment

Generally, comparison of all experimental data with the model well confirm that the spin relaxation in Si/SiGe layers is dominated by the Dyakonov–Perel mechanism caused by the Bychkov–Rashba field. The expressions for the spin relaxation rates (7.24), (7.25) describe the anisotropy of the relaxation times (see Fig. 7.6) showing the importance of the cyclotron modulation in the Dyakonov–Perel spin relaxation. This effect is important in high mobility samples as it is seen in Fig. 7.5 where for weak momentum scattering both spin relaxation rates, measured for perpendicular orientation of the applied field, approach very small values. This narrowing does not occur, when the magnetic field is oriented in-plane, and $\omega_c = 0$. As it is shown in Fig. 7.7, in the high mobility range the strongest anisotropy (and thus the lowest anisotropy ratio) of the linewidth is observed.

For high mobility quantum wells, the fitting of (7.24) and (7.25) to the angular dependencies of the relaxation rates allows for a sensitive evaluation of Ω_{BR}^2 and τ_p. The lines in Fig. 7.6 correspond to the dependencies described by (7.24) and (7.25) and the best fit parameter values are listed in the figure. The magnitude of the Bychkov–Rashba field, $B_{BR} = 6\,\text{mT}$, for the sample with $n_s = 2 \times 10^{11}\,\text{s}^{-1}$ fits well to the Bychkov–Rashba field evaluated from the g-factor anisotropy (see Figs. 7.2 and 7.3). For low mobility samples the relaxation rates and their weak anisotropy are determined by the product $\Omega_{BR}^2 \tau_p$ and these two factors cannot be determined independently from the experimentally observed linewidth. In that mobility range, the linewidth anisotropy becomes independent of momentum scattering and tends to a constant value of 2/3 (see Fig. 7.7).

In the high mobility range, for $\Omega_{BR}^2 \tau_p \gg 1$, and for perpendicular orientation ($\theta = 0$), the spin relaxation and the anisotropy ratio, caused by the Dyakonov–Perel mechanism, tend to zero. This allows us to investigate other spin relaxation mechanisms. As it is seen in Fig. 7.7, the linewidth for ($\theta = 0$) is larger than the value expected for the Dyakonov–Perel mechanism, indicating the occurrence of other spin relaxation mechanisms. As mentioned in Sect. 7.3.1, most likely the Elliott–Yafet relaxation becomes visible. Generally, a high anisotropy of the linewidth reflects high quality of the two-dimensional layer with high electron mobility and lack of other mechanisms, e.g., those originating from non-uniformity of the sample.

In principle, the structure inversion asymmetry in these samples could be avoided using symmetrical doping. This asymmetry in our samples limits the spin lifetime to a few microseconds at a magnetic field of 0.34 T. In practice, symmetrical doping

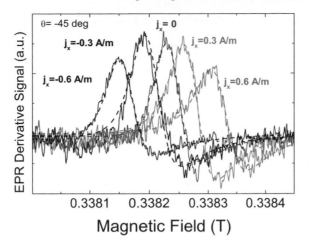

Fig. 7.8. ESR spectra of a two-dimensional electron gas in a Si quantum well for various values of an electric current density passing a 3 mm wide sample. Measurements were performed with B perpendicular to the current and tilted by $\theta = -45°$ from the direction normal to the sample plane, at a microwave frequency of 9.4421 GHz

does not improve the situation because donors "float" at the surface during the MBE growth and any attempt to dope the deeper barrier results in donor proliferation into the channel causing low mobility. Possibly symmetry could be achieved applying some gate voltage to an asymmetrical sample or by some other technological trick. In that case we can extrapolate the Elliott–Yafet rate for high mobilities, lets say 400,000 cm^2/Vs, and we would expect then a spin lifetime of 25 µs.

7.4 Current Induced Spin–Orbit Field

Recently we found more direct evidence for the Bychkov–Rashba field in our one-sided modulation doped quantum well samples [40]: when we pass a moderate dc current through our two-dimensional electron gas we see a shift of the ESR line position and the shift changes its sign when either the current or the magnetic field direction is reversed (see Fig. 7.8). A similar effect, namely the influence of the current-induced Rashba field on the Hanle depolarization, has been observed by Kalevich and Korenev [47]. The current-induced Bychkov–Rashba field is a direct consequence of the $k \times \sigma$ term in the Hamiltonian. The application of an in-plane electric field (as needed to obtain the electric current, j_x, see Fig. 7.9) causes a shift of the Fermi circle and thus an additional δk_x for each electron. This results in the additional Zeeman splitting which manifests itself in a shift of the spin resonance field. The corresponding Bychkov–Rashba field is perpendicular to both δk_x and the z direction in which the symmetry is distorted. Therefore an additional δB_{BR} is caused

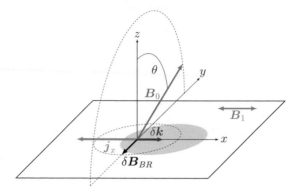

Fig. 7.9. A current j_x passes through a two-dimensional electron gas (xy plane). The Fermi circle shifts by an amount δk_x. Within this approximation, each electron experiences a Bychkov–Rashba field, $\delta \boldsymbol{B}_{\mathrm{BR}}$, in addition to that resulting from its thermal momentum. The static field $\boldsymbol{B}_0$ (drawn not to scale) is applied in the yz plane to enable ESR measurements. After [40]

by δk_x, which is perpendicular to δk_x, and it is in-plane:

$$\delta \boldsymbol{B}_{\mathrm{BR}} = \frac{\beta_{\mathrm{BR}}}{n_s e}(\boldsymbol{n} \times \boldsymbol{j}). \tag{7.26}$$

Here $\beta_{\mathrm{BR}} = a_{\mathrm{BR}} m^* / g \mu_{\mathrm{B}} \hbar$. When the static field is applied in-plane and perpendicular to the dc current, the additional δB_{BR} is either parallel or anti-parallel to the static field (see Fig. 7.9) which is seen by the shift in ESR.

The observed angular dependence of the ESR field is shown in Fig. 7.10 with and without current. When the current direction is inverted, the shift in the resonance field, which reflects directly the current induced Bychkov–Rashba field, is also reversed. Altogether this angular dependence shows all features expected from the $\boldsymbol{k} \times \boldsymbol{\sigma}$ term.

This finding implies a number of effects and consequences. It shows directly that the ESR field can be adjusted by a current. This kind of "ESR tuning" could be utilized, e.g., for selective spin manipulation: in a system of wires in a resonator, spins can be manipulated in one particular wire by bringing that wire into resonance by some current pulse. Once in resonance, the electron spins precess at the Rabi frequency (determined by the microwave intensity) and a specific rotation angle can be achieved by choice of the current pulse duration. A Rabi frequency of 100 MHz can been obtained in spin–echo experiments and thus up to about 100 rotation periods can be achieved within the dephasing time of more than 1 μs.

The observation of a dc Bychkov–Rashba field implies also the possibility to generate ac fields by applying ac currents. This should be possible up to frequencies of the order of the inverse momentum scattering time, τ_p. (Beyond that frequency we have ballistic oscillations which still should give rise to spin rotation.) Experimental evidence for such effects are described in the next section.

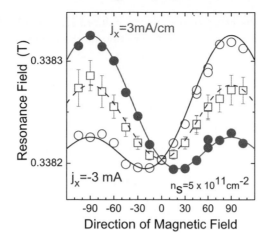

Fig. 7.10. Angular dependence of the ESR field for a current of $J = 0$ (*squares*) and $J = \pm 1$ mA (*open* and *full circles*, respectively). The electron concentration is $n_s = 5 \times 10^{11}$ cm^{-2} and the sample width $w = 3$ mm. After [40]

7.5 ESR Excited by an ac Current

7.5.1 Electric Dipole vs. Magnetic Dipole Spin Excitation

It has been shown by Rashba in his early theoretical work that the spin–orbit coupling causes electric dipole transitions between the states of a spin doublet [17]. The transition probability of electric dipole transitions, when allowed, can be by orders of magnitude higher than that of the magnetic dipole transitions [17, 48, 49]. Therefore even a small admixture of excited states, which would allow electric dipole transitions, can lead to a substantial increase of the ESR amplitude. The investigation of the absolute ESR amplitude is experimentally difficult. Therefore experimental evidence of electric dipole ESR usually is based on a specific angular dependence of the ESR amplitude. In contrast to magnetic dipole transitions, the efficiency of electric dipole transitions strongly depends on the experimental geometry. In particular, the electric dipole transition probability vanishes for specific relative orientations of the sample, the external static magnetic field and the direction of the microwave electric field. Dobrowolska et al. demonstrated the occurrence of such effects specific for the electric dipole ESR of conduction electrons (originally called electric dipole spin resonance) in a broad class of semiconductors and metals [50].

Recent papers, investigating spin properties of high mobility two-dimensional electrons, indicate some unsolved problems in understanding the details of electric dipole ESR. Originally in Rashba's model dissipation mechanisms were not taken into account [48, 49]. On the other hand, it is experimentally well evidenced now that the electric dipole ESR scales with the electron mobility [40]. A general theoretical treatment, taking into account momentum and spin dissipation, has been proposed by Duckheim and Loss [51].

The finding of the effect of the current induced Bychkov–Rashba field, as described in Sect. 7.4, may give new insight into the mechanisms of electric dipole ESR. The main idea is that the electric microwave field induces an ac electric current which leads to an effective ac magnetic field that stimulates spin–flip transitions. The ac current is proportional to the electron conductivity, and the dependence of the effective field on electron dissipation becomes thus obvious within this approach. In that sense this approach of current induced transitions is different from the electric dipole model. Both approaches describe, however, the same phenomenon, namely, the excitation of the spin resonance by an electric field via spin–orbit coupling. As it is shown by the recent papers [41, 51] two different limits should be considered. The model of current-induced transitions describes the case when a low frequency field induces a drift current while the electric dipole model stands for a high frequency range were a displacement current is induced, equivalent to the dissipationless oscillation of an electric dipole.

7.5.2 The ESR Signal Strength in Two-dimensional Si/SiGe Structures— Experimental Results

Sensitivity of ESR in Two-dimensional Si/SiGe

The ESR of two-dimensional electrons in Si can be easily observed in spite of the fact that the number of spins is rather small. For a typical sample area of $10\,mm^2$ the number of spins is of the order of 10^8. Nevertheless, the ESR can be easily observed with a reasonable signal to noise ratio.

The narrow linewidth of the two-dimensional electron resonance in Si/SiGe structures is only one reason for the high sensitivity of ESR. A comparison of the observed sensitivity with the instrument specifications (only a rough estimation is possible) shows that the experimental sensitivity is at least by an order of magnitude higher than expected for magnetic dipole ESR, indicating the role of the spin–orbit coupling in the resonance excitation and detection mechanisms. The ESR signal is also big enough to allow for spin echo experiments [24].

Temperature Dependence

With increasing temperature, the line shape and the width of the ESR in Si/SiGe does not change significantly. In the low temperature range the width slightly decreases with increasing temperature reflecting the increase of the momentum relaxation rate and thus less effective Dyakonov–Perel spin relaxation. In the high temperature range a weak increase of the linewidth is observed, probably due to spin–phonon coupling [52].

Investigations of ESR above $40\,K$ are very difficult because the signal amplitude strongly decreases. As it was shown recently, the decrease of ESR signal amplitude is correlated with momentum scattering and scales with the square of the electron mobility [18, 41]. This observation clearly indicates that the ESR in Si/SiGe structures is induced by an ac electric current via spin–orbit coupling.

Power Dependence of the Line Shape and Amplitude

The line shape of the two-dimensional electron resonance in Si/SiGe layers is very different from the classical ESR shape which is usually described by the imaginary part of the dynamic magnetic susceptibility, $\chi''(\omega)$, and the corresponding Lorentzian shape function (see Fig. 7.4). The experimentally observed line shape is of the so-called Dysonian type [53], i.e., it can be described by a combination of real (dispersive), $\chi'(\omega)$, and imaginary (absorption), $\chi''(\omega)$, contributions [22, 40]. The amplitudes of both contributions depend on the experimental geometry, microwave power, temperature, etc. A deconvolution of the experimental line shape into these two contributions is easily achievable and all dependencies can be independently investigated for both types of signals.

Angular Dependence of the Amplitudes of ESR Signals

As discussed above, a rotation of the sample in the external magnetic field leads to a variation of the resonance position and the linewidth. The amplitude of the signal additionally depends on the position of the sample in the microwave cavity, i.e., on the magnitude of the magnetic, B_1, and electric, E_1, components of the microwave field and on the direction of both components [18, 40].

The complex angular dependence of the ESR amplitude shows that the ESR is not of the magnetic dipole type but rather that it is excited by an effective spin–orbit field. In that case, when the ESR can be excited by both magnetic dipole (i.e., by B_1) and by the current induced or electric dipole field (i.e., by E_1) the signal amplitude depends on the experimental geometry in a really complex way [40, 42, 54].

7.5.3 Modeling the Current Induced Excitation and Detection of ESR

Two different types of electrical excitations can be distinguished [40, 51]. In the low frequency range, when $\omega\tau_p \ll 1$, the drift current is limited by momentum scattering. In the high frequency limit, when $\omega\tau_p \gg 1$, the displacement current dominates and the amplitude of the current becomes independent of momentum dissipation. This type of electric dipole ESR has been analyzed by Rashba and coworkers [17, 48, 49] and there is ample experimental evidence for it [50].

A detailed comparison of current induced and electric dipole ESR will be presented elsewhere [42]. The effective Bychkov–Rashba field is given by (7.26), where now B_{BR} and j are complex quantities describing amplitudes and phase shifts. Within the Drude model, the current $j = \hat{\sigma}(\omega)E_1$ is induced by the ac electric field and described by the complex conductivity:

$$\hat{\sigma}(\omega) = \frac{n_s e^2 \tau_p}{m^*} \frac{1}{1 - i\omega\tau_p}. \tag{7.27}$$

The driving force for ESR is obtained from the transverse component of the ac Bychkov–Rashba field, $|B_{BR\perp}|$. In general, this field depends in a complex way on

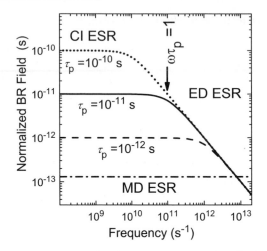

Fig. 7.11. Dependence of the ESR driving field on frequency. The effective ac field is normalized by $\beta_{BR} e E / m^*$ and evaluated for the material parameters of Si/SiGe layers

geometry and frequency. For a specific geometry, however, when microwave electric field, E_1, and external magnetic field B_0 are parallel and both are oriented in the sample plane, the expression for $|B_{BR\perp}|$ has a simple form:

$$|B_{BR\perp}| = \frac{\beta_{BR} e E_1}{m^*} \frac{\tau_p}{\sqrt{1 + \omega^2 \tau_p^2}}. \tag{7.28}$$

This dependence is plotted in Fig. 7.11 as a function of frequency for various values of τ_p. The vertical axis is normalized by $\beta_{BR} e E_1 / m^*$ and evaluated for the material parameters of Si/SiGe layers. To compare $|B_{BR\perp}|$ with the driving force in magnetic dipole excitation, caused by a microwave field B_1, we assume that the sample is placed at the respective positions with maximum amplitude of E_1 and B_1 in the microwave cavity.

In Fig. 7.11 one can distinguish two ranges of frequency with different characteristic dependencies. At low frequency, $\omega \tau_p \ll 1$, the driving force of current induced ESR is independent of frequency and proportional to the electron mobility. Even for Si/SiGe, where the spin–orbit interaction is rather small, its driving force can strongly exceed that of magnetic dipole transitions. A typical value of τ_p for high quality Si/SiGe is on the order of 10^{-11} s. Therefore $|B_{BR\perp}|$ is two orders of magnitude higher than B_1. Because the absorption signal scales with the square of the ac field, the current induced ESR signal can be 4 orders of magnitude stronger than the magnetic dipole ESR signal.

In the high frequency range, for $\omega \tau_p \gg 1$, the driving force of the electric dipole ESR decreases with frequency and becomes independent of the electron mobility. This originates from the fact that the amplitude of the charge oscillations decreases with frequency. As a result, for very high frequency, the electric dipole ESR signal becomes smaller than that caused by magnetic dipole transitions.

7.5.4 Power Absorption, Line Shape

As long as the effect of the ac electric field can be expressed by the time dependent Bychkov–Rashba field, the power absorption originating from spin–orbit interaction can be described by the Bloch equations, which lead to the known expression for the power absorption. The power absorbed per unit area is described by [44]

$$\frac{dP}{dA} = \frac{1}{2}\mu_0\omega\chi''(\omega)|H_{BR\perp}|^2.$$

(7.29)

This power is proportional thus to the imaginary part of the dynamic susceptibility, $\chi''(\omega)$, which in turn is proportional to the Lorentz shape function, $f_L(\omega)$. For the discussed experimental geometry, the final expression for the frequency dependence of the ESR, caused by spin–orbit interaction, is

$$\frac{dP}{dA} = \frac{1}{2\mu_0}\left(\frac{\beta_{BR}eE}{m^*}\right)^2 \frac{\omega\tau_p^2}{1+\omega^2\tau_p^2}\pi\gamma M_0 f_L(\omega).$$

(7.30)

The discussion presented above is based on the simultaneous solution of the Bloch equations for the precession of the magnetic moment and the Boltzmann equation for the motion of electrical charges. The solution leads to the conclusion that the line shape is expected to be of pure absorption type. The experimentally observed line shape of the absorption spectra shows, however, not only the absorption-like contribution, proportional to $\chi''(\omega)$, but also a pronounced dispersive contribution, proportional to the dispersive component $\chi'(\omega)$. The occurrence of this component can neither be explained by the model described above, which discusses a channel of energy transfer from the ac electric field via excitation of charge motion, to the spin system, nor by the Dyson effect [53], which for two-dimensional system also predicts a pure absorption line shape.[3]

Therefore, the experimentally observed line shape indicates that other channels for the energy transfer are spin dependent leading to the occurrence of a dispersive component of the absorption signal. We infer that the dispersive component reflects the spin dependent Joule heating, namely, the dependence of the electron velocity on the precession phase and precession angle.

7.6 Spin Relaxation under Lateral Confinement

Interest in the spin properties of electrons, confined in all three dimensions, was stimulated recently by the fact that the spin–orbit driven spin relaxation processes can be suppressed by confinement [55–57] and long spin coherence times are necessary

---

[3] Dyson considered the ESR line shape in metals. He showed that spin diffusion beyond the skin depth and the effect of the induction of eddy currents by the oscillating magnetization leads to a dispersive component of the absorption signal. In two-dimensional samples, however, both mechanisms are not effective.

for quantum computers. Present technology allows lateral confinement by chemical composition to a radius down to about 50 nm. Much smaller confinement radii occur in impurity bound electron states. Shallow donors have a ground state Bohr radius of about 3 to 7 nm, whereas deep impurity states are confined essentially to the unit cell containing the defect. At present, mostly artificial quantum dots and shallow impurity states are considered as candidates in quantum computing concepts.

Shallow and deep defect states in Si were investigated most thoroughly in the 1960s and ESR was a most successful method to reveal microscopic details of the electronic defect states. Right from the beginning it was clear that the low temperature spin coherence was limited by hyperfine interaction of the bound electron with nuclear spins of the host material [3]. Hyperfine interaction is described by the Fermi contact term, i.e., the probability of finding an electron at the site of a nucleus with a spin I. For delocalized electrons in two-dimensional or three-dimensional structures, this probability is small and the wave function extends over a huge number of such sites, so that the effect is averaged out.

For the highly localized defect states, on the other hand, only a small number of nuclear spins are seen by the bound electron and their discrete arrangement and their finite number of quantized orientations $(2I + 1)$ may lead to well resolved spectra which yield very detailed information on the defect composition and its geometry [3]. Si was the best investigated material right from the beginning, mostly because of its importance as the main material in electronics, but also because of the low abundance ($<5\%$) of ^{29}Si ($I = 1/2$) which allows for well resolved hyperfine splittings in the spectra in contrast to III–V compounds where each host atom carries nuclear spin.

In the first part of this section we recall early and recent work on shallow donors in Si which demonstrate the probably longest spin coherence times seen in solids. Work on quantum dots in Si is still in its infancy in contrast to III–V compounds, mostly because of technological issues. In the second part of this section we discuss first results which do not show the expected quenching of spin dephasing, possibly because of electron–hole interaction of photogenerated carriers. We find that effects of hyperfine dephasing also become appreciable.

7.6.1 Shallow Donors

Hyperfine Interaction in Shallow Donors

The basic spin properties of shallow donors in Si were investigated 50 years ago [3, 52, 58]. A full comprehensive picture was experimentally established by Feher verifying the theoretical description of Kohn and Luttinger [59]. Using the method of electron–nuclear double resonance (ENDOR) in which the effect of a nuclear spin transition on the ESR is detected, Feher [3] was able to map out the wave function of donor-bound electrons in Si: the hyperfine splitting caused by a particular host atom with nuclear spin within the Bohr radius of the donor bound electron is proportional to the probability of finding the electron at this site, $\psi^2(r_l)$, where ψ is the wave function of the bound electron and r_l a vector pointing from the donor nucleus to the ^{29}Si site. The hyperfine interaction with the nucleus at r_l thus depends on the orientation of the vector r_l with respect to the external static magnetic field. Analysis of the NMR induced spectra allows thus to determine $\psi^2(r_l)$.

Feher was able to show in this way that $\psi^2(r_l)$ does not fall off monotonically with increasing r_l as one might expect for s states. Due to the six-fold degeneracy of the Si conduction band minimum, the donor wave function is a linear combination of six terms related to the Bloch waves at the six minima, multiplied by hydrogen-like envelope functions. For finite r_l, the six oscillatory terms interfere and cause additional maxima of $\psi^2(r_l)$. In addition, he obtained also a value for the position of the conduction band minima which occur at 85% of the radius of the Brillouin zone in good agreement with band structure calculations [3].

In this most influential paper, Feher also analyzed the observed ESR line shape and thus the spin coherence time. In the case of dilute, noninteracting donors in pure Si, the line shape results from the actual distribution of the host atoms with nuclear spins within the orbit of the bound electron and their spin orientation. The observed Gaussian ESR line shape with a linewidth of 2.3 G for the ESR of Sb, P, As donors in Si at 1.25 K could be explained by summing up the contributions of hyperfine coupling to all the ^{29}Si host atoms situated within reach of the donor electron, identified before by the ENDOR experiments [3].

Recently, in the context of developing concepts for quantum computers, interest in the spin properties of shallow donors has revived. Feher had obtained a spin relaxation time of $T_1 = 3\,000$ s at 1.25 K at 0.32 T, falling off rapidly with temperature [58]. In view of the inhomogeneous broadening due to hyperfine interaction, it appeared natural to apply the spin–echo techniques to distinguish homogeneous and inhomogeneous contributions to spin dephasing of the donors. Tyryshkin et al. investigated isotopically enriched ^{28}Si, in which the concentration of ^{29}Si was less than 0.1%. In that material, they found a donor spin coherence time of 60 ms which they consider as *the intrinsic decoherence time for an isolated phosphorus donor bound electron in Si* at 6.9 K. They seem to be the *longest electron spin coherence times ever observed* [7, 60].

Summarizing, for strongly confined electrons the decoherence rate of an ensemble of electron spins is dominated by the hyperfine coupling to nuclear spins. The dynamics of the relaxation is additionally complicated due to spin diffusion, which results in non-exponential relaxation processes. A smart application of the spin–echo technique by Tyryshkin et al. [7, 60] allows to distinguish decoherence of the ensemble of donors in the presence of spin diffusion from the transverse spin relaxation of a single donor. The main contribution to the decoherence rate originates from the variety of configurations of nuclear spins. The decoherence rate of individual spins is much smaller, determined by fluctuations of nuclear spins [61] and by the longitudinal relaxation mechanisms. This fact stimulated Stoneham et al. to propose a Si based quantum computer making use of deep donor qubits which can be spectroscopically distinguished due to their surrounding [62].

Using spin–echo methods, the effect of dephasing due to hyperfine interaction can be repaired as long as the time required to invert the spin orientation (by a so-called π-pulse) is short as compared to the spin coherence time. In addition, the principle of Stark tuning of the ESR of ^{121}Sb donors as a method for selective spin manipulation has been demonstrated recently [63].

Longitudinal Spin Relaxation in Donors

In contrast to the Dyakonov–Perel relaxation in metallic layers, which is almost temperature independent (apart from the effect of a temperature dependent scattering rate) down to very low temperature ($k_B T \ll \mu B$), the longitudinal rate of the donor spins is temperature activated in a wide range [7, 60]. For donors in Si the activation energy corresponds to the valley splitting of the donor ground state. The longitudinal relaxation rate $1/T_1$ is thus attributed to the well-known Orbach process [64] in which thermal excitation to real excited states stimulate spin flip processes due to spin–orbit interaction. In the high temperature range ($T > 10$ K), for ^{28}Si:P, the longitudinal relaxation rate therefore exceeds the transverse one caused by the hyperfine coupling. The spin coherence becomes lifetime-limited which constitutes the real limitation for spin coherence in donors in Si.

7.6.2 From the Two-dimensional Electron Gas to Quantum Dots

The spin relaxation mechanisms in two-dimensional layers and in a donor are completely different. In the two-dimensional limit at low temperature the longitudinal and transverse relaxation of the spin ensemble are dominated by Elliott–Yafet or by Dyakonov–Perel relaxation, where the spin–orbit field is modulated by momentum scattering and in the latter case also by the cyclotron motion. In the other extreme, for a donor bound spin, decoherence is caused by hyperfine coupling, while the (longitudinal) spin–flip rate is governed by thermal excitations from the ground to excited state within the ground state multiplet caused by valley splitting.

We expect thus a qualitative change in the mechanism of spin relaxation when the effective confinement radius, r, becomes shorter than the electron mean free path, λ_p. For an electron mobility of 10^5 cm^2/Vs in Si and an electron concentration $n_s = 5 \times 10^{11}$ cm^{-2} the mean free path is $\lambda_p \cong 0.8\,\mu$m. For quantum dots of smaller size, $r < \lambda_p$, the modulation frequency in the Dyakonov–Perel mechanism is determined rather by the confinement frequency, but not by the momentum relaxation rate. In this stage, scattering at the dot surface dominates momentum scattering imposing a new contribution to the modulation of the spin–orbit field. As a consequence, the Dyakonov–Perel spin relaxation rate is reduced and other mechanisms become dominant.

Finally, when the states become localized to an extent such that sharp levels exist which are separated by more than $k_B T$, spin–orbit effects are suppressed [55]. In the quantum mechanical description, the continuous density of state, characteristic for two-dimensional systems becomes quantized, with an energy splitting of $\hbar\omega_{\mathrm{conf}} \cong 4\hbar^2\pi^2/r^2$. The confinement frequency ω_{conf} is inversely proportional to the square of the quantum dot diameter, and thus the Dyakonov–Perel rate is strongly reduced in small quantum dots. For example, for $r = 20$ nm the reduction amounts to 3 orders of magnitude.

7.6.3 Spin Relaxation and Dephasing in Si Quantum Dots

Extremely long spin lifetimes (>ms) [56] but short dephasing times (10 ns) [57] have been already found in III–V quantum dots. For Si the situation is less favorable to start with: the higher effective mass requires smaller structures which are difficult to make. Only recently gate defined single electron transistor structures and double dots made of Si were reported in the literature [65, 66] whereas in III–V compounds they have been used for many years already.

Another approach to achieve localization of electrons in Si is based on the self-organized or the seeded growth of Ge dots on Si [67, 68]. Due to the large lattice mismatch of Si and Ge, the latter shows two-dimensional growth only for 2–3 monolayers and switches spontaneously to the 3D Stranski–Krastanow growth mode. These islands are under compressive strain and cause dilatational strain in the adjacent Si layer. Dilatation in Si causes a lowering of the conduction band and thus localization of electrons. The strain distribution is essential for the splitting of the sixfold degenerate conduction band states. Usually the lowest state derives from the doubly degenerate Δ_z valley which is oriented in the growth direction.

In order to reduce the fluctuations in size and location of self-organized grown dots, the method of seeded growth was developed in which prestructured surfaces are used to obtain controllable nucleation of quantum dots. Using various nanolithography techniques, regular arrays of pits can be produced on a Si (100) surface. Typically, the period is 200 nm, the pit diameter less than 100 nm [67, 68].

Both in self-organized dot samples and seeded growth samples spin resonance has been observed: a single narrow line is seen during illumination with sub-bandgap light [69]. Typically the ESR linewidth is of the order of a few tenths of a Gauss, one order of magnitude smaller than that of shallow donors, but one order of magnitude wider than that of the two-dimensional electrons. Surprisingly there was no big difference between self-organized and seeded samples in spite of a quite substantial improvement in the size distribution of the latter ones.

Spin–echo experiments were performed in order to measure the spin lifetime and coherence in these quantum dots [69]. As a result, we saw practically no improvement in the spin lifetime as compared to two-dimensional electrons. Probably the presence on the holes in Ge islands in the direct neighborhood of the Si quantum dots opens a fast spin relaxation channel, similar to the Bir–Aronov–Pikus mechanism [33, 70]. This mechanism is known to be very effective because the strong spin–orbit interaction mixes holes states leading to strong spin lattice relaxation of holes. The electron–hole coupling is evident from the effective photoluminescence of electron–hole pairs seen in the discussed type of quantum dots.

The concept of electron spin relaxation via coupling to holes is supported by the angular dependence of the spin relaxation rate. In the Dyakonov–Perel relaxation the cyclotron motion, which appears for perpendicular magnetic field, leads to a faster modulation and thus to narrowing of the resonance linewidth (see Sect. 7.3.2). For the Si quantum dots, however, the linewidth and the relaxation rate are higher for perpendicular than for in-plane orientation of the applied magnetic field. This opposite angular dependence of the linewidth could be accounted for by noting that the perpendicular magnetic field leads to an additional localization of states amplifying electron–hole coupling. The main relaxation effect is expected to occur for hole states in the Ge quantum dots which are less localized than the electron states in Si and comparable with the magnetic length.

In these experiments we obtained a spin coherence time of about 300 ns, much longer than the coherence time observed for III–V quantum dots (10 ns) but shorter than in Si quantum wells. We expect a shorter spin coherence time because of the hyperfine interaction of the localized electrons with the ^{29}Si nuclei within the dot. A detailed modeling of the effect of hyperfine interaction on spin coherence, as was done by Feher for the shallow donors, is hardly possible here since these quantum dots show a wide variation in size and shape in addition to the statistical arrangement of ^{29}Si nuclei within the dot.

At present there is also no way to model the wave functions of the confined electrons reliably. Therefore we adopted a most simple model to estimate the hyperfine broadening ΔB_{hf}. We assume that the rms value of the hyperfine field due to ^{29}Si scales with the inverse volume V of the quantum dot and that each ^{29}Si contributes in the same way to the hyperfine broadening. In this way we obtain a broadening of: $\Delta B_{hf} = c_{hf}(a_{29}/V)^{1/2}$, where c_{hf} is a phenomenological constant describing the strength of the hyperfine interaction which can be estimated by comparison to the effect for a shallow donor, and a_{29} stands for the abundance of nuclear spin isotopes. We obtain thus a broadening of 0.1 G for a dot with a diameter of 60 nm and a height of 5 nm which is the order of magnitude for the size of our dots, and the value obtained is comparable also to the observed linewidth.

7.7 Conclusions

Silicon definitely is a candidate for a material with long spin lifetime and spin coherence. The spin coherence time of donors in isotopically purified ^{28}Si was found to be at least 60 ms which seems to be the longest ever seen. In a modulation doped two-dimensional electron gas with high electron mobility there is ample evidence for the dominating effect of the Bychkov–Rashba spin–orbit field which limits both the spin lifetime and coherence in spite of the weak spin–orbit interaction. The Bychkov–Rashba field also enables interaction of currents and spins which manifests itself in a shift of the ESR, when a dc current is applied to the two-dimensional electron gas. This effect could be used to tune the resonance to a given frequency. It is also responsible for a microwave current-induced contribution to the ESR which can be by orders of magnitude more efficient than the usual magnetic dipole transition.

First experiments on the spin coherence in laterally confined Si structures did not show the anticipated prolongation of the spin lifetime—for strain induced quantum dots it is still on the order of 0.3 µs or less. We attribute this to the fact that in these undoped structures electron–hole pairs were generated by optical interband excitation and the resulting electron–hole pair spin interaction reduces the spin lifetime. Residual effects of spin–orbit interaction are unlikely since the linewidth anisotropy shows the opposite behavior as compared to two-dimensional layers.

It is also not simple to estimate the spin–flip relaxation rate in Si quantum dots without this effect. By analogy to the longitudinal relaxation of donor electron spins, spin relaxation is caused by thermal excitation within the valley multiplet. For strained two-dimensional Si layers the six-fold degeneracy of the bulk Si (and donors in bulk Si) is reduced to a two-fold valley degeneracy. The remaining degeneracy could be a serious problem in the construction of a quantum computer but the degeneracy can be lifted by a lateral confinement. Recent experimental data show that atomic steps cause a splitting of the ground doublet. Modification of the potential by a point contact leads to a splitting in Si/SiGe structures by up to 1.5 meV, corresponding to an activation temperature of 17 K [8]. A higher valley splitting can in principle be obtained by a construction of a low symmetry quantum dot.

Acknowledgements. It is a pleasure to thank Hans Malissa for his continuous help and years of fruitful collaboration. Work supported by the "Fonds zur Förderung der Wissenschaftlichen Forschung", Projects P-16631 and N1103, and ÖAD and GMe, all Vienna and in Poland by a MNiSW Project granted for the years 2007–2010.

References

[1] L. Liu, Phys. Rev. Lett. **6**, 683 (1961)
[2] Y. Yafet, in *g Factors and Spin-Lattice Relaxation of Conduction Electrons*, ed. by F. Seitz, D. Turnbull. Solid State Physics, vol. 1414 (Academic, San Diego, 1963)
[3] G. Feher, Phys. Rev. **114**, 1219 (1959)
[4] B.E. Kane, Nature **393**, 133 (1998)
[5] R. Vrijen, E. Yablonovitch, K. Wang, H.W. Jiang, A. Balandin, V. Roychowdhury, T. Mor, D. DiVincenzo, Phys. Rev. A **62**, 012306 (2000)
[6] D. Loss, D.P. DiVincenzo, Phys. Rev. A **57**, 120 (1998)
[7] A.M. Tyryshkin, S.A. Lyon, A.V. Astashkin, A.M. Raitsimring, Phys. Rev. B **68**, 193207 (2003)
[8] S. Goswami, K.A. Slinker, M. Friesen, L.M. Mcguire, J.L. Truitt, C. Tahan, L.J. Klein, J.O. Chu, P.M. Mooney, D.W. Van Der Weide, R. Joynt, S.N. Coppersmith, M.A. Eriksson, Nat. Phys. **3**, 41 (2007)
[9] G.W. Ludwig, H.H. Woodbury, in *Electron Spin Resonance in Semiconductors*, ed. by F. Seitz, D. Turnbull. Solid State Physics, vol. 13 (Academic, San Diego, 1992), p. 223
[10] H.D. Fair Jr., R.D. Ewing, F.E. Williams, Phys. Rev. Lett. **15**, 355 (1965)
[11] B.C. Cavenett, Adv. Phys. **30**, 475–538 (1981)
[12] N. Nestle, G. Denninger, M. Vidal, C. Weinzierl, K. Brunner, K. Eberl, K. von Klitzing, Phys. Rev. B **56**, R4359 (1997)
[13] C.F.O. Graeff, M.S. Brandt, M. Stutzmann, M. Holzmann, G. Abstreiter, F. Schäffler, Phys. Rev. B **59**, 13242 (1999)
[14] W. Jantsch, Z. Wilamowski, N. Sandersfeld, F. Schäffler, Phys. Stat. Sol. (b) **210**, 643 (1998)
[15] J. Matsunami, M. Ooya, T. Okamoto, Phys. Rev. Lett. **97**, 066602 (2006)
[16] D.R. McCamey, H. Huebl, M.S. Brandt, W.D. Hutchison, J.C. McCallum, R.G. Clark, A.R. Hamilton, Appl. Phys. Lett. **89**, 182115 (2006)
[17] E.I. Rashba, V.I. Sheka, in *Landau Level Spectroscopy*, ed. by G. Landwehr, E.I. Rashba (Elsevier Science, Amsterdam, 1991)

[18] M. Schulte, J.G.S. Lok, G. Denninger, W. Dietsche, Phys. Rev. Lett. **94**, 137601 (2005)
[19] A. Wolos, W. Jantsch, K. Dybko, Z. Wilamowski, C. Skierbiszewski, AIP Conf. Proc. **893**, 1313 (2007)
[20] A. Wolos, W. Jantsch, K. Dybko, Z. Wilamowski, C. Skierbiszewski, Phys. Rev. B **76**, 045301 (2007)
[21] Z. Wilamowski, W. Jantsch, H. Malissa, U. Rössler, Phys. Rev. B **66**, 195315 (2002)
[22] Z. Wilamowski, N. Sandersfeld, W. Jantsch, D. Többen, F. Schäffler, Phys. Rev. Lett. **87**, 026401 (2001)
[23] A. Schweiger, G. Jeschke, *Principles of Pulse Electron Paramagnetic Resonance* (Oxford University Press, London, 2001)
[24] A.M. Tyryshkin, S.A. Lyon, W. Jantsch, F. Schäffler, Phys. Rev. Lett. **94**, 126802 (2005)
[25] E.I. Rashba, Fiz. Tverd. Tela (Leningrad) **2**, 1224 (1960); Sov. Phys. Solid State **2**, 1109 (1960)
[26] Y.A. Bychkov, E.I. Rashba, J. Phys. C **17**, 6039 (1984)
[27] G. Dresselhaus, Phys. Rev. **100**, 580 (1955)
[28] Z. Wilamowski, W. Jantsch, Phys. Rev. B **69**, 035328 (2004)
[29] M.M. Glazov, E.L. Ivchenko, JETP **99**, 1279 (2004)
[30] P. Vogl, J.A. Majewski, in *Institute of Physics Conference Series*, vol. 171, ed. by A.R. Long, J.H. Davies (Institute of Physics, Bristol, 2003), P3.05, ISBN7503-0924-5
[31] F. Schäffler, Semicond. Sci. Technol. **12**, 1515 (1997)
[32] R.J. Elliott, Phys. Rev. **96**, 266 (1954)
[33] G.L. Bir, A.G. Aronov, G.E. Pikus, Sov. Phys. JETP **42**, 705 (1976)
[34] M.I. Dyakonov, V.I. Perel, Sov. Phys. Solid State **13**, 3023 (1972)
[35] R.M. White, *Quantum Theory of Magnetism* (McGraw–Hill, New York, 1970)
[36] A. Abragam, *The Principles of Nuclear Magnetism* (Clarendon Press, Glasgow, 1961)
[37] E.L. Ivchenko, Sov. Phys. Solid State **15**, 1048 (1973)
[38] M. Glazov, Phys. Rev. B **70**, 195314 (2004)
[39] R. Freedman, D.R. Fredkin, Phys. Rev. B **11**, 4847 (1975)
[40] Z. Wilamowski, H. Malissa, F. Schäffler, W. Jantsch, Phys. Rev. Lett. **98**, 187203 (2007)
[41] W. Ungier, W. Jantsch, Z. Wilamowski, Acta Phys. Pol. A **112**(2), 345 (2007)
[42] Z. Wilamowski, W. Ungier, W. Jantsch, to be published
[43] A. Abragam, B. Bleaney, *Electron Paramagnetic Resonance of Transition Ions* (Clarendon Press, Oxford, 1970)
[44] C.P. Poole, *Electron Spin Resonance*, 2nd edn. (Wiley, New York, 1983)
[45] F.G. Pikus, G.E. Pikus, Phys. Rev. B **51**, 16928 (1995)
[46] C. Tahan, R. Joynt, Phys. Rev. B **71**, 07315 (2005)
[47] V.K. Kalevich, V.L. Korenev, JETP Lett. **52**, 230 (1990)
[48] E.I. Rashba, A.L. Efros, Appl. Phys. Lett. **83**, 5295 (2003)
[49] A.L. Efros, E.I. Rashba, Phys. Rev. B **73**, 165325 (2006)
[50] M. Dobrowolska, Y.F. Chen, J.K. Furdyna, S. Rodriguez, Phys. Rev. Lett. **51**, 134 (1983)
[51] M. Duckheim, D. Loss, Nat. Phys. **2**, 195 (2006)
[52] G. Lampel, PhD Thesis, University of Orsay (1968)
[53] G. Feher, A.F. Kip, F.J. Dyson, Phys. Rev. **98**, 337 (1955)
[54] Z. Wilamowski, W. Jantsch, Physica E **10**, 17 (2001)
[55] A.V. Khaetskii, Y.V. Nazarov, Phys. Rev. B **61**, 12639 (2000)
[56] M. Kroutvar, Y. Ducommun, D. Heiss, M. Bichler, D. Schuh, G. Abstreiter, J.J. Finley, Nature **432**, 81 (2004)
[57] J.R. Petta, A.C. Johnson, J.M. Taylor, E.A. Laird, A. Yacoby, M.D. Lukin, C.M. Marcus, M.P. Hanson, A.C. Gossard, Science **309**, 2180 (2005)
[58] G. Feher, E.A. Gere, Phys. Rev. **114**, 1245 (1959)

[59] W. Kohn, J.M. Luttinger, Phys. Rev. **97**, 1721 (1955)
[60] A.M. Tyryshkin et al., Physica E **35**, 257 (2006)
[61] J.P. Gordon, K.D. Bowers, Phys. Rev. Lett. **1**, 368 (1958)
[62] A.M. Stoneham, A.J. Fisher, P.T. Greenland, J. Phys. Condens. Matter **15**, L447 (2003)
[63] F.R. Bradbury, A.M. Tyryshkin, G. Sabouret, J. Bokor, T. Schenkel, S.A. Lyon, Phys. Rev. Lett. **97**, 176404 (2006)
[64] R. Orbach, Proc. Phys. Soc. Lond. **77**, 821 (1961)
[65] D.S. Gandolfo, D.A. Williams, H. Qin, J. Appl. Phys. **101**, 013701 (2007)
[66] S.J. Shin et al., Appl. Phys. Lett. **91**, 053114 (2007)
[67] Z. Zhong, G. Chen, J. Stangl, T. Fromherz, F. Schäffler, G. Bauer, Physica E **21**, 588 (2004)
[68] G. Chen, H. Lichtenberger, G. Bauer, W. Jantsch, F. Schäffler, Phys. Rev. B **74**, 035302 (2006)
[69] H. Malissa, W. Jantsch, G. Chen, H. Lichtenberger, T. Fromherz, F. Schäffler, G. Bauer, A. Tyryshkin, S. Lyon, Z. Wilamowski, *Proceedings of the Material Research Society Fall Meeting 2006*. Boston and AIP Conf. Proc., vol. 893 (2007), p. 1317
[70] G. Lampel, Phys. Rev. Lett. **20**, 491 (1968)

8

Spin Hall Effect

M.I. Dyakonov and A.V. Khaetskii

Since the electrons have an internal degree of freedom, spin, they are characterized not only by charge density and electric current, but also by spin density and spin current. The spin current is described by a tensor q_{ij}, where the first index indicates the direction of flow, while the second one says which component of the spin is flowing. Thus, if all electrons with concentration n are completely spin-polarized along z and move with a velocity v in the x direction, the only non-zero component of q_{ij} is $q_{xz} = nv$.[1]

Both the charge current and the spin current change sign under space inversion (because spin is a pseudovector). In contrast, they behave differently with respect to time inversion: while the electric current changes sign, the spin current does not (because spin, like velocity, changes sign under time inversion).

This chapter is devoted to transport phenomena, predicted a long time ago [1, 2] and originating from the mutual transformation of the charge and spin currents due to spin–orbit interaction. Recently, this has become a subject of considerable interest and intense research, both experimental and theoretical.

8.1 Background: Magnetotransport in Molecular Gases

Like much of the spin physics in semiconductors, the subject of this chapter has its roots in atomic physics. Atoms and molecules also have internal degrees of freedom, their orbital and/or spin angular momentum, which can have an influence (albeit a rather weak one) on transport properties of gases. Historically, the first phenomenon of this kind was the *Senftleben effect*. In 1930, Senftleben discovered [3, 4] that the

---

[1] Since $s = 1/2$, it might be more natural to define the spin current density for this case as $(1/2)nv$. We find it more convenient to omit $1/2$, because this allows to avoid numerous factors $1/2$ and 2 in other places. It would be more correct to describe our definition of q_{ij} as the spin *polarization* current density tensor. Below we will use the shorthand "spin current".

electrical resistance of a thin platinum wire, immersed in oxygen gas, slightly (less than 1%) increases when a weak magnetic field is applied. The effect was not seen for other gases. Senftleben understood that the observed resistance increase is due to the magnetic field induced *decrease* of the thermal conductivity of oxygen.[2]

The influence of the magnetic field on the transport properties of oxygen was explained by Gorter [5]. The O_2 molecule is one of only two diatomic molecules (the other one is NO), that are diamagnetic: in its ground state it has a non-compensated total electron spin, equal to 1, as well as an associated magnetic moment. Because of the spin–orbit interaction, this spin in the ground state is directed along the orbital momentum related to the rotation of the molecule.

Qualitatively, one might say that there is a mixture of two species with angular momenta parallel and perpendicular to the velocity. The important point is that the transport cross section depends on the orientation of the axis of rotation with respect to the translational velocity, the collisions being more effective when the angular momentum and the velocity are aligned. As a result of this asymmetry, the two species have thermal resistances that are slightly different, and the total resistance will be *less* than it would be in the absence of asymmetry.[3]

The role of the magnetic field is to produce a precession of the rotation axis of the molecule (or its angular momentum) with a frequency Ω, depending on the magnetic moment of the molecule. If the magnetic field is strong enough for the angular momentum to make many rotations during the time τ_p between collisions, $\Omega \tau_p \gg 1$, then the effect of asymmetry in scattering will be averaged out. As a result, the two species become identical, and the thermal resistivity will increase. Since τ_p is inversely proportional to the gas pressure, P, this explains the experimental observation that the effect scales as B/P.

More recently, Gorelik [6] experimentally observed that the thermal conductivity of oxygen shows spectacular irregular oscillations as a function of the magnetic field. The oscillation amplitude is very small, on the order of 10^{-5}, but well reproducible. An explanation proposed in [7] relates these oscillations to small changes in the transport cross section which occur at numerous level crossing in the rotational spectrum in the presence of the spin–orbit interaction. It is a pity, that those experiments were never reproduced.

Thus, there is a variety of transport phenomena in molecular gases, related to the existence of internal degrees of freedom, which can be influenced by the application of a magnetic field.

---

[2] The resistance of a metal increases with temperature, while the temperature depends on the balance between the Joule heating and the cooling via thermal conductivity of the surrounding media. Thus, a decrease of the thermal conductivity will result in an increase of the wire temperature, and hence of its resistance.

[3] Consider two parallel resistances $R + \Delta R$ and $R - \Delta R$. The total resistance will be *smaller* by an amount $\sim (\Delta R/R)^2$ compared to the case when both resistances are equal to R.

8.2 Phenomenology (with Inversion Symmetry)

We will now discuss, from pure symmetry considerations, what phenomena of spin-charge coupling are, in principle, possible. For the moment, we restrict ourselves to an isotropic media with inversion symmetry. This does not mean that the results obtained below are not valid when inversion symmetry is absent. Rather, it means that we will not take into account additional specific effects, which are due entirely to the lack of inversion symmetry.

It will be shown, that the phenomenological approach allows to describe a number of interesting physical effects by introducing a single dimensionless parameter.

8.2.1 Preliminaries

To begin, consider spin-up and spin-down (with respect to the z axis) electrons and suppose that we force our electrons to flow in the direction x. Let $q_x^\pm$ be the corresponding flow densities, which are not necessarily equal.

The crucial point is that because of the spin–orbit interaction these currents will induce currents of *opposite signs* for the two spin species in the y-direction[4]:

$$q_y^\pm = \mp\gamma q_x^\pm, \tag{8.1}$$

where γ is a dimensionless parameter proportional to the strength of the spin–orbit interaction. We assume that $|\gamma|$ is small, the sign of γ is a priori unknown. Note, that under time inversion we have: $q^\pm \to -q^\mp$. Consequently, γ *changes sign under time inversion.*

We now introduce the total (charge) flow density $q = q^+ + q^-$, and the spin current $q_{iz} = q_i^+ - q_i^-$. It follows from (8.1) that

$$q_y = -\gamma q_{xz}, \qquad q_{yz} = -\gamma q_x. \tag{8.2}$$

These equations demonstrate the mutual transformations of spin and charge currents.

8.2.2 Spin and Charge Current Coupling

More accurately, the transport phenomena related to coupling of the spin and charge currents can be described phenomenologically in the following simple way [8]. We introduce the charge and spin currents, $q^{(0)}$ and $q_{ij}^{(0)}$, which would exist in the absence of a spin–orbit interaction:

$$q^{(0)} = -\mu n E - D\nabla n, \tag{8.3}$$

$$q_{ij}^{(0)} = -\mu E_i P_j - D\frac{\partial P_j}{\partial x_i}, \tag{8.4}$$

---

[4] This is reminiscent of the *Magnus effect*: a spinning tennis ball deviates from its straight path in air in a direction depending on the sign of its rotation. From the point of view of symmetry, this effect is described by (8.1).

where μ and D are the mobility and the diffusion coefficient, connected by the Einstein relation, $\boldsymbol{P}$ is the vector of spin polarization density.[5]

Equation (8.3) is the standard drift–diffusion expression for the electron flow. Equation (8.4) describes the spin current of polarized electrons, which may exist even in the absence of a spin–orbit interaction simply because spins are carried by the electron flow. We ignore possible dependence of mobility on spin polarization, which is assumed to be small.

If there are other sources for currents, like, for example, a temperature gradient, the corresponding terms should be included in (8.3) and (8.4).

A spin–orbit interaction couples the two currents and gives corrections to the values $\boldsymbol{q}^{(0)}$ and $q_{ij}^{(0)}$. For an isotropic material with inversion symmetry, the only possibility is[6]

$$q_i = q_i^{(0)} + \gamma \epsilon_{ijk} q_{jk}^{(0)}, \tag{8.5}$$

$$q_{ij} = q_{ij}^{(0)} - \gamma \epsilon_{ijk} q_k^{(0)}, \tag{8.6}$$

where q_i and q_{ij} are the corrected currents, ϵ_{ijk} is the unit antisymmetric tensor[7] and γ is the small dimensionless parameter introduced above. The difference in signs in (8.5) and (8.6) is consistent with the Onsager relations and is due to the different properties of charge and spin currents with respect to time inversion. One can check that (8.2) follows from these equations.

8.2.3 Phenomenological Equations

Explicit phenomenological expressions for the two currents follow from (8.3)–(8.6) (the electric current density $\boldsymbol{j}$ is related to $\boldsymbol{q}$ by $\boldsymbol{j} = -e\boldsymbol{q}$):

$$\boldsymbol{j}/e = \mu n \boldsymbol{E} + D \nabla n + \beta \boldsymbol{E} \times \boldsymbol{P} + \delta \operatorname{curl} \boldsymbol{P}, \tag{8.7}$$

$$q_{ij} = -\mu E_i P_j - D \frac{\partial P_j}{\partial x_i} + \epsilon_{ijk} \left(\beta n E_k + \delta \frac{\partial n}{\partial x_k} \right). \tag{8.8}$$

Here

$$\beta = \gamma \mu, \qquad \delta = \gamma D, \tag{8.9}$$

so that the coefficients β and δ, similar to μ and D, satisfy the Einstein relation. However, since γ changes sign under time inversion (Sect. 8.2.1), β and δ are *non-dissipative* kinetic coefficients, unlike μ and D. Equations (8.7) and (8.8) should be complemented by the continuity equation for the vector of spin polarization:

$$\frac{\partial P_j}{\partial t} + \frac{\partial q_{ij}}{\partial x_i} + \frac{P_j}{\tau_s} = 0, \tag{8.10}$$

where τ_s is the spin relaxation time.

---

[5] It is convenient to use this quantity, instead of the normal spin density $S = \boldsymbol{P}/2$, see footnote 1.

[6] Not exactly, see Sect. 8.2.5 below.

[7] This tensor is defined by: $\epsilon_{xyz} = \epsilon_{zxy} = \epsilon_{yzx} = -\epsilon_{yxz} = -\epsilon_{zyx} = -\epsilon_{xzy} = 1$.

While (8.7)–(8.10) are written for a three-dimensional sample, they are equally applicable to the 2D case, with obvious modifications: the electric field, space gradients, and all currents (but not the spin polarization vector) should have components in the 2D plane only.

Another remark concerns the equilibrium situation when obviously all currents should vanish. If an inhomogeneous magnetic field $B(r)$ exists, the equilibrium spin polarization will be space dependent, however this by itself should produce neither spin, nor charge currents. To assure this, an additional counter-term should be introduced into the right-hand side of (8.4), proportional to $\partial B_j / \partial x_i$, which takes care of the force acting on the electron with a given spin in an inhomogeneous magnetic field. Corresponding terms will appear in (8.7) and (8.8). We ignore these terms assuming that, if present, the magnetic field is homogeneous.

8.2.4 Physical Consequences of Spin–Charge Coupling

Equations (8.7)–(8.10), which appeared for the first time in [1, 2], describe all the physical consequences of spin–charge current coupling. The effects of spin–orbit interaction are contained in the additional terms with the coefficients β and δ.

Anomalous Hall Effect

The term $\beta E \times P$ in (8.7) describes the anomalous Hall effect, which was first observed in ferromagnets by Hall himself [9, 10]. The measured Hall voltage contains a part, which is proportional to magnetization, but cannot be explained as being due to the magnetic field produced by magnetization (it is much greater than that, especially at elevated temperatures). It took 70 years to understand [11–14] that the anomalous Hall effect is due to a spin–orbit interaction.

The anomalous Hall effect can also be seen in nonmagnetic semiconductors, where the spin polarization is created by application of a magnetic field. The spin-related anomalous effect can be separated from the much larger ordinary Hall effect by magnetic resonance of the conduction electrons, which results in a resonant change of the Hall voltage [15]. Non-equilibrium spin polarization, produced either by optical means or by spin injection, should also result in an anomalous Hall voltage. Such an experiment was recently done by Miah [16] with GaAs illuminated by circularly polarized light.

Electric Current Induced by curl P

The term δ curl P in (8.7) describes an electrical current induced by an inhomogeneous spin density (now referred to as the Inverse Spin Hall Effect). A way to measure this current under the conditions of optical spin orientation was proposed in [17]. The circularly polarized exciting light is absorbed in a thin layer near the surface of the sample. As a consequence, the photo-created electron spin density is inhomogeneous, however curl $P = 0$, since P is perpendicular to the surface and it varies in the same direction. By applying a magnetic field parallel to the surface,

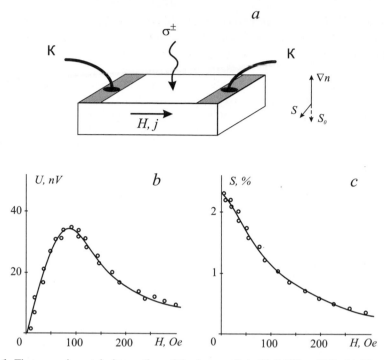

Fig. 8.1. First experimental observation of the inverse Spin Hall Effect [18]. (**a**) The experimental setup. (**b**) Voltage measured between the contacts K as a function of magnetic field. (**c**) Measured degree of circular polarization of luminescence, equal to the normal component of the average electron spin, as a function of magnetic field. The solid line in (**b**) is calculated using the results in (**c**)

one can create a parallel component of P, thus inducing a non-zero curl P and the corresponding surface electric current (or voltage).

This effect was found by Bakun et al. [18], thus providing the first experimental observation of the Inverse Spin Hall Effect, see Fig. 8.1. In a later publication by Tkachuk et al. [19] observed very clear manifestations of the nuclear magnetic resonance in the surface current induced by curl P.

Current-Induced Spin Accumulation, or Spin Hall Effect

The term $\beta n \epsilon_{ijk} E_k$ (and its diffusive counterpart $\delta \epsilon_{ijk} \partial n / \partial x_k$) in (8.8), describes what is now called (after Hirsch [20]) the Spin Hall Effect: an electrical current induces a transverse spin current, resulting in spin accumulation near the sample boundaries [1, 2]. This phenomenon was observed experimentally only in recent years [21, 22] and has attracted widespread interest. Spin accumulation can be seen by solving (8.10) in the steady state ($\partial P_j / \partial t = 0$) and using (8.8) for the spin current. Since the spin polarization will be proportional to the electric field, terms $E P$ can be neglected. Also, the electron concentration should be considered uniform.

We take the electric field along the x axis and look at what happens near the boundary $y = 0$ of a wide sample situated at $y > 0$ (when the sample size is greater than the spin diffusion length, spin accumulation near the other boundary can be considered independently). The boundary condition obviously should correspond to a vanishing of the normal to the boundary component of the spin current, $q_{yj} = 0$.

The solution of the diffusion equation

$$D\frac{d^2 P}{dy^2} = \frac{P}{\tau_s} \tag{8.11}$$

with the boundary conditions at $y = 0$, following from (8.8),

$$\frac{dP_x}{dy} = 0, \qquad \frac{dP_y}{dy} = 0, \qquad \frac{dP_z}{dy} = \frac{\beta n E}{D}, \tag{8.12}$$

gives the result:

$$P_z(y) = P_z(0) \exp\left(-\frac{y}{L_s}\right), \qquad P_z(0) = -\frac{\beta n E L_s}{D}, \qquad P_x = P_y = 0, \tag{8.13}$$

where $L_s = \sqrt{D\tau_s}$ is the spin diffusion length.

Thus the current-induced spin accumulation exists in thin layers (the *spin layers*) near the sample boundaries. The width of the spin layer is given by the spin–diffusion length, L_s, which is typically on the order of 1 μm. The polarization within the spin layer is proportional to the driving current, and the signs of spin polarization at the opposing boundaries are opposite. For a cylindrical wire the spins wind around the surface.

It should be stressed that all these phenomena are closely related and have their common origin in the coupling between spin and charge currents given by (8.5) and (8.6). Any mechanism that produces the anomalous Hall effect will also lead to the spin Hall effect and vice versa.

It is remarkable that there is a single dimensionless parameter, γ, that governs the resulting physics. The calculation of this parameter should be the objective of a microscopic theory, see Sect. 8.4 below.

The Degree of Polarization in the Spin Layer

Using (8.13) and (8.9), the *degree* of polarization in the spin layer, $\mathcal{P} = P_z(0)/n$, can be rewritten as follows:

$$\mathcal{P} = -\gamma \frac{v_d}{v_F}\left(\frac{3\tau_s}{\tau_p}\right)^{1/2}, \tag{8.14}$$

where we have introduced the electron drift velocity $v_d = \mu E$ and used the conventional expression for the diffusion coefficient of degenerate 3D electrons $D = v_F^2 \tau_p/3$, v_F is the Fermi velocity, τ_p is the momentum relaxation time.[8]

---

[8] For 2D electrons, the factor $1/3$ should be replaced by $1/2$. If the electrons are not degenerate, v_F should be replaced by the thermal velocity.

In materials with inversion symmetry, like Si, where both the spin–charge coupling and spin relaxation via the Elliott–Yaffet mechanism are due to spin asymmetry in scattering by impurities, the strength of the spin–orbit interaction cancels out in (8.14), since $\tau_s \sim \gamma^{-2}$.

Thus the most optimistic estimate for the degree of polarization within the spin layer is $\mathcal{P} \sim v_d/v_F$ [1]. In semiconductors, this ratio may be, in principle, on the order of 1. In the absence of inversion symmetry, usually the Dyakonov–Perel mechanism makes the spin relaxation time considerably shorter, which is unfavorable for an appreciable spin accumulation.

8.2.5 Related Problems

Here we briefly discuss some more subtle points related to our subject.

The Validity of the Approach Based on the Diffusion Equation

The diffusion equation is valid, when the scale of spatial variation of concentration (in our case, of spin polarization density) is large compared to the mean free path $\ell = v_F \tau_p$. The variation of P occurs on the spin diffusion length, so the condition $L_s \gg \ell$ should be satisfied. Since $L_s/\ell \sim (\tau_s/\tau_p)^{1/2}$, this condition can be equivalently rewritten as $\tau_s \gg \tau_p$.

Thus, if the spin relaxation time becomes comparable to the momentum relaxation time (which is the case of the so-called "clean limit", when the spin band splitting is greater than $\hbar/\tau_p$, see Chap. 1, Sect. 1.4.2), the diffusion equation approach breaks down.

The diffusion equation still can be derived for spatial scales much greater than ℓ, but it will be of no help for the problem at hand, because neither this equation, nor the boundary conditions in (8.12) can any longer be used to study spin accumulation. Surface spin effects will occur on distances less than ℓ from the boundaries and will crucially depend on the properties of the interfaces (e.g., flat or rough interface, etc.).

How the Spin Current Should Be Defined

There was a discussion concerning the correct microscopic definition of spin currents, and the form of the boundary conditions, see [23]. Our point of view is the following. One should distinguish two situations: (1) the case when $\tau_s \gg \tau_p$ and (2) the case, when τ_s is comparable to τ_p.

In the first case, spin accumulation can be studied by the diffusion equation approach. Microscopically, one has to *derive* (8.10). The quantity entering the $\partial q_{ij}/\partial x_i$ term will be the true spin current, and it is *this* quantity, whose normal component should vanish at the boundary.

In the second case, the diffusion equation approach cannot be used (see above), because all the spin effects near the boundary occur on the spatial scale ℓ or even less.[9] To understand what happens near the boundaries, one must address the quantum-mechanical problem of electrons reflecting from the boundary in the presence of an electric field and spin–orbit interaction. Under such circumstances the definition of the bulk spin current (which cannot be measured directly), as well of the "Spin Hall conductivity", β, is immaterial, since it has no bearing on the possible spin polarization near the boundaries, see Sect. 8.4.3.

Apart from the boundary spin accumulation, there are also bulk effects of second order in the spin–orbit interaction (see Sect. 8.2.6). The "normally" defined spin current (as an anticommutator of the spin and velocity operators) shows up in such effects.

Additional Terms in (8.6)

In fact, symmetry considerations allow additional terms in one of our basic equations (8.5), (8.6). Namely, it is possible to complement the rhs of (8.6) for the spin current by additional terms of the type: $q_{ji}^{(0)}$ (note the transposition of the indices i and j!) and $\delta_{ij} q_{kk}^{(0)}$ (the sum over repeating indices is assumed).

These terms will result [1, 2] in corresponding additional terms in (8.8): $\mu E_j P_i + D \partial P_i / \partial x_j$ (note again the transposition of i and j with respect to (8.8)) and $\delta_{ij}(\mu \boldsymbol{E} \boldsymbol{P} + D \operatorname{div} \boldsymbol{P})$ with some new coefficients.

These additional terms are of no importance for all the effects considered so far. Their origin and their physical meaning will be discussed in Sect. 8.4.1.

8.2.6 Electrical Effects of Second Order in Spin–Orbit Interaction

An electrical current produces a spin current, which in turn gives an additional electric current. Thus there should be a correction to the sample resistance, which is second order in the coupling parameter γ. This effect *reduces* the bulk conductivity [24, 25], which can be seen from Fig. 8.2, illustrating the double scattering effect known in atomic physics [26]. Since there is no way to determine experimentally what is the uncorrected value of resistance, such effects can be revealed only by influencing the intermediate link, the spin current. This can be done by applying a magnetic field. Below we consider separately corrections to the bulk conductivity and surface effects related to spin accumulation.

---

[9] In the absence of inversion symmetry, spin relaxation is usually related to the spin band splitting. If the splitting at the Fermi level, $\hbar\Omega(\boldsymbol{p})$ is such that $\Omega(p)\tau_\mathrm{p} \ll 1$, then $\tau_\mathrm{s} \gg \tau_\mathrm{p}$. In the opposite case, spin relaxation goes through two stages (see Sect. 1.4.2). The first one has a duration $\sim 1/\Omega(p)$ and the second one is characterized by the time τ_p. Thus there are also *two* characteristic spatial scales: $v_\mathrm{F}/\Omega(p)$ and $v_\mathrm{F}\tau_\mathrm{p} = \ell$, and the first one is much smaller than the second one. Obviously, the physics on these scales can not be treated by the diffusion equation.

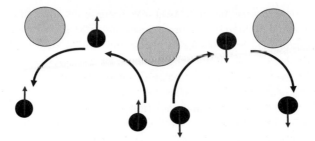

Fig. 8.2. Appearance of a negative correction to electric current due to double spin-asymmetric scattering

Bulk Effects

One can easily understand what happens by using the Drude-like equations taking into account the spin–charge current coupling [25]:

$$\frac{dq_i}{dt} = -\frac{en}{m}E_i - (q \times \omega_c)_i - \frac{q_i}{\tau_p} + \frac{\gamma}{\tau_p}\epsilon_{ijk}q_{jk}, \tag{8.15}$$

$$\frac{dq_{ij}}{dt} = \epsilon_{ikl}\omega_{ck}q_{lj} + \epsilon_{jkl}\Omega_k q_{il} - \frac{q_{ij}}{\tau_p} - \frac{\gamma}{\tau_p}\epsilon_{ijk}q_k. \tag{8.16}$$

Here $\omega_c = eB/mc$ and $\Omega = g\mu_B B/\hbar$, ω_c and Ω are the cyclotron and spin precession frequencies, respectively, n is the electron concentration. The two first terms in the right side of (8.16) describe the influence of the magnetic field on the spin current due to the rotation of the velocity (with frequency ω_c) and of the spin (with frequency Ω). The third term describes the spin current relaxation due to scattering. Strictly speaking, the spin relaxation rate should be also taken into account. We have ignored it, assuming that the spin–orbit interaction is weak and $\tau_s \gg \tau_p$.

In the steady state and in the absence of a magnetic field (8.15), (8.16) reduce to (8.5) and (8.6), and the conductivity is given by

$$\sigma = \sigma_0(1 - 2\gamma^2), \quad \sigma_0 = \frac{ne^2\tau_p}{m}, \tag{8.17}$$

where σ_0 is the normal Drude conductivity. In the magnetic field (directed along the z axis), the steady state solution of (8.15), (8.16) gives the following results [25] for the resistivity components:

$$\rho_{zz} = \rho_0\left(1 + \frac{2\gamma^2}{1 + (\omega_c - \Omega)^2\tau_p^2}\right), \tag{8.18}$$

$$\rho_{xx} = \rho_0\left[1 + \gamma^2\left(\frac{1}{1 + (\omega_c\tau_p)^2} + \frac{1}{1 + (\Omega\tau_p)^2}\right)\right], \tag{8.19}$$

$$\rho_{xy} = -\frac{B}{nec}\left[1 - \gamma^2\left(\frac{1}{1 + (\omega_c\tau_p)^2} + \frac{\Omega}{\omega_c}\frac{1}{1 + (\Omega\tau_p)^2}\right)\right], \tag{8.20}$$

where $\rho_0 = 1/\sigma_0$ is the Drude resistivity.

The most unusual result is the magnetic field dependence of the ρ_{zz} component. While q_z is not influenced directly by the Lorentz force, it is coupled to the spin current component q_{xy}. By changing the relative orientation of spin and velocity, the magnetic field destroys the spin current. As a result, the negative correction to the conductivity in (8.17) disappears. The scale of the relevant magnetic fields is defined by the classical condition $\omega_c \tau_p \sim 1$.

Note, that there are cases (Sect. 8.4.3) when the parameter γ is not necessarily small, so that the magnetic effects due to the spin–charge coupling can be really important. In particular, as seen from (8.20), the Hall constant may even change sign!

Surface Effects

These second order effects are related to the surface (or edge) spin accumulation and can be visible if the sample lateral width is comparable to or less than the spin diffusion length L_s. Now an additional correction to the electric current comes from the curl P term in (8.7).

While being a small correction, it hopefully can be measured because (i) the intermediate link, the spin density, can be influenced by application of a magnetic field, and (ii) the precision of electric measurements is far greater than that of optical measurements, usually used to reveal the spin polarization (see Sect. 8.5 below). One might say that these kinds of phenomena are a combination of the direct and inverse spin Hall effects, and the Hanle effect.

One of them was proposed in [8]. Near the boundary, the z component of spin polarization changes in the direction perpendicular to the sample boundary (the y direction). Thus curl $P \neq 0$, and according to (8.7), a correction to the electric current should exist within the spin layer. It turns out that this correction is always positive,[10] i.e., it leads to a slight decrease of the sample resistance. By applying a magnetic field in the xy plane, we can destroy the spin polarization and thus observe a positive magnetoresistance. For wide samples ($L \gg L_s$), the relevant field scale corresponds to $\Omega \tau_s \sim 1$, where Ω is the spin precession frequency.[11] However, for narrow samples ($L \ll L_s$), the width of the Hanle curve is determined by the condition $\Omega \tau_d \sim 1$, where $\tau_d = L^2/D \ll \tau_s$ is the time of diffusion on a distance L. Similar results for the Hanle effect in the case when spin diffusion is important were obtained previously [27].

Another interesting effect is predicted in [28], where a narrow 2D strip with two pairs of point contacts A, B and C, D is considered. The two pairs are separated by a distance x that is much larger than the strip width, w, but comparable to the spin diffusion length, L_s. Normally, if $x \gg w$ a current passing across the strip through

---

[10] The sign is opposite to that of the correction in the bulk. The reason is that in the spin layers the spin current induced by electric field is compensated by the opposing diffusion spin current due to the polarization gradient. It is this diffusion spin current that causes the surface correction to the electric current.

[11] To take account of the magnetic field, an additional term $\Omega \times P$ should be added to (8.10).

the A, B contacts will not induce any noticeable voltage between C and D. However, the spin current and the spin density will appear on distances on the order of L_s from the current source. As a result, a voltage, which is second order in γ, should appear between C and D. This spin-mediated non-local charge transport can be again destroyed by applying a magnetic field, which will diminish the spin polarization.

8.3 Phenomenology (without Inversion Symmetry)

If inversion symmetry is absent, whether in a bulk crystal, or in a two-dimensional structure, effects additional to those considered above can arise. In gyrotropic crystals a current can be induced by a *homogeneous* non-equilibrium spin density, as it was shown theoretically by Ivchenko and Pikus [29] and by Belinicher [30, 31]. The first experimental demonstration of this effect was reported in [32]. Inversely, an electric current will generate a uniform spin polarization.

Phenomenologically, this sort of effects can be described by a second rank tensor Q_{ij}, which connects the pseudovector of spin polarization P with the polar vector of electric current j:

$$j_i = Q_{ik} P_k, \qquad P_i = R_{ik} j_k, \qquad (8.21)$$

where the tensor R_{ik} is the inverse of Q_{ik}. Note, that the left- and right-hand sides of (8.21) behave similarly with respect to time inversion, which means that these equations describe *non-dissipative* phenomena. We refer the reader to Chap. 9 for a detailed description of these effects.

The physical reason for this interconnection can be illustrated by the case of the (110) quantum well (see Sect. 1.4.2), where the spin band splitting term is proportional to $p_x s_z$, the z and x axes being chosen along the [110] and [1$\bar{1}$0] direction, respectively [33].

In this case, the energy spectrum consists of two parabolic bands with $s_z = \pm 1/2$, which are oppositely shifted in p-space in the x direction. Each band has a non-zero average p_x, however, in equilibrium both bands are equally filled, so that on the average neither current, nor spin polarization exists. It is now obvious that a non-equilibrium net spin polarization along z means that one band is populated more than the other one, and consequently $\langle p_x \rangle \neq 0$. Thus, spin polarization along z will produce a current along x and vice versa.

For a 2D electron gas with the Bychkov–Rashba splitting [34] the tensor Q_{ij} can be constructed using the Rashba field E^R, a vector pointing in the growth direction z: $Q_{ij} \sim \epsilon_{ijk} E_k^R$, so that (8.21) reduces simply to

$$j \sim E^R \times P. \qquad (8.22)$$

Concerning the spin current, in the latter case it may contain two additional terms. The first one is quadratic in the Rashba field and proportional to the in-plane electric field E:

$$q_{ij} \sim \left(E^R \times E \right)_i E_j^R. \qquad (8.23)$$

If we do not care about the dependence on E^R, this term has the same symmetry properties as the previously considered $\epsilon_{ijk} E_k$ term, with $i, k = x, y$.

The second term was first derived by Kalevich, Korenev, and Merkulov [35]. It is linear in the Rashba field and proportional to the spin polarization:

$$q_{ij} \sim P_i E_j^R - \delta_{ij}(\boldsymbol{PE}^R), \quad i = x, y, \; j = x, y, z. \tag{8.24}$$

For this term, the non-zero components are

$$q_{xz} \sim P_x, \qquad q_{yz} \sim P_y, \qquad q_{xx} = q_{yy} \sim -P_z. \tag{8.25}$$

Thus, the most important new phenomena, that may exist in the absence of inversion symmetry, are the generation of both charge and spin currents by a *uniform* non-equilibrium spin polarization and the inverse effect of producing a bulk spin polarization by charge or spin current.

8.4 Microscopic Mechanisms

The microscopic mechanisms responsible for the spin–charge coupling, and their relative role are still not sufficiently well understood, in spite of a half a century of theoretical efforts, and especially in recent years. We refer the reader to Sinitsyn's review [36] providing the history of this research and recent developments, see also [37]. Here we will limit ourselves to qualitative considerations. The originally proposed mechanism for the spin Hall effect [1, 2] is related to the spin asymmetry in electron scattering due to the spin–orbit interaction (the Mott effect [26, 38]), which was previously used to explain the anomalous Hall effect [11–13]. It is likely that this mechanism accounts for the existing experimental observations (see Sect. 8.5). Also related to scattering is the side jump mechanism proposed by Berger [39, 40] in the contest of the anomalous Hall effect in ferromagnets. Another, "intrinsic", mechanism was first considered by Karplus and Luttinger [14] and proposed recently for specific cases [46, 47], causing much excitement. It is related exclusively to the spin band splitting and does not involve spin asymmetry in scattering.

8.4.1 Spin Asymmetry in Electron Scattering

Mott has shown [26, 38] that the spin–orbit interaction results in an asymmetric scattering of polarized electrons. If a polarized electron beam hits a target, it will deviate in a direction depending on the sign of polarization (similar to a rotating tennis ball in air). The Mott detectors based on this effect are used in high-energy facilities to analyze the electron spin polarization.[12]

---

[12] Interestingly, the *sources* of polarized electrons use optical spin orientation in semiconductors. A GaAs sample pumped by circularly polarized light serves as a photocathode and emits polarized electrons coming from the conduction band, which are then accelerated to high energy. To date, this is the most important practical application of semiconductor spin physics.

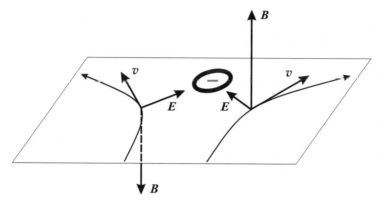

Fig. 8.3. Schematics of electron scattering by a negative charge. The electron spin sees a magnetic field $B \sim v \times E$ perpendicular to the plane of the electron trajectory. Note that the magnetic field has opposite directions for electrons scattered to the left and to the right

The scattering of electrons by a charged center is schematically depicted in Fig. 8.3. The most important element for us is the magnetic field B existing in the electron's moving frame and seen by the electron spin, as explained in Sect. 1.2.3. This field is perpendicular to the plane of the electron trajectory and has opposite signs for electrons moving to the right and to the left of the charged center.

Simply looking at Fig. 8.3, one can make the following observations:

Electron Spin Rotates

If the electron spin is not exactly perpendicular to the trajectory plane, it will make a precession around B during the time of collision. The angle of spin rotation during an individual collision depends on the impact parameter and on the orientation of the trajectory plane with respect to spin. This precession is at the origin of the Elliott–Yafet mechanism of spin relaxation.

The Scattering Angle Depends on Spin

The magnetic field B in Fig. 8.3 is inhomogeneous in space because the electric field E is non-uniform and also because the velocity v changes along the trajectory. For this reason there is a spin-dependent force (proportional to the gradient of the Zeeman energy $2\mu_B(BS)$), which acts on the electron. As a consequence, a left-right asymmetry in scattering of electrons with a given spin appears. This is the Mott effect, or *skew scattering*, resulting, among other things, in the anomalous Hall effect.

If the incoming electrons are *not* polarized the same spin asymmetry in scattering will result in separation of spin-up and spin-down electrons. Spin-ups will go to the right, while spin-downs will go to the left, which means that a spin current in the direction perpendicular to the incoming flux will appear (the Spin Hall Effect).

In quantum mechanics, the spin-dependent asymmetry in scattering appears only beyond the Born approximation.

Spin Rotation is Correlated with Scattering

As seen from Fig. 8.3, the spin rotation around the field B is *correlated* with scattering. If the spin on the right trajectory (corresponding to scattering to the right) is rotated clockwise, then the spin on the left trajectory (scattered to the left) is rotated counterclockwise. Let us see what happens if the incoming beam (x axis) is polarized along the y axis in the trajectory plane, i.e., characterized by a spin current q_{xy}. After scattering, the electrons going to the right will have some spin component along the x axis, while the electrons going to the left will have an x component of the opposite sign! This means that scattering transforms the initial spin current q_{xy} to q_{yx}. Similarly, q_{xx} will transform to $-q_{yy}$.

Such an analysis shows that during the scattering process the initial spin current $q_{ij}^{(0)}$ generates a new spin current q_{ij} according to the rule:

$$q_{ji}^{(0)} - \delta_{ij}q_{kk}^{(0)} \rightarrow q_{ij}. \tag{8.26}$$

Thus the correlation between spin rotation and the direction of scattering gives a physical reason for the additional terms described in Sect. 8.2.5. At present is not clear, under what conditions the two terms can enter not only in the special combination given by (8.26), but with arbitrary coefficients. The experimental consequences of this effect are also not obvious.

The Value of γ for Skew Scattering

The general expressions for the kinetic coefficients in (8.7), (8.8) through the scattering amplitude were derived in [2]. In the presence of the spin–orbit interaction, the scattering amplitude is a matrix in spin indices of the form [41]:

$$\hat{F}_{p'}^{p} = A(\theta)\hat{I} + B(\theta)\hat{\sigma} \cdot n, \tag{8.27}$$

where θ is the scattering angle, $n = p' \times p/|p' \times p|$ is the unit vector normal to the scattering plane, $\hat{I}$ is the unit 2×2 matrix, and $\hat{\sigma}$ are Pauli matrices. The solution of the kinetic equation (Appendix A) gives the coefficients β and β_1, which are the real and imaginary parts of a single expression [2]:

$$\beta + i\beta_1 = 4\pi \frac{e}{m} N \left\langle v\tau_p^2 \int_0^\pi AB^* \sin^2 \theta \, d\theta \right\rangle, \tag{8.28}$$

where N is the concentration of scatterers, τ_p is the momentum relaxation time for a given energy, brackets stand for averaging over the equilibrium distribution. The coefficient β_1 defines the magnitude of the additional terms in the expression for the spin current q_{ij}, which were discussed in Sect. 8.2.5 and above: $\beta_1(E_j P_i - \delta_{ij}EP)$.

In this case γ depends only on the form of the scattering potential, the electron energy, and the strength of the spin–orbit interaction. In the Born approximation the functions $A(\theta)$ and $B(\theta)$ have a phase difference of $\pi/2$, so that in order to obtain a non-zero β one has to go beyond this approximation. In contrast, the coefficient β_1, originating from the correlation between spin rotation and the scattering direction, exists already within the Born approximation.

Abakumov and Yassievich [42] have considered spin-dependent scattering by a charged center in the first order beyond the Born approximation. From their results one can deduce

$$\gamma \sim \frac{\lambda k_F}{a_B}, \tag{8.29}$$

where $a_B = \hbar^2 \epsilon / me^2$ is the Bohr radius, k_F is the Fermi wave vector, ϵ is the dielectric constant, and the constant λ determines the strength of the spin–orbit interaction, see (8.31) below.

For a 3D bulk concentration of $10^{17} \, \text{cm}^{-3}$ (or a 2D surface concentration of $3 \times 10^{11} \, \text{cm}^{-2}$) one gets the estimates $\gamma \sim 4 \times 10^{-4}$ for GaAs and $\gamma \sim 2 \times 10^{-2}$ for InSb. It should be noted that for the same electron concentration the value of the Born parameter $e^2/(\epsilon \hbar v_F)$, where v_F is the Fermi velocity, for GaAs is $\simeq 1.2$.[13] Thus the Born expansion is practically not always justified.

We equivalently rewrite (8.29) in the form:

$$\gamma \sim \lambda k_F^2 \left(\frac{e^2}{\epsilon \hbar v_F} \right), \tag{8.30}$$

which explicitly displays the Born parameter, considered as small in [42]. If this parameter becomes equal to 1, we have $\gamma \sim \lambda k_F^2$, or $\gamma \sim E_F/E_g$ (for $\Delta \gg E_g$) and $\gamma \sim (E_F/E_g)(\Delta/E_g)$ (for $\Delta \ll E_g$).[14] It is likely that these estimates will hold also when the Born parameter becomes large.

8.4.2 The Side Jump Mechanism

The side jump mechanism for the anomalous Hall effect was proposed by Berger [39, 40] and studied in detail by Nozières and Lewiner [43], see also [44]. It is described as a spin-dependent lateral displacement of the electron wave packet during each scattering event. In our opinion, the role of the side jump effect is still not very well understood. Here we will discuss this effect from the point of view of classical mechanics, which allows to achieve clarity and transparency lacking so far in the quantum-mechanical approach.

---

[13] For a degenerate electron gas, the Born parameter coincides with the parameter r_s, the ratio of the mean potential energy to the kinetic energy, which defines whether the gas is ideal, or not. At relatively low electron concentration, this parameter may be quite large.

[14] These formulas are analogous to the estimate $\gamma \sim (v/c)^2$ for the scattering of an electron by a proton in vacuum, the velocity of light, c, being replaced by $(E_g/m)^{1/2}$ for the case $\Delta \gg E_g$ or by $(E_g^2/(\Delta m))^{1/2}$ for $\Delta \ll E_g$.

The effective mass Hamiltonian describing the spin–orbit interaction is conventionally written as

$$H_{so} = 2\lambda(\boldsymbol{k} \times \nabla V) \cdot \boldsymbol{s}, \tag{8.31}$$

where $\boldsymbol{k}$ is the electron wave vector, $\boldsymbol{s} = \boldsymbol{\sigma}/2$ is the electron spin operator, and $V(\boldsymbol{r})$ is the electron potential energy. In vacuum $\lambda = -\hbar^2/(4m_0^2 c^2)$, m_0 being the free electron mass. In semiconductors with the band structure of GaAs, in the limit when the effective mass m is much smaller than m_0, the Kane model gives [45]

$$\lambda = \frac{\hbar^2}{4m\,E_{\rm g}} \quad \text{for } \Delta \gg E_{\rm g}, \tag{8.32}$$

$$\lambda = \frac{\hbar^2}{3m\,E_{\rm g}} \frac{\Delta}{E_{\rm g}} \quad \text{for } \Delta \ll E_{\rm g}, \tag{8.33}$$

where Δ is the spin–orbit splitting of the valence band (Sect. 1.3.6).

Classical Mechanics of a Spinning Particle

We can eliminate the Planck constant by rewriting (8.31) in the form

$$H_{so} = \mathcal{A}(\boldsymbol{p} \times \nabla V) \cdot \boldsymbol{S}. \tag{8.34}$$

Here we have introduced the constant $\mathcal{A} = 2\lambda/\hbar^2$ with the dimension $(\text{momentum})^{-2}$ and the *dimensional* intrinsic angular momentum of the electron $\boldsymbol{S} = \hbar \boldsymbol{s}$, $\boldsymbol{p} = \hbar \boldsymbol{k}$ is the electron momentum. We can now write down the classical Hamiltonian equations, corresponding to the Hamiltonian function $H = p^2/(2m) + V(\boldsymbol{r}) + H_{so}$:

$$\dot{\boldsymbol{r}} = \frac{\boldsymbol{p}}{m} + \mathcal{A}(\nabla V \times \boldsymbol{S}), \tag{8.35}$$

$$\dot{\boldsymbol{p}} = -\nabla\big[V + \mathcal{A}(\boldsymbol{p} \times \nabla V) \cdot \boldsymbol{S}\big], \tag{8.36}$$

$$\dot{\boldsymbol{S}} = \mathcal{A}(\boldsymbol{p} \times \nabla V) \times \boldsymbol{S}. \tag{8.37}$$

These equations can be applied to a classical object with an internal angular momentum $\boldsymbol{S}$ (e.g., a tennis ball, with an appropriate choice of the constant $\mathcal{A}$). Obviously, they are identical to the quantum-mechanical operator equations for $\hat{\boldsymbol{r}}$, $\hat{\boldsymbol{p}}$, and $\hat{\boldsymbol{S}}$. Note, that the observable quantities are $\boldsymbol{r}$ and $\boldsymbol{v} = \dot{\boldsymbol{r}}$, not the canonical momentum $\boldsymbol{p}$. Therefore, it may be useful to rewrite these equations in form of Newton's law for the variables $\boldsymbol{r}$ and $\boldsymbol{v}$. In the two-dimensional case one obtains

$$m\dot{\boldsymbol{v}} = -\nabla V + m\mathcal{A}(\boldsymbol{v} \times \boldsymbol{S})\Delta_2 V, \tag{8.38}$$

where Δ_2 stands for the two-dimensional Laplacian.[15] A similar, but more complicated, equation can be easily derived for the three-dimensional case.

---

[15] Note that unlike the full Laplacian of V, $\Delta_2 V$ is *not* related to the charge density by the Poisson equation.

One can see from (8.38) that in the two-dimensional case the role of the spin–orbit interaction for the particle motion reduces to the action of an effective inhomogeneous magnetic field directed along S and proportional to $\Delta_2 V$. One of the consequences of (8.38) is that the accelerated motion of an electron in a uniform electric field ($V = e\boldsymbol{E}\boldsymbol{r}$) is not modified by the spin–orbit interaction.[16]

Nozières and Lewiner [43] derive the side jump in the following simple way. To the first order in the spin–orbit interaction, on can replace ∇V in the second term of (8.35) by $-\dot{\boldsymbol{p}}$. Then, considering S as being constant and integrating this term over time from $-\infty$ to $+\infty$, one obtains the generally accepted result for the spin-dependent displacement during an individual collision:

$$\delta\boldsymbol{r} = \mathcal{A}(\boldsymbol{S} \times \delta\boldsymbol{p}), \qquad \delta\boldsymbol{p} = \boldsymbol{p}' - \boldsymbol{p}, \qquad (8.39)$$

where $\boldsymbol{p}$ and $\boldsymbol{p}'$ are the electron momenta before and after the collision. However, it is noted in [43] that (8.39) might be not generally true, because the first term in (8.35) may also contribute to this displacement. We will see below that this is indeed the case.

Reflection from a Flat Wall

We start with the simplest problem: the reflection from a flat wall located in the plane $y = 0$. We chose the potential energy to be zero for $y > 0$, the actual form of the repulsive potential at $y < 0$ is not really important. The incident velocity of the particle lies in the xy plane, while S is directed along z, so that S remains constant.

For $y < 0$ (8.35), (8.36) yield

$$\dot{x} = \frac{p_x}{m} + \mathcal{A}S\frac{\partial V}{\partial y}, \qquad \dot{y} = \frac{p_y}{m}, \qquad \dot{p}_x = 0, \qquad \dot{p}_y = -\frac{\partial V}{\partial y} - \mathcal{A}Sp_x\frac{\partial^2 V}{\partial y^2}. \tag{8.40}$$

The trajectories are schematically depicted in Fig. 8.4. Since energy and p_x are conserved, the reflection angle is equal to the incidence angle, like in the absence of spin–orbit coupling. However, because of spin–orbit interaction there is an additional spin-dependent shift of the outgoing part of the trajectory by an amount δx. This value can be calculated exactly for a parabolic potential $V(y) = ky^2/2$. To the first order in $\mathcal{A}$ one finds[17]

$$\delta x = \delta x^{(0)}\left(1 - \tan^2\theta\right), \tag{8.41}$$

---

[16] The opposite statement can be found in the literature. By looking at the "anomalous velocity" (the second term in (8.35)) one can be tempted to claim the existence of a transverse velocity $e\mathcal{A}(\boldsymbol{E} \times \boldsymbol{S})$. This is an illusion: the transverse to the electric field component of $\boldsymbol{p}$ being conserved (see (8.36)), the transverse component of velocity is a constant. This constant is equal to the initial value of the transverse velocity, exactly like in the absence of the spin–orbit interaction.

[17] This result was derived by Maria Lifshits (to be published).

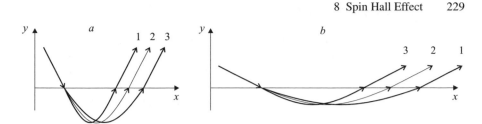

Fig. 8.4. Reflection of a spinning particle from a flat wall. (**a**) Angle of incidence $\theta < \pi/4$. (**b**) $\theta > \pi/4$. (*1*) With spin–orbit interaction for $S > 0$. (*2*) Without spin–orbit interaction. (*3*) With spin–orbit interaction for $S < 0$. Note that the spin-dependent shifts of the outgoing trajectory have opposite signs in (**a**) and (**b**)

where $\delta x^{(0)}$ is given by (8.39) and θ is the incidence angle. Thus the expression (8.39) is true for normal incidence only. The shift changes sign at $\theta = \pi/4$. It can be shown that (8.41) does not depend on the actual form of $V(y)$.[18]

The difference between (8.41) and (8.39) can be explained as follows. Let us integrate the first of (8.40) over time from the moment $t = 0$ when the particle hits the wall to the moment t_0, when it reappears at $y = 0$ on its way backwards. One finds $\delta x = \delta x^{(0)} + (p_x/m)t_0$ (since $p_x = $ const). The time t_0 is modified by a spin–orbit interaction term in the last of (8.40). To the first order in $\mathcal{A}$ this gives the correction in (8.41), proportional to $\tan^2 \theta$.

We must also remember, that the spin–orbit term in (8.31) or (8.34) should always be considered as a small perturbation to the main Hamiltonian, i.e., it should only *slightly* modify the particle trajectory. This forbids us to consider spin–orbit effects in the limit of an absolutely rigid wall: the spin-dependent shift in Fig. 8.4 should always remain a small correction.

Scattering by a Hard Sphere

If the sphere radius r_0 is much greater than the particle penetration depth (and hence, than δr), its surface may be locally considered as flat and we can use the previous result to calculate not only the side jump (which remains the same), but also the spin-dependent correction to the scattering angle $\delta\theta$. Let again S be perpendicular to the scattering plane. Then the orbital angular momentum is conserved, which means that the minimal distance from the sphere center to the continuation of the outgoing trajectory should be equal to the impact parameter ρ, see Fig. 8.5. This consideration gives the relation between the side jump δr and the correction to the scattering angle: $\delta\theta = \delta r/r_0$.

A similar relation should exist for classical scattering by an arbitrary potential $V(r)$, with r_0 replaced by the effective scattering radius.

---

[18] The vicinity of $\theta = \pi/2$, where (8.41) diverges, requires a more accurate approach. It can be shown that in fact at $\pi/2 - \theta \sim \mathcal{A}S(km)^{1/2} \ll 1$ the value of δx as a function of θ has a sharp extremum and becomes zero at $\theta = \pi/2$. The details of this behavior depend on the form of $V(y)$.

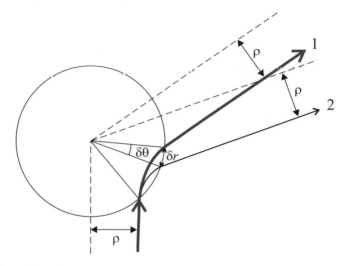

Fig. 8.5. Scattering of a spinning particle by a hard sphere of radius r_0. (*1*) With spin–orbit interaction. (*2*) Without spin–orbit interaction. The shift of the outgoing trajectory is due to the side jump δr. The angle between (*1*) and (*2*) is $\delta\theta = \delta r / r_0$

Side Jump versus Skew Scattering

The relative role of the two spin-dependent effects can be understood as follows. During the time between collisions τ_p the displacement of spin-up and spin-down particles will be different. Because of skew scattering this difference is on the order of $\delta\theta\ell$, where ℓ is the mean free path. This should be compared to the side jump during one collision, δr. Using the relation $\delta\theta = \delta r / r_0$, we find that the side jump to skew scattering ratio is $\sim r_0/\ell$ for scattering by hard spheres, and most probably for any kind of classical scattering (when the De Broglie wavelength is smaller than the scattering diameter).

If one wants to use the kinetic theory, the ratio r_0/ℓ must be small. Thus, within the validity of the kinetic theory and the Boltzmann equation, the contribution of the side jump effect is always small compared to that of skew scattering, *if* scattering is classical.

Apparently, the situation is different for quantum scattering. For example, in the Born approximation the skew scattering does not exist, while the side jump seems to remain the same [39, 40, 43]. The value of the parameter γ for the side jump mechanism can be estimated as $\gamma_{SJ} \sim \delta r/\ell \sim \lambda k_F/\ell \sim \lambda m/(\hbar\tau_p)$. Using (8.29) for skew scattering (γ_{SS}), one obtains $\gamma_{SJ}/\gamma_{SS} \sim a_B/\ell$. *This* ratio may be large even within the validity of the kinetic theory, since the Bohr radius a_B may be quite large (10^{-6} cm for electrons in GaAs). The relatively large value of this ratio is due to the assumed small value of the Born parameter. If this parameter is on the order of 1, then we should use the estimate $\gamma_{SS} \sim \lambda k_F^2$ (see (8.30)). This gives $\gamma_{SJ}/\gamma_{SS} \sim (k_F\ell)^{-1}$, a value, that again should be small for the kinetic theory to be valid.

We believe that this important issue needs further clarification.

8.4.3 Intrinsic Mechanism

The so-called "intrinsic" or "Berry-phase" mechanism of generating spin current by electric current is related entirely to the spin band splitting and does not rely on spin-dependent effects in scattering (in the clean limit, when the band splitting $\hbar\Omega(p)$ at the Fermi level is large, $\Omega(p)\tau_p \gg 1$). Essentially, it was first put forward by Karplus and Luttinger [14].

Spin Current of Bulk $J = 3/2$ Holes

The first recent proposal was made by Murakami, Nagaosa, and Zhang [46] for bulk holes described by the Luttinger Hamiltonian (see Sect. 1.3.6). They found the spin Hall conductivity, $\beta n \sim ek_F/\hbar$, corresponding to $\gamma \sim (k_F\ell)^{-1}$, which can be also presented as $\gamma \sim (\Omega(p)\tau_p)^{-1}$.[19]

Theoretically, it is certainly very interesting that such a spin current exists. However, the question arises, what could be the experimental consequences? Since the spin current cannot be measured directly, it could manifest itself either by bulk second order effects ($\sim\gamma^2$, see Sect. 8.2.6) or by the surface spin accumulation. These observable effects were not calculated. The spin relaxation time for holes being on the order of τ_p (see Sect. 1.4.2), the diffusion equation approach cannot be applied (Sect. 8.2.5). Still, we can expect that (8.14) may be used as an estimate for the degree of polarization near the surface, if we put $\tau_s \sim \tau_p$ and $\gamma \sim (k_F\ell)^{-1}$. This gives

$$\mathcal{P} \sim \frac{eE\lambda_F}{E_F}. \tag{8.42}$$

The degree of spin polarization is on the order of the ratio of the voltage drop on the distance of the Fermi wavelength λ_F to the Fermi energy. In order to achieve $\mathcal{P} = 1\%$ for a hole concentration of 10^{17} cm^{-3} in GaAs one would need an electric field $E \sim 10$ V/cm. However, there is also a question about the width of the surface layer where this polarization exists. It is likely that this width is determined by λ_F, which can hardly be resolved by optical measurements.

We note that the splitting into light and heavy hole bands is *not* due to spin–orbit interaction, but rather to the p-state nature of the Bloch functions in the valence band. Indeed, even in the absence of spin–orbit interaction, holes are still particles with internal angular momentum $L = 1$ (see Sect. 1.3.6) and the splitting into light and heavy holes still exists. Thus, for the case of holes the spin–orbit interaction is not of primary importance.

Note also that the $J = 3/2$ holes may not be described by the simple equations (8.5), (8.6) because for higher spins the number of coupled macroscopic quantities increases compared to spin 1/2 particles. In 1984 we studied [24] the mutual transformation of spin and charge currents for holes (and for carriers in a gapless semiconductor like HgTe) due to elastic scattering, by finding a general solution of the

---

[19] Since the difference between the light and heavy hole masses is large, this splitting is on the order of the Fermi energy E_F. It is assumed that $\Omega(p)\tau_p \gg 1$, which is normally the case.

kinetic equation describing correlations between $\boldsymbol{J}$ and $\boldsymbol{p}$. Here we present some results, which are relevant for our discussion.

1. There are two contributions to spin current. The first one comes from states with helicity $\pm 1/2$ in the light hole band, and the second one is due to the non-diagonal in the band indices (light/heavy) elements of the density matrix. The heavy hole $\pm 3/2$ band does not support a spin current.
2. In [24] we have neglected the non-diagonal elements on the grounds that they contain the small parameter $(\Omega(p)\tau_p)^{-1}$. It is this contribution that is responsible for the intrinsic mechanism proposed in [46] with γ on the order of the above small parameter.
3. Due to scattering, the charge and spin currents in the light hole band (or for electrons in a gapless semiconductor) are coupled. This is described by the Drude-like equations similar to (8.15), (8.16), which can be written in the form (we skip the action of magnetic field):

$$\frac{dq_i}{dt} = \frac{en}{m}E_i - \frac{q_i}{\tau_1} + \frac{\gamma}{\sqrt{\tau_1 \tau_2}}\epsilon_{ijk}q_{jk}, \qquad (8.43)$$

$$\frac{dq_{ij}}{dt} = -\frac{q_{ij}}{\tau_2} - \frac{\gamma}{\sqrt{\tau_1 \tau_2}}\epsilon_{ijk}q_k, \qquad (8.44)$$

the only difference being that the relaxation times for charge and spin currents, τ_1 and τ_2, are now different (by a numerical factor), n is the concentration of light holes.

It is important that unlike the case of spin–orbit interaction for electrons, for holes the coupling between their spin and momentum is very strong. Because of this, the parameter γ, which was very small in all previous considerations, now is generally on the order of 1! It can be reduced only due to the smallness of the Born parameter $e^2/\epsilon\hbar v_F$ (we remind that in the Born approximation $\gamma = 0$). As mentioned previously, at low concentrations the Born parameter may easily be on the order of 1. These results suggest that the spin current generated in the light hole band by scattering might provide a strong competition with the intrinsic current. It seems that the $J = 3/2$ holes, and especially electrons in a gapless semiconductor, may be the best candidates for observing strong spin-charge coupling effects, which could be revealed by the anomalous magnetic field dependence of the resistivity tensor (Sect. 8.2.6) and by an anomalous frequency dependence of conductivity.

4. The frequency dependence of conductivity in an n-type gapless semiconductor can be easily derived from (8.43), (8.44) [24]:

$$\sigma(\omega) = \frac{ne^2}{m}\frac{\tau_1(1 + i\omega\tau_2)}{(1 + i\omega\tau_1)(1 + i\omega\tau_2) + 2\gamma^2}. \qquad (8.45)$$

In the Born approximation $\gamma = 0$ and (8.45) reduces to the usual Drude result.

Intrinsic Mechanism for 2D Electrons and Holes

Sinova et al. [47] proposed an intrinsic mechanism of spin current due to the Bychkov–Rashba spin splitting of the conduction band for 2D electrons. They claimed that in the clean limit there is a universal spin Hall conductivity $\beta n = e/(8\pi\hbar)$.[20] A lively discussion followed with many papers confirming this result by sophisticated theoretical techniques, and many others disproving it. Finally, it was understood that the "universal spin conductivity" is actually *zero* (see for example [48, 49]). However, this annulation is specific for the case when the spin band splitting is linear in momentum [33, 34]. The current consensus is that the intrinsic mechanism may exist for any type of spin band splitting, *except* if it is linear in p. An example is the two-dimensional system of heavy holes in a (100) quantum well. In this case the spin–orbit interaction is cubic in the two-dimensional momentum [50]:

$$\hat{H}_{\text{hh}} = \frac{\alpha_{\text{hh}}}{2}\left(\hat{\sigma}_+ p_-^3 + \hat{\sigma}_- p_+^3\right), \tag{8.46}$$

where $p_\pm = p_x \pm ip_y$, etc. In a quantum well, the sub-band of heavy holes is doubly degenerate at $p = 0$, therefore the holes can be attributed a "pseudospin" $1/2$, the Pauli matrices in (8.46) act in the basis of the two degenerate states.

Calculations [51, 52] give an electrical field induced intrinsic spin current with $\gamma \sim (k_F\ell)^{-1}$. However, even in the clean limit, the numerical factor depends on the nature of the scattering potential, i.e., whether it allows momentum transfer corresponding to inter-sub-band transitions (hard potential), or not (soft potential). In this sense there is no universality of the intrinsic spin current.

Spin band splitting can occur only in the absence of inversion symmetry. In this case, there are other, maybe more important effects, mentioned in Sect. 8.3, see also Chap. 9. And, of course, there is always the question about the relation of the calculated spin current to the experimentally observable quantities.

Spin Accumulation in the Ballistic Regime

Is there any relation between the mean value of the spin current operator calculated in the bulk to the local spin polarization near the boundary in the clean limit? The answer is negative, and the reason for that is the fast spin precession and spin relaxation due to the strong spin–orbit coupling.

Consider the case of Bychkov–Rashba splitting characterized by the energy Δ_R. The characteristic spatial scale near the boundary is the spin precession length, $L_s = \hbar v_F/\Delta_R$.[21] The region in the interior of the sample which is beyond the strip of width L_s is disconnected from the surface layer.

---

[20] Note the difference between the 3D and 2D cases stemming from the different dimensionality of concentration: $ek_F/\hbar$ in 3D becomes $e/\hbar$ in 2D. In both cases $\gamma \sim (k_F\ell)^{-1}$.

[21] It is interesting that this expression is valid both in the clean and in the dirty limits. In the latter case $\Delta_R\tau_p/\hbar \ll 1$ and L_s may be equivalently presented as $L_s = (D\tau_s)^{1/2}$ with $1/\tau_s \sim \Delta_R^2\tau_p/\hbar^2$.

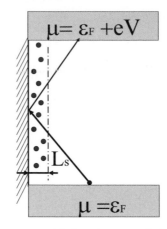

Fig. 8.6. Ballistic structure with the spin accumulation near the boundary

To illustrate the point, let us consider a mesoscopic ballistic structure with a mean free path much greater than the sample sizes. This ballistic region is attached to the leads which serve as source–drain contacts, see Fig. 8.6. A voltage V is applied between the leads which causes a charge current through the structure.

We consider specular scattering at the boundaries. Since the structure is ballistic, we can trace every electron trajectory and find the exact analytical solution [53, 54] for the spin polarization density P_z (the component perpendicular to the two-dimensional plane). The spin accumulation exists over the spin precession length L_s near the boundary and decays as a power law of the distance from the boundary. Note, that the edge spin polarization in a ballistic system appears not as a result of acceleration of electrons by an electric field and associated spin precession. The electric field is absent inside an ideal ballistic conductor. Edge spin accumulation in a ballistic system is due to the difference in populations of left-moving and right-moving electrons. Combined with boundary scattering and spin precession, this results in oscillatory edge polarization.

The characteristic magnitude of the accumulated polarization density is found to be $P_z \sim eV \Delta_R/(\hbar v_F)^2$, which corresponds to the degree of spin polarization within the spin layer $\mathcal{P} \sim (eV/E_F)(\Delta_R/E_F)$. For GaAs with typical ratio $\Delta_R/E_F \approx 10^{-2}$ one obtains $\mathcal{P} = 1\%$ when $eV \sim E_F = 10\,\mathrm{meV}$. The direction of $\boldsymbol{P}$ is defined by the vector $\boldsymbol{n} \times \boldsymbol{j}$, where $\boldsymbol{n}$ is the normal to the boundary and $\boldsymbol{j}$ is the charge current density. Thus the directions of the accumulated spins are opposite for the opposing boundaries, like in the diffusive spin Hall effect.

Since the sample width is much larger than L_s, there is no connection with the spin current in the bulk. In the ballistic case considered here we cannot use the diffusion equation for $\boldsymbol{P}$. Therefore, the results are not obtained by the use of the boundary condition of the type that some spin current is equal to zero. Instead, the boundary condition consists of requiring that the wave function, i.e., the sum of the incident and reflected waves equals zero for both helicities.

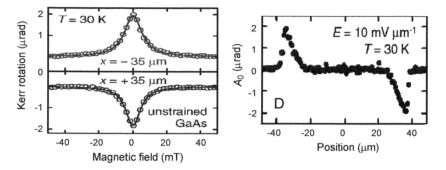

Fig. 8.7. First observation of the spin Hall effect [21]. *Left panel*: Measurements of the Kerr rotation as a function of magnetic field at the opposing sample edges. *Right panel*: Spatial dependence of the Kerr rotation across the channel

The conclusion that the width of the spin layer is the spin precession length L_s (which in the clean limit is much less than the mean free path ℓ) holds also in the non-ballistic case, when the sample size is greater than ℓ.

8.5 Experiments

Here we present some experimental data on the spin Hall effect. The number of experimental papers is still about two orders of magnitude less than the number of theoretical ones. Most of the results on semiconductors were obtained by Awschalom's group in Santa Barbara. Some important work was also done in metals.

First Observation of the Spin Hall Effect

Kato et al. [21] report on the optical detection of the spin Hall effect in thin films of GaAs and InGaAs. A linearly polarized beam is tuned to the absorption edge of the semiconductor. The rotation of the polarization axis of the reflected beam provides a measure of the electron spin polarization along the beam direction. In Fig. 8.7 typical Kerr rotation data for scans of external magnetic field are shown for two edges of the sample. These curves give the projection of the spin polarization along the z-axis, which diminishes with an applied transverse magnetic field because of spin precession (the Hanle effect). It is seen that the spin polarization has opposite signs at the two edges and falls off rapidly with the distance from the edge, disappearing at the center of the channel.

The effect in both unstrained and strained samples is investigated. No marked crystal direction dependence is observed in the strained samples, which suggests the dominance of the scattering mechanism. (see also a later publication of the same group [55]).

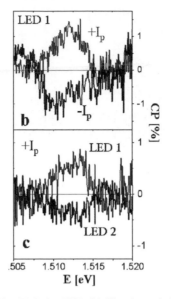

Fig. 8.8. The spin Hall effect for 2D holes [22]. (**b**) Circular polarization measured at one edge for two opposite current directions. (**c**) Circular polarization measured at opposing edges with fixed current

Spin Hall Effect for 2D Holes

Wunderlich et al. [22] (see also [56]) report the experimental observation of the spin Hall effect in a 2D hole system. The 2D hole layer is a part of a p–n junction light-emitting diode with a specially designed coplanar geometry which allows an angle-resolved polarization detection at opposite edges of the sample. The detection of spin is done by measuring the circular polarization of the light. A p-channel current is applied along the x direction and the circular polarization of light emitted along the z axis is measured, while biasing one of the light-emitting diodes at either side of the p-channel.

The occurrence of the spin Hall effect is demonstrated in Fig. 8.8(b). When biasing, a circular polarization of light emitted from the region near the corresponding p–n junction step edge appears, and its sign flips upon reversing the current direction. Figure 8.8(c) compares the circular polarizations ($\approx 1\%$) obtained with a fixed current, when either the first light-emitting diode or the second one is activated. The opposite sign of these two signals confirms the opposite directions of P_z at the two edges.

The authors estimate that the conditions are close to the clean limit and interpret the edge spin accumulation as resulting from the intrinsic mechanism, arguing that the observed degree of polarization is consistent with the estimate (8.42). However there is no direct experimental evidence of the intrinsic origin of the observed effect.

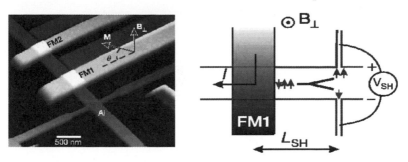

Fig. 8.9. Experimental observation of the inverse spin Hall effect [58]. *Left panel*: Atomic force microscope image of the device. A thin aluminium Hall cross is contacted with two ferromagnetic electrodes, FM1 and FM2. A magnetic field perpendicular to the substrate, $B_\perp$ sets the orientation of magnetization of FM1 and FM2, which is characterized by an angle θ. *Right panel*: A current is injected out of FM1 into the Al film and away from the Hall cross. A Hall voltage, V_{SH}, is measured between the two Hall probes at a distance L_{SH} from the injection point

Spin Hall Effect for 2D Electrons

The current-induced spin polarization in a two-dimensional electron gas confined in (110) AlGaAs quantum wells was observed by Sih et al. [57], with the goal to determine which mechanism of the spin Hall effect was dominant. For the [110] growth direction the Dresselhaus field is perpendicular to the 2D plane [33] and should not contribute to the intrinsic spin current. The Rashba splitting was measured as a shift of the Hanle curve in magnetic field-dependent Kerr rotation when a dc voltage is applied along the [001] direction. The thus determined splitting happens to be very small. The authors conclude that the observed spin Hall effect is dominated by the spin-dependent scattering mechanism.

Observation of the Inverse Spin Hall Effect in Metals

Valenzuela and Tinkham [58] report electrical measurements of the inverse spin Hall effect in aluminium, using a ferromagnetic electrode in combination with a tunnel barrier to inject a spin-polarized current. They observe an induced voltage that results exclusively from the conversion of the injected spin current into charge imbalance. This voltage is proportional to the component of the injected spins that is perpendicular to the plane defined by the spin current direction and the voltage probes.

A ferromagnetic electrode FM1 is used to inject the spin-polarized electrons via a tunnel barrier in one of the arms of an aluminium Hall cross (Fig. 8.9). The injected current is driven away from the Hall cross, where only a pure spin current flows as a result of the spin injection.

Because of spin–orbit interaction, the spin current induces a transverse charge imbalance and generates a measurable voltage, V_{SH}. Spin imbalance in the Al film occurs with a spin direction defined by the magnetization orientation of the FM1

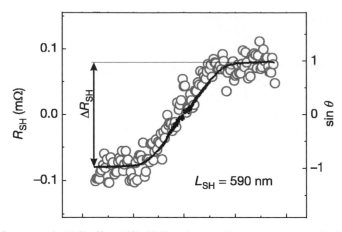

Fig. 8.10. Inverse spin Hall effect [58]. Hall resistance R_{SH} versus perpendicular magnetic field $B_\perp$ for $L_{SH} = 590$ nm compared to the measured value of $\sin\theta$

electrode. Thus V_{SH} is expected to vary when a magnetic field perpendicular to the substrate, $B_\perp$, is applied and the magnetization of the electrode $\boldsymbol{M}$ is tilted out of the substrate plane by an angle θ. Then V_{SH} should be proportional to $\sin\theta$, correlating with the component of $\boldsymbol{M}$ normal to the substrate.

Figure 8.10 shows $R_{SH} = V_{SH}/I$ as a function of $B_\perp$. The saturation in R_{SH} for large $B_\perp$ strongly suggests that the result is related to the magnetization orientation of the ferromagnetic electrode and thus is a manifestation of the inverse spin Hall effect. This is further supported by comparing $R_{SH}(B_\perp)$ with the magnetization component perpendicular to the substrate (line in Fig. 8.10). The agreement is excellent. The experimental data allow to evaluate the spin-Hall conductivity and to determine $\gamma \approx (1\text{–}3) \times 10^{-4}$.

Both the direct and the inverse spin Hall effects in a platinum wire, detected by electrical measurements involving spin injection are reported in a very interesting article by Kimura et al. [59]. Both effects persist even at room temperature. The parameter γ for Pt is deduced to be 3.7×10^{-3}, much larger than that for Al obtained in [58], a difference which is naturally explained by the stronger spin–orbit coupling in Pt.

Room Temperature Spin Hall Effect in Semiconductors

Stern et al. [60] report observation of the spin Hall effect in n-type ZnSe epilayers with electron concentration from 5×10^{16} to 9×10^{18} cm^{-3}. Based on the value and the sign of the effect, they attribute it to the scattering mechanism. The most important finding is that spin Hall effect is observable (by the Kerr rotation technique) up to room temperature!

8.6 Conclusion

The spin Hall effect is a new transport phenomenon, predicted a long time ago but observed only in recent years. It was experimentally studied in three- and two-dimensional semiconductors samples. The inverse spin Hall effect was seen in semiconductor structures, as well as in metals. Finally, it is important that these effects are observable not only at cryogenic, but also at room temperature. At present, it is difficult to predict whether this effect will have any practical applications, as many people believe, or it will belong only to fundamental research as a tool for studying spin interactions in solids.

The spin Hall effect shares with the long-studied anomalous Hall effect an uncertainty about its microscopic origin. Let us hope that future, primarily experimental, but also theoretical work will help to elucidate this problem.

Appendix A: The Generalized Kinetic Equation

The Boltzmann equation, generalized for the existence of spin and taking account of the spin–orbit interaction is probably the best, most simple, and economic theoretical tool for studying spin phenomena in conditions when the orbital motion can be considered classically (when, for example, the Landau quantization is not important). It can be derived from the quantum equation for the one-particle density matrix $\hat{\rho}_{\boldsymbol{p}_1,\boldsymbol{p}_2}(t)$. We use the momentum representation, hats indicate matrices in spin indices:

$$i\hbar \frac{\partial \hat{\rho}}{\partial t} = [\hat{H}, \hat{\rho}].   \tag{8.47}$$

To approach classical physics, one introduces the Wigner density matrix $\hat{\rho}(\boldsymbol{r}, \boldsymbol{p}, t)$, by putting $\boldsymbol{p} = (\boldsymbol{p}_1 + \boldsymbol{p}_2)/2$, $\boldsymbol{\kappa} = \boldsymbol{p}_1 - \boldsymbol{p}_2$ and doing the Fourier transform over $\boldsymbol{\kappa}$:

$$\hat{\rho}(\boldsymbol{r}, \boldsymbol{p}, t) = \int \exp(i\boldsymbol{\kappa} \boldsymbol{r}/\hbar)\hat{\rho}_{\boldsymbol{p}+\boldsymbol{\kappa}/2, \boldsymbol{p}-\boldsymbol{\kappa}/2}(t) \frac{d^3\kappa}{(2\pi\hbar)^3}.   \tag{8.48}$$

Consider a Hamiltonian $\hat{H}(\boldsymbol{p})$ describing the electron spin band splitting and the interaction with an external electric field $\boldsymbol{E}$ (e is the absolute value of the electron charge):

$$\hat{H}(\boldsymbol{p}) = \frac{p^2}{2m} + \hbar \boldsymbol{\Omega}(\boldsymbol{p})\hat{s} + e\boldsymbol{E}\boldsymbol{r}.   \tag{8.49}$$

Using (8.47) and (8.49) one can derive the kinetic equation for $\hat{\rho}(\boldsymbol{r}, \boldsymbol{p}, t)$:

$$\frac{\partial \hat{\rho}}{\partial t} + \{\hat{v}, \nabla\hat{\rho}\} - e\boldsymbol{E}\frac{\partial \hat{\rho}}{\partial \boldsymbol{p}} + i[\boldsymbol{\Omega}(\boldsymbol{p})\hat{s}, \hat{\rho}] = \hat{I}\{\hat{\rho}\},   \tag{8.50}$$

where $\{A, B\} = (AB + BA)/2$ and $\hat{v} = \partial \hat{H}(\boldsymbol{p})/\partial \boldsymbol{p}$ is the electron velocity, which is a matrix in spin indices. The left-hand side of (8.50) follows exactly from (8.47). The right-hand side is the collision integral, which is added "by hand" (like in the conventional Boltzmann equation).[22] One has to evaluate the change of the density matrix, $\delta\hat{\rho}$, during one collision and make a sum over all collisions per unit time.

Spin–orbit interaction (8.31) during the act of an individual collision makes the integral operator $\hat{I}$ a matrix with four spin indices. For elastic collisions with impurities and *if the effect of band splitting can be neglected while considering individual collisions*, we have [61]

$$\hat{I}\{\hat{\rho}(\boldsymbol{p})\}_{\mu\mu'} = \int d\Omega_{p'}\left\{ W^{\mu\mu'}_{\mu_1\mu_1'} \cdot \rho_{\mu_1\mu_1'}(\boldsymbol{p}') \right.$$
$$\left. - \frac{1}{2}\left[W^{\mu\mu_1}_{\mu_2\mu_2} \cdot \rho_{\mu_1\mu'}(\boldsymbol{p}) + \rho_{\mu\mu_1}(\boldsymbol{p}) \cdot W^{\mu_1\mu'}_{\mu_2\mu_2} \right] \right\}. \qquad (8.51)$$

Here the summation over repeated indices is implied. This expression is the generalization of the Boltzmann collision integral taking account of spin–orbit effects. If spin–orbit interaction during collisions is neglected, it reduces to the conventional Boltzmann term. The transition probability matrix $\hat{W}$ can be expressed through the scattering amplitude $\hat{F}^p_{p'}$ given by (8.27):

$$W^{\mu\mu'}_{\mu_1\mu_1'} = N v F^{\mu p}_{\mu_1 p'}\left(F^{\mu' p}_{\mu_1' p'}\right)^{\star}, \qquad (8.52)$$

where N is the impurity concentration, $v = p/m = p'/m$ is the electron velocity, and $F^{\mu p}_{\mu_1 p'}$ is the scattering amplitude for the transition from the initial state $\mu_1 p'$ to the final state μp.

The question of whether or not the spin band splitting can be neglected, while considering collisions, is a subtle one, especially when dealing with non-diagonal in band indices elements of the density matrix.[23]

Equation (8.50) resembles the usual Boltzmann equation, the main differences being the form of the collision integral and the additional commutator term due to the spin band splitting. It can be separated into two coupled equations by putting $\hat{\rho} = (1/2)f\hat{I} + 2S\hat{s}$, where the particle and spin distributions in phase space are related to $\hat{\rho}$ by the relations:

$$f(\boldsymbol{r}, \boldsymbol{p}, t) = \text{Tr}(\hat{\rho}), \qquad S(\boldsymbol{r}, \boldsymbol{p}, t) = \text{Tr}(\hat{s}\hat{\rho}). \qquad (8.53)$$

The spin polarization density $\boldsymbol{P}(\boldsymbol{r})$ used in this chapter is related to the distribution S by $\boldsymbol{P}(\boldsymbol{r}) = 2\int S(\boldsymbol{r}, \boldsymbol{p})\, d^3p$. The equations for f and S can be derived from (8.50):

$$\frac{\partial f}{\partial t} + \frac{p_i}{m}\frac{\partial f}{\partial x_i} - eE_i\frac{\partial f}{\partial p_i} + \frac{\partial \Omega_j}{\partial p_i}\frac{\partial S_j}{\partial x_i} = \text{Tr}(\hat{I}), \qquad (8.54)$$

---

[22] Like in the case of the classical Boltzmann equation, this term can be derived by making the usual assumptions of the kinetic theory.

[23] This point might be important for the "intrinsic" mechanism of spin currents, which arise because the applied electric field mixes the states in different bands. So does the impurity potential, and great care should be taken to be sure that *all* corrections on the order of $(\Omega(p)\tau_p)^{-1}$ have been picked up. The lesson of the "universal spin Hall conductivity" calls for extreme caution in these matters.

$$\frac{\partial S_k}{\partial t} + \frac{p_i}{m} \frac{\partial S_k}{\partial x_i} - eE_i \frac{\partial S_k}{\partial p_i} + \frac{1}{4} \frac{\partial \Omega_k}{\partial p_i} \frac{\partial f}{\partial x_i} - \left[\boldsymbol{\Omega}(\boldsymbol{p}) \times \boldsymbol{S} \right]_k = \mathrm{Tr}(\hat{s}_k \hat{I}). \qquad (8.55)$$

These equations are further simplified in the spatially homogeneous situation.

They contain most of the relevant spin physics in semiconductors: spin relaxation, spin diffusion, coupling between spin and charge currents (due to skew scattering, as well as intrinsic), etc. After including a magnetic field in the usual manner, they can be used to study magnetic effects in spin relaxation and spin transport. Similar, but more complicated, equations were derived and analysed for $J = 3/2$ holes in the valence band and carriers in a gapless semiconductor [24]. Compared to other, more sophisticated techniques, the approach based on the kinetic equation, has the advantage of being much more transparent and of allowing to use the physical intuition accumulated in dealing with the Boltzmann equation. It also avoids invoking quantum mechanics where it is not really necessary.

Acknowledgement. We thank Maria Lifshits for very helpful discussions.

References

[1] M.I. Dyakonov, V.I. Perel, Pis'ma Z. Eksp. Teor. Fiz. **13**, 657 (1971); JETP Lett. **13**, 467 (1971)
[2] M.I. Dyakonov, V.I. Perel, Phys. Lett. A **35**, 459 (1971)
[3] H. Senftleben, Z. Phys. **31**, 822 (1930)
[4] H. Senftleben, Z. Phys. **31**, 961 (1930)
[5] C.J. Gorter, Naturwissenschaften. **26**, 140 (1938)
[6] L. Gorelik, Pis'ma Z. Eksp. Teor. Fiz. **33**, 403 (1981); Sov. Phys. JETP Lett. **33**, 387 (1981)
[7] N.S. Averkiev, M.I. Dyakonov, Pis'ma Z. Eksp. Teor. Fiz. **35**, 196 (1982); Sov. Phys. JETP Lett. **35**, 242 (1982)
[8] M.I. Dyakonov, Phys. Rev. Lett. **99**, 126601 (2007)
[9] E.H. Hall, Philos. Mag. **10**, 301 (1880)
[10] E.H. Hall, Philos. Mag. **12**, 157 (1881)
[11] J. Smit, Physica. **17**, 612 (1951)
[12] J. Smit, Physica. **21**, 877 (1955)
[13] J. Smit, Physica. **24**, 29 (1958)
[14] R. Karplus, J.M. Luttinger, Phys. Rev. **95**, 1154 (1954)
[15] J.N. Chazalviel, I. Solomon, Phys. Rev. Lett. **29**, 1676 (1972)
[16] M.I. Miah, J. Phys. D: Appl. Phys. **40**, 1659 (2007)
[17] N.S. Averkiev, M.I. Dyakonov, Fiz. Tekh. Poluprovodn. **17**, 629 (1983); Sov. Phys. Semicond. **17**, 393 (1983)
[18] A.A. Bakun, B.P. Zakharchenya, A.A. Rogachev, M.N. Tkachuk, V.G. Fleisher, Pis'ma Z. Eksp. Teor. Fiz. **40**, 464 (1984); Sov. Phys. JETP Lett. **40**, 1293 (1984)
[19] M.N. Tkachuk, B.P. Zakharchenya, V.G. Fleisher, Z. Eksp. Teor. Fiz. Pis'ma **44**, 47 (1986); Sov. Phys. JETP Lett. **44**, 59 (1986)
[20] J.E. Hirsh, Phys. Rev. Lett. **83**, 1834 (1999)
[21] Y.K. Kato, R.C. Myers, A.C. Gossard, D.D. Awschalom, Science **306**, 1910 (2004)
[22] J. Wunderlich et al., Phys. Rev. Lett. **94**, 047204 (2005)

[23] J. Shi, P. Zhang, D. Xiao, Q. Niu, Phys. Rev. Lett. **96**, O76604 (2006)

[24] M.I. Dyakonov, A.V. Khaetskii, Z. Eksp. Teor. Fiz. **86**, 1843 (1984); Sov. Phys. JETP **59**, 1072 (1984)

[25] A.V. Khaetskii, Fiz. Tekh. Poluprovodn **18**, 1744 (1984); Sov. Phys. Semicond. **18**, 1091 (1984)

[26] N.F. Mott, H.S.W. Massey, *The Theory of Atomic Collisions*, 3rd edn. (Clarendon Press, Oxford, 1965)

[27] M.I. Dyakonov, V.I. Perel, Fiz. Tekh. Poluprov. **10**, 350 (1976); Sov. Phys. Semicond. **10**, 208 (1976)

[28] D.A. Abanin, A.V. Shytov, L.S. Levitov, B.I. Halperin, Nonlocal charge transport mediated by spin diffusion in the Spin-Hall Effect, arXiv:0708.0455 (2007)

[29] E.L. Ivchenko, G.E. Pikus, Sov. Phys. JETP Lett. **27**, 604 (1978)

[30] V.I. Belinicher, Phys. Lett. A **66**, 213 (1978)

[31] V.I. Belinicher, B.I. Sturman, Usp. Fiz. Nauk **130**, 415 (1980)

[32] V.M. Asnin, A.A. Bakun, A.M. Danishevskii, E.L. Ivchenko, G.E. Pikus, A.A. Rogachev, Solid State Commun. **30**, 565 (1979)

[33] M.I. Dyakonov, V.Yu. Kachorovskii, Sov. Phys. Semicond. **20**, 110 (1986)

[34] Y.A. Bychkov, E.I. Rashba, J. Phys. C **17**, 6039 (1984)

[35] V.K. Kalevich, V.I. Korenev, I.A. Merkulov, Solid State Commun. **91**, 559 (1994)

[36] N.A. Sinitsyn, J. Phys. Condens. Matter. **20**, 023201 (2008); arXiv:0712.0183

[37] H.-A. Engel, E.I. Rashba, B.I. Halperin, *Handbook of Magnetism and Advanced Magnetic Materials*, vol. 5 (Wiley, New York, 2006), p. 2858; arXiv:cond-mat/0603306

[38] N.F. Mott, Proc. R. Soc. A **124**, 425 (1929)

[39] L. Berger, Phys. Rev. B **2**, 4559 (1970)

[40] L. Berger, Phys. Rev. B **6**, 1862 (1972)

[41] L.D. Landau, E.M. Lifshits, *Quantum Mechanics. Non-relativistic Theory*, 3rd edn. (Elsevier Science, Oxford, 1977)

[42] V.N. Abakumov, I.N. Yassievich, Z. Eksp. Teor. Fiz. **61**, 2571 (1971); Sov. Phys. JETP **34**, 1375 (1971)

[43] P. Nozières, C. Lewiner, J. Phys. (Paris) **34**, 901 (1973)

[44] S.K. Lyo, T. Holstein, Phys. Rev. Lett. **29**, 423 (1972)

[45] V.F. Gantmakher, Y.B. Levinson, *Carrier Scattering in Metals and Semiconductors* (North-Holland, Amsterdam, 1987)

[46] S. Murakami, N. Nagaosa, S.-C. Zhang, Science **301**, 1348 (2003)

[47] J. Sinova, D. Culcer, Q. Niu, N.A. Sinitsyn, T. Jungwirth, A.H. MacDonald, Phys. Rev. Lett. **92**, 126603 (2004)

[48] E.G. Mishchenko, A.V. Shytov, B.I. Halperin, Phys. Rev. Lett. **93**, 226602 (2004)

[49] O.V. Dimitrova, Phys. Rev. B **71**, 245327 (2005)

[50] R. Winkler, *Spin-Orbit Coupling Effects in Two-Dimensional Electron and Hole Systems* (Springer, Berlin, 2003)

[51] A.V. Shytov, E.G. Mishchenko, H.-A. Engel, B.I. Halperin, Phys. Rev. B **73**, 075316 (2006)

[52] A. Khaetskii, Phys. Rev. B **73**, 115323 (2006)

[53] V.A. Zyuzin, P.G. Silvestrov, E.G. Mishchenko, Phys. Rev. Lett. **99**, 106601 (2007)

[54] A.V. Khaetskii, E.V. Sukhorukov, to be published

[55] V. Sih, V.H. Lau, R.C. Myers, V.R. Horowitz, A.C. Goassard, D.D. Awschalom, Phys. Rev. Lett. **97**, 096605 (2006)

[56] K. Nomura, J. Wunderlich, J. Sinova, B. Kaestner, A.H. MacDonald, T. Jungwirth, Phys. Rev. B **72**, 245330 (2005)

[57] V. Sih, R.C. Myers, Y.K. Kato, W.H. Lau, A.C. Gossard, D.D. Awschalom, Nat. Phys. **1**, 31 (2005)

[58] S.O. Valenzuela, M. Tinkham, Nature **442**, 176 (2006)

[59] T. Kimura, Y. Otani, T. Sato, S. Takahashi, S. Maekawa, Phys. Rev. Lett. **98**, 156601 (2007)

[60] N.P. Stern, S. Ghosh, G. Xiang, M. Zhu, N. Samarth, D.D. Awschalom, Phys. Rev. Lett. **97**, 126603 (2006)

[61] M.I. Dyakonov, V.I. Perel, Decay of atomic polarization moments, in: *Proc. 6th Intern. Conf. on Atomic Physics*, Riga, 1978, p. 410

Spin–Photogalvanics

E.L. Ivchenko and S.D. Ganichev

9.1 Introduction. Phenomenological Description

The spin of electrons and holes in solid state systems is an intensively studied quantum mechanical property showing a large variety of interesting physical phenomena. One of the most frequently used and powerful methods of generation and investigation of spin polarization is optical orientation [1]. Besides purely optical phenomena like circularly-polarized photoluminescence, the optical generation of an unbalanced spin distribution in a semiconductor may lead to spin photocurrents.

Light propagating through a semiconductor and acting upon mobile carriers can generate a dc electric current, under short-circuit condition, or a voltage, in the case of open-circuit samples. In this chapter we consider only the photogalvanic effects (PGE) which, by definition, appear neither due to inhomogeneity of optical excitation of electron–hole pairs nor due to inhomogeneity of the sample. Moreover, we focus the attention here on the spin–photogalvanics and discuss spin-related mechanisms of the following effects: the circular PGE, spin–galvanic effect, inverse spin–galvanic effect or spin polarization by electric current, generation of pure spin photocurrents and magneto-gyrotropic photogalvanic effect.

The macroscopic features of all spin-dependent photogalvanic effects discussed in this chapter, e.g., the possibility to generate a helicity-dependent current, its behavior upon variation of radiation helicity, crystallographic orientation, experimental geometry, etc., can be described in the frame of a phenomenological theory which operates with conventional vectors, or *polar* vectors, and pseudovectors, or *axial* vectors describing rotation, and does not depend on details of microscopic mechanisms. Below we consider one by one the phenomenological theory of various spin–photogalvanic effects.

Circular Photogalvanic Effect

The signature of photocurrent due to the circular PGE is that it appears only under illumination with circularly polarized light and reverses its direction when the sign of circular polarization is changed. Physically, the circular PGE can be considered as a transformation of the photon angular momenta into a translational motion of free charge carriers. It is an electronic analog of mechanical systems which transmit rotatory motion to linear one. In general there exist two different possibilities for such transmission. The first is based on the wheel effect: when a wheel being in mechanical contact with the plane surface rotates it simultaneously moves as a whole along the surface. The second transformation is based on the screw effect and exemplified by a screw thread or a propeller. In the both cases the rotation inversion results in the reversal of motion, like the circular photogalvanic current changes its sign following the inversion of the photon helicity described by the degree of circular polarization P_c.

Phenomenologically, the circular photogalvanic current j is described by a second-order pseudotensor

$$j_\lambda = I\gamma_{\lambda\mu}\mathrm{i}(e \times e^*)_\mu = IP_c\gamma_{\lambda\mu}n_\mu, \qquad (9.1)$$

where n is the unit vector pointing in the direction of the exciting beam, I and e are the light intensity and polarization unit vector. Hereafter a repeated subscript is understood to imply summation over the range of this subscript. In a bulk semiconductor or superlattice the index λ runs over all three Cartesian coordinates x, y, z. In quantum well structures the free-carrier motion along the growth direction is quantized and the index λ enumerates two in-plane coordinates. In quantum wires the free movement is allowed only along one axis, the principal axis of the structure, and the coordinate λ is parallel to this axis. On the other hand, the light polarization unit vector e and directional unit vector n can be arbitrarily oriented in space and, therefore, $\mu = x, y, z$. The tensor γ in (9.1) relates components of the polar vector j and the axial vector $e \times e^*$ which means that it is nonzero for point groups that allow optical activity or *gyrotropy*. We remind the reader that the gyrotropic point group symmetry makes no difference between components of polar vectors, like current or electron momentum, and axial vectors, like a magnetic field or spin. Among 21 crystal classes lacking inversion symmetry, 18 are gyrotropic. Three nongyrotropic noncentrosymmetric classes are T_d, C_{3h}, and D_{3h}.

Spin–Galvanic and Inverse Spin–Galvanic Effects

Another root to spin–photogalvanics is provided by optical spin orientation. A uniform nonequilibrium spin polarization obtained by any means, including optical, yields an electric current if the system is characterized by the gyrotropic symmetry. The current j and spin S are also related by a second-order pseudotensor

$$j_\lambda = Q_{\lambda\mu}S_\mu. \qquad (9.2)$$

This equation shows that the direction of the electric current is coupled to the orientation of the nonequilibrium spin which is given by the radiation helicity. The effect inverse to the spin–galvanic effect is an electron spin polarization induced by a dc electric current, namely,

$$S_\mu = R_{\mu\lambda}j_\lambda. \qquad (9.3)$$

We note the similarity of (9.1)–(9.3) characteristic for effects due to gyrotropy: all three equations linearly couple a polar vector with an axial vector.

Pure Spin Photocurrents

While describing spin-dependent phenomena, one needs, in addition to electric currents, introduce spin currents. Hereafter, similarly to the notations of Chap. 8, we use the notation $q_{\lambda\mu}$ for the spin flux density with μ indicating the spin orientation and λ indicating the flow direction. Of special interest is the generation of pure spin currents in which case a charge current is absent but at least one of the $q_{\lambda\mu}$ components is nonzero. A fourth-order tensor P relating the pseudotensor q with the light intensity and polarization as follows:

$$q_{\lambda\mu} = I P_{\lambda\mu\nu\eta} e_\nu e_\eta^* \tag{9.4}$$

has nonzero components in all systems lacking a center of inversion symmetry. However an equivalence between polar and axial vector components in the gyrotropic point groups suggests new important mechanisms of pure spin currents connected with the spin–orbit splitting of electronic bands.

Magneto-Photogalvanic Effects

The variety of effects under consideration is completed by magnetic-field induced photocurrents gathered in the class of magneto-photogalvanic effects represented by the phenomenological equation

$$j_\lambda = I \Phi_{\lambda\mu\nu\eta} B_\mu e_\nu e_\eta^*, \tag{9.5}$$

where B is an external magnetic field. The symmetry properties of the tensor Φ coincide with those of P. In this chapter we will consider a magneto-gyrotropic photocurrent induced by a linearly polarized radiation which can be directly connected with the pure spin current generated at zero magnetic field.

9.2 Circular Photogalvanic Effect

9.2.1 Historical Background

The circular photogalvanic effect was independently predicted by Ivchenko and Pikus [2] and Belinicher [3]. It was first observed and studied in tellurium crystals by Asnin et al. [4], see more references in the book [5]. In tellurium the current arises due to spin splitting of the valence band edge at the boundary of the first Brillouin-zone ("camel back" structure). While neither bulk zinc-blende materials like GaAs and related compounds nor bulk diamond crystals like Si and Ge allow this effect, in quantum well structures the circular PGE is possible due to a reduction of symmetry. The circular PGE in gyrotropic quantum wells was observed by Ganichev et al. applying terahertz radiation [6]. In this chapter we discuss the circular PGE in quantum well structures grown along the [001], [113], and [110] directions, present experimental data for demonstration and outline the microscopic theory of the effect under inter-sub-band and interband optical transitions.

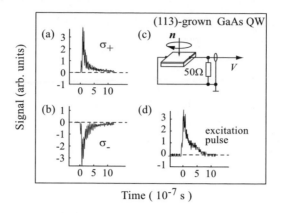

Fig. 9.1. Oscilloscope traces obtained for pulsed excitation of (113)-grown n-type GaAs quantum wells at $\lambda = 76\,\mu$m. (**a**) and (**b**) show circular PGE signals obtained for circular σ_+ and σ_- polarization, respectively. For comparison, in (**d**) a signal pulse for a fast photon drag detector is plotted. In (**c**) the measurement arrangement is sketched. After [7]

9.2.2 Basic Experiments

With illumination of quantum well structures by polarized radiation a current signal proportional to the helicity P_c is generated in unbiased samples. The irradiated structure represents a current source wherein the current flows in the quantum well, see a scheme in Fig. 9.1(c). Figures 9.1(a) and (b) show measurements of the voltage drop across a 50 Ohm load resistor in response to 100 ns laser pulses at $\lambda = 76\,\mu$m. Signal traces are plotted for right-handed (a) and left-handed circular polarization (b), in comparison to a reference signal shown in Fig. 9.1(d) and obtained from a fast photon drag detector [7, 8]. The width of the current pulses is about 100 ns which corresponds to the THz laser pulses duration.

Figure 9.2(a) presents results of measurements carried out at room temperature on (113)-grown p-GaAs/AlGaAs multiple quantum wells under normal incidence and (001)-grown n-InAs/AlGaSb single quantum well structure under oblique incidence. Optical excitation was performed by a high-power THz pulsed NH$_3$ laser operating at wavelength $\lambda = 76\,\mu$m. The linearly polarized light emitted by the laser could be modified to an elliptically polarized radiation by applying a $\lambda/4$ plate and changing the angle φ between the optical axis of the plate and the polarization plane of the laser radiation. Thus the helicity P_c of the incident light varies from -1 (left handed, σ_-) to $+1$ (right handed, σ_+) according to

$$P_c = \sin 2\varphi. \tag{9.6}$$

One can see from Fig. 9.2(a) that the photocurrent direction is reversed when the polarization switches from right-handed circular, $\varphi = 45°$, to left-handed, $\varphi = 135°$. The experimental points are well fitted by the equation

$$j_\lambda(\varphi) = j_\lambda^0 \sin 2\varphi \tag{9.7}$$

with one scaling parameter.

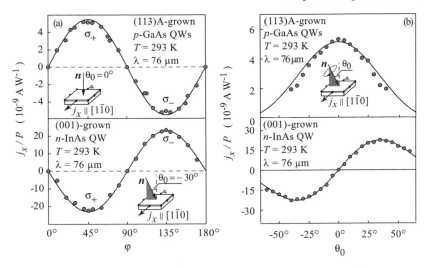

Fig. 9.2. (a) Photocurrent in quantum wells normalized by the light power P as a function of the phase angle φ defining helicity. *The insets* show the geometry of the experiment. *Upper panel*: Normal incidence of radiation on p-type (113)-grown GaAs/AlGaAs quantum wells (symmetry class C_s). The current j_x flows along the $[1\bar{1}0]$ direction perpendicular to the mirror plane. *Lower panel*: Oblique incidence of radiation with an angle of incidence $\theta_0 = -30°$ on n-type (001)-grown InAs/AlGaSb quantum wells (symmetry class C_{2v}). *Full lines* are fitted using one parameter according to (9.7). **(b)** Photocurrent as a function of the incidence angle θ_0 for right-circularly polarized radiation σ_+ measured perpendicularly to light propagation. *Upper panel*: p-type (113)A-grown GaAs/AlGaAs quantum wells. *Lower panel*: n-type (001)-grown InAs/AlGaSb quantum wells. *Full lines* represent theoretical fit. After [6]

In Fig. 9.2(b) closer look is taken at the dependence of the photocurrent on the angle of incidence θ_0 in configuration with the incidence plane normal to the axis $x \parallel [1\bar{1}0]$.

For (113)-oriented quantum wells belonging to the symmetry class C_s the current retains its sign for all θ_0 and achieves its maximum at normal incidence (see upper panel of Fig. 9.2(b)). In contrast, in asymmetric (001)-oriented samples (C_{2v}-symmetry) a variation of θ_0 in the plane of incidence normal to x changes the sign of the current j_x for normal incidence, $\theta_0 = 0$, as can be seen in the lower panel of Fig. 9.2(b). Solid curves in this figure show a fit with phenomenological equation (9.1) adapted to a corresponding symmetry and being in a good agreement with experiment.

Further experiments demonstrate that circular PGE can be generated by the radiation of wide range of frequencies from terahertz to visible light. Applying light of various frequencies the photocurrent due to interband, inter-sub-band and free carrier absorption was detected. Absorption of radiation in the range of 9 to 11 μm in n-type GaAs quantum well samples of well widths 8–9 nm is dominated by resonant direct inter-sub-band optical transitions between the first ($e1$) and the second

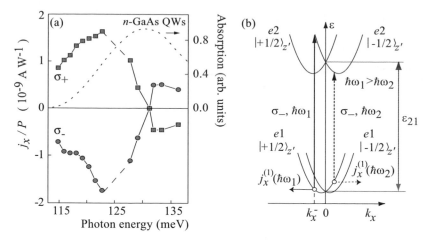

Fig. 9.3. (a) Photocurrent in quantum wells normalized by the light power P as a function of the photon energy $\hbar\omega$. Measurements are presented for n-type (001)-grown GaAs/AlGaAs quantum wells of 8.2 nm width at room temperature. Oblique incidence of σ_+ (*squares*) and σ_- (*circles*) circularly polarized radiation with an angle of incidence $\theta_0 = 20°$ was used. The current j_x was measured perpendicular to the light incidence plane (y, z). *The dotted line* shows the absorption measurement using a Fourier transform infrared spectrometer. After [9]. (b) Microscopic picture describing the origin of the circular photogalvanic current and its spectral inversion in C_s point group samples. The essential ingredient is the splitting of the conduction band due to k-linear terms. Left handed circularly polarized radiation σ_- induces direct spin–flip transitions (*vertical arrows*) from the $e1$ sub-band with $s = 1/2$ to the $e2$ sub-band with $s' = -1/2$. As a result an unbalanced occupation of the k_x states occurs yielding a spin polarized photocurrent. For transitions with k_x^- lying left to the minimum of $e1$ ($s = 1/2$) sub-band the current indicated by j_x is negative. At smaller ω the transition occurs at a value of k_x on the right-hand side of the sub-band minimum, and the current reverses its sign

($e2$) size-quantized sub-bands. Applying mid-infrared (MIR) radiation from the CO_2 laser, which causes direct transitions in GaAs quantum wells, a current signal proportional to the helicity P_c has been observed at normal incidence in (113)-oriented samples and (110)-grown asymmetric quantum wells and at oblique incidence in (001)-oriented samples, indicating a spin orientation induced circular PGE [9, 10]. In Fig. 9.3(a) the data are presented for a (001)-grown n-type GaAs quantum well of 8.2 nm width measured at room temperature. It is seen that the current for both left- and right-handed circular polarization changes sign at the frequency of the absorption peak. Spectral inversion of the photocurrent direction at inter-sub-band resonance has also been observed in (113)-oriented samples and (110)-grown asymmetric quantum wells [9, 10]. In the next subsection we start the microscopic consideration from the inter-sub-band mechanism of circular PGE.

9.2.3 Microscopic Model for Inter-Sub-Band Transitions

Microscopically, as shown below, a conversion of photon helicity into a current as well as a number of effects described in this chapter is due to a removal of spin degeneracy in the k-space resulting in a shift of two spin sub-bands as sketched in Fig. 9.3(b). Thus before a discussion of the photocurrent origin we briefly describe the band spin splitting.

9.2.4 Relation to k-Linear Terms

The linear in k terms are given by the term

$$\mathcal{H}_{k}^{(1)} = \beta_{\mu\lambda}\sigma_{\mu}k_{\lambda} \tag{9.8}$$

in the expansion of the electron effective Hamiltonian in powers of the wave vector k. The coefficients $\beta_{\mu\lambda}$ form a pseudotensor subjected to the same symmetry restriction as the current-to-spin tensor R or the transposed pseudotensors γ and Q. The coupling between the electron spin and momentum described by products of the Pauli spin matrices σ_{μ} and the wave vector components k_{λ} as well as spin-dependent selection rules for optical transitions yield a net current sensitive to circularly polarized optical excitation. The sources of k-linear terms are the bulk inversion asymmetry resulted in the Dresselhaus term [11], including a possible interface inversion asymmetry [12, 13], and a structure inversion asymmetry causing the Rashba term [14] (for reviews see [7, 15, 16]).

9.2.5 Circular PGE Due to Inter-Sub-Band Transitions

Figure 9.3(b) illustrates the inter-sub-band transitions $e1 \rightarrow e2$ resulting in the circular PGE. In order to make the physics more transparent we will first consider the inter-sub-band circular photogalvanic current generated under normal incidence in quantum wells of the C_s symmetry, say, in (113)-grown quantum wells, and use the relevant coordinate system $x \parallel [1\bar{1}0]$, $y' \parallel [33\bar{2}]$, $z' \parallel [113]$. In the linear-k Hamiltonian we retain only the term $\sigma_{z'}k_x$ because other terms make no contribution to the photocurrent under the normal incidence. Therefore, the energy dispersion in the νth electron sub-band depicted in Fig. 9.3(b) is taken as

$$E_{e\nu,k,s} = E_{\nu}^0 + \frac{\hbar^2 k^2}{2m_c} + 2s\beta_{\nu}k_x, \tag{9.9}$$

where $s = \pm 1/2$ is the electron spin component along the z' axis, $\beta_{\nu} = \beta_{z'x}^{(\nu)}$ and, for the sake of simplicity, we neglect nonparabolicity effects assuming the effective mass m_c to be the same in both sub-bands.

For the direct $e1$–$e2$ transitions shown in Fig. 9.3 by vertical arrows, the energy and momentum conservation laws read

$$E_{21} + 2(s'\beta_2 - s\beta_1)k_x = \hbar\omega,$$

where E_{21} is the Γ-point gap $E_2^0 - E_1^0$ and $s', s = \pm 1/2$. As a result of optical selection rules the circular polarization, e.g., left-handed, under normal incidence induces direct optical transitions between the sub-band $e1$ with spin $s = +1/2$ and the sub-band $e2$ with spin $s' = -1/2$. For monochromatic radiation with photon energy $\hbar\omega_1 > E_{21}$ optical transitions occur only at a fixed value of k_x^- where the energy of the incident light matches the transition energy as indicated by the arrow in Fig. 9.3(b). Therefore, optical transitions induce an imbalance of the momentum distribution in both sub-bands yielding an electric current along the x-direction with the $e1$ and $e2$ contributions, antiparallel ($j^{(1)}$) or parallel ($j^{(2)}$) to x, respectively. Since in n-type quantum wells the energy separation between the $e1$ and $e2$ sub-bands is typically larger than the energy of longitudinal optical phonons $\hbar\omega_{LO}$, the nonequilibrium distribution of electrons in the $e2$ sub-band relaxes rapidly due to emission of phonons. As a result, the electric current $j^{(2)}$ vanishes and the current magnitude and direction are determined by the group velocity and the momentum relaxation time τ_p of uncompensated electrons in the $e1$ sub-band with $s = +1/2$, i.e., by $j^{(1)}$. By switching circular polarization from left- to right-handed due to selection rules light excites the spin down sub-band only. Thus the whole picture mirrors and the current direction reverses. Spectral inversion of the photocurrent at fixed helicity also follows from the model picture of Fig. 9.3(b). Indeed decreasing the photon frequency to $\hbar\omega_2 < E_{21}$ shifts the transitions toward positive k_x and the direction of the current reverses (horizontal dashed arrow).

Formally this process is described as follows. In the polarization $e \perp z'$, the direct inter-sub-band absorption is weakly allowed only for the spin–flip transitions, $(e1, -1/2) \to (e2, 1/2)$ for σ_+ photons and $(e1, 1/2) \to (e2, -1/2)$ for σ_- photons. Particularly, under the σ_- photoexcitation the electrons involved in the transitions have the fixed x-component of the wave vector

$$k_x^- = -\frac{\hbar\omega - E_{21}}{\beta_2 + \beta_1} \tag{9.10}$$

and velocity

$$v_x^{(e\nu)} = \frac{\hbar k_x^-}{m_c} + (-1)^{\nu+1}\frac{\beta_\nu}{\hbar}. \tag{9.11}$$

The circular photogalvanic current can be written as

$$j_x^{(e1)} = e\left(v_x^{(e2)}\tau_p^{(2)} - v_x^{(e1)}\tau_p^{(1)}\right)\frac{\eta_{21}I}{\hbar\omega}P_c, \tag{9.12}$$

where $\tau_p^{(\nu)}$ is the electron momentum relaxation time in the ν sub-band, η_{21} is the absorbance or the fraction of the energy flux absorbed in the quantum well due to the transitions under consideration, and minus in the right-hand side means that the $e1$-electrons are removed in the optical transitions.

We assume that inhomogeneous broadening, δ_{21}, of the resonance E_{21} exceeds, by far, the sub-band spin splitting. In this case the convolution of the current given by (9.12) with the inhomogeneous distribution function leads to [9]

$$j_x = \frac{e}{\hbar}(\beta_2 + \beta_1)\left[\tau_{\mathrm{p}}^{(2)}\eta_{21}(\hbar\omega) + \left(\tau_{\mathrm{p}}^{(1)} - \tau_{\mathrm{p}}^{(2)}\right)\langle E\rangle \frac{\mathrm{d}\eta_{21}(\hbar\omega)}{\mathrm{d}\hbar\omega}\right]\frac{I\,P_{\mathrm{c}}}{\hbar\omega}, \qquad (9.13)$$

where η_{21} is the absorbance in the polarization $e \perp z'$ calculated neglecting the linear-k terms but taking into account the inhomogeneous broadening, $\langle E\rangle$ is the mean value of the two-dimensional (2D) electron energy, namely, half of the Fermi energy E_{F} for a degenerate 2D electron gas and $k_{\mathrm{B}}T$ for a nondegenerate gas. Since the derivative $\mathrm{d}\eta_{21}/\mathrm{d}(\hbar\omega)$ changes its sign at the absorption peak frequency and usually the time $\tau_{\mathrm{p}}^{(1)}$ is much longer than $\tau_{\mathrm{p}}^{(2)}$, the circular photogalvanic current given by (9.13) exhibits the sign-inversion behavior within the resonant absorption contour, in agreement with the experimental observations [9, 10].

Similarly to the previously discussed case of the C_{s} symmetry the circular photogalvanic current for the $e1$–$e2$ transitions in (001)-grown quantum wells exhibits the sign-inversion behavior within the resonant absorption contour [9], in agreement with the experimental data presented in Fig. 9.3(a).

9.2.6 Interband Optical Transitions

For direct optical transitions between the heavy-hole valence sub-band hh1 and conduction sub-band $e1$, the circular PGE is also most easily conceivable in quantum wells of the C_{s} symmetry which allows the spin–orbit term $\beta_{z'x}\sigma_{z'}k_x$. For simplicity, for a while we take into account the linear-k terms only in the conduction sub-band assuming the following parabolic dispersion in the $e1$ and hh1 sub-bands:

$$E_{e1,k,\pm 1/2} = E_{\mathrm{g}}^{\mathrm{QW}} + \frac{\hbar^2 k^2}{2m_{\mathrm{c}}} \pm \beta_{\mathrm{e}}k_x, \qquad E_{\mathrm{hh1},k,\pm 3/2}^{\mathrm{v}} = -\frac{\hbar^2 k^2}{2m_{\mathrm{v}}} \pm \beta_{\mathrm{h}}k_x, \qquad (9.14)$$

where m_{v} is the hole in-plane effective mass, $\beta_{\mathrm{e}} = \beta_{z'x}^{(e1)}$, $\beta_{\mathrm{h}} = \beta_{z'x}^{(\mathrm{hh1})}$, $E_{\mathrm{g}}^{\mathrm{QW}}$ is the band gap renormalized because of the quantum confinement of free carriers and the energy is referred to the valence-band top. In Fig. 9.4(a) the allowed optical transitions are from $j = -3/2$ to $s = -1/2$ for the σ_+ polarization and from $j = 3/2$ to $s = 1/2$ for the σ_- polarization. Under circularly polarized radiation with a photon energy $\hbar\omega$ and for a fixed value of $k_{y'}$ the energy and momentum conservation allow transitions only from two values of k_x labeled k_x^- and k_x^+. The corresponding transitions are shown in Fig. 9.4(a) by the solid vertical arrows with their "center-of-mass" shifted from the point $k_x = 0$. Thus the average electron velocity in the excited state is nonzero and the contributions of $k_x^\pm$ photoelectrons to the current do not cancel each other as in the case $\beta_{\mathrm{e}} = \beta_{\mathrm{h}} = 0$. Changing the photon helicity from $+1$ to -1 inverts the current because the "center-of-mass" for these transitions is now shifted in the opposite direction. The asymmetric distribution of photoelectrons in the k-space decays within the momentum relaxation time. However, under steady-state optical excitation new photocarriers are generated resulting in a dc photocurrent. The photohole contribution is considered in a similar way. The final result for the interband circular photogalvanic current can be presented as

$$j_x = -e\left(\tau_{\mathrm{p}}^{\mathrm{e}} - \tau_{\mathrm{p}}^{\mathrm{h}}\right)\left(\frac{\beta_{\mathrm{e}}}{m_{\mathrm{v}}} + \frac{\beta_{\mathrm{h}}}{m_{\mathrm{c}}}\right)\frac{\mu_{\mathrm{cv}}}{\hbar}\frac{\eta_{\mathrm{eh}}I}{\hbar\omega}P_{\mathrm{c}}, \qquad (9.15)$$

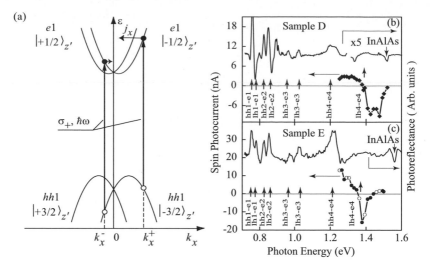

Fig. 9.4. (a) Microscopic picture describing the origin of interband circular PGE. The essential ingredient is the spin splitting of the electron and/or hole states due to linear-k terms. (b), (c) Spectral response of the circular photogalvanic current observed in two (001)-grown $In_xGa_{1-x}As$/InAlAs quantum well structures, sample D and sample E, under oblique incidence. To enhance the structure inversion asymmetry, sample E was grown with a graded indium composition from 0.53 to 0.75 for the quantum well, instead of the uniform indium composition of 0.70 for sample D. The photoreflectance spectra of two samples are also shown to determine the electronic structures of the samples. The arrows indicate the heavy hole (hh) and light hole (lh) related transitions. Data are from [19]

where η_{eh} is the fraction of the photon energy flux absorbed in the quantum well due to the hh1 $\rightarrow$ e1 transitions and τ_p^e, τ_p^h are the electron and hole momentum relaxation times.

The circular PGE at interband absorption was observed in GaAs-, InAs- and GaN-based quantum well structures [17–20]. In Figs. 9.4(b) and (c) diamonds and circles present the spectral dependence of the circular photogalvanic current measured under interband optical transitions between the higher valence and conduction sub-bands. The photoreflectance spectra of the samples are shown (solid curves) to clearly indicate the quantized energy levels of electrons and holes as marked by the arrows. For the both samples the photocurrent spectral contour exhibits a change in sign, in a qualitative agreement with the theoretical prediction [21].

9.2.7 Spin-Sensitive Bleaching

Application of high intensities results in saturation (bleaching) of PGE. This effect was observed for direct inter-sub-band transitions in p-type GaAs quantum wells and gave an experimental access to spin relaxation times [22, 23]. The method is based on the difference in nonlinear behavior of circular and linear PGE. The linear PGE is an another photogalvanic effect allowed in GaAs structures and can be induced by

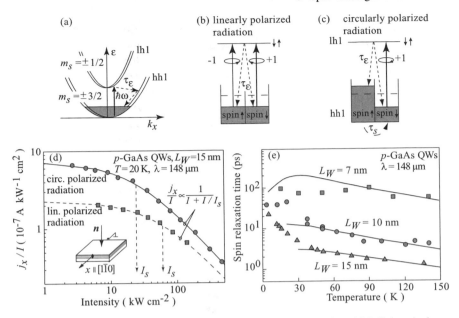

Fig. 9.5. (a)–(c) Microscopic picture of spin sensitive bleaching: Direct hh1–lh1 optical transitions (a) and process of bleaching for two polarizations, linear (b) and circular (c). *Dashed arrows* indicate energy (τ_ε) and spin (τ_s) relaxation. (d) Circular (*squares*) and linear PGE (*circles*) currents j_x normalized by the intensity as a function of the intensity for circularly and linearly polarized radiation. *The inset* shows the geometry of the experiment. The measurements are fitted to $j_x/I \propto 1/(1 + I/I_s)$ with one parameter I_s for each state of polarization. Data are given for (113)-grown samples from [22]. (e) Spin relaxation times of holes for three different widths of (113)-grown GaAs/AlGaAs quantum wells as a function of temperature. *The solid lines* show a fit according to the Dyakonov–Perel relaxation mechanism. From [23]

linearly polarized light [5, 8, 24]. Both currents are proportional to absorption and their nonlinear behavior reflects the nonlinearity of absorbance.

Spin sensitive bleaching can be analyzed in terms of a simple model taking into account both optical excitation and nonradiative relaxation processes. Excitation with THz radiation results in direct transitions between heavy-hole (hh1) and light-hole (lh1) sub-bands (Fig. 9.5(a)). This process depopulates and populates selectively spin states in the hh1 and lh1 sub-bands. The absorption is proportional to the difference of populations of the initial and final states. At high intensities the absorption decreases since the photoexcitation rate becomes comparable to the nonradiative relaxation rate to the initial state. Absorption of linearly polarized light is not spin selective and the saturation is controlled by energy relaxation (see Fig. 9.5(b)). In contrast, absorption of circularly polarized light is spin selective due to selection rules, and only one type of spin is excited, Fig. 9.5(c). Note that during energy relaxation the hot holes lose their photoinduced orientation due to rapid relaxation so that the spin orientation occurs only within the bottom of the hh1 sub-band. Thus the absorption bleaching of circularly polarized radiation is governed by energy relaxation of photoexcited carriers and spin relaxation within the sub-band hh1. These processes are characterized by energy and spin relaxation times, τ_ε and τ_s, respectively. If τ_s is longer than τ_ε the bleaching of absorption becomes spin sensitive and the saturation intensity of circularly polarized radiation drops below the value of linear polarization.

Bleaching of absorption with increasing the intensity of linearly polarized light is described phenomenologically by the function

$$\eta(I) = \frac{\eta_0}{1 + I/I_{se}}, \tag{9.16}$$

where $\eta_0 = \eta(I \to 0)$ and I_{se} is the characteristic saturation intensity controlled by energy relaxation of the 2D hole gas. Since the photocurrent of linear PGE, j_{LPGE}, induced by the linearly polarized light is proportional to ηI, one has

$$\frac{j_{LPGE}}{I} \propto \frac{1}{1 + I/I_{se}}. \tag{9.17}$$

The circular photogalvanic current j_{CPGE} induced by the circularly polarized radiation is proportional to the degree of hole spin polarization and given by [22]

$$\frac{j_{CPGE}}{I} \propto \frac{1}{1 + I(I_{se}^{-1} + I_{ss}^{-1})}, \tag{9.18}$$

where $I_{ss} = p_s \hbar \omega / (\eta_0 \tau_s)$, p_s is the 2D hole density.

The measurements illustrated in Fig. 9.5(d) indicate that the photocurrent j_x at a low power level depends linearly on the light intensity and gradually saturates with increasing intensity, $j_x \propto I/(1 + I/I_s)$, where I_s is the saturation parameter. One can see from Fig. 9.5(d) that the measured saturation intensity for circular polarized radiation is smaller than that for linearly polarized light. Using the measured values of I_s and (9.17) and (9.18) one can estimate the parameter I_{ss} and even the time τ_s [22, 23].

Figure 9.5(e) presents spin relaxation times extracted from experiment (points) together with a theoretical fit assuming that the Dyakonov–Perel mechanism of hole spin relaxation is dominant.

9.3 Spin–Galvanic Effect

The mechanisms of circular PGE discussed so far are linked with the asymmetry in the momentum distribution of carriers excited in optical transitions which are sensitive to the light circular polarization due to selection rules. Now we discuss an additional possibility to generate a photocurrent sensitive to the photon helicity. In a system of free carriers with nonequilibrium spin-state occupation but equilibrium energy distribution within each spin branch, the spin relaxation can be accompanied by generation of an electric current. This effect, predicted by Ivchenko et al. [25], was observed by Ganichev et al. applying THz radiation and named the spin–galvanic effect [26]. If the nonequilibrium spin is produced by optical orientation proportional to the degree of light circular polarization P_c the current generation can be reputed just as another mechanism of the circular PGE. However the nonequilibrium spin S can be achieved both by optical and non-optical methods, e.g., by electrical spin injection, and, in fact, (9.2) presents an independent effect. Usually the circular PGE and spin–galvanic effect are observed simultaneously under illumination by circularly polarized light and do not allow an easy experimental separation. However, they can be separated in time-resolved measurements. Indeed, after removal of light or under pulsed photoexcitation the circular photogalvanic current decays within the momentum relaxation time τ_p whereas the spin–galvanic current decays with the spin relaxation time. Another method which, on the one hand, provides a uniform distribution in spin sub-bands and, on the other hand, excludes the circular PGE was proposed in [26]. It is based on the use of optical excitation and the assistance of an external magnetic field to achieve an in-plane polarization in (001)-grown low-dimensional structures.

9.3.1 Microscopic Mechanisms

For (001)-grown asymmetric quantum wells characterized by the C_{2v} symmetry only two linearly independent components, Q_{xy} and Q_{yx}, of the tensor Q in (9.2) are nonzero so that

$$j_x = Q_{xy}S_y, \qquad j_y = Q_{yx}S_x. \tag{9.19}$$

Hence, a spin polarization driven current needs a spin component lying in the quantum well plane. For the C_s symmetry of (hhl)-oriented quantum wells, particularly, (113) and asymmetric (110), an additional tensor component $Q_{xz'}$ is nonzero and the spin–galvanic current may be caused by nonequilibrium spins oriented normally to the quantum well plane.

Figure 9.6 illustrates the generation of a spin–galvanic current. As already addressed above, it arises due to k-linear terms in the electron effective Hamiltonian, see (9.8). For a 2D electron gas system, these terms lead to the situation sketched in Fig. 9.6(a). More strictly, the scattering changes both k_x and k_y components of the electron wave vector as shown Fig. 9.6(b) by dashed lines. However, the one-dimensional sketch in Fig. 9.6(a) conveys the interpretation in a simpler and clearer way. In the figure the electron energy spectrum along k_x with allowance for the spin-dependent term $\beta_{yx}\sigma_y k_x$ is shown. In this case $s_y = \pm 1/2$ is a good quantum number. The electron energy band splits into two sub-bands which are shifted in the k-space, and each of the bands comprises states with spin up or down. Spin orientation in the y-direction causes the unbalanced population in spin-down and spin-up sub-bands. As long as the carrier distribution in each sub-band is symmetric around the sub-band minimum point no current flows.

As illustrated in Fig. 9.6(a) the current flow is caused by k-dependent spin–flip relaxation processes. Spins oriented in the y-direction are scattered along k_x from the more populated spin sub-band, e.g., the sub-band $|+1/2\rangle_y$, to the less populated sub-band $|-1/2\rangle_y$. Four different spin–flip scattering events are sketched in Fig. 9.6(a)

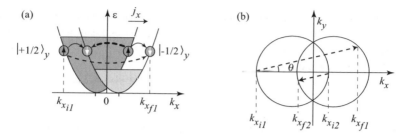

Fig. 9.6. Microscopic origin of the spin–galvanic current. (**a**) One-dimensional sketch: the $\sigma_y k_x$ term in the Hamiltonian splits the conduction band into two parabolas with the spin $s_y = \pm 1/2$ pointing in the y-direction. If one of the spin–split sub-bands is preferentially occupied, e.g., by spin injection ($|{+}1/2\rangle_y$-states in the figure), spin–flip scattering results in a current in the x-direction. The scattering rate depends on values of the initial and final electron wave vectors. Thus, the transitions sketched by dashed arrows yield an asymmetric filling of the sub-bands and, hence, a current flow. If instead of the spin-up sub-band the spin-down sub-band is preferentially occupied, the current direction is reversed. (**b**) The spin–flip transitions in the two dimensions at scattering angle θ different from 0

by bent arrows. The spin–flip scattering rate depends on the values of wave vectors of the initial and final states [27]. Therefore, the spin–flip transitions shown by solid arrows in Fig. 9.6(a) have the same rates. They preserve the symmetric distribution of carriers in the sub-bands and, thus, do not yield a current. However, two other scattering processes shown by broken arrows are inequivalent and generate an asymmetric carrier distribution around the sub-band minima in both sub-bands. This asymmetric population results in a current flow along the x-direction. The occurrence of a current is due to the spin dependence of the electron scattering matrix elements $\hat{M}_{k'k} = A_{k'k}\hat{I} + \boldsymbol{\sigma} \cdot \boldsymbol{B}_{k'k}$, where $A^*_{k'k} = A_{kk'}$, $B^*_{k'k} = B_{kk'}$ due to hermicity of the interaction and $A_{-k',-k} = A_{kk'}$, $\boldsymbol{B}_{-k',-k} = -\boldsymbol{B}_{kk'}$ due to the symmetry under time inversion. Within the model of elastic scattering the current is not spin polarized since the same number of spin-up and spin-down electrons move in the same direction with the same velocity. The spin–galvanic current can be estimated by [28]

$$j_x = Q_{xy} S_y \sim e n_s \frac{\beta_{yx}}{\hbar} \frac{\tau_p}{\tau'_s} S_y, \tag{9.20}$$

and the similar equation for j_y, where n_s is the 2D electron density, τ'_s is the spin relaxation time due to the Elliott–Yafet mechanism [1]. Since spin–flip scattering is the origin of the current given by (9.20), this equation is valid even if Dyakonov–Perel mechanism [1] of spin relaxation dominates. The Elliott–Yafet relaxation time τ'_s is proportional to the momentum relaxation time τ_p. Therefore the ratio τ_p/τ'_s in (9.20) does not depend on the momentum relaxation time. The in-plane average spin, e.g., S_y in (9.20), decays with the total spin relaxation time τ_s and, hence, the time decay of the spin–galvanic current following pulsed photoexcitation is described by the exponential function $\exp(-t/\tau_s)$. In contrast, the circular PGE current induced by a short-pulse decays within the momentum relaxation time τ_p allowing to distinguish these two effects in time resolved measurements.

In general, in addition to the kinetic contribution to the current there exists the so-called relaxational contribution which arises due to the linear-k terms neglecting the Elliott–Yafet spin relaxation, i.e., with allowance for the Dyakonov–Perel mechanism only. This contribution has the form

$$j = -en_s\tau_p \nabla_k \left(\boldsymbol{\Omega}_k^{(1)} \dot{S} \right), \qquad (9.21)$$

where the spin rotation frequency $\boldsymbol{\Omega}_k^{(1)}$ is defined by $\mathcal{H}^{(1)} = (\hbar/2)\boldsymbol{\sigma}\boldsymbol{\Omega}_k^{(1)}$, i.e., $\hbar\Omega_{k,\mu}^{(1)} = 2\beta_{\mu\lambda}k_\lambda$, and $\dot{S}$ is the spin generation rate.

For optical transitions excited under oblique incidence of the light in n-type zinc-blende-lattice quantum wells of the C_{2v} symmetry, the spin–galvanic effect coexists with the circular PGE described in Sect. 9.2.3. In the case of inter-sub-band transitions in (001)-grown quantum wells, the spin orientation is generated by resonant spin-dependent and spin-conserving photoexcitation followed by energy relaxation of the photoelectrons from the sub-band $e2$ to $e1$ and their further thermalization within the sub-band $e1$. The resulting spin generation rate is given by a product of the optical transition rate times the factor of depolarization, ξ, of the thermalizing electrons, and the current j_x is estimated as

$$j_x \sim e \frac{\beta_{yx}}{\hbar} \frac{\tau_p \tau_s}{\tau_s'} \frac{\eta_{21} I}{\hbar\omega} P_c \xi n_y, \qquad (9.22)$$

where η_{21} is the absorbance under the direct transitions $e1 \rightarrow e2$. Equation (9.22) shows that the spin–galvanic current is proportional to the absorbance and determined by the spin splitting constant in the first sub-band, β_{yx} or β_{xy}. This is in contrast to the circular PGE which is proportional to the absorbance derivative, see (9.13).

Finally we note that besides spin–flip mechanisms of the current generation the spin–galvanic effect can be caused by the interference of spin-preserving scattering and spin relaxation processes in a system of spin-polarized two-dimensional carriers [29].

9.3.2 Spin–Galvanic Photocurrent Induced by the Hanle Effect

The spin–galvanic effect can be investigated by pure optical spin orientation due to absorption of circularly polarized radiation in quantum wells. However, the irradiation of quantum wells with circularly polarized light also results in the circular PGE, and an indivisible mixture of both effects may be observed since phenomenologically they are described by the tensors, $\boldsymbol{\gamma}$ and $\boldsymbol{Q}$, equivalent from the symmetry point of view. Nevertheless, microscopically these two effects are definitely inequivalent. Indeed, the spin–galvanic effect is caused by asymmetric spin–flip scattering of spin polarized carriers and determined by the spin relaxation processes. If spin relaxation is absent the spin–galvanic current vanishes. In contrast, the circular PGE is a result of selective photoexcitation of carriers in the k-space with circularly polarized light due to optical selection rules, it is independent of the spin relaxation if $\tau_s \gg \tau_p$.

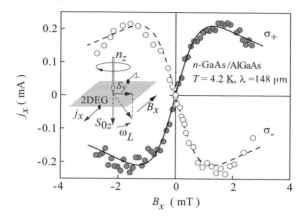

Fig. 9.7. Spin–galvanic current j_x as a function of magnetic field $\boldsymbol{B} \parallel x$ for normally incident right-handed (*open circles*) and left-handed (*solid circles*) circularly polarized radiation. *Solid* and *dashed curves* are fitted after (9.19) and (9.23) using the same value of the spin relaxation time τ_s and scaling of the ordinate. *Inset* shows optical scheme of generating a uniform in-plane spin polarization which causes a spin–galvanic current. Electron spins are oriented normal to the quantum well plane by circularly polarized radiation and rotated into the plane by the Larmor precession in an in-plane magnetic field B_x. From [26]

Here we describe a method which, on the one hand, achieves a uniform distribution of nonequilibrium spin polarization by optical means and, on the other hand, excludes the circular PGE [26]. The polarization is obtained by absorption of circularly polarized radiation at normal incidence on (001)-grown quantum wells as depicted in the inset in Fig. 9.7. For normal incidence the spin–galvanic effect as well as the circular PGE vanish because both $S_x = S_y = 0$ and $n_x = n_y = 0$. Thus, the spin orientation S_{0z} along the z-axis is achieved but no spin-induced photocurrent is generated. We note that similar method have been applied in experiments of Bakun et al. [30] carried out on bulk AlGaAs excited by interband absorption and demonstrating spin photocurrents caused by the inhomogeneous spin distribution predicted in [31, 32]. The crucial difference to the spin–galvanic effect is that in the case of surface photocurrent caused by optical orientation a gradient of spin density is needed. Naturally this gradient is absent in quantum wells where the spin–galvanic effect has been investigated because quantum wells are two-dimensional and have no 'thickness'.

An in-plane spin component, necessary for the spin–galvanic effect, arises in a magnetic field $\boldsymbol{B} \parallel x$. The field perpendicular to the initially oriented spins rotates them into the plane of the 2D electron gas due to the Larmor precession (Hanle effect). The nonequilibrium spin polarization S_y is given by

$$S_y = -\frac{\omega_L \tau_{s\perp}}{1 + (\omega_L \tau_s)^2} S_{0z}, \qquad (9.23)$$

where $\tau_s = \sqrt{\tau_{s\parallel} \tau_{s\perp}}$, $\tau_{s\parallel}$ and $\tau_{s\perp}$ are the longitudinal and transversal electron spin relaxation times, and ω_L is the Larmor frequency. Since in the experimental set-up $\dot{\boldsymbol{S}}$ is parallel to z, the scalar product $\boldsymbol{\Omega}_k^{(1)} \dot{\boldsymbol{S}}$ vanishes and, according to (9.21), the spin–

galvanic effect is not contributed by the relaxational mechanism and arises only due to the kinetic mechanism described by (9.20). The observation of the Hanle effect, see Fig. 9.7 demonstrates that free carrier intra-sub-band transitions can polarize the spins of electron systems. The measurements allow one to extract the spin relaxation time τ_s from the peak position of the photocurrent reached at $\omega_L \tau_s = 1$ providing experimental access to investigation of spin relaxation times for monopolar spin orientation where only one type of charge carriers is involved in the excitation–relaxation process [26, 33]. This condition is close to that of electrical spin injection in semiconductors.

For optical excitation of the spin–galvanic effect mid-infrared, far-infrared (terahertz frequencies) and visible laser radiation has been used [7, 8, 24]. Most of the measurements were carried out in the long wavelength range with photon energies less than the energy gap of investigated semiconductors. The advantage is that, in contrast to interband excitation resulting in the valence-band–conduction-band transitions, there are no spurious photocurrents due to other mechanisms like the Dember effect, photovoltaic effects at contacts and Schottky barriers, etc.

In contrast to the circular PGE the spin–galvanic effect induced by the Hanle effect under inter-sub-band transitions excited by the mid-infrared radiation does not change its sign with frequency radiation and follows the spectral behavior of direct inter-sub-band absorption [33]. This result is in agreement with the mechanism of the spin–galvanic effect discussed in the previous subsection, see (9.22), and clearly demonstrates that this effect has different microscopic origin. The observation of the mid-infrared and terahertz radiation excited spin–galvanic effect, which is due to spin orientation, gives clear evidence that direct inter-sub-band and Drude absorption of circularly polarized radiation result in a monopolar spin orientation. Mechanisms of the monopolar spin orientation were analyzed in [33, 34]. We would like to emphasize that spin-sensitive $e1$–$e2$ inter-sub-band transitions in (001)-grown n-type quantum wells have been observed at normal incidence when there is no component of the electric field of the radiation normal to the plane of the quantum wells.

9.3.3 Spin–Galvanic Effect at Zero Magnetic Field

In the experiments described above an external magnetic field was used for reorientation of an optically generated spin polarization. The spin–galvanic effect can also be generated at optical excitation only, without application of an external magnetic field. The necessary in-plane component of the spin polarization can be obtained by oblique incidence of the exciting circularly polarized radiation but in this case the circular PGE may also occur interfering with the spin–galvanic effect. However, the spin–galvanic effect due to pure optical excitation was demonstrated making use the difference in spectral behavior of these two effects excited by inter-sub-band transitions in n-type GaAs quantum wells [28]. Experiments have been carried out making use of the spectral tunability of the free electron laser "FELIX". The helicity dependent photocurrent closely following the absorption spectrum was detected demonstrating dominant contribution of the spin–galvanic effect.

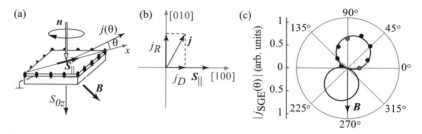

Fig. 9.8. The separation of the Dresselhaus and Rashba contributions to the spin–galvanic effect observed in an *n*-type InAs single quantum well at room temperature for the case of the electron spin $S_\parallel \parallel [100]$. (**a**) Geometry of the experiment. (**b**) The direction of Dresselhaus and Rashba contributions to the photocurrent. (**c**) The spin galvanic current measured as a function of the angle ϑ between the pair of contacts and the *x* axis. From [35]

9.3.4 Determination of the Rashba/Dresselhaus Spin Splitting Ratio

An important application of the spin–galvanic was addressed in [35]. It was demonstrated that angular dependent measurements of spin photocurrents allow us to separate the Dresselhaus and Rashba terms.

Experiments were carried out on (001)-oriented quantum wells for which the Hamiltonian (9.8) for the first sub-band reduces to

$$\mathcal{H}_k^{(1)} = \alpha(\sigma_{x_0}k_{y_0} - \sigma_{y_0}k_{x_0}) + \beta(\sigma_{x_0}k_{x_0} - \sigma_{y_0}k_{y_0}), \qquad (9.24)$$

where the parameters α and β result from the structure-inversion and bulk-inversion asymmetries, respectively, and x_0, y_0 are the crystallographic axes [100] and [010]. Note that, in the coordinate system with $x \parallel [1\bar{1}0]$, $y \parallel [110]$, the matrix $\mathcal{H}_k^{(1)}$ gets the form $\beta_{xy}\sigma_x k_y + \beta_{yx}\sigma_y k_x$ with $\beta_{xy} = \beta+\alpha$, $\beta_{yx} = \beta-\alpha$. According to (9.20) the current components j_x, j_y are proportional, respectively, to β_{xy} and β_{yx} and, therefore, angular dependent measurements of spin photocurrents allow one to separate the Dresselhaus and Rashba terms. By mapping the magnitude of the spin photocurrent in the quantum well plane the ratio of both terms can directly be determined from experiment and does not rely on theoretically obtained quantities. The relation between the photocurrent and spin directions can be conveniently expressed in the following matrix form:

$$j \propto \begin{pmatrix} \beta & -\alpha \\ \alpha & -\beta \end{pmatrix} S_\parallel, \qquad (9.25)$$

where j and $S_\parallel$ are two-component columns with the in-plane components along the crystallographic axes $x_0 \parallel [100]$ and $y_0 \parallel [010]$. The directions of the Dresselhaus and Rashba coupling induced photocurrents are shown in Fig. 9.8(b) for the particular case $S_\parallel \parallel [100]$.

Figure 9.8(c) shows the angular dependence of the spin–galvanic current $j(\vartheta)$ measured on an *n*-type (001)-grown InAs/Al$_{0.3}$Ga$_{0.7}$Sb single quantum well of 15 nm width at room temperature. Because of the admixture of photon helicity-independent magneto-gyrotropic effects (see Sect. 9.5.2) the spin–galvanic effect is extracted after eliminating current contributions which are helicity-independent: $j = (j_{\sigma_+} - j_{\sigma_-})/2$.

The sample edges are oriented along the [1$\bar{1}$0] and [110] crystallographic axes. Eight pairs of contacts on the sample allow one to probe the photocurrent in different directions, see Fig. 9.8(a). The optical spin orientation was performed by using a pulsed molecular NH_3 laser. The photocurrent j is measured in the unbiased structure in a closed circuit configuration. The nonequilibrium in-plane spin polarization $S_\parallel$ is prepared as described in Sect. 9.3.2 (see also Fig. 9.8(a)). The angle between the magnetic field and $S_\parallel$ can in general depend on details of the spin relaxation process. In these particular InAs quantum well structures the isotropic Elliott–Yafet spin relaxation mechanism dominates. Thus, the in-plane spin polarization $S_\parallel$ is always perpendicular to B and can be varied by rotating B around z as illustrated in Fig. 9.8(a). The circle in Fig. 9.8(c) represents the angular dependence $\cos(\vartheta - \vartheta_{max})$, where ϑ is the angle between the pair of contacts and the x axis and $\vartheta_{max} = \arctan j_R/j_D$. The best fit in this sample is achieved for the ratio $j_R/j_D = \alpha/\beta = 2.1$. The method was also used for investigation of Rashba/Dresselhaus spin–splitting in GaAs heterostructures where spin relaxation is controlled by Dyakonov–Perel mechanism [36]. These experiments demonstrate that growth of structures with various delta-doping layer position accompanied by experiments on spin–galvanic effect makes possible a controllable variation of the structure inversion asymmetry and preparation of samples with equal Rashba and Dresselhaus constants or with a zero Rashba constant.

9.4 Inverse Spin–Galvanic Effect

The effect inverse to the spin–galvanic effect is the electron spin polarization generated by a charge current j. First it was predicted in [2] and observed in bulk tellurium [37]. Aronov, Lyanda-Geller [38], Edelstein [39], and Vas'ko [40] demonstrated that spin orientation by current is also possible in quantum well systems. This study was extended in [41–46]. Most recently the first direct experimental proofs of this effect were obtained in semiconductor quantum wells [47, 48] as well as in strained bulk material [49]. At present inverse spin–galvanic effect has been observed in various low-dimensional structures based on GaAs, InAs, ZnSe, and GaN.

Phenomenologically, the averaged nonequilibrium free-carrier spin S is linked to j by (9.3). Microscopically, the spin polarization can be found from the kinetic equation for the electron spin density matrix ρ_k which can conveniently be presented in the form

$$\rho_k = f_k + s_k \sigma, \qquad (9.26)$$

where $f_k = \text{Tr}\{\rho_k/2\}$ is the distribution function and $s_k = \text{Tr}\{\rho_k \sigma/2\}$ is the average spin in the k state. In the presence of an electric field F the kinetic equation reads

$$\frac{eF}{\hbar} \frac{\partial \rho_k}{\partial k} + \frac{i}{\hbar} \left[\mathcal{H}_k^{(1)}, \rho_k \right] + Q_k\{\rho\} = 0, \qquad (9.27)$$

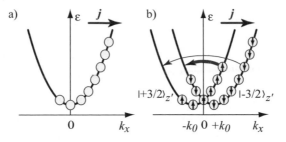

Fig. 9.9. Comparison of current flow in (**a**) spin-degenerate and (**b**) spin–split sub-bands. (**a**) Hole distribution at a stationary current flow due to acceleration in an electric field and momentum relaxation. (**b**) Spin polarization due to spin–flip scattering. Here only $\beta_{z'x}\sigma_{z'}k_x$ term is taken into account in the Hamiltonian which splits the valence sub-band into two parabolas with spin-up $|+3/2\rangle_{z'}$ and spin-down $|-3/2\rangle_{z'}$ in the z'-direction. Biasing along the x-direction causes an asymmetric in k-space occupation of both parabolas

where $Q_k\{\rho\}$ is the collision integral and $\mathcal{H}_k^{(1)}$ is the linear-k Hamiltonian. Similarly to the spin–galvanic effect there exist two different mechanisms of the current-to-spin transformation, namely, spin–flip mediated and precessional.

9.4.1 Spin-Flip Mediated Current-Induced Polarization

In the spin–flip mediated mechanism, a value of the spin generation rate is calculated neglecting the commutator in (9.27) and taking into account the spin–flip processes in the collision integral and the linear-k terms in the electron dispersion. Microscopic illustration of this mechanism is sketched in Fig. 9.9(b) for a 2D hole gas in a system of the C_s symmetry, a situation relevant for the experiments of [47]. In the simplest case the electron kinetic energy in a quantum well depends quadratically on the in-plane wave vector k. In equilibrium, the spin degenerate k states are symmetrically occupied up to the Fermi energy E_F. If an external electric field is applied, the charge carriers drift in the direction of the resulting force. The carriers are accelerated by the electric field and gain kinetic energy until they are scattered, Fig. 9.9(a). A stationary state forms where the energy gain and the relaxation are balanced resulting in an asymmetric distribution of carriers in the k-space. The holes acquire the average quasimomentum

$$\hbar\bar{k} = -e\tau_p F = -\frac{m_c}{en_s}j, \qquad (9.28)$$

where τ_p is the momentum relaxation time, j the electric current density, m_c the effective mass, and n_s the 2D carrier concentration. As long as the energy band is spin degenerated in the k-space a current is not accompanied by spin orientation. However, in zinc-blende-lattice quantum wells or strained bulk semiconductors the spin degeneracy is lifted due to the linear-k terms given by (9.8). To be specific for the mechanism depicted in Fig. 9.9(b) we consider solely spin–orbit interaction of the form $\beta_{z'x}\sigma_{z'}k_x$. Then the parabolic energy band splits into two parabolic sub-bands of opposite spin directions, $s_{z'} = 3/2$ and $s_{z'} = -3/2$, with minima symmetrically shifted in the k-space along the k_x axis from the point $k = 0$ into the points $\pm k_0$, where $k_0 = m_c \beta_{z'x}/\hbar^2$. The corresponding dispersion is sketched in Fig. 9.9(b). In the presence of an in-plane electric field $F \parallel x$ the distribution of carriers in the k-space gets shifted yielding an electric current. Until the spin relaxation is switched off the spin branches are equally populated and equally contribute to the current. Due to the band splitting, spin–flip relaxation processes $\pm 3/2 \to \mp 3/2$ are different because of the difference in quasimomentum transfer from initial to final states. In Fig. 9.9(b) the k-dependent spin–flip scattering processes are indicated by arrows of different lengths and thicknesses. As a consequence different amounts of spin-up and spin-down carriers contribute to the spin–flip transitions causing a stationary spin orientation. Thus, in this picture we assume that the origin of the current induced spin orientation is, as sketched in Fig. 9.9(b), exclusively due to scattering and hence dominated by the Elliott–Yafet spin relaxation processes.

9.4.2 Precessional Mechanism

The precessional mechanism resulting in the current induced spin orientation is based on the Dyakonov–Perel spin relaxation. In this mechanism of spin polarization the contribution of spin–flip scattering to the collision integral is ignored and the spin appears taking into account the linear-k Hamiltonian, both in the collision integral and the commutator $[\mathcal{H}_k^{(1)}, \rho_k]$. For example, we present here the collision integral for elastic scattering

$$Q_k\{\rho\} = \frac{2\pi}{\hbar} N_i \sum_{k'} |A_{k'k}|^2 \{\delta(E_k + \mathcal{H}_k^{(1)} - E_{k'} - \mathcal{H}_{k'}^{(1)}), \rho_k - \rho_{k'}\}, \quad (9.29)$$

where $E_k = \hbar^2 k^2/(2m_c)$, N_i is the density of static defects acting as the scatterers, $A_{k'k}$ is the scattering matrix element and the braces mean the anticommutator, $\{AB\} = (AB + BA)/2$ for two arbitrary 2×2 matrices A and B. Similar equation can be written for electron-phonon scattering.

In the equilibrium the electron spin density matrix is given by

$$\rho_k^0 = f^0(E_k + \mathcal{H}_k^{(1)}) \approx f^0(E_k) + \frac{\partial f^0}{\partial E_k} \mathcal{H}_k^{(1)}, \quad (9.30)$$

where $f^0(E) = \{\exp[(E - \mu)/k_B T] + 1\}^{-1}$ is the Fermi–Dirac distribution function, μ is the electron chemical potential, k_B is the Boltzmann constant and T is the temperature.

Neglecting the spin splitting we can write the solution of (9.27) in the textbook form

$$f_k = f^0(E_k) - e F_x v_x \tau_1(E_k) \frac{\partial f^0}{\partial E_k} \quad (9.31)$$

with $s_k = 0$. Here $v_x = \hbar k_x/m_c$, τ_1 is the time describing the relaxation of a distribution–function harmonic with the angular dependence of the functions k_x or k_y. If we substitute ρ_k with the distribution function (9.31) into the collision integral and

ρ_k^0 into the first term of (9.27) we obtain an equation for s_k. By solving this equation one arrives at the estimation for the spin density

$$s_\mu = \sum_k s_{k,\mu} \sim \beta_{\mu\lambda} \bar{k}_\lambda g_{2d}, \qquad (9.32)$$

where $g_{2d} = m_c/(\pi \hbar^2)$ is the 2D density of states and $\hbar \bar{k}_\lambda/m_c$ is the electron drift velocity. The exact formula can be found in [43, 46].

Spin orientation by electric current in low-dimensional structures has been observed applying various experimental techniques, comprising transmission of polarized THz-radiation, polarized luminescence and space resolved Faraday rotation [19, 47–52]. Here we briefly sketch results of experiments on the THz-transmission and polarized photoluminescence in which the spin orientation by electric current in quantum well structures was initially observed.

9.4.3 Current Induced Spin Faraday Rotation

In order to observe current induced spin polarization, in [47] the circular dichroism and Faraday rotation of THz radiation transmitted through samples containing multiple quantum wells were studied. This method allows one to detect spin polarization at normal incidence in the growth direction. The materials chosen for studies were (113)- and miscut (001)-oriented p-type GaAs multiple quantum wells of the C_s point group symmetry. The transmission measurements were carried out at room temperature using linearly polarized $\lambda = 118\,\mu m$ radiation as shown in Fig. 9.10(a): the sample is placed between two metallic grid polarizers and the cw-terahertz radiation is passed through this optical arrangement. Using modulation technique the Faraday rotation was observed only for the current flowing in the x-direction. This is in agreement with the phenomenological equation $S_{z'} = R_{z'x} j_x$ relating the induced spin with the current density, the spin polarization can be obtained only for the current flowing along the direction normal to the mirror reflection plane which is perpendicular to the x-axis. The signal ΔV caused by rotation of polarization plane is shown in Fig. 9.10(a) as a function of the current strength. Experiment shows that, in agreement with (9.32), the spin polarization increases with the decreasing temperature.

Current induced spin polarization has also been detected by the Faraday rotation of infrared radiation applying a mode-locked Ti:sapphire laser. Figure 9.10(b) demonstrates an optical detection of current-induced electron spin polarization in strained InGaAs epitaxial layers [49]. The heterostructure studied consists of 500 nm of n-In$_{0.07}$Ga$_{0.93}$As (Si doped for $n = 3 \times 10^{16}$ cm^{-3}) grown on (001) semi-insulating GaAs substrate and capped with 100 nm of undoped GaAs. The n-InGaAs layer is strained due to the lattice mismatch. An alternating electric field $\boldsymbol{F}$ is applied along either of the two crystal directions [110] and [1$\bar{1}$0], the in-plane magnetic field $\boldsymbol{B}$ is parallel to $\boldsymbol{F}$. A linearly polarized probe beam is directed along the z axis, normally incident and focused on the sample. The polarization axis of the transmitted beam rotates by an angle θ_F that is proportional to the z component of the spins S_z (the

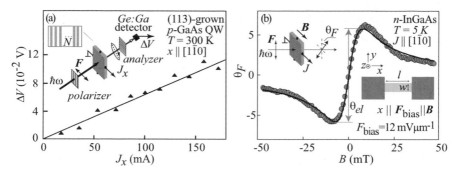

Fig. 9.10. (a) Polarization dependent signal for current in the active direction as a function of current strength for two samples. From [47]. (b) Voltage-induced angle θ_F as a function of the magnetic field B for $F = 12\,\mathrm{mV\,\mu m^{-1}}$ ($F \parallel [1\bar{1}0]$) (from [49]). *Open circles* are data, and lines are fits according to (9.33). *Insets to both panels* show experimental set-up: (a) The sample is placed between crossed polarizer and analyzer blocking optical transmission at zero current through the sample. Injecting a modulated current in the sample yields a signal at the detector which is recorded by the box-car technique. (b) The current yields an in-plane spin polarization which applying magnetic field B is rotated out off plane yielding Faraday rotation of the probe light. *Second inset* shows sample geometry. Here *dark areas* are Ni/GeAu contacts and *the light grey area* is the InGaAs channel

spin Faraday rotation). The current-induced angle is lock-in detected at the modulation frequency as a function of the applied magnetic field. The experiment data in Fig. 9.10(b) can be explained by assuming a constant orientation rate $\dot{s}_{k,y}$ for spins polarized along the y axis. The rotation of the spins around the magnetic field yields the z spin component given by

$$S_z(B) = \frac{\omega_L \tau_{s\parallel}}{1 + (\omega_L \tau_s)^2} S_{0y}, \quad S_{0y} = \tau_{s,\perp} \sum_k \frac{\dot{s}_{k,y}}{n_s}, \tag{9.33}$$

where the notations for the spin relaxation times and Larmor frequency are introduced in (9.23). The high sensitivity of the Faraday rotation technique allows detection of 100 spins in an integration time of about 1 s, unambiguously revealing the presence of a small spin polarization due to laterally applied electric fields.

9.4.4 Current Induced Polarization of Photoluminescence

In [48, 50] in order to detect the inverse spin–galvanic effect the degree of circular polarization of the 2D hole gas photoluminescence (PL) was measured. This experimental procedure has become a proven method for probing spin polarization [1, 53]. A (001)-grown sample cleaved into bars was studied with the current flowing along the long side cleaved parallel to the [1$\bar{1}$0] direction. Lately (113)-grown samples were also studied [50]. The PL was excited with 633 nm line from a helium-neon laser. In (001)-oriented samples the PL was collected from the cleaved (110) facet of the sample. On the other hand, at the (113)-oriented heterojunctions, because of

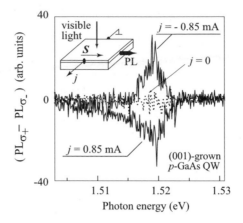

Fig. 9.11. Differential spectra of polarized PL for two current directions (from [48]). Base line is taken with the current turned off. Inset shows experimental geometry

the C_s symmetry, the mean spin density will have a component along the growth direction. Therefore, the PL in this case was detected in the back scattering geometry; the degree of PL circular polarization P_c was analyzed with a $\lambda/4$ plate and a linear polarizer. Inset in Fig. 9.11 shows the experimental arrangement for measuring the current induced polarization and differential spectra, $(PL_{\sigma_+} - PL_{\sigma_-})$, for the two opposite current directions. The observation of the circularly polarized radiation and, in particular, the reversal of helicity upon the inversion of the current direction demonstrate the effect of current induced spin polarization. The observed degree of polarization in (001)-grown samples yields a maximum of 2.5% [48]. In (113)-grown samples even higher polarization of 12% at 5.1 K is achieved [50].

9.5 Pure Spin Currents

Pure spin current represents a nonequilibrium distribution where free carriers, electrons or holes, with the spin up propagate mainly in one direction and an equal number of spin-down carriers propagates in the opposite direction. This state is characterized by zero charge current because electric currents contributed by spin-up and spin-down quasiparticles cancel each other, but leads to separation of spin-up and spin-down electron spatial distributions and accumulation of the opposite spins at the opposite edges of the sample. Spin currents in semiconductors can be driven by an electric field acting on unpolarized free carriers (the spin Hall effect, not considered here). They can be induced as well by optical means under interband or intraband optical transitions in non-centrosymmetric bulk and low-dimensional semiconductors [54–58].

9.5.1 Pure Spin Current Injected by a Linearly Polarized Beam

In general, the spin current density pseudotensor $q_{\lambda\mu}$ describes the flow of the μ component of the spin polarization in the spatial direction λ. Phenomenologically, the spin photocurrent $q_{\lambda\mu}$ is related with bilinear products $I e_\nu e_\eta^*$ by a fourth-rank tensor, in (9.4) it is the tensor $P_{\lambda\mu\nu\eta}$. Here we assume the light to be linearly polarized. In this particular case the product $e_\nu e_\eta^* \equiv e_\nu e_\eta$ is real and the tensor $P_{\lambda\mu\nu\eta}$ is symmetric with respect to interchange of the third and fourth indices.

Among microscopic mechanisms of the pure spin photocurrent we first discuss those related to the k-linear terms in the electron effective Hamiltonian [55]. Let us consider the $e1$–$hh1$ interband absorption of linearly polarized light under normal incidence on (001)-grown quantum wells. In this case linear-k splitting of the $hh1$ heavy-hole valence sub-band is negligibly small. For the sake of simplicity, but not at the expense of generality, in the $e1$ conduction sub-band we take into account the spin–orbit term $\beta_{yx}\sigma_y k_x$ only, the contribution to the tensor P coming from the term $\beta_{xy}\sigma_x k_y$ is considered similarly. Then the conduction-electron spin states are eigenstates of the spin matrix σ_y. For the light linearly polarized, say, along x, all the four transitions from each heavy-hole state $\pm 3/2$ to each $s_y = \pm 1/2$ state are allowed. The energy and momentum conservation laws read

$$E_g^{QW} + \frac{\hbar^2(k_x^2 + k_y^2)}{2\mu_{cv}} + 2s_y\beta_{yx}k_x = \hbar\omega,$$

where we use the same notations for E_g^{QW} and μ_{cv} as in Sect. 9.2.6. For a fixed value of k_y the photoelectrons are generated only at two values of k_x labeled $k_x^{\pm}$. The average electron velocity in the s_y spin sub-band is given by

$$\bar{v}_{e,x} = \frac{\hbar(k_x^+ + k_x^-)}{2m_c} + 2s_y\frac{\beta_{yx}}{\hbar} = \frac{2s_y\beta_{yx}}{\hbar}\frac{m_c}{m_c + m_v}.$$

The spin fluxes $i_{\pm 1/2}$ are opposite in sign, the electric current $j = e(i_{1/2} + i_{-1/2})$ is absent but the spin current $j_s = (1/2)(i_{1/2} - i_{-1/2})$ is nonzero. This directional movement decays in each spin sub-band within the momentum relaxation time τ_p^e. However under the cw photoexcitation the electron generation is continuous which results in the spin current

$$q_{xy} = \frac{\beta_{yx}\tau_p^e}{2\hbar}\frac{m_c}{m_c + m_v}\frac{\eta_{eh}I}{\hbar\omega}.$$

Under the normal incidence it is independent of the light polarization plane.

In (110)-grown quantum well structures the spin component along the normal $z' \parallel [110]$ is coupled with the in-plane electron wave vector due to the term $\beta_{z'x}\sigma_{z'}k_x$ in the conduction band and the term proportional to $J_{z'}k_x$ in the heavy-hole band where $J_{z'}$ is the 4×4 matrix of the angular momentum 3/2. The coefficient $\beta_{z'x}^{(e1)}$ is relativistic and can be ignored compared with the nonrelativistic constant $\beta_{z'x}^{(hh1)}$ describing the spin splitting of heavy-hole states. The allowed direct optical transitions from the valence sub-band hh1 to the conduction sub-band $e1$ are $|+3/2\rangle \rightarrow |+1/2\rangle$ and $|-3/2\rangle \rightarrow |-1/2\rangle$, where $\pm 1/2$, $\pm 3/2$ indicate the z' components of the electron spin and hole angular momentum. Under linearly polarized photoexcitation the charge photocurrent is not induced, and for the electron pure spin photocurrent one has

$$q_{xz'} = \frac{\beta_{z'x}^{(hh1)}\tau_p^e}{2\hbar} \frac{m_v}{m_c + m_v} \frac{\eta_{eh}I}{\hbar\omega}. \tag{9.34}$$

The similar hole spin current can be ignored in the spin separation experiments because of the much shorter spin relaxation time for holes as compared to the conduction electrons.

Another contribution to spin photocurrents comes from k-linear terms in the matrix elements of the interband optical transitions [59]. Taking into account $k \cdot p$ admixture of the remote Γ_{15} conduction band to the Γ_{15} valence-band states $X_0(k)$, $Y_0(k)$, $Z_0(k)$ and the Γ_1 conduction-band states $S(k)$, one derives the interband matrix elements of the momentum operator for bulk zinc-blende-lattice semiconductors [60, 61]

$$\langle iS(k)|e \cdot p|X_0(k)\rangle = \mathcal{P}[e_{x_0} + i\chi(e_{y_0}k_{z_0} + e_{z_0}k_{y_0})], \tag{9.35}$$

$\langle iS(k)|e \cdot p|Y_0(k)\rangle$ and $\langle iS(k)|e \cdot p|Z_0(k)\rangle$ are obtained by the cyclic permutation of indices, the coefficient χ is a material parameter dependent on the interband spacings and interband matrix elements of the momentum operator at the Γ point, and we use here the crystallographic axes $x_0 \parallel [100]$, etc. For GaAs band parameters [62] the coefficient χ can be estimated as $0.2\,\text{Å}$. Calculation shows that, in (110)-grown quantum wells (QW), the spin photocurrent caused by k-linear terms in the interband matrix elements has the form

$$q_{xz'} = \varepsilon(e_{y'}^2 - e_x^2)\frac{\chi\tau_p^e}{\hbar}\frac{\eta_{cv}}{\hbar\omega}I, \qquad q_{y'z'} = \varepsilon e_x e_{y'}\frac{\chi\tau_p^e}{\hbar}\frac{\eta_{ev}}{\hbar\omega}I \tag{9.36}$$

with $\varepsilon = (\hbar\omega - E_g^{QW})m_v/(m_c + m_v)$ being the kinetic energy of the photoexcited electrons and $y' \parallel [00\bar{1}]$. In contrast to (9.34), this contribution depends on the polarization plane of the incident light and vanishes for unpolarized light. From comparison of (9.34) and (9.36) one can see that depending on the value of $\hbar\omega - E_g^{QW}$ the two contributions to $q_{xz'}$ can be comparable or one of them can dominate over the other.

The injection and control of pure spin currents in (110)-oriented GaAs quantum wells at room temperature by one-photon absorption of a linearly polarized optical pulse was demonstrated by Zhao et al. [56]. Spatially resolved pump–probe technique was used. The pump pulse excited electrons from the valence to the conduction band with an excess energy of $\sim 148\,\text{meV}$ large enough for the polarization-dependent contribution (9.36) to dominate over the polarization independent contribution (9.34). The probe was tuned near the band edge. The σ_+ component of the linearly polarized probe interacts stronger with the spin-down electrons of the density $n_{-1/2}$, while the σ_- component interacts stronger with the spin-up electrons of

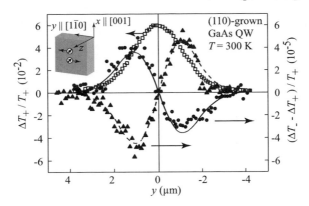

Fig. 9.12. Measurement of $\Delta T_+/T_+ \propto n_{-1/2}(y)$ *(open squares)* and $\Delta T_-/T_+ - \Delta T_+/T_+ \propto$ $n_{1/2}(y) - n_{-1/2}(y)$ for an x polarized *(solid circles)* and a y polarized *(solid triangles)* pump pulse at room temperature for a pump fluence of $10\,\mu\mathrm{J/cm}^2$. *The lines are the fits to the data* for spin separation $d = 2.8$ nm. From [56]

the density $n_{1/2}$. Consequently, the net spin polarization of the carriers present in the sample at the position of the probe can be readily deduced from the difference in the transmission $T_\pm$ of the σ_+ and σ_- components of the probe. The results of measuring $n_{-1/2}(y) \propto \Delta T_+/T_+$ and $\Delta n(y) \equiv n_{1/2}(y) - n_{-1/2}(y) \propto \Delta T_-/T_+ - \Delta T_+/T_+$ for the x and y polarized pump are shown in Fig. 9.12. Note that here we retain the notations $x \parallel [001]$, $y \parallel [1\bar{1}0]$, $z \parallel [110]$ as they are introduced in the original paper [56] while in (9.34) and (9.36) we use the Cartesian frame $x \parallel [1\bar{1}0]$, $y' \parallel [00\bar{1}]$, $z' \parallel [110]$. Clearly, the $\Delta n(y)$ signal is consistent with a pure spin current. It can be well fitted by the product of the spatial derivative of the original Gaussian profile and the separation d of the order of the photoelectron mean free path. The solid curve in Fig. 9.12 corresponds to a fit for $d = 28\,\text{Å}$. In agreement with (9.36) the $\Delta n(y)$ signal has opposite signs for the x and y linear polarization of the pump.

It is worth adding that a pure spin current may be generated at simultaneous one- and two-photon coherent excitation of proper polarization as demonstrated in bulk GaAs [63] and GaAs/AlGaAs quantum wells [64]. This phenomenon may be attributed to a photogalvanic effect where the reduced symmetry is caused by the coherent two-frequency excitation [65].

9.5.2 Pure Spin Currents Due to Spin-Dependent Scattering

Light absorption by free carriers, or the Drude-like absorption, is accompanied by electron scattering by acoustic or optical phonons, static defects, etc. Scattering-assisted photoexcitation with unpolarized light also gives rise to a pure spin current [57, 58]. However, in contrast to the direct transitions considered above, the spin splitting of the energy spectrum leads to no essential contribution to the spin current induced by free-carrier absorption. The more important contribution comes from asymmetry of the electron spin-conserving scattering. In gyrotropic low-dimensional

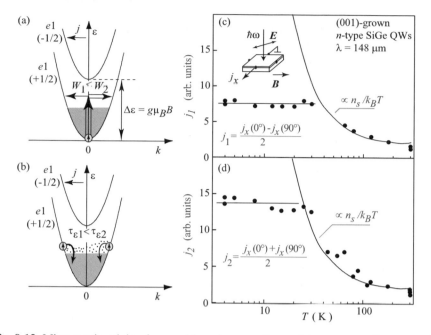

Fig. 9.13. Microscopic origin of a zero-bias spin separation and the corresponding magnetic field-induced photocurrent for excitation (**a**) and relaxation (**b**) models corresponding to currents j_1 and j_2 (see explanation in the text). Temperature dependencies of the contributions j_1 (**c**) and j_2 (**d**) to the photocurrent j_x in a magnetic field $B \parallel y$. Full lines are fits to $An_s/k_B T$ with a single fitting parameter A, and to a constant, respectively. From [58]

structures, spin–orbit interaction adds an asymmetric spin-dependent term to the scattering probability. This term in the scattering matrix element is proportional to components of $[\sigma \times (k + k')]$, where σ is the vector composed of the Pauli matrices, k and k' are the initial and scattered electron wave vectors. Figure 9.13(b) sketches the process of energy relaxation of hot electrons for the spin-up sub-band $(s = +1/2)$ in a quantum well containing a two-dimensional electron gas. Energy relaxation processes are shown by curved arrows. Due to the spin-dependent scattering, transitions to positive and negative k'_x states occur with different probabilities. In Fig. 9.13(b) the difference is indicated by curved arrows of different thickness. This asymmetry causes an imbalance in the distribution of carriers in both sub-bands $(s = \pm 1/2)$ between positive and negative k_x-states. This in turn yields a net electron flows, $i_{\pm 1/2}$, within each spin sub-band. Since the asymmetric part of the scattering amplitude depends on spin orientation, the probabilities for scattering to positive or negative k'_x-states are inverted for spin-down and spin-up sub-bands. Thus, the charge currents, $j_+ = ei_{+1/2}$ and $j_- = ei_{-1/2}$, where e is the electron charge, have opposite directions because $i_{+1/2} = -i_{-1/2}$ and therefore they cancel each other. Nevertheless, a finite pure spin current $J_s = \frac{1}{2}(i_{+1/2} - i_{-1/2})$ is generated since electrons with spin-up and spin-down move in opposite directions [57].

Similarly to the relaxation mechanism, optical excitation of free carriers by Drude absorption, also involving electron scattering, is asymmetric and yields spin separation. Figure 9.13(a) sketches the process of Drude absorption via virtual states for the spin-up sub-band. Vertical arrow indicates optical transitions from the initial state with $k_x = 0$ while the horizontal arrows describe an elastic scattering event to a final state with either positive or negative electron wave vector. Due to the spin dependence of scattering, transitions to positive and negative k_x states occur with different probabilities. This is indicated by the different thickness of the horizontal arrows. The asymmetry causes an imbalance in the distribution of photoexcited carriers in the spin sub-band between positive and negative k_x-states. This in turn yields electron flow.

Magneto-Gyrotropic Effects

A pure spin current and zero-bias spin separation can be converted into a measurable electric current by application of a magnetic field. Indeed, in a Zeeman spin-polarized system, the two fluxes $i_{\pm 1/2}$, whose magnitudes depend on the free carrier densities in spin-up and spin-down sub-bands, $n_{\pm 1/2}$, respectively, no longer compensate each other and hence yield a net electric current. Since the fluxes $i_{\pm 1/2}$ are proportional to the carrier densities $n_{\pm 1/2}$ the charge current is given by

$$j = e(i_{1/2} + i_{-1/2}) = 4eSj_s, \tag{9.37}$$

where $x \parallel [1\bar{1}0]$, $y \parallel [110]$, $S = (1/2)(n_{1/2} - n_{-1/2})/(n_{1/2} + n_{-1/2})$ is the average spin per particle and j_s is the pure spin current in the absence of magnetic field. An external magnetic field B results in different equilibrium populations of the two spin sub-bands due to the Zeeman effect. We remind the reader that in equilibrium the average spin is given by $S = -g\mu B/4k_B T$ for a nondegenerate two-dimensional electron gas and $S = -g\mu B/4E_F$ for a degenerate one.

In the structures of the C_{2v} symmetry, the phenomenological equation (9.5) for the magneto-photogalvanic effects induced by normally-incident linearly polarized radiation reduces to [66]

$$\begin{aligned} j_x &= S_1 B_y I + S_2 B_y (e_x^2 - e_y^2)I + 2S_3 B_x e_x e_y I, \\ j_y &= S_1' B_x I + S_2' B_x (e_x^2 - e_y^2)I + 2S_3' B_y e_x e_y I. \end{aligned} \tag{9.38}$$

Here the parameters S_1 to S_3 and S_1' to S_3' are linearly independent components of the tensor $\Phi_{\lambda\mu\nu\eta}$ in (9.5) and only in-plane components of the magnetic field are taken into account. For $B \parallel y$ we have

$$j_x = j_1 \cos 2\alpha + j_2, \qquad j_y = j_3 \sin 2\alpha, \tag{9.39}$$

where $j_1 = S_2 B_y I$, $j_2 = S_1 B_y I$, $j_3 = S_3' B_y I$.

Right panels of Fig. 9.13 show the temperature dependence of the currents j_1 and j_2 corresponding to the excitation and relaxation mechanisms depicted in Figs. 9.13(a) and (d), respectively. The data are obtained in an n-type SiGe quantum well structure for the magnetic field 0.6 T under excitation with a THz molecular laser ($\lambda = 148\,\mu$m). The analysis shows [57, 58] that the temperature dependence of the current can be reduced to $n_s S$, the current becomes independent of temperature at low temperatures and is proportional to n_s / T at high temperatures, in agreement with the experimental data. Thus, the application of an external magnetic field gives experimental access to investigations of pure spin currents. Like circular PGE and spin–galvanic effect, magneto-gyrotropic effect provides an efficient tool for investigation of inversion asymmetry as it is demonstrated for (110)-grown GaAs quantum wells [67].

9.6 Concluding Remarks

The two main fields of study in the modern physics of semiconductors are transport phenomena and optical effects. Sometimes an impression arises that, founded on much the same basis of tremendous successes achieved in technology, these fields are developing independently of each other. It is also true for the extensive studies of spin physics in semiconductors. One of the aims of this chapter is to show that the spin–photogalvanics builds a solid bridge between the two fields and sets up a base for the reciprocation of ideas. Indeed, the spin-dependent photogalvanic effects, including charge and spin photocurrents, as well as the inverse effects allowing optical detection of current-induced spin polarization need a thorough knowledge in both the transport physics and the polarized optical spectroscopy. As a result the different concepts supplement each other and provide a deeper insight in the spin-dependent microscopic processes.

As for future work, one of problems still existing is that the circular photogalvanic effect, in addition to the spin-dependent mechanisms, can also arise from orbital effects as shown, e.g., in [61, 68, 69]. Therefore, some experimental data on the circular PGE, especially those performed under intra-sub-band excitation, need independent and direct experiments aimed at separation of the spin-dependent and spin-independent contributions to the circular PGE. In this respect future experiments on time-resolved photogalvanics under short-pulsed circularly polarized photoexcitation, with the pulse duration being comparable with the free-carrier momentum and spin relaxation times, would be desirable and informative. Such experiments would also reveal a great deal about the momentum, energy, and spin relaxation of nonequilibrium photoexcited carriers.

References

[1] F. Meier, B.P. Zakharchenya, in *Optical Orientation*, ed. by V.M. Agranovich, A.A. Maradudin. Modern Problems in Condensed Matter Sciences, vol. 8 (Elsevier Science, Amsterdam, 1984)
[2] E.L. Ivchenko, G.E. Pikus, Pis'ma Z. Eksp. Teor. Fiz. **27**, 640 (1978); JETP Lett. **27**, 604 (1978)

[3] V.I. Belinicher, Phys. Lett. A **66**, 213 (1978)

[4] V.M. Asnin, A.A. Bakun, A.M. Danishevskii, E.L. Ivchenko, G.E. Pikus, A.A. Rogachev, Pis'ma Z. Eksp. Teor. Fiz. **28**, 80 (1978); JETP Lett. **28**, 74 (1978)

[5] B.I. Sturman, V.M. Fridkin, *The Photovoltaic and Photorefractive Effects in Non-Centrosymmetric Materials* (Gordon and Breach Science Publishers, Philadelphia, 1992)

[6] S.D. Ganichev, E.L. Ivchenko, S.N. Danilov, J. Eroms, W. Wegscheider, D. Weiss, W. Prettl, Phys. Rev. Lett. **86**, 4358 (2001)

[7] S.D. Ganichev, W. Prettl, J. Phys. Condens. Matter **15**, R935 (2003)

[8] S.D. Ganichev, W. Prettl, *Intense Terahertz Excitation of Semiconductors* (Oxford University Press, Oxford, 2006)

[9] S.D. Ganichev, V.V. Bel'kov, P. Schneider, E.L. Ivchenko, S.A. Tarasenko, D. Schuh, W. Wegscheider, D. Weiss, W. Prettl, Phys. Rev. B **68**, 035319 (2003)

[10] V.A. Shalygin, H. Diehl, Ch. Hoffmann, S.N. Danilov, T. Herrle, S.A. Tarasenko, D. Schuh, Ch. Gerl, W. Wegscheider, W. Prettl, S.D. Ganichev, Pis'ma Z. Eksp. Teor. Fiz. **84**, 666 (2006); JETP Lett. **84**, 570 (2006)

[11] M.I. Dyakonov, V.Yu. Kachorovskii, Fiz. Tekh. Poluprovodn. **20**, 178 (1986); Sov. Phys. Semicond. **20**, 110 (1986)

[12] O. Krebs, P. Voisin, Phys. Rev. Lett. **77**, 1829 (1996)

[13] U. Rössler, J. Keinz, Solid State Commun. **121**, 313 (2002)

[14] Y.A. Bychkov, E.I. Rashba, Pis'ma Z. Eksp. Teor. Fiz. **39**, 66 (1984); JETP Lett. **39**, 78 (1984)

[15] R. Winkler, *Spin–Orbit Coupling Effects in Two-Dimensional Electron and Hole Systems.* Springer Tracts in Modern Physics, vol. 191 (Springer, Berlin, 2003)

[16] W. Zawadzki, P. Pfeffer, Semicond. Sci. Technol. **19**, R1 (2004)

[17] V.V. Bel'kov, S.D. Ganichev, P. Schneider, C. Back, M. Oestreich, J. Rudolph, D. Hägele, L.E. Golub, W. Wegscheider, W. Prettl, Solid State Commun. **128**, 283 (2003)

[18] M. Bieler, N. Laman, H.M. van Driel, A.L. Smirl, Appl. Phys. Lett. **86**, 061102 (2005)

[19] C.L. Yang, H.T. He, L. Ding, L.J. Cui, Y.P. Zeng, J.N. Wang, W.K. Ge, Phys. Rev. Lett. **96**, 186605 (2006)

[20] K.S. Cho, Y.F. Chen, Y.Q. Tang, B. Shen, Appl. Phys. Lett. **90**, 041909 (2007)

[21] L.E. Golub, Phys. Rev. B **67**, 235320 (2003)

[22] S.D. Ganichev, S.N. Danilov, V.V. Bel'kov, E.L. Ivchenko, M. Bichler, W. Wegscheider, D. Weiss, W. Prettl, Phys. Rev. Lett. **88**, 057401 (2002)

[23] P. Schneider, J. Kainz, S.D. Ganichev, V.V. Bel'kov, S.N. Danilov, M.M. Glazov, L.E. Golub, U. Rössler, W. Wegscheider, D. Weiss, D. Schuh, W. Prettl, J. Appl. Phys. **96**, 420 (2004)

[24] E.L. Ivchenko, *Optical Spectroscopy of Semiconductor Nanostructures* (Alpha Science Int., Harrow, 2005)

[25] E.L. Ivchenko, Yu.B. Lyanda-Geller, G.E. Pikus, Pis'ma Z. Eksp. Teor. Fiz. **50**, 156 (1989); JETP Lett. **50**, 175 (1989)

[26] S.D. Ganichev, E.L. Ivchenko, V.V. Bel'kov, S.A. Tarasenko, M. Sollinger, D. Weiss, W. Wegscheider, W. Prettl, Nature (Lond.) **417**, 153 (2002)

[27] N.S. Averkiev, L.E. Golub, M. Willander, J. Phys. Condens. Matter **14**, R271 (2002)

[28] S.D. Ganichev, P. Schneider, V.V. Bel'kov, E.L. Ivchenko, S.A. Tarasenko, W. Wegscheider, D. Weiss, D. Schuh, D.G. Clarke, M. Merrick, B.N. Murdin, P. Murzyn, P.J. Phillips, C.R. Pidgeon, E.V. Beregulin, W. Prettl, Phys. Rev. B **68**, 081302 (2003)

[29] L.E. Golub, Pis'ma Z. Eksp. Teor. Fiz. **85**, 479 (2007); JETP Lett. **85**, 393 (2007)

[30] A.A. Bakun, B.P. Zakharchenya, A.A. Rogachev, M.N. Tkachuk, V.G. Fleisher, Pis'ma Z. Eksp. Teor. Fiz. **40**, 464 (1984); Sov. JETP Lett. **40**, 1293 (1984)

[31] N.S. Averkiev, M.I. D'yakonov, Fiz. Tekh. Poluprov. **17**, 629 (1983); Sov. Phys. Semi-cond. **17**, 393 (1983)
[32] M.I. Dyakonov, V.I. Perel, Pis'ma Z. Eksp. Teor. Fiz. **13**, 206 (1971); Sov. JETP Lett. **13**, 144 (1971)
[33] S.A. Tarasenko, E.L. Ivchenko, V.V. Bel'kov, S.D. Ganichev, D. Schowalter, P. Schneider, M. Sollinger, W. Prettl, V.M. Ustinov, A.E. Zhukov, L.E. Vorobjev, cond-mat/301393 (2003); See also J. Supercond.: Incorporating Novel Magn. **16**, 419 (2003)
[34] E.L. Ivchenko, S.A. Tarasenko, Z. Eksp. Teor. Fiz. **126**, 426 (2004); JETP **99**, 379 (2004)
[35] S.D. Ganichev, V.V. Bel'kov, L.E. Golub, E.L. Ivchenko, P. Schneider, S. Giglberger, J. Eroms, J. De Boeck, G. Borghs, W. Wegscheider, D. Weiss, W. Prettl, Phys. Rev. Lett. **92**, 256601 (2004)
[36] S. Giglberger, L.E. Golub, V.V. Bel'kov, S.N. Danilov, D. Schuh, Ch. Gerl, F. Rohlfing, J. Stahl, W. Wegscheider, D. Weiss, W. Prettl, S.D. Ganichev, Phys. Rev. B **75**, 035327 (2007)
[37] L.E. Vorob'ev, E.L. Ivchenko, G.E. Pikus, I.I. Farbstein, V.A. Shalygin, A.V. Sturbin, Pis'ma Z. Eksp. Teor. Fiz. **29**, 485 (1979); JETP Lett. **29**, 441 (1979)
[38] A.G. Aronov, Yu.B. Lyanda-Geller, Pis'ma Z. Eksp. Teor. Fiz. **50**, 398 (1989); JETP Lett. **50**, 431 (1989)
[39] V.M. Edelstein, Solid State Commun. **73**, 233 (1990)
[40] F.T. Vasko, N.A. Prima, Fiz. Tverd. Tela **21**, 1734 (1979); Sov. Phys. Solid State **21**, 994 (1979)
[41] A.G. Aronov, Yu.B. Lyanda-Geller, G.E. Pikus, Z. Eksp. Teor. Fiz. **100**, 973 (1991); Sov. Phys. JETP **73**, 537 (1991)
[42] A.V. Chaplik, M.V. Entin, L.I. Magarill, Physica E **13**, 744 (2002)
[43] F.T. Vasko, O.E. Raichev, *Quantum Kinetic Theory and Applications* (Springer, New York, 2005)
[44] S.A. Tarasenko, Pis'ma Z. Eksp. Teor. Fiz. **84**, 233 (2006); JETP Lett. **84**, 199 (2006)
[45] M. Trushin, J. Schliemann, Phys. Rev. B **75**, 155323 (2007)
[46] O.E. Raichev, Phys. Rev. B **75**, 205340 (2007)
[47] S.D. Ganichev, S.N. Danilov, P. Schneider, V.V. Bel'kov, L.E. Golub, W. Wegscheider, D. Weiss, W. Prettl, cond-mat/0403641 (2004); See also J. Magn. Magn. Mater. **300**, 127 (2006)
[48] A.Yu. Silov, P.A. Blajnov, J.H. Wolter, R. Hey, K.H. Ploog, N.S. Averkiev, Appl. Phys. Lett. **85**, 5929 (2004)
[49] Y.K. Kato, R.C. Myers, A.C. Gossard, D.D. Awschalom, Phys. Rev. Lett. **93**, 176601 (2004)
[50] A.Yu. Silov, P.A. Blajnov, J.H. Wolter, R. Hey, K.H. Ploog, N.S. Averkiev, in *Proc. 13th Int. Symp. Nanostructures: Phys. and Technol.* (St. Petersburg, Russia, 2005)
[51] V. Sih, R.C. Myers, Y.K. Kato, W.H. Lau, A.C. Gossard, D.D. Awschalom, Nat. Phys. **1**, 31 (2005)
[52] N.P. Stern, S. Ghosh, G. Xiang, M. Zhu, N. Samarth, D.D. Awschalom, Phys. Rev. Lett. **97**, 126603 (2006)
[53] D.D. Awschalom, D. Loss, N. Samarth, in *Semiconductor Spintronics and Quantum Computation*, ed. by K. von Klitzing, H. Sakaki, R. Wiesendanger. Nanoscience and Technology (Springer, Berlin, 2002)
[54] R.D.R. Bhat, F. Nastos, A. Najmaie, J.E. Sipe, Phys. Rev. Lett. **94**, 096603 (2005)
[55] S.A. Tarasenko, E.L. Ivchenko, Pis'ma Z. Eksp. Teor. Fiz. **81**, 292 (2005); JETP Lett. **81**, 231 (2005)
[56] H. Zhao, X. Pan, A.L. Smirl, R.D.R. Bhat, A. Najmaie, J.E. Sipe, H.M. van Driel, Phys. Rev. B **72**, 201302 (2005)

[57] S.D. Ganichev, V.V. Bel'kov, S.A. Tarasenko, S.N. Danilov, S. Giglberger, Ch. Hoffmann, E.L. Ivchenko, D. Weiss, W. Wegscheider, Ch. Gerl, D. Schuh, J. Stahl, J. De Boeck, G. Borghs, W. Prettl, Nat. Phys. **2**, 609 (2006)

[58] S.D. Ganichev, S.N. Danilov, V.V. Bel'kov, S. Giglberger, S.A. Tarasenko, E.L. Ivchenko, D. Weiss, W. Jantsch, F. Schäffler, D. Gruber, W. Prettl, Phys. Rev. B **75**, 155317 (2007)

[59] S.A. Tarasenko, E.L. Ivchenko, Proc. ICPS-28 (Vienna, 2006). AIP Conf. Proc. **893**, 1331 (2007)

[60] E.L. Ivchenko, A.A. Toropov, P. Voisin, Fiz. Tverd. Tela **40**, 1925 (1998); Phys. Solid State **40**, 1748 (1998)

[61] J.B. Khurgin, Phys. Rev. B **73**, 033317 (2006)

[62] J.-M. Jancu, R. Scholz, E.A. de Andrada e Silva, G.C. La Rocca, Phys. Rev. B **72**, 193201 (2005)

[63] M.J. Stevens, A.L. Smirl, R.D.R. Bhat, J.E. Sipe, H.M. van Driel, J. Appl. Phys. **91**, 4382 (2002)

[64] M.J. Stevens, A.L. Smirl, R.D.R. Bhat, A. Najimaie, J.E. Sipe, H.M. van Driel, Phys. Rev. Lett. **90**, 136603 (2003)

[65] M.V. Entin, Fiz. Tekh. Poluprov. **23**, 1066 (1989); Sov. Phys. Semicond. **23**, 664 (1989)

[66] V.V. Bel'kov, S.D. Ganichev, E.L. Ivchenko, S.A. Tarasenko, W. Weber, S. Giglberger, M. Olteanu, P. Tranitz, S.N. Danilov, P. Schneider, W. Wegscheider, D. Weiss, W. Prettl, J. Phys. Condens. Matter **17**, 3405 (2005)

[67] V.V. Bel'kov, P. Olbrich, S.A. Tarasenko, D. Schuh, W. Wegscheider, T. Korn, C. Schüller, D. Weiss, W. Prettl, S.D. Ganichev, Phys. Rev. Lett. **100**, 176806 (2008)

[68] E.L. Ivchenko, B. Spivak, Phys. Rev. B **66**, 155404 (2002)

[69] S.A. Tarasenko, Pis'ma Z. Eksp. Teor. Fiz. **85**, 216 (2007); JETP Lett. **85**, 182 (2007)

10

Spin Injection

M. Johnson

10.1 Introduction

The magnetic dipole moment that is associated with a particle's spin state is a fundamental property with manifestations in many branches of physics. In Condensed Matter Physics, electron (and hole) spin offers a rich phenomenology and this volume gives an appreciation of many of these topics. This chapter is concerned with transport phenomena that involve carrier spin as well as charge. Since electric currents in solids typically utilize carriers within a thermal energy range of the Fermi level E_F, our focus is on the spin state of conduction electrons near E_F and we examine how spin states can affect charge flow in the form of current and voltage distributions.

A unique feature of this study is that interface effects are important, as can be understood with a simple description of basic concepts. An electric current in a ferromagnetic metal (F) is spin polarized, a fact known since the early part of the last century [1]. An electric current in a nonmagnetic metal (N) is not spin polarized. When a ferromagnet is in interfacial contact with a nonmagnetic metal, the current crossing the F/N interface is spin polarized. This phenomenon is broadly known as *spin injection* [2]. It results in a nonequilibrium population of spin polarized electrons in N, a spin accumulation. These nonequilibrium spins spread by diffusion, and this can result in a small current of oriented spins in N and/or F.

The phenomenology of spin injection, accumulation and detection was developed for metals. Research on a variety of topics in the late 1990s caused high interest in the plausibility of spin injection in semiconductors. Recent results [3] have demonstrated the effect, and have confirmed that the models of spin injection, accumulation and detection developed for metals are also valid for semiconductors.

10.1.1 History

Our knowledge of spin dependent transport in the solid state has derived from several key experiments and theoretical insights. Tedrow and Meservey [4, 5] fabricated planar $F/I/S$ tunnel junctions, where S was superconducting aluminum, F was a transition metal ferromagnet, and I was an aluminum oxide tunnel barrier. After applying a field of order 1 tesla in the film plane, tunnel conductance spectroscopy was used to demonstrate that the current tunneling into the quasiparticle states of the aluminum was spin polarized. These experiments gave the first empirical estimate of the fractional polarization, $\mathcal{P}$, of such currents. Shortly thereafter, Julliere [6] extended the work of Meservey et al. to form a structure that has become important for applications. For his Ph.D. thesis, he made a tunneling structure $F1/I/F2$ in which he substituted a second ferromagnetic film for the aluminum electrode, thereby inventing the magnetic tunnel junction. He succeeded in measuring a tunnel magnetoresistance only at low temperature, but 25 years later his invention would become a technological success.

The prevailing opinion in the 1970s was that any spin polarized current that crossed a F/N interface would rapidly decay inside N, on a length scale similar to that of Ruderman–Kittel–Kasaya–Yosida interactions (a few angstroms). Aronov, however, took a contrary view. In a series of short theory papers [7–9], he predicted that a current crossing an F/N interface would be spin polarized for a distance of an electron mean free path or longer for three specific cases where N was a nonmagnetic metal, a superconductor, or a semiconductor. At about the same time, Silsbee et al. [10] had been studying nonequilibrium populations of spin polarized electrons using transmission electron spin resonance. They fabricated N foil samples with thin F films on both surfaces and observed an enhanced transmission electron spin resonance (a larger population of spin polarized electrons) in comparison with samples having no F films. Silsbee [11], arguing that the effect could be extrapolated to zero frequency, predicted that a current crossing a F/N interface would remain spin polarized for a long distance *and* that it would generate a nonequilibrium population of spin polarized electrons in N having a spatial distribution identically the same as the spin diffusion length L_s that characterized transmission electron spin resonance. This nonequilibrium population is commonly called a spin accumulation, $\tilde{M}$. Silsbee further predicted a converse effect: the presence of spin accumulation in N would generate an electric voltage at a separate N/F interface, and the magnitude and sign of the voltage would depend on the relative orientations of the magnetizations $M1$ and $M2$ of the two ferromagnetic films, $F1$ and $F2$.

The detailed predictions of Silsbee were successfully demonstrated by the Spin Injection Experiment [2, 12]. The original experiment used a bulk aluminum "wire" as the N material, and the ferromagnetic metal injectors and detectors were Permalloy. The experiment was the first to demonstrate a resistance modulation ΔR in a $F1/N/F2$ structure that changed when the orientations of $M1$ and $M2$ changed from parallel to antiparallel, an effect that is now generally called "magnetoresistance." It also used the Hanle effect [13] to show that spin dynamics and relaxation (and the spin accumulation) at zero frequency were identically the same as at the high frequencies of transmission electron spin resonance. Johnson and Silsbee formalized Silsbee's model with a theory [14], and also derived an original theory of spin-dependent transport using nonequilibrium thermodynamics [15]. Some of the details of this calculation were later confirmed independently by Wyder et al. [16].

Following the discoveries of spin injection and spin dependent transport in $F/N/F$ structures, the giant magnetoresistance effect was reported in 1988 [17, 18]. The giant magnetoresistance and spin valve effects are closely related to spin injection and detection, but the resistance modulation (magnetoresistance) is determined by spin dependent scattering at F/N interfaces. Spin accumulation is negligible in current-in-the-plane spin valves, but both interfacial spin scattering and spin accumulation are relevant to current-perpendicular-to-the-plane spin valves. Interest in giant magnetoresistance was high for a few years because spin valves were used as read heads in magnetic recording technology from 1999–2005. More recently, magnetic tunnel junctions have become the dominant device family for magnetoelectronic devices.

The theory of spin injection, accumulation and detection was developed first in systems where F and N are metals [11, 14]. These systems provide a relatively complete understanding of the phenomenology, and early experiments using metals provide an empirical confirmation. These models were then extended to describe systems in which N is a nonmagnetic semiconductor. Roughly speaking, the first half of this chapter (Sect. 10.2) will develop the models and theoretical framework of spin injection. The second half (Sect. 10.3 and Sect. 10.4) will begin with a brief review of experimental techniques, originally applied to metals, that confirmed the theory. The remainder of the chapter will review recent experiments in the area of spin injection in semiconductors. Very recent experiments [3] have confirmed the validity of the spin injection models of Sect. 10.2 for semiconductors, as well as metals.

10.2 Theoretical Models of Spin Injection and Spin Accumulation

As noted in Sect. 10.1 above, models of spin injection, accumulation and detection preceded experiments. In a similar way, we begin the chapter by developing a theoretical framework. The first model provides a simple but intuitively correct picture. This heuristic model is next made quantitative, and then a fully rigorous theory based on nonequilibrium thermodynamics is given.

10.2.1 Heuristic Introduction

This microscopic transport model can be used to explain the basic physical principles of electrical spin injection, nonequilibrium spin accumulation, and electrical spin detection [2, 12, 14]. The following discussion uses the pedagogical three-terminal geometry shown in Fig. 10.1(a), along with simplified density of states diagrams to describe the transport processes that are shown in Figs. 10.1(b) and (c). When the switch next to the battery is closed, a bias current I is driven through a single domain ferromagnetic film $F1$ and into a nonmagnetic metal layer N. It carries magnetization across the interface (with area A) and into N at the rate $I_M = \eta_1 \mu_B I / e$, where μ_B is the Bohr magneton and I/e is the number current. Here η_1 is the fractional polarization of carriers driven across the interface. It is derived and defined, in

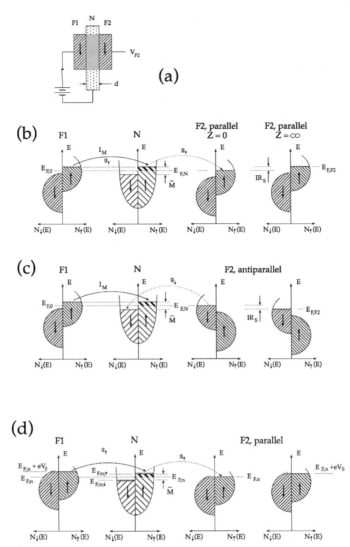

Fig. 10.1. (a) Cross-section sketch of a pedagogical spin accumulation device. (b)–(d) Simplified density of states diagrams to describe the microscopic transport model of the spin injection/accumulation/detection device in (a). (b) $M1$ and $M2$ parallel. (c) $M1$ and $M2$ antiparallel. (d) $M1$ and $M2$ parallel, using band diagrams in which the Fermi level of $F1$ and $F2$ intersects both spin sub-bands

Sect. 10.2.2 below, as the ratio $\eta_1 = (g_\uparrow - g_\downarrow)/(g_\uparrow + g_\downarrow)$, where $g_\uparrow$ and $g_\downarrow$ are the up- and down-spin sub-band conductances. The simplified model of Fig. 10.1 uses half-metal ferromagnets for $F1$ and $F2$, having perfect spin polarization, $\eta_1 = \eta_2 = 1$. In the more realistic model of Sect. 10.2.2, the Fermi level intersects both the up-and down-spin sub-bands of each ferromagnet and $\eta_1, \eta_2 < 1$. Elsewhere in this volume, the symbol $\mathcal{P}$ is used to denote the fractional polarization of a current. Here we use η_i to specifically refer to the fractional polarization of current crossing an interface, as measured at the interface. In this simple one-dimensional model, we use I_M to represent an interfacial magnetization current. Note that more generally, a current of magnetization (equivalently a spin polarized current), $\mathbf{J_M}$, is described by a second rank tensor [15] (see also Chap. 8). The current density $\mathbf{J}$ has three vector components and the direction of spin orientation has three vector components. In this chapter, we will simplify notation for the sake of convenience. For three dimensional systems we will choose a convenient axis for spin orientation (e.g., the $\hat{z}$ axis), so that $\mathbf{J_M}$ can be treated as a vector.

The sample thickness d (Fig. 10.1(a)) is larger than an electron mean free path but smaller than a spin diffusion length, $d < L_s \equiv \sqrt{DT_2}$, where T_2 is the spin relaxation time ($L_s \equiv \delta_s$ and both δ_s and L_s are commonly used in the literature). In metals, the transverse spin relaxation time T_2 is the same as the longitudinal time, $T_2 = T_1$. This time is also called the "spin flip time," τ_s, elsewhere in this volume. Our use of T_2 reinforces that this model is based in TESR phenomenology, but $\tau_s \equiv T_2$ will be used for the remainder of this chapter. In the steady state, I_M is the source rate that spin magnetization is added to the sample region, and relaxation at the sink rate $1/\tau_s$ is steadily removing oriented spins by spin relaxation and randomization. The resulting nonequilibrium magnetization,

$$\tilde{M} = I_M \tau_s / \text{Vol}, \qquad (10.1)$$

is a balance between these source and sink rates and is called *spin accumulation*. It represents a difference in spin sub-band chemical potential in N, $\tilde{M} \propto N(E_F)(E_{F,N\uparrow} - E_{F,N\downarrow})$ (Fig. 10.1(b)), and is depicted as gray shading in Fig. 10.1(a). In (10.1), the volume $\text{Vol} = A \cdot d$ is the volume occupied by the nonequilibrium spins.

A second ferromagnetic film $F2$ that is in interfacial contact with the sample region acts as a spin detector. An output terminal is attached to $F2$, labeled V_{F2} in Fig. 10.1(a). When this terminal is connected to ground through a low impedance current meter, a positive current $I_d \propto (E_{F,N\uparrow} - E_{F,N})$ (where $E_{F,N}$ is the average chemical potential of the two spin sub-bands) is driven across the $N/F2$ interface and through a current detector when the magnetizations $M1$ and $M2$ are parallel (Fig. 10.1(b)). When $M1$ and $M2$ are antiparallel, the current $I_d \propto (E_{F,N} - E_{F,N,\downarrow})$ is negative. Conceptually, this induced electric current is the converse of the injection process and is an interface effect: A gradient of spin sub-band electrochemical potential across the $N/F2$ interface (a thermodynamic force) causes an interfacial electric field (an *emf* source) that drives an electrical current, either positive or negative depending on the sign of the gradient, across the interface.

In real devices it is most common to measure a voltage at the output terminal. Consider V_{F2} to be a floating voltage, separated from ground by infinite impedance. Then a positive (negative) voltage

$$V_{F2} = V_s = \pm \frac{\eta_2 \mu_B}{e} \frac{\tilde{M}}{\chi}, \qquad (10.2)$$

is developed at the $N/F2$ interface [2, 14] when $M1$ and $M2$ are parallel (antiparallel). Here χ is the Pauli susceptibility, $\tilde{M}/\chi \equiv -H^*$ is the effective magnetic field associated with the nonequilibrium spin accumulation, and $\mu_B \tilde{M}/\chi$ is the effective Zeeman energy of each nonequilibrium spin. The voltage V_s is directly related to the interfacial, spin sub-band electrochemical potential gradient described above. The expression for V_s can be combined with that for the magnitude of $\tilde{M}$, (10.1), to give the spin-coupled transresistance R_s that is observed in a spin injection/detection experiment. In Fig. 10.1(a), $\tilde{M}$ is confined to a volume Ad and the transresistance is [19]

$$R_s = \frac{\eta_1 \eta_2}{\chi} \frac{\mu_B^2}{e^2} \frac{\tau_s}{\text{Vol}} = \eta_1 \eta_2 \frac{\rho L_s^2}{\text{Vol}} = \eta_1 \eta_2 \frac{\rho L_s^2}{\text{Ad}}. \tag{10.3}$$

Note that floating voltage V_{F2} is bipolar, $V_{F2} = \pm IR_s$, with positive (negative) voltage observed when $M1$ and $M2$ are parallel (antiparallel).

Equation (10.3) was derived for a device of a geometry similar to Fig. 10.1(a), and having an N layer that is thin relative to the spin diffusion length, $d \ll L_s$. When the N layer thickness is larger than L_s, the value of $\tilde{M}$ decays exponentially away from the $F1/N$ interface and V_s is smaller than its value in the thin limit, $V_s(d) = V_{s,0}e^{-d/L_s}$. Experimentally, manipulating the magnetization orientations $M1$ and $M2$ between parallel and antiparallel gives a measurement of $\Delta R = 2R_s$. Furthermore, measuring $\Delta R(d)$ for a number of nominally identical samples, having different thicknesses d, provides a determination of the spin diffusion length, L_s.

The spin injection phenomenology is distinctly different from giant magnetoresistance associated with spin dependent interfacial scattering in several ways. First, the resistance V_{F2}/I measured at $F2$ is relative high (low) when the magnetization orientations $M1$ and $M2$ are parallel (antiparallel), which is opposite the case for giant magnetoresistance. Second, the resistance V_{F2}/I is truly negative when $M1$ and $M2$ are antiparallel. By contrast, giant magnetoresistance always has a positive resistance of variable magnitude. As discussed below, the negative resistance associated with spin accumulation can be empirically observed when floating voltage V_{F2} is measured with respect to an appropriate reference ground. Third, the voltage associated with spin accumulation is distinctly different from giant magnetoresistance in the response to a transverse field. Analogous with the phenomenology of transmission electron spin resonance, $\tilde{M}$ can be destroyed by application of an external magnetic field applied perpendicular to the plane of the spin polarized electrons, $H_\perp$. Discussed in detail below, the amplitude of a Lorentzian feature observed in a spin injection structure, $\Delta V(H_\perp)$, is proportional to the spin-coupled voltage of the accumulated spins, $\Delta V(H_\perp) = V_s \propto \tilde{M} \propto \tau_s$. Finally, magnetoresistive spin valve devices scale according to the scaling rules of the resistivities of the materials. By contrast, the output voltage of spin injection device structures is proportional to $\tilde{M}$, $V_{F2} \propto \tilde{M}$, and $\tilde{M} \propto 1/\text{Vol}$. This is an important feature: A spin injection device observes *inverse scaling*. The output voltage is inversely proportional to the volume of the N base. As discussed briefly in Sect. 10.5, the inverse scaling law has been experimentally confirmed over ten decades of sample volume, and this offers technological promise for spin injection devices that might be fabricated with nanometer dimensions.

10.2.2 Microscopic Transport Model

The heuristic picture described with Fig. 10.1 can be derived more formally [14]. A treatment of spin injection and spin diffusion in nonmagnetic metals (N) begins with the realization that scattering events that alter the spin state of a carrier in N are rare. Transport in ferromagnetic (F) metals has long been modeled by using independent conductances for the up-spin and down-spin sub-bands [1]. Because populations of up-spin and down-spin carriers in N don't easily mix, Johnson and Silsbee introduced separate up-spin and down-spin conductances to describe transport in a nonmagnetic metal N, as well as in F [14]. Referring to Fig. 10.1(d), generalizing to the more realistic model in which the Fermi level intersects both up- and down-spin sub-bands of $F1$ and $F2$, and using the simplifying assumption that spin relaxation in the ferromagnet is rapid so that its magnetization remains in equilibrium, $E_{F;F1,\uparrow} = E_{F;F1,\downarrow} = E_{F;N} + eV_0$, the electric current from $F1$ to N is

$$J_e = (1/e)\big[g_\uparrow(E_{F;F1,\uparrow} - E_{F;N}) + g_\downarrow(E_{F;F1,\downarrow} - E_{F;N})\big]$$
$$= (g_\uparrow + g_\downarrow)V_0.$$

In the simplified sketch of Fig. 10.1(d), only the up-spin conductance $g_\uparrow$ is shown but both spin conductances contribute. The magnetization current is

$$J_M = (\mu_B/e)\big[g_\uparrow(E_{F;F1,\uparrow} - E_{F;N}) - g_\downarrow(E_{F;F1,\downarrow} - E_{F;N})\big]$$
$$= (\mu_B/e)(g_\uparrow - g_\downarrow)V_0.$$

The ratio of J_M to J_e is

$$\frac{J_M}{J_e} = \frac{g_\uparrow - g_\downarrow}{g_\uparrow + g_\downarrow}\frac{\mu_B}{e} \equiv \eta_1\frac{\mu_B}{e} \tag{10.4}$$

and this defines the interfacial spin polarization coefficient η_1, under the assumptions of no interfacial spin scattering.

A current is driven across the $N/F2$ interface for the low impedance case ($Z = 0$), because of the gradient of spin sub-band electrochemical potential associated with the spin accumulation:

$$J_e = (1/e)\big[g_\uparrow(E_{F;N,\uparrow} - E_{F;F2}) + g_\downarrow(E_{F;N,\downarrow} - E_{F;F2})\big]$$
$$= \frac{1}{e}\left[(E_{F;N} - E_{F;F2})(g_\uparrow + g_\uparrow) + \frac{\mu_B\tilde{M}}{\chi}(g_\uparrow - g_\uparrow)\right]. \tag{10.5}$$

For the high impedance case, the spin-coupled voltage V_s that is a measure of the spin accumulation is found by setting $J_e = 0$ in (10.5),

$$V_s = \frac{\eta_2\mu_B}{e}\frac{\tilde{M}}{\chi}, \tag{10.6}$$

where η_2 is defined in analogy with η_1 as the ratio of the difference and sum of spin-up and spin-down conductances.

The microscopic transport model reviewed above was used in the earliest discussion of spin injection phenomenology [2, 12], and provides several useful features: (i) The idea of using separate and independent conductances for up- and down-spin conduction electron populations had been used to describe transport in ferromagnets, but this model extended the idea and applied it to nonmagnetic metals. This enabled the concept of a spin polarized current in N, an idea with numerous ramifications, such as spin torque switching. (ii) The prediction of a negative "resistance" for the case where $M1$ and $M2$ are antiparallel is intuitively explained. (iii) An output transresistance R_s that obeys "inverse scaling" with sample volume is predicted. As discussed in Sect. 10.5 below, "inverse scaling" may be important for device applications. (iv) The fractional polarization η_1 (η_2) that is characteristic of the injector (detector) interface is defined.

Of high importance to this volume, it will be seen in Sect. 10.4 that the microscopic transport model is qualitatively and quantitatively valid for spin injection in semiconductors.

10.2.3 Thermodynamic Theory of Spin Transport

Nonequilibrium thermodynamics offers a powerful theoretical framework for the description of transport phenomena [20]. While the microscopic transport model offers a good qualitative and quantitative model of spin injection and spin accumulation, the tools of thermodynamics can be used to derive a fully complete and rigorous theory of spin dependent transport [15]. In particular, subtle issues that involve details of spin dependent transport across each F/N interface can be studied and explained with this approach.

Thermodynamic Equations of Motion

Recognizing that gradients of nonequilibrium magnetization $\tilde{M}$ can drive currents of spin and charge across metal-metal interfaces, Johnson and Silsbee developed a thermodynamic theory [15] that can be used to derive the equations of motion of charge and spin in F/N systems. These are particularly useful for understanding interface effects, such as spin injection across a tunnel barrier or with a "resistance mismatch" that characterizes the F and N materials. The formal approach uses an entropy production calculation, where a *flux* J_N of a thermodynamic parameter N (charge, heat, and spin magnetization) is associated with a generalized force, or affinity, F_N (gradients of voltage, temperature, magnetization potential). Each flux can, in general, be driven by each of the generalized forces, so that J_N can be expanded in powers of the F_N. Only the first order terms are kept for linear response theory, and the coefficients are known as the kinetic coefficients L_{mn}.

To summarize the derivation, electronic transport inside a bulk conductor, either ferromagnetic or nonmagnetic, is described by the linear dynamic transport equations [15]:

$$
\begin{pmatrix} \mathbf{J_q} \\ \mathbf{J_Q} \\ \mathbf{J_M} \end{pmatrix} = -\sigma \begin{pmatrix} 1 & a''\frac{k_B^2 T}{eE_F} & \frac{p\mu_B}{e} \\ \frac{a''k_B^2 T^2}{eE_F} & \frac{a'k_B^2 T}{e^2} & p'\frac{\mu_B}{E_F}\left[\frac{k_B T}{e}\right]^2 \\ p\frac{\mu_B}{e} & p'\frac{\mu_B T}{E_F}\left[\frac{k_B}{e}\right]^2 & \zeta\frac{\mu_B^2}{e^2} \end{pmatrix} \begin{pmatrix} \nabla V \\ \nabla T \\ \nabla(-H^*) \end{pmatrix}. \tag{10.7}
$$

Recall that $\mathbf{J_M}$ is generally a second rank tensor, but is treated as a vector for simple assumptions about the spin orientation axis. The kinetic coefficients $L_{m,n}$ can be provided phenomenologically, or estimated within a specific transport model. For example, in a ferromagnet $L_{1,3} = L_{3,1} = p_f(\mu_B/e)$ describes the flow of a magnetization current associated with an electric current, with fractional polarization p_f. Values of p_f are $p_f \approx 0.35$ to 0.45, according to experimental measurements [21]. Note that $L_{3,3} = \zeta(\mu_B/e)^2$ describes self-diffusion of nonequilibrium spins, and $\zeta \approx 1$ is an excellent approximation. In most cases, gradients of temperature are small, heat flow is minimal, and all terms except $L_{1,1}$, $L_{3,1}$, $L_{1,3}$, and $L_{3,3}$ are negligible. In particular, thermal transport coefficients a'' and p' [15] are vanishingly small.

Similar equations are derived for the case of two metals separated by an interface characterized by electrical conductance G:

$$
\begin{pmatrix} I_q \\ I_Q \\ I_M \end{pmatrix} = -G \begin{pmatrix} 1 & \frac{k_B^2 T}{e\varepsilon} & \frac{\eta\mu_B}{e} \\ \frac{k_B^2 T^2}{e\varepsilon} & \frac{ak_B^2 T}{e^2} & \eta'\frac{\mu_B}{s}\left[\frac{k_B T}{e}\right]^2 \\ \frac{\eta\mu_B}{e} & \eta'\frac{\mu_B T}{\varepsilon}\left[\frac{k_B}{e}\right]^2 & \xi\frac{\mu_B^2}{e^2} \end{pmatrix} \begin{pmatrix} \Delta V \\ \Delta T \\ \Delta(-H^*) \end{pmatrix}. \tag{10.8}
$$

Similar to (10.7), differences of temperature and flow of heat across the interface are typically small. Interfacial thermal transport parameters η' and $1/\varepsilon$ are small [15] and the terms $L_{1,2}$, $L_{2,1}$, $L_{2,3}$, and $L_{3,2}$ are negligible.

Consider transport inside a nonmagnetic material. Driven by a gradient of voltage, the currents of charge and of polarized spins are given by (10.7):

$$
\mathbf{J_q} = -\sigma \nabla V, \tag{10.9}
$$

$$
\mathbf{J_M} = -\sigma (p_n\mu_B/e)\nabla V = 0, \tag{10.10}
$$

where $p_n = 0$ in a nonmagnetic material: There is no current of polarized electrons associated with an electric current in a nonmagnetic material. Spin polarized currents may, however, exist in N. The currents of spin polarized electrons are driven by self-diffusion and by the $L_{3,3}$ term of (10.7),

$$
\mathbf{J_M} = -(\sigma\mu_B^2/e^2)\nabla(-H^*). \tag{10.11}
$$

Spin injection and diffusion in N can be described as follows. Spin-charge coupling occurs on the length scale of an electron mean-free path, and an interfacial current of polarized electrons, I_M, generates a nonequilibrium distribution of polarized carriers in N, near the F/N interface. The spatial dependence of $\tilde{M}$ in N is thereafter determined by self diffusion of the spins. In particular it can be shown [22] that flows of spin accumulation are governed by the diffusion equation

$$\nabla^2 \mu_s = \frac{\mu_s}{L_s^2}, \tag{10.12}$$

where $\mu_s \equiv (E_{F;N;\uparrow} - E_{F;N;\downarrow})/2$, while flows of charge are governed by the Laplace equation

$$\nabla^2 \mu_q = 0, \tag{10.13}$$

where $\mu_q \equiv (E_{F;N;\uparrow} + E_{F;N;\downarrow})/2 = E_{F;N}$.

Boundary Conditions for Charge and Spin Diffusion

Spin injection, accumulation, diffusion and detection were first studied in the quasi-one-dimensional geometry of Fig. 10.2(b). Although this geometry will be discussed below, it can be introduced here in order to make some remarks about boundary conditions. Ferromagnetic injector $F1$ spans the width of the sample at $x = 0$, and detector $F2$ spans the sample at $x = L_x \sim L_s$. Spin polarized current enters the sample at the $F1/N$ interface, and a ground for the charge current is provided at $x = -b$, where b is a distance much larger than L_x. It may seem trivial to state that electrical charge and charge current are conserved. The charge current that enters at the injector is equal to that which exits at ground, there is no current flow out the sides of the wire, and the solution to (10.13) is a linear voltage drop from $x = 0$ to $x = -b$. The portion of the wire $x > 0$ is a single equipotential surface.

It is equally important to note that spin orientation is *not conserved*. If a carrier with an oriented spin is injected at $x = 0$, there is no constraint on the orientation of the carrier spin removed at ground. This can be stated another way by recalling from (10.7) and (10.8), $I_M = \eta(\mu_B/e)I_q$ at the F/N interface, but $\mathbf{J}_M$ is unrelated to $\mathbf{J}_q$ in N because $p_n = 0$ in a nonmagnetic material.

Detailed Model of an F/N Interface

The Johnson–Silsbee thermodynamic theory can be used to provide a detailed description of charge and spin transport across a F/N interface. Referring to Fig. 10.3(a), a ferromagnetic metal F and nonmagnetic material N are in interfacial contact. Considering isothermal flow, a constant current J_q is imposed and the solution for the resultant magnetization current J_M is calculated. Equations (10.7) are used to relate currents to potential gradients and to describe steady state flows in each of the materials F and N. The analogous discrete equations, (10.8), are used to relate interfacial currents with differences of potential across the interface. Boundary conditions demand that the magnetization currents of all three regions are equal at the interface ($x = 0$), $J_{M,F} = J_M = J_{M,N}$ (where J_M is the interfacial magnetization current), and that the electric currents are also equal, $J_{q,F} = J_q = J_{q,N}$.

In the general case, the flow of J_M into N generates a spin accumulation, $\tilde{M}$, and an associated effective field, $-H^* = \tilde{M}/\chi$, in N (Fig. 10.3(b)), and $-H^*(x)$ decreases as x increases away from the F/N interface. The nonequilibrium spin population can also diffuse *backwards*, along $-x$, going back across the F/N interface and into F. The effective field $-H^*$ in F is not constrained to match that in N,

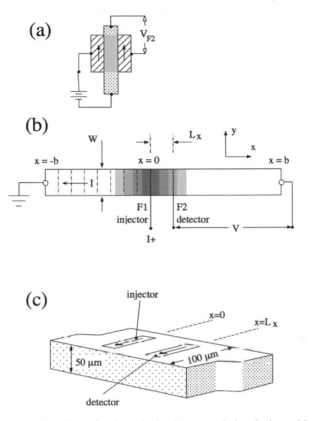

Fig. 10.2. (a) Cross section view of pedagogical spin accumulation device, with floating voltage V_{F2} grounded to a region of N that is remote from the path of bias current. (b) Schematic top view of nonlocal quasi-one-dimensional geometry, the lateral spin valve, used in original spin injection experiment as well as in recent semiconductor samples. *Dotted lines* represent equipotentials characterizing electrical current flow. *Gray shading* represents diffusing population of nonequilibrium spin polarized electrons injected at $x = 0$, with *darker shades* corresponding to higher density of polarized electrons. (c) Perspective sketch of bulk Al sample used in original spin injection experiment

$-H_f^*(x = 0) \neq -H_n^*(x = 0)$ (Fig. 10.3(b)) because, for example, the susceptibilities χ_f and χ_n can be quite different.

The backflow of diffusing, spin polarized electrons across the interface must be overcome by the imposed current. The interface has some intrinsic resistance, $R_i = 1/G$, and the backflow acts as an additional, effective interface resistance (Fig. 10.3(c)). The spatial extent of the nonequilibrium spin population in F is described by the spin diffusion length in F, $\delta_{s,f}$. An estimate for transition metal ferromagnetic films is $\delta_{s,f} = 14.5$ nm [23]. The backwards diffusion of polarized spins near the F/N interface effectively cancels a portion of the forward flowing polarized current $J_{M,f}$. The result is that the fractional polarization of the magnetization

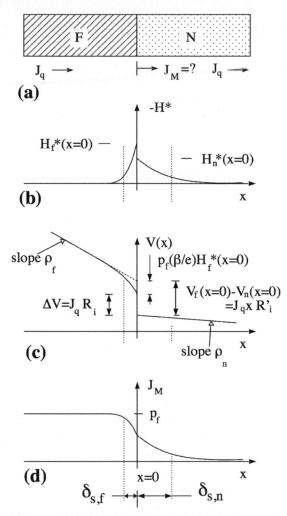

Fig. 10.3. (a) Model for flow of charge and spin currents, J_q and J_M, at the interface between a ferromagnetic metal and nonmagnetic material. $x = 0$ at the interface. (b) Magnetization potential. The nonequilibrium spin population decays in F and N with characteristic lengths $\delta_{s,f}$ and $\delta_{s,n}$, respectively. (c) Voltage. (d) Current of spin magnetization, J_M

current that reaches and crosses the interface, J_M, is reduced relative to the bulk value, $J_M < J_{M,f}$ (Fig. 10.3(d)). This reduction of polarization arises from the $L_{3,3}$ self-diffusion term in (10.8).

After algebraic manipulation, a general form for the interfacial magnetization current is found to be [15, 24]

$$J_M = \frac{\eta \mu_B}{e} J_q \left[\frac{1 + G(p_f/\eta) r_f (1 - \eta^2)/(1 - p_f^2)}{1 + G(1 - \eta^2)[r_n + r_f/(1 - p_f^2)]} \right], \qquad (10.14)$$

where $r_f = \delta_{s,f} \rho_f = \delta_{s,f}/\sigma_f$, $r_n = \delta_{s,n} \rho_n = \delta_{s,n}/\sigma_n$, $G = 1/R_i$. It is important to note that spin transport is governed by the relative values of the intrinsic interface resistance, $R_i = 1/G$, the resistance of a length of normal material equal to a spin depth, r_n, and the resistance of a length of ferromagnetic material equal to a spin depth, r_f. Typical values of these resistances are easily estimated [23]; $r_f \sim 10^{-11} \,\Omega\mathrm{cm}^2$, $r_n \sim 2 \times 10^{-11}$ to $2 \times 10^{-10} \,\Omega\mathrm{cm}^2$, and $R_i = R_c \approx 10^{-11} \,\Omega\mathrm{cm}^2$. Since all of the characteristic values fall within a range of a factor of ten, all of the terms in (10.14) are important for the general case.

Resistance Mismatch at an F/N Interface

One limiting case of (10.14) is that of low interfacial resistance, $R_i \to 0$. An appropriate experimental system is a multilayer, current-perpendicular-to- the-plane giant magnetoresistance sample grown in ultrahigh vacuum [25]. In this case, $R_i \approx 3 \times 10^{-12} \,\Omega\mathrm{cm}^2 \ll r_f$ may justify the high conductance approximation. Equation (10.14) reduces to the simpler form [15, 24]:

$$J_M = p_f \frac{\mu_B}{e} J_q \frac{1}{1 + (r_n/r_f)(1 - p_f^2)}. \tag{10.15}$$

The polarization of the injected current is reduced from that in the bulk ferromagnet by the *resistance mismatch* factor $(1 + M')^{-1} = [1 + (r_n/r_f)(1 - p_f^2)]^{-1}$. Using the above estimates for r_f and r_n, the mismatch factor can be expected to be as large as $M' \sim 20$.

Another limiting case is that of high interfacial resistance. Spin accumulation in N can be large, but the resistive barrier prevents back diffusion. The nonequilibrium spin population in F remains small, and the voltage drop across the interface is dominated by R_i. The interfacial magnetization current is now given by

$$J_M = \eta \frac{\mu_B}{e} J_q, \tag{10.16}$$

and the fractional polarization is determined by the interface parameter η. The resistive barrier may be asymmetric with respect to spin, and, in general, the limit $\eta \le p_f$ is imposed.

The transport effects related to "resistance mismatch" (equivalently called "conductance mismatch") are not likely to be observed in systems where N is a non-magnetic metal because the relevant resistances are so small that even a "negligible" interface resistance, such as a contact resistance, can dominate interfacial transport. The condition $R_i \ll r_f, r_n$ is almost never realized, and the general expression, (10.14), should be used. A prediction [26] that "resistance mismatch" would limit injection efficiency to less than 1% when F is a metal and N is a semiconductor was based on a calculation using the infinite interface conductance limit, (10.15). However, such an F/N interface is always characterized by a Schottky barrier, tunnel barrier, or low conductance "Ohmic contact." In each case, the intrinsic interface resistance is high and mediates interfacial spin transport [27]. As discussed in

Sects. 10.4.1 and 10.4.2 below, experiments have confirmed this and demonstrated typical injection polarizations of 20% to 50% for ferromagnetic metal/nonmagnetic semiconductor interfaces.

10.2.4 Hanle Effect

The Johnson–Silsbee spin injection experiment [2] introduced a novel technique, employing the Hanle effect [13], to detect spin accumulation and measure spin relaxation times, and to demonstrate that spin injection and accumulation are fundamentally related to nonequilibrium spin resonance phenomena. This technique is often cited as an infallible proof of spin injection, in both metals and semiconductors [3], and the physics is briefly reviewed in this section.

In-plane magnetic fields can be used to make magnetoresistance measurements of R_s, and this is discussed below in Sect. 10.3. However, using the Hanle effect [2, 12, 13], the zero frequency analogue of transmission electron spin resonance, permits a quantitative measurement of both R_s and spin relaxation time T_2 (in this section, T_2 is proper and will be used instead of τ_s) using only a single data set. In a simple picture, spin polarized electrons diffuse across a distance L from the injector to the detector. Consider the thin limit, $L < L_s$. Under the influence of a perpendicular magnetic field B, each spin precesses by a phase angle that is proportional to the time it takes to reach the detector. Since the electrons are moving diffusively, there is a distribution of arrival times. In the limit of zero external magnetic field, all the diffusing, polarized electrons that reach the detector have the same phase, as long as they have spent a time less than T_2 in the sample. For sufficiently large field, the spin phase angles of the electrons reaching the detector at any one time are completely random. If T_2 is long, there is a large distribution of arrival times at the detector, and a very small field is needed to randomize the distribution of phases. The characteristic field B_{hw} is given by the condition that the product of the precessional frequency and T_2 is a complete phase rotation angle of 2π. Thus, the characteristic field is $B_{hw} = 1/\gamma T_2$, where γ is the gyromagnetic ratio for electrons. A plot of the number of spin polarized electrons, or of the voltage V_s at the detector $F2$, as a function of the perpendicular field would have a Lorentzian (absorptive) line shape. The detailed field dependence of the spin-coupled voltage V_s detected at a second ferromagnetic electrode $F2$ can be found by solving the Bloch equations with a diffusion term [12, 14]. The advantage of using the Hanle effect is that a single measurement gives relaxation time T_2 from the width of the Hanle feature, and leaves polarization η as the sole parameter to be fit to the amplitude of the feature.

10.3 Spin Injection Experiments in Metals

Experimentally, a reference ground must be provided for any real measurement of the floating voltage V_{F2} (refer to Fig. 10.1(a)). An ideal ground will not introduce Ohmic voltage contributions to the measurement of spin accumulation. An appropriate ground is provided in the schematic sketch of Fig. 10.2(a). Because electric bias current flows across the $F1/N$ interface and through the bottom of N, the region of N near the top is remote from the current path and the Ohmic voltage between $F2$ and the ground at the top is minimal. An experimental realization of this idea is shown in Fig. 10.2(b).

This quasi-one-dimensional, nonlocal geometry was introduced in the original spin injection experiment [2]. It has become widely adopted for studies of spin injection in mesoscopic systems [28], and in semiconductors [3]. Describing the details of the original experiment by referring to Fig. 10.2(b), an aluminum wire extends along the $\hat{x}$ axis. A narrow ferromagnetic electrode $F1$ spans the width of the wire near its center, at $x = 0$. When bias current I is injected at $F1$ and grounded at the left end of the wire, $x = -b$, there is a linear voltage drop from $x = 0$ to $-b$. This is depicted by the regularly spaced equipotential (dot-dash) lines in Fig. 10.2(b). However, there is no net current flow in the region $x > 0$ and the wire is at a constant potential from $x = 0$ to $x = b$. A voltage measurement between the end of the wire $x = b$ and a narrow electrode that spans the wire at $x = L_x$ is necessarily a null measurement, $V = 0$.

Spin polarized electrons injected at $F1$ diffuse equally along $\pm\hat{x}$. Self-diffusion of the nonequilibrium spin population is symmetric, with each polarized electron performing a random walk along the $\hat{x}$ axis [2]. The density of diffusing spin polarized electrons is depicted in Fig. 10.2(b) by the shaded region, with darker shades representing higher density. When the electrode at $x = L_x$ is also a ferromagnetic film, $F2$, a potentiometric measurement V records a spin dependent voltage that is relatively high (low) when the magnetization orientation $M2$ is parallel (antiparallel) with $M1$. Since $V = 0$ in the absence of nonequilibrium spin effects, this measurement uniquely discriminates against any background voltages. The ideal geometry of Fig. 10.2(b) is achieved when the electrodes have negligible width (they are drawn as lines in Fig. 10.2(b)) and have uniform contact across the width of the wire. Departures from ideality result in a small, spin independent baseline voltage $V \neq 0$ [29]. The geometry of Fig. 10.2(b) has been called "nonlocal," and structures of this kind are sometimes called "lateral spin valves" to distinguish them from thin film sandwich structures typically used in giant magnetoresistance.

In the original spin injection experiment, the N sample was a "wire" of bulk, high purity aluminum about $100\,\mu m$ wide and $50\,\mu m$ thick, sketched in a perspective view in Fig. 10.2(c). An array of ferromagnetic pads, about $15\,\mu m$ wide by $45\,\mu m$ long, were fabricated by photolithography and liftoff as electrodes on the top surface, with interprobe spacings x in multiples of $50\,\mu m$. The F electrodes were e-beam deposited from a single source of $Ni_{0.8}Fe_{0.2}$ after cleansing the Al surface with an Ar ion mill.

Before describing details of the experimental methods, some clarification about the notation used for magnetic field can be offered. The symbol $\mathbf{B}$ is often called the "magnetic flux density," "magnetic induction," or simply the "magnetic field." The symbol $\mathbf{H}$ is called the "magnetic field strength." In a magnetizable medium, $\mathbf{B}$ and $\mathbf{H}$ are related by $\mathbf{B} = \mathbf{H} + 4\pi\mathbf{M}$, where $\mathbf{M}$ is the volume magnetization of the medium. In free space, $\mathbf{B}$ and $\mathbf{H}$ are identically the same. Ferromagnetic materials have spontaneous magnetization, and experiments involving F/N material structures

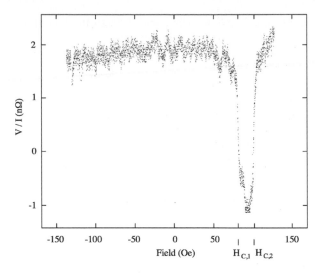

Fig. 10.4. Example of spin injection, accumulation, diffusion and detection in aluminum, using in-plane magnetic fields. External field is applied along $\hat{y}$ axis, in the plane of $F1$ and $F2$, starting at $H_y = -140$ Oe, sweeping up and stopping at $H_y = 125$ Oe. The dip occurs when $M1$ and $M2$ change their relative orientation from parallel to antiparallel and shows V/I as a negative resistance. A similar dip occurs at symmetric negative field when H_y is swept from positive to negative. $L_x = 50\,\mu m$, $T = 27$ K

typically use **H** to refer to externally applied magnetic fields. This convention is adopted for this chapter, although other chapters in this volume correctly use **B**.

For magnetoresistance measurements with magnetic fields in the film plane, the magnetizations $M1$ and $M2$ must have slightly different coercivities, $H_{C,1} \neq H_{C,2}$. The detected voltage V as a function of externally applied field H should be positive whenever the magnetizations of $F1$ and $F2$ are aligned and negative over a field range $H_{C,2} - H_{C,1}$, when anti-aligned.

Figure 10.4 shows magnetotransport data for a sample with $L_x = 50\,\mu m$ using in-plane fields, $H = H_y$. As the field is swept along the easy axis of both $F1$ and $F2$ from negative to positive values, the positive voltage for $H < H_{C,1}$ is a measure of spin accumulation, $R_s \propto \tilde{M}$, with $M1$ and $M2$ parallel. The region $H_{C,1} < H < H_{C,2}$ represents the region where magnetizations $M1$ and $M2$ are reorienting between parallel and antiparallel and the detected voltage V_s drops from positive to negative. In the region $H > H_{C,2}$ the orientations $M1$ and $M2$ return to parallel and the original, positive voltage is regained. A field sweep from positive to negative values shows the same feature in the field range -100 Oe $< H < -80$ Oe, as expected for the hysteresis of the ferromagnetic films. Note that the voltage V_s is positive for parallel magnetizations and negative for antiparallel magnetizations, as discussed in Sect. 10.2.1.

As a demonstration of the Hanle effect, a magnetic field $H_\perp$ is applied transverse to the orientation of the polarized spins. An example of the Hanle effect from an

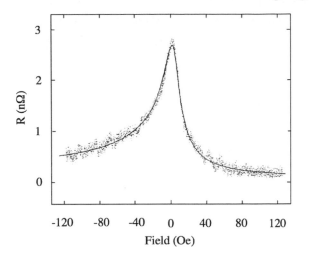

Fig. 10.5. Example of spin injection in aluminum. Hanle data showing absorptive lineshape, with small admixture of dispersive character. $L_x = 50\,\mu m$, $T = 21\,K$. Solid line is fit to equations in the text: $T_2 = 7.0\,nsec$, $\eta = 0.075$

Al wire sample is shown in Fig. 10.5 presented in units of resistance, $R = V/I$. Figure 10.5 presents typical data, primarily absorptive in appearance, for a sample with $L_x = 50\,\mu m$, $T = 21\,K$, and the deduced fitting parameters are $\eta = 7.5\%$, $T_2 = 7.0\,ns$.

Finally, it can be noted that the spin injection experiment offered an early demonstration of a pure spin current. In the nonlocal geometry of Fig. 10.2(b), voltage measurements confirm that there is zero current of charged particles. Measurements of spin accumulation using a ferromagnetic detector, however, show a spin population that decreases with distance, thereby showing a nonzero diffusion current of spins.

10.4 Spin Injection in Semiconductors

Measurement of the spin lifetimes of carriers in semiconductors has been studied optically and with electron spin resonance (ESR) for many years [30–33]. Although Aronov had proposed the possibility of spin injection into semiconductors [8], interest in the topic was triggered by the proposal to fabricate a spin injected field effect transistor [34]. In the Datta–Das structure (Fig. 10.6), a ferromagnetic source and drain were connected by a two dimensional electron gas channel, with the source-drain distance L on the order of an electron ballistic mean free path. The magnetizations of both source and drain were oriented along the axis of the channel, the $\hat{x}$ axis. An intrinsic electric field E_z perpendicular to the two dimensional electron gas plane (discussed below) transformed, in the rest frame of the carriers, as an effective magnetic field H_y^*. The notation H^* is commonly used for an effective magnetic field of any kind, and H_y^* is not related to the thermodynamic effective field discussed

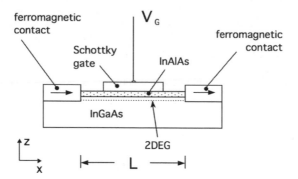

Fig. 10.6. Cross sectional schematic of the spin injected field effect transistor proposed by Datta and Das

in Sects. 10.2.1 and 10.2.3 and Fig. 10.3. Carriers were injected at the source with their spin axes oriented along $\hat{x}$, precessed under the influence of H_y^*, and arrived at the drain with a spin phase ϕ that depended on their time of transit, $\phi \propto L/v_F$, where v_F is the Fermi velocity. The source-drain conductance was predicted to be proportional to the projection of the carrier spin on the magnetization orientation of the drain, and therefore to be a function of ϕ. By applying a gate voltage to the channel, the field E_z, the effective field H_y^*, and ϕ could be varied and a modulation of source-drain conductance would result.

The Datta–Das device was designed as a field effect transistor. The geometry is similar to the lateral spin valve, but two conceptual differences are important. First, the Johnson–Silsbee idea applied to metals involves diffusive charge and spin transport and the development of spin accumulation. By contrast, the Datta–Das device was designed to involve ballistic charge and spin transport, and no spin accumulation was expected. Second, the transport of spin polarized carriers in the Datta–Das device is determined by internal electric fields (the Rashba effect), and these fields can be modulated by a gate voltage. This has no analogue in metals-based devices.

Some researchers perceive a spin injected field effect transistor to have technological impact for a variety of new applications in digital semiconductor technology. Possible advantages are unclear at best and, furthermore, no viable prototype device has yet been demonstrated successfully. However, research is active and addresses individual topics related to device development. One class of topics relates to basic issues of spin injection across a F/NS interface (where NS is a nonmagnetic semiconductor). The dominant experimental methodology involves a light emitting diode sample structure that has a ferromagnetic electrode, and optical measurements are used. A second class of topics relates to spin dependent scattering and spin-flip lifetimes in nonmagnetic semiconductors. The dominant experimental technique is again optical, using measurements of Faraday rotation to infer the spin orientations of carriers. Although experiments using spin injected light emitting diodes suggest that the fractional polarization η of injected current can be relatively large, transport experiments have not succeeded in demonstrating devices with large η. Since the spin dependent voltage modulations are small, the use of transport measurements to characterize the dynamics of spin dependent transport is marginal. Since spin injected light emitting diodes have no technological applications, the few results from transport experiments have high importance. For these reasons, the experimental review is broken into two portions. Optical experiments are discussed in Sect. 10.4.1 and transport experiments are discussed in Sect. 10.4.2.

10.4.1 Optical Experiments

Spin Injection

The spin injected field effect transistor requires the transmission of a spin-polarized current at a F/NS (and NS/F) interface, and the transport of spin polarized carriers from a source to a drain. A successful device relies on predictions about the efficiency of spin injection and about the dynamics of spin-polarized transport in a two dimensional electron gas channel. Spin-polarized tunneling experiments using ferromagnetic metals and semiconductors were the earliest experiments performed to examine the issue of spin-polarized transport at a F/NS interface. In a typical experiment [35], luminescence in a GaAs sample is used as a detector of polarized carriers injected across a vacuum tunnel barrier. In the reverse of optical pumping, circularly polarized light is emitted in proportion to the degree of spin polarization of the recombining minority carriers at the fundamental band gap. GaAs is an ideal material because of the relatively large spin orbit splitting at the valence band. Room temperature experiments used nickel as the ferromagnet, and the polarization of the tunnel current was found to vary between 5 and 30%.

Optical measurements using spin injected light emitting diodes have become a popular technique for inferring the polarization of injected current. The experiments use a variety of materials, but a typical device structure is sketched in cross section in Fig. 10.7. Structures are prepared by molecular beam epitaxy. A quantum well heterostructure is grown between two spacer layers. A ferromagnetic material is grown on top of the top spacer and is often capped with a thin metal film for passivation. When biased by a voltage, an electric current flows from F into the nonmagnetic semiconductor NS and the interfacial current is characterized by a polarization η. Recombination occurs in the heterostructure region, accompanied by photon emission (luminescence). The optical polarization of the emitted photons is related to the polarization of carriers that reach the quantum well. The luminescence may be detected at the edges of the structure, or at the back side of the substrate. Photons may also escape through the top of the structure, and the metal cap must be sufficiently thin such that transmission of the luminescence is substantial. The experimental methodology involves measuring the polarization of the luminescence as a function of bias voltage and wavelength.

Structures have been fabricated using III–V compound semiconductors with ferromagnetic injectors that may be magnetic semiconductors (e.g., GaMnAs) or transition metal ferromagnets. In this case, an external magnetic field may be applied along $\hat{x}$ or $\hat{z}$ in order to saturate the magnetization of the F layer. Structures also have been

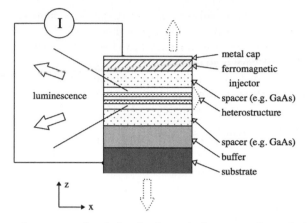

Fig. 10.7. Schematic cross sectional sketch of a typical structure for optical spin injection experiments

fabricated using II–VI compound semiconductors, in which case the ferromagnetic injector is a diluted magnetic semiconductor also from the II–VI family. For these materials, a large external magnetic field applied along $\hat{z}$ is required to saturate the magnetization of the diluted magnetic semiconductor layer.

In one of the earliest optical studies of spin injection [37], the nonmagnetic semiconductor was a 1.6 micron thick layer of the II–VI compound CdTe and the ferromagnetic injector was a 360 nm thick layer of the diluted magnetic semiconductor $Cd_{0.98}Mn_{0.02}Te$. This structure was not electrically biased, rather pulses of circular polarized light that were incident on the metal cap generated spin polarized carriers in the diluted magnetic semiconductor. The polarized carriers diffused into NS and the polarization of the photoluminescence was measured as functions of time (relative to the excitation pulse) and energy. The experimental temperature, $T = 5\,K$, was slightly above the spin glass temperature of the $Cd_{0.98}Mn_{0.02}Te$ and external fields of approximately 2 tesla along $\hat{z}$ were used to create the necessary Zeeman splitting. This early study demonstrated the feasibility of using a magnetic semiconductor for spin injection, and the authors used a semiquantitative analysis to estimate that the polarization of interfacial F/NS current was 50% or higher.

An important variation of this technique [38] measured the circular polarization of the electroluminescence from a GaAs/AlGaAs light emitting diode (heterostructure layer in Fig. 10.7). A diluted magnetic semiconductor, n-doped $Be_xMn_yZn_{1-x-y}Se$, was used as the ferromagnetic injector. External fields applied along the $\hat{z}$ axis with magnitude of a few tesla were used. Under forward bias, spin polarized electrons were injected into the GaAs with an efficiency estimated at 90%. The effect was robust at fields as low as 1.5 tesla and temperatures as high as 10 K. The resistivities of the magnetic semiconductor F and the nonmagnetic semiconductor NS are of the same order, and therefore effects of "resistance mismatch" do not diminish the high values of η. As discussed in Sect. 10.2.3, this means that spin injection structures using only semiconductors can be fabricated without incorporating low transmission barriers between F and NS.

This variation was also performed [39] using the ferromagnetic semiconductor $Ga_{1-x}Mn_xAs$ ($x = 0.045$) as the spin injector. Spin polarized holes were electrically injected across a F/NS interface and into a spacer layer of undoped GaAs. The spacer layer, which had different thicknesses d ranging from 20 nm to 220 nm for a variety of samples, was grown on top of an InGaAs/GaAs light emitting quantum well heterostructure. The F layer GaMnAs has in-plane magnetic anisotropy. An external magnetic field H_x in the film plane (along $\hat{x}$ in Fig. 10.7) changed the magnetization of F from orientation along $+\hat{x}$ to $-\hat{x}$. The polarization of the electroluminescence was measured as functions of H_x, d and temperature T. The switching field of F was about 45 Oe, and a strong spin injection effect was observed for temperatures as high as 35 K.

These studies [38, 39] offer an important demonstration of interfacial spin-polarized transport and generated tremendous interest in the topical area of magnetic semiconductors. However, experiments with magnetic semiconductors have some limitations worth noting. Structures using a diluted magnetic semiconductor as the magnetic material require large magnetic fields, of order several tesla. Structures using the ferromagnetic semiconductor GaMnAs have shown relatively small effects. Both categories of materials require cryogenic temperatures, and only weak effects have been seen at temperatures above 40 K. For these reasons, magnetic semiconductors may not be relevant for integrated electronic device applications.

Some of these limitations may be avoided by using a transition metal ferromagnet as the ferromagnetic injector. Using the same basic technique introduced by Oestreich [37], two InGaAs quantum wells, separated by a GaAs barrier and sandwiched between two GaAs spacer layers (refer to Fig. 10.7) were used as a light emitting diode [40]. A 70 nm thick n-GaAs layer was grown on the light emitting diode, followed by a 20 nm thick Fe layer and a 10 nm Al cap. As discussed in Sect. 10.2.3 above, efficient spin transport can occur between a ferromagnetic metal and a nonmagnetic semiconductor when there is a sufficiently large interfacial resistance. Here a Schottky barrier at the Fe/GaAs interface provides the necessary resistance. Electroluminescence measurements were performed in the Faraday geometry, requiring a field along $\hat{z}$. Fields of about 2 tesla were sufficient to saturate the Fe magnetization along $\hat{z}$, perpendicular to the in-plane easy axis. Spin polarized injection with interfacial polarization of about 2% was observed at room temperature. While the polarization efficiency was not very large, this experiment demonstrated that transition metal ferromagnets could be used and that room temperature spin injection is possible. More recent experiments have shown that injection using ferromagnetic metal films can have interfacial polarization as large as 50% [41].

Optical studies of spin injection in a light emitting diode have been popular, but the converse measurement also has been performed [42]. Using a structure similar to that of Fig. 10.7, a thin Fe layer is grown on n-doped GaAs (100). Some samples include a tunnel barrier of aluminum oxide and other samples only have the native Schottky barrier between F and NS. The Fe layers are sufficiently thin that incident polarized light is transmitted to the GaAs, where spin polarized photocurrent is induced. The photocurrent is measured for a variety of voltage bias conditions. The results were highly sensitive to the morphology of the F/NS interface, but spin asymmetries of about 1.7% were observed at room temperature at a bias of about 0.04 V. This magnitude is comparable with the early studies of spin injection from Fe to GaAs [40].

The study of spin injection in semiconductors is not yet characterized by the successful correspondence between theory and experiment of spin injection in metals. The preponderance of empirical results implies that electrical spin injection across a F/NS interface should be quite possible at room temperature, and may have efficiency as large as tens of percent. Other research suggests that novel techniques for spin injection may be viable. As one example, there is evidence that carriers in a NS may become spin polarized by proximity with a proximal interface with a ferromagnetic film [43].

Spin Dynamics and Lifetimes

The dynamics of spin-polarized carriers in a semiconductor have also been studied by optical techniques. Purely optical techniques for probing transport have been used for decades [33] and reviews appear in other chapters of this volume. A few experiments relevant to this chapter's topic of spin injection can be noted.

The recent interest in semiconductor spintronics was ignited, in part, by an experiment that demonstrated the diffusion of spin polarized carriers in GaAs [44]. A thick layer of intrinsic GaAs (about 5 microns thick) was grown on top of a thin GaInAs (8 nm) quantum well, and was capped by an optically semitransparent layer of Ni/Cr. The structure was biased by a voltage. Circularly polarized pulses of laser light created an electron-hole plasma near the top surface of the GaAs. The electrons, which have a preferential spin orientation because of optical selection rules, diffuse under the influence of the electric field. Electrons reaching the quantum well recombine with holes and the polarization of the photoluminescence is related to the fractional spin polarization of the recombining electrons. From their measurements, the authors deduced that the electron spin polarization of about 30% is almost completely conserved during diffusion over a length of 4 microns ($T = 10$ K). These results implied that the electron spin diffusion length in GaAs, in an electric field, is much longer than 10 microns. This was important because it demonstrated that spin orientation could be maintained on length scales longer than those characteristic of electronic devices.

Long spin relaxation times and spin diffusion lengths were confirmed in a series of experiments [36, 45]. The optical technique involves the resonant pumping of optically generated spin-polarized excitations and subsequent time-dependent optical detection by Faraday rotation. This approach has measured spin relaxation times of order several nsec in n-type GaAs at 5 K, and spin diffusion lengths of order tens of microns at low temperature.

10.4.2 Transport Experiments

The spin injected field effect transistor proposed by Datta and Das was not designed for use in digital electronics technology. By contrast, the authors proposed to use some novel characteristics of the carriers in a two dimensional electron gas of a III–V compound semiconductor heterostructure. Small variations of the Datta–Das structure result in a device that may have digital electronics applications, and these will be mentioned below. A large portion of semiconductor spintronics research has studied the unusual transport characteristics of III–V semiconductors, however, and we begin with a brief review of the Rashba Hamiltonian.

Large Spin–Orbit Effects

An asymmetry in the confinement potential of a quantum well can generate an intrinsic field E_z (where $\hat{z}$ is the direction of film growth, normal to the film plane). The electric fields $\pm E_{z,c}$ associated with the confining walls of the well are very large. They cancel for a perfectly symmetric well and the net electric field is zero. If there is any asymmetry in the well, however, cancellation is imperfect and a large, intrinsic field E_z is the result. Spin–orbit effects can cause a large splitting Δ_{so} of the conduction band, as noted by Rashba [46]. For a 2D carrier moving with a weakly relativistic Fermi velocity in the $\hat{x}$ $(-\hat{y})$ direction, E_z transforms, in the rest frame of the carrier, as an effective magnetic field H_y^* (H_x^*). The coupling of field H^* with the carrier spin results in spin eigenstates.

With this understanding of a Rashba spin system, the Datta–Das device can be reviewed again. Referring to Fig. 10.6, the effective magnetic field H^* is along the $\hat{y}$ axis. Spin polarized carriers injected from the source with orientation along the $\hat{x}$ axis are not in a spin eigenstate. They precess under the influence of field H_y^*. When the distance L_x corresponds to a transit time that develops a spin precession of π (2π), the spin orientation of the carrier is antiparallel (parallel) with the dominant orientation of the drain and the source-drain conductance is relatively low (high). The gate voltage modulates the strength of the effective field. Therefore, as the gate voltage is monotonically increased, the spin orientation of a carrier ballistically reaching the drain varies periodically from antiparallel to parallel and the source-drain conductance varies periodically between relatively high and low values. It can be noted that the original idea of Datta and Das [34] relied on ballistic transport. The spin orbit interaction in a 2D electron gas with Rashba Hamiltonian is so strong that a few scattering events might be expected to randomize the carrier spin orientation, and transport by diffusion would show no interesting effects.

The device of Datta and Das can be modified for integrated electronics applications. For example, the magnetization orientation of one of the contacts (e.g., the source) can be chosen to be fixed in one direction along the $\hat{y}$ axis, and that of the other (the drain) can be free to be manipulated to be parallel or antiparallel with the first. Such a device would operate as a memory cell in a nonvolatile magnetic random access memory. A datum would be stored as a parallel or antiparallel magnetization state of the drain. Sensing a high or low channel conductance would correspond to readout of the binary state. The gate voltage would not be used to modulate the spin transport. However, it could be used to isolate any single device from an array of devices.

Experimental Progress

All of the measurements described in Sect. 10.4.1 have been purely optical, and it is not valid to extrapolate plausible transport results from optical measurements. Instead, experiments on device prototypes that utilize electrical injection and electrical detection are necessary. Empirically observed voltage and/or current modulations have been small, and the number of transport experiments is relatively few. From existing data, the basic models of spin injection and detection (refer to Sect. 10.2) seem to be valid.

Several early experiments used a geometry similar to that of Datta and Das. In one example, NS was an InAs quantum well and the F electrodes were permalloy. The device was configured such that magneto-optic Kerr effect measurements of the permalloy electrodes could be performed in situ. A small but reproducible resistance modulation could be observed at low temperature ($T = 0.3$ K) [47].

An experiment using the nonlocal geometry of Fig. 10.2 can be reviewed with some detail [48]. Two F/NS junctions, were fabricated on a common two dimensional electron gas channel. In the actual device, six separate channels were connected in parallel for improved signal-to-noise. Narrow channels, about 900 nm wide, were defined on an InAs single quantum well heterostructure using optical lithography and an Ar ion mill dry etch. The chip was backfilled with SiN to planarize the surface at the level of the mesa and to cover the side edges of the two dimensional electron gas. The interelectrode spacing L_x was the order of magnitude of the carrier mean free path, a few microns.

Referring to the nonlocal geometry of Fig. 10.2, spin polarized electrons were injected from $F1$ into the two dimensional electron gas and the injected current was grounded at the left end of the "wire." Detecting electrode $F2$ was grounded at the right end and acts as a spin-sensitive potentiometer. Ballistic and quasi-ballistic spin polarized electrons that are injected at the $F1$/NS interface have initial trajectories to the left and right in equal numbers. Carriers with initial trajectories to the right eventually scatter, and all the current is drained at ground. Injector $F1$ was fabricated with permalloy and had a relatively small coercivity, $H_{C1} \approx 30$ Oe. Detector $F2$ was fabricated with FeCo and had a relatively large coercivity, $H_{C2} \approx 70$ Oe. By applying an external field H_y in the film plane and parallel to the easy magnetization axis of the ferromagnetic films, the relative magnetization orientation of injector and detector was manipulated between parallel and antiparallel. The top barrier layer of the InAs SQW remained in tact during processing and formed a low transmission barrier between the ferromagnetic metal electrodes and NS. The junction resistance of a few $K\Omega$ was sufficiently large that effects of resistance mismatch (refer to Sect. 10.2.3) were not relevant.

An example of electrical spin injection and detection is seen in the data shown in Fig. 10.8, taken at a temperature of 4.2 K using a sample with $L_x = 10.6$ μm. The

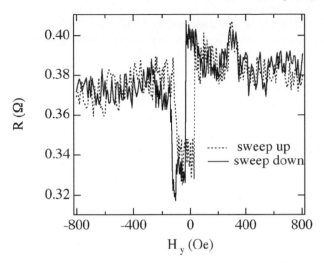

Fig. 10.8. Examples of data showing detection of electrical spin injection in an InAs quantum well. The sample has a separation of $L_x = 10.6\,\mu$m between injector and detector. *Solid lines*: sweep field down. *Dotted lines*: sweep field up. The hysteretic dips are characteristic of spin injection and detection

baseline resistance is nearly zero, demonstrating the effectiveness of the nonlocal geometry. The overlapping dips that appear in the range $-200\,\text{Oe} < H_y < +200\,\text{Oe}$ have the qualitative shape that is characteristic of spin injection. Data were also taken on a sample with an interprobe separation of $L_x = 3.2$ microns. Hysteretic dips that were qualitatively similar to those of Fig. 10.8 were observed, and the amplitude was substantially larger. From the amplitude dependence of these two probe separations, an upper bound of the spin dependent mean free path was estimated to be $\Lambda_s = 4$ microns. The magnitude of the spin injection effects diminished by about 20% at an elevated temperature of 150 K. The magnitude and temperature dependence agree with a recent theory of spin dependent transport in quantum wells of III–V heterostructures [49]. It should be noted that this experiment can not distinguish spin accumulation effects (diffusive transport) from spin-dependent voltage modulation arising from ballistic transport.

The most definitive experiments have also used the nonlocal geometry [3] and, significantly, the measurements included observation of the Hanle effect as well as magnetoresistance with in-plane magnetic fields. The NS material is lightly n-doped GaAs (Si doped at $2 \times 10^{16}\,\text{cm}^{-3}$ to $3.5 \times 10^{16}\,\text{cm}^{-3}$ for three samples). It was epitaxially grown, with a thickness of 1.5 microns, on top of a 300 nm thick buffer layer on a GaAs (100) substrate. A thin (15 nm thick) layer of n^+ GaAs ($5 \times 10^{18}\,\text{cm}^{-3}$), on top of a thin transition layer of equal thickness, was grown on NS. The 5 nm thick epitaxial Fe electrodes were grown on top of the n^+ GaAs. The NS sample was patterned to form a channel about 70 microns wide and 0.35 mm long. The Fe films were patterned to form F electrodes that nearly spanned the NS channel, had width of about 10 microns, and had spacing of $L_x = 12$ microns between $F1$ and $F2$. A narrow Schottky barrier formed at the interface between Fe and n^+ GaAs. As discussed in Sect. 10.2.3, the Schottky barrier provides a dominant interfacial resistance that permits efficient spin injection, avoiding issues related to the mismatch of resistivities of Fe and GaAs.

Using these samples, a variety of transport measurements were performed and all of the results were self-consistent. The current-voltage $(I - V)$ characteristic of each device is nonlinear, but a summary of the results can be given for a typical injection current of $+1$ mA. First, in-plane magnetic fields (along the $\hat{y}$ axis in Fig. 2) were used to manipulate the magnetization orientations $M1$ and $M2$ between parallel and antiparallel, and hysteretic changes of resistance, ΔR, were observed in a small field range around $|H_y| \approx 250$ Oe, representing the field range for which the coercivities H_{C1} and H_{C2} differed. The magnitude of ΔR was about 16 mΩ at 50 K. It's noteworthy that the magnitude of ΔR is roughly the same as $\Delta R \approx 7$ mΩ for the InAs single quantum well data (Fig. 10.8), also noting that both samples had comparable spacing L_x. A spin diffusion length of $L_s = 6$ microns was deduced, similar to the estimate of $L_s = 4$ microns for the InAs samples.

Second, the Hanle effect was observed by saturating the magnetizations of both injector $M1$ and detector $M2$ along the $\hat{y}$ axis, and then sweeping an external magnetic field along the $\hat{z}$ axis. Measurements were made for $M1$ and $M2$ parallel (antiparallel), and a positive (negative) Hanle feature was observed. The Hanle feature was fit to theory [3, 14], and spin relaxation times were deduced, ranging from $\tau_s = 24$ nsec $(T = 10$ K$)$ to $\tau_s = 4$ nsec $(T = 70$ K$)$. The magnitude of the Hanle feature was consistent with the magnitude of ΔR measured with in-plane fields.

Third, optical Kerr measurements were performed to measure the spin polarization of carriers in NS independently. These measurements were particularly helpful in explaining details of the bias dependence.

Data were also taken for a "crossed probe" geometry. Referring to Fig. 10.2, current injected at $F1$ was grounded at $x = +b$ and the voltage measured at $F2$ was referenced to a ground at $x = -b$. These measurements necessarily included a resistive voltage in proportion to L_x, and the carrier diffusion was affected by the associated electric field. The results were qualitatively the same, and the quantitative differences were small. Semiconductors differ from metals in that relatively large internal electric fields may exist and spin diffusion in the nonlocal geometry may not be isotropic.

The results of these experiments are important because the Hanle effect is the most rigorous and unequivocal test for the existence of electrical spin injection, detection and spin accumulation. As such, the results represent a definitive demonstration of spin injection in a semiconductor. They also demonstrate the existence of spin accumulation in NS, confirming that a diffusive transport model is valid (for thin film and/or bulk samples, but not necessarily for 2D samples). It is significant that the authors [3] analyzed their results using Johnson–Silsbee theory [14]. In particular, the NS sample was modeled as a Pauli metal with carrier density $n = 3 \times 10^{16}$ cm^{-3} and effective mass $m^* = 0.07 m_e$, where m_e is the free electron mass. The interfacial injection polarization was estimated to be about 20%, comparable with values observed for injection into metal or InAs samples [12, 48].

While these results represent convincing proof of electrical spin injection and detection, the magnitude of the observed effects is quite small at cryogenic temperatures and vanishes at temperatures above 80 K. The plausibility of a Datta–Das kind of spin injected field effect transistor has therefore been shown, but it remains unlikely that a prototype device can be fabricated having device characteristics that would be competitive with those of dominant complimentary metal oxide semiconductor (CMOS) technology.

10.5 Related Topics

Discussions of numerous topics involving spin injection are readily found in the literature. Without summarizing these varied aspects, a few comments can be made. As a first example, it was noted in Sect. 10.3 that spin injection can be used as a technique to create a pure spin current. This may be relevant for subfields related to quantum measurement or quantum computing. Recalling (10.1), spin accumulation is directly proportional to the spin relaxation time of the carrier in the nonmagnetic material. As a technique, spin injection has been useful for directly measuring spin relaxation times in a variety of materials systems including bulk metals, thin film metals, and semiconductors. Dynamic nuclear polarization is the subject of Chap. 11. It has been predicted that the spin polarized currents [7] and nonequilibrium spin populations [50] that result from spin injection can cause dynamic nuclear polarization.

The possibility of using spin for novel functionality in new generations of devices has been a large motivation in the popularity of spintronics research. Semiconductor based devices, such as the spin injected field effect transistor, have been a focus of promising new applications. After a decade of research, spin dependent transport effects in semiconductors have been demonstrated unequivocally. However, their magnitude is relatively small, observations have been constrained to a cryogenic temperature range, and the prospects of a competitive, integrated electronic device are remote. Magneto-optic effects have been more robust, but no applications for a spin injected light emitting diode are known.

Spintronic devices based on metals have enjoyed considerable success. The magnetic tunnel junction is the dominant device family, and magnetic tunnel junctions are used as sensors in disk drive read heads and as memory cells in nonvolatile, magnetic random access memories. The continued development of both of these applications involves scaling device structures to smaller feature sizes for future generations. A magnetic tunnel junction fabricated with a traditional aluminum oxide barrier faces scaling problems for devices with a minimum feature size f below 50 nm because the device impedance may be too large, and a new device may soon replace the magnetic tunnel junction. The lateral spin valve, based on spin injection and spin accumulation, is a candidate to replace the magnetic tunnel junction for several reasons. First, spin accumulation scales inversely with the volume of N [refer to (10.1)]. As discussed below, this has been verified over ten decades of sample volume. Second, for magnetoresistive device families such as the magnetic tunnel junction, the device impedance R and output resistance modulation ΔR are intrinsically related as the magnetoresistance ratio, $MR = \Delta R/R$. This is not the case for lateral spin valves: the output impedance and output modulation are independent. In principle, R can be reduced to an acceptable level without diminishing the output modulation ΔR. Third, this all-metal device family uses the same ferromagnetic and nonmagnetic materials as spin valves, so that materials and processing issues have been largely solved.

Recent experiments have confirmed the inverse scaling law remarkably well. In order to make a quantitative analysis, (10.3) can be written as

$$\frac{\Delta R \cdot \text{Vol}}{\eta_1 \eta_2 \tau_s} \approx \text{constant}, \tag{10.17}$$

where $\Delta R = 2R_s$ is the parameter typically measured and reported. This relationship is approximately correct because the susceptibility χ may vary for different materials. Equation (10.17) is independent of temperature, although τ_s may have intrinsic temperature dependence. A value of $\Delta R \cdot \text{Vol}/\eta_1 \eta_2 \tau_s$ of 5.9×10^{-4} $\Omega \text{cm}^3/\text{sec}$ was measured for a thin film Al sample having a volume given by transverse dimensions of 120 nm by 400 nm and a film thickness of 6 nm [28]. This compares very well with the value 5.4×10^{-4} $\Omega \text{cm}^3/\text{sec}$ that was measured in the original spin injection experiment [2]. The comparison may be fortuitous, but a variety of recent experiments on Al, Cu and Ag thin films have given values in the range 5.9×10^{-4} $\Omega \text{cm}^3/\text{sec}$ to 1.1×10^{-3} $\Omega \text{cm}^3/\text{sec}$. Thus, the inverse scaling rule has been demonstrated to an accuracy of a factor of 2 for N volumes that vary by a factor of 10^{10}. Should devices continue to follow the inverse scaling rule for another decade of volume, equivalently a factor of $10^{1/3}$ in linear dimensions, the lateral spin valve might well become the next generation device for sensing and magnetic random access memory applications.

References

[1] N.F. Mott, Proc. R. Soc. A **153**, 699 (1936)
[2] M. Johnson, R.H. Silsbee, Phys. Rev. Lett. **55**, 1790 (1985)
[3] X. Lou, C. Adelmann, S.A. Crooker, E.S. Garlid, J. Zhang, S.M. Reddy, S.D. Flexner, C.J. Palmstrm, P.A. Crowell, Nat. Phys. **3**, 197 (2007)
[4] P.M. Tedrow, R. Meservey, Phys. Rev. Lett. **25**, 1270 (1970)
[5] P.M. Tedrow, R. Meservey, Phys. Rev. Lett. **26**, 192 (1971)
[6] M. Julliere, Phys. Lett. **54**, 225 (1975)
[7] A.G. Aronov, JETP Lett. **24**, 32 (1976)
[8] A.G. Aronov, Sov. Phys. Semicond. **10**, 698 (1976)
[9] A.G. Aronov, Sov. Phys. JETP **44**, 193 (1976)
[10] R.H. Silsbee, A. Janossy, P. Monod, Phys. Rev. B **19**, 4382 (1979)
[11] R.H. Silsbee, Bull. Mag. Res. **2**, 284 (1980)
[12] M. Johnson, R.H. Silsbee, Phys. Rev. B **37**, 5326 (1988)
[13] W. Hanle, Z. Phys. **30**, 93 (1924)

[14] M. Johnson, R.H. Silsbee, Phys. Rev. B **37**, 5312 (1988)
[15] M. Johnson, R.H. Silsbee, Phys. Rev. B **35**, 4959 (1987)
[16] P.C. van Son, H. van Kempen, P. Wyder, Phys. Rev. Lett. **58**, 2271 (1987)
[17] M.N. Baibich et al., Phys. Rev. Lett. **61**, 2472 (1988)
[18] G. Binasch, P. Grnberg, F. Saurenbach, W. Zinn, Phys. Rev. B **39**, 4828 (1989)
[19] M. Johnson, J. Appl. Phys. **75**, 6714 (1994)
[20] H.B. Callen, *Thermodynamics* (Wiley, New York, 1960)
[21] B. Nadgorny, R.J. Soulen, M.S. Osofsky, I.I. Mazin, G. LaPrade, R.J.M. van de Veerdonk, A.A. Smits, S.F. Cheng, E.F. Skelton, S.B. Qadri, Phys. Rev. B **61**, R3788 (2000)
[22] M. Johnson, J. Byers, Phys. Rev. B **67**, 125112 (2003)
[23] R. Godfrey, M. Johnson, Phys. Rev. Lett. **96**, 136601 (2006)
[24] M. Johnson, R.H. Silsbee, Phys. Rev. Lett. **60**, 377 (1988)
[25] Q. Yang, P. Holody, S.-F. Lee, L.L. Henry, R. Lolee, P.A. Schroeder, W.P. Pratt Jr., J. Bass, Phys. Rev. Lett. **72**, 3274 (1994)
[26] G. Schmidt, D. Ferrand, L.W. Mollenkamp, A.T. Filip, B.J. van Wees, Phys. Rev. B **62**, R4790 (2000)
[27] E.I. Rashba, Phys. Rev. B **62**, R16267 (2000)
[28] S.O. Valenzuela, M. Tinkham, Appl. Phys. Lett. **85**, 5914 (2004)
[29] M. Johnson, R.H. Silsbee, Phys. Rev. B **76**, 153107 (2007)
[30] R.R. Parsons, Phys. Rev. Lett. **23**, 1152 (1969)
[31] A.I. Ekimov, V.I. Safarov, JETP Lett. **12**, 198 (1970)
[32] D. Stein, K. v. Klitzing, G. Weimann, Phys. Rev. Lett. **551**, 130 (1983)
[33] F. Meier, B.P. Zakharchenya (eds.), *Optical Orientation* (North-Holland, New York, 1984)
[34] S. Datta, B. Das, Appl. Phys. Lett. **56**, 665 (1990)
[35] S.F. Alvarado, P. Renaud, Phys. Rev. Lett. **68**, 1387 (1992)
[36] J.M. Kikkawa, D.D. Awschalom, Phys. Rev. Lett. **80**, 4313 (1998)
[37] M. Oestreich, J. Hübner, D. Hägale, P.J. Klar, W. Heimbrodt, W.W. Rühle, D.E. Ashenford, B. Lunn, Appl. Phys. Lett. **74**, 1251 (1999)
[38] R. Fiederling, M. Keim, G. Reuscher, W. Ossau, G. Schmidt, A. Waag, L.W. Molenkamp, Nature **402**, 787 (1999)
[39] Y. Ohno, D.K. Young, B. Beschoten, F. Matsukura, H. Ohno, D.D. Awschalom, Nature **402**, 790 (1999)
[40] H.J. Zhu, M. Ramsteiner, H. Kostial, M. Wassermeier, H.-P. Schonherr, K.H. Ploog, Phys. Rev. Lett. **87**, 016601 (2001)
[41] T. Manago, H. Akinaga, Appl. Phys. Lett. **81**, 694 (2002)
[42] T. Taniyama, G. Wastlbauer, A. Ionescu, M. Tselepi, J.A.C. Bland, Phys. Rev. B **68**, 134430 (2003)
[43] C. Ciuti, J.P. McGuire, L.J. Sham, Appl. Phys. Lett. **81**, 4781 (2002)
[44] D. Hägale, M. Oestreich, W.W. Rühle, N. Nestle, K. Eberl, Appl. Phys. Lett. **73**, 1580 (1998)
[45] J. Kikkawa, D.D. Awschalom, Nature **397**, 139 (1999)
[46] Yu.A. Bychkov, E.I. Rashba, JETP Lett. **39**, 78 (1984)
[47] W.Y. Lee, S. Gardelis, B.-C. Choi, Y.B. Xu, C.G. Smith, C.H.W. Barnes, D.A. Ritchie, E.H. Linfield, J.A.C. Bland, J. Appl. Phys. **85**, 6682 (1999)
[48] P.R. Hammar, M. Johnson, Phys. Rev. Lett. **88**, 066806 (2002)
[49] K.C. Hall et al., Phys. Rev. B **68**, 115311 (2003)
[50] M. Johnson, Appl. Phys. Lett. **77**, 1680 (2000)

11

Dynamic Nuclear Polarization and Nuclear Fields

V.K. Kalevich, K.V. Kavokin, and I.A. Merkulov

In memory of our teachers V.I. Perel and B.P. Zakharchenya

The optical polarization of nuclear spins may develop in a semiconductor when it is illuminated by circularly polarized light [1, 2]. The ultimate reason of this phenomenon is the hyperfine interaction of electron and nuclear spins, enabling the transfer of angular momentum (spin) from optically oriented electrons to the lattice nuclei. This process is called dynamic polarization. The dynamically polarized nuclei, in turn, produce a mean effective magnetic field (called Overhauser field) that can substantially change the spin polarization of electrons. In that way, a strongly coupled electron–nuclear spin system is formed, where the polarization of nuclei is not simply determined by the spin state of electrons, but exerts a back influence upon the electron polarization.

The technically simplest way to trace the behavior of the electron–nuclear spin system is to detect the circular polarization of photoluminescence, because it is proportional to the mean electron spin [2]. This method has been widely and successfully used from earliest experiments on optical orientation until present. Alternative methods for detection of the electron polarization using Faraday and Kerr rotation of linearly-polarized probe beams have also been developed [3, 4]. Lately, the direct spectroscopic observation of the splitting of electron spin levels in the Overhauser field has become possible [5–8].

The experiments on optical orientation of the electron–nuclear spin system, started in 1968 [1], have resulted in the discovery of many spectacular spin-related effects and to the development of many subtle methods for characterization of semiconductor structures. The basics of the phenomenon were investigated in bulk crystals and covered in several reviews, the most complete of which is the book *Optical Orientation* [2], published in 1984. The progress of growth technology, which resulted in the appearance of high-quality low-dimensional structures (quantum wells, wires and dots), gave a new impetus to this research. In the last years, much experimental work has been done on these new objects, often using newly developed measurement techniques. Recent research is reviewed, for instance, in [9–13].

In this chapter, we analyze, from our point of view, the most important results of the investigation of the electron–nuclear spin system in low-dimensional structures. We will show how the most part of these results can be understood within the existing theoretical framework, and where and why the old theory is insufficient.

The chapter starts with an overview of basic interactions of electron and nuclear spins (Sect. 11.1). Section 11.2 shows, both theoretically and on demonstrative experimental examples, how the electron spin relaxation by nuclei changes with the increase of the electron spin correlation time. This introductory part is followed by Sects. 11.3 and 11.4 devoted to the dynamic polarization of nuclei under various experimental conditions. The main properties of the electron–nuclear spin system, theoretically described in the approximation of short correlation time, are illustrated by new experimental results, which are not covered by the book *Optical Orientation* [2]. Most of them make use of the reduced symmetry and spatial confinement of electron states in quantum wells and quantum dots. These include, for instance, bistability of the nuclear polarization, induced by modification of the electron spectrum by the Overhauser field, anisotropy of the electron g-factor and spin relaxation time, or strong uniaxial exchange interaction with the hole in a quantum-dot exciton. Section 11.5 considers optical detection of multispin nuclear magnetic resonance, and also optically induced NMR.

In the last part of the chapter (Sect. 11.6), we discuss the long-lived spin memory in quantum dots, with a particular stress on the behavior of the electron–nuclear spin system at a long correlation time of the electron spin, which is a developing field in both experiment and theory. New entities, such as the nuclear spin polaron, a correlated electron–nuclear spin complex, appear on the scene. In the case of high nuclear polarization, the memory time for the direction of the total spin of the polaron in zero magnetic field can by orders of magnitude exceed the dipole–dipole relaxation time of the nuclei not coupled with an electron.

11.1 Electron–Nuclear Spin System of the Semiconductor: Characteristic Values of Effective Fields and Spin Precession Frequencies

Here we consider the basics of the theory of electron–nuclear spin system, using a quantum dot as an example. The estimates will be done for the dot containing $N = 10^5$ nuclei.

11.1.1 Zeeman Splitting of Spin Levels

The Zeeman energy of an electron spin s in the magnetic field B [14] is

$$H_{Ze} = \mu_B g_e (Bs), \tag{11.1}$$

where $\mu_B = 9.27 \times 10^{-24}$ J/T is the Bohr magneton, and g_e is the electron g-factor. Not only the value of g_e, but also its sign can be different, depending on the composition of the semiconductor and parameters of the heterostructure.

The Zeeman energy of a system of nuclear spins $\mathbf{I}_n$ [14] is equal to

$$H_{ZN} = -\mu_N \sum_n g_n (\mathbf{B} \mathbf{I}_n), \tag{11.2}$$

where the sum is over all the nuclei in the system, $\mu_N = 5.05 \times 10^{-27}$ J/T is the nuclear magneton, and g_n is the g-factor of the nth nucleus (in the nuclear radiospectroscopy the Zeeman energy is often expressed through the nuclear gyromagnetic ratio of the nucleus, $\gamma_n = \mu_N g_n / \hbar$). As the Bohr magneton is approximately 2 000 times larger than the nuclear magneton, the splittings of electron and nuclear spin levels differ by three orders of magnitude. In the magnetic field of 1 T, the Larmor precession periods of electron and nuclear spins equal, respectively, $T_{Le} = h / \mu_B g_e B \approx 0.7 \times 10^{-10}$ s and $T_{LN} \approx 1.3 \times 10^{-7}$ s for $g_e = g_n = 1$.

11.1.2 Quadrupole Interaction

The quadrupole interaction of the nuclear spin with gradients of the local electric field in the crystal is described by the following Hamiltonian [15]:

$$H_Q = V_{\alpha\beta} I_\alpha I_\beta. \tag{11.3}$$

Here the coefficients $V_{\alpha\beta}$ are proportional to the electric field gradient at the nucleus and to the quadrupole moment of the nucleus, which is nonzero for nuclei with $I \geq 1$. Nuclear spin levels remain partially degenerate because of the time-inversion symmetry of the Hamiltonian, forming a few Kramers doublets. The electric-field gradients may be caused by redistribution of valence electrons of the atom if one or more of its neighbors are replaced by different species (in alloys like GaAlAs [16]), by deformation of the crystal, by charged impurities, etc. The range of characteristic energies is very wide. For example, in [17] the quadrupole splitting of As nuclei in GaAlAs, caused by the replacement of Ga by Al, corresponds to the Larmor frequency of 17 MHz. In [18] the quadrupole splitting of In nuclei due to the strain in the InP/InGaP heterostructure is estimated to be of 1 MHz.

11.1.3 Hyperfine Interaction

In most semiconductors, the interaction between electron and nuclear spins is dominated by the Fermi contact interaction[1] [15]:

$$H_{hf} = \sum_n a_n (\mathbf{s} \mathbf{I}_n), \tag{11.4}$$

---

[1] Holes, having p-type Bloch wave functions, are not coupled with nuclei by the Fermi interaction. For this reason, their hyperfine interaction is 4–5 orders weaker than that of electrons [19].

where

$$a_n = v_0 A_n |\Psi(\boldsymbol{r}_n)|^2, \tag{11.5}$$

v_0 is the unit cell volume, $\Psi(\boldsymbol{r}_n)$ is the envelope wave function of the electron at the position of the nth nucleus, $\boldsymbol{I}_n$ is the spin operator of that nucleus, and $A_n \propto g_n$ is the hyperfine constant. In GaAs-type semiconductors, A_n is of the order of $100\,\mu\text{eV}$.

Overhauser Field

If there exists a mean polarization of the lattice nuclei, the electron is affected by the hyperfine (Overhauser) field. This field does not depend on the electron localization volume:

$$\boldsymbol{B}_{\mathrm{N}} = \frac{v_0 \sum_{\eta} A_{\eta} \langle \boldsymbol{I}_{\eta} \rangle}{\mu_{\mathrm{B}} g_e} = \sum_{\eta} B_{\eta\,\mathrm{max}} \langle \boldsymbol{I}_{\eta} \rangle / I_{\eta}. \tag{11.6}$$

Here, the sum is over all the nuclear species, not over individual nuclei. $\langle \boldsymbol{I}_{\eta} \rangle$ is the mean spin of the nuclei of the type η. The total field of completely polarized nuclei $B_{\mathrm{N\,max}} = \sum_{\eta} B_{\eta\,\mathrm{max}}$ is of the order of a few Tesla. For example, $B_{\mathrm{N\,max}} \approx 5.3\,\text{T}$ in GaAs [20, 54].

The Field of Nuclear Spin Fluctuation

Apart of the mean hyperfine field, the electron is affected by a fluctuating field, B_{NF}, proportional to the square root of the number of nuclei in the electron's localization region. If nuclear spins are not strongly polarized,

$$\langle B_{\mathrm{NF}}^2 \rangle = \frac{v_0^2 \sum_n A_n^2 |\Psi(\boldsymbol{r}_n)|^4 I_n (I_n + 1)}{\mu_{\mathrm{B}}^2 g_e^2} \approx \frac{B_{\mathrm{N\,max}}^2}{N}. \tag{11.7}$$

Usually N is of the order of 10^5, i.e., the fluctuation field is about 0.3% of the maximum nuclear field, that is $B_{\mathrm{NF}} \sim 100\,\text{G}$. The Larmor period of the electron spin in this field is about 10^{-8} s.

Knight Field

Polarized electrons, in turn, produce a hyperfine field on nuclei (Knight field). In the case of free electrons this field is proportional to their concentration and usually is weak.

For a localized electron, the Knight field is inversely proportional to its localization volume:

$$\boldsymbol{B}_{en} = -\frac{v_0 A_n}{\mu_{\mathrm{N}} g_n} |\Psi(\boldsymbol{r}_n)|^2 \boldsymbol{s}. \tag{11.8}$$

This field is determined by the electron spin direction. It changes when the electron is captured or released by the localization center or quantum dot, because of the exchange interaction between the localized and delocalized electrons, precession of the electron spin in the Overhauser field, etc.

The time averaged Knight field is

$$\boldsymbol{B}_{en} = -F \frac{v_0 A_n}{\mu_N g_n} |\Psi(\boldsymbol{r}_n)|^2 \boldsymbol{S}, \tag{11.9}$$

where $\boldsymbol{S}$ is the mean electron spin and $F \leq 1$ is the filling factor characterizing the occupation of the localization center by electrons.

The Knight field is spatially inhomogeneous. Apart of a relatively weak dependence on the envelope-function coordinate $\boldsymbol{r}_n$, it may considerably change from one nuclear species to another within one crystal cell.

For a quantum dot containing N nuclei, the splitting of nuclear spin levels in the Knight field is N times smaller than that of the electron in the Overhauser field of fully polarized nuclei. Correspondingly, the maximum Knight field is $B_{e\,max} = (B_{N\,max}/N)(\mu_B g_e/\mu_N g_n)$. For $N = 10^5$, $B_{e\,max} \approx 10^{-1}$ T. The Knight field is the largest at the center of the dot, where the Larmor period of the nuclear spin is of the order of microseconds. The field decreases towards the dot periphery, where the interaction of nuclear spins with the electron may become weaker than their interaction with each other.

11.1.4 Nuclear Dipole–Dipole Interaction

The main interaction between nuclear spins is the dipole–dipole interaction [15]:

$$H_{dd} = \frac{\mu_N^2}{2} \sum_{n \neq n'} \frac{g_n g_{n'}}{r_{nn'}^3} \left((\boldsymbol{I}_n \boldsymbol{I}_{n'}) - 3 \frac{(\boldsymbol{I}_n \boldsymbol{r}_{nn'})(\boldsymbol{I}_{n'} \boldsymbol{r}_{nn'})}{r_{nn'}^2} \right), \tag{11.10}$$

where $\boldsymbol{r}_{nn'}$ is the translation vector between the nuclei n and n'. The strength of the fluctuating local magnetic field created at the nucleus by the other nuclei, B_L, is a few Gauss (for GaAs, $B_L \approx 1.5$ G [20]). Correspondingly, the Larmor period of the nuclear spin in the local field is in the sub-millisecond range.

The magneto–dipole interaction does not conserve the total angular momentum of interacting nuclei, which is transferred to the crystal as a whole. The corresponding relaxation time of nuclear polarization, T_2, is of the order of the Larmor period in the local field, 10^{-4} s. In weak magnetic fields, $B \ll B_L$, all the spin components relax with this characteristic time. In strong fields, $B \gg B_L$, the time T_2 is the relaxation time of transversal (perpendicular to $\boldsymbol{B}$) nuclear spin components. In this latter case, the longitudinal (parallel to $\boldsymbol{B}$) spin component decays with the time $T_1 \gg T_2$, because the relaxation process requires dissipation of a large Zeeman energy.

If the nuclear polarization is spatially inhomogeneous, the magneto-dipole interaction results in diffusional spin flows with the diffusion coefficient [21]

$$D \approx d^2 T_2^{-1}, \tag{11.11}$$

where d is the distance between adjacent nuclei. The diffusion slows down if large gradients of magnetic field (e.g., Knight field) are present, so that the difference of its strength at the nearest nuclei $\delta B \gg B_L$.

11.2 Electron Spin Relaxation by Nuclei: from Short to Long Correlation Time

As mentioned in Chap. 1, random hyperfine fields produced by disordered nuclear spins may cause spin relaxation of electrons. For free electrons, both in three and two dimensions, this mechanism of spin relaxation is ineffective because of small root-mean-square value and very short correlation time of the Overhauser field [22–24]. For localized electrons, the situation is quite different. Here nuclear fluctuation fields may be of the order of hundreds of Gauss (see Sect. 11.1.3), corresponding to electron Larmor periods of about 1 ns. The correlation time τ_c varies depending on the nature of the localization center (see the general discussion of spin relaxation at short and long correlation time in Chap. 1).

If the localization center is a donor impurity, τ_c at helium temperatures is limited either by tunneling to empty donor-bound states [22] or by exchange interaction with electrons localized at other donors nearby [25]. Naturally, in this case τ_c strongly depends on the donor concentration n_d. In GaAs with $n_d \approx 10^{14}\,\mathrm{cm}^{-3}$, τ_c is about 20 ns [26], and the regime of long correlation time is realized: 2/3 of the electron spin polarization decays within the time $1/\Omega_{NF} \approx 5$ ns. With increasing n_d, τ_c decreases and eventually gets shorter than $1/\Omega_{NF}$ (short-correlation-time regime). In this regime, the electron spin relaxation time τ_{eN} is given by the expression:

$$\tau_{eN} = \left(\langle \Omega_{NF}^2 \rangle \tau_c/3\right)^{-1} = \left(\tau_c I(I+1)\frac{\sum_n a_n^2}{3\hbar^2}\right)^{-1}. \qquad (11.12)$$

It becomes longer and longer, reaching nearly 200 ns^2 at $n_d \approx 4 \times 10^{15}\,\mathrm{cm}^{-3}$. The crossover from long- to short-correlation-time regime was also realized by injecting free electrons to a GaAs layer containing donors [26]. In this case, τ_c of donor-bound electrons was limited by exchange scattering of delocalized electrons.

In quantum dots, the extrinsic mechanisms that limit τ_c by affecting the *electron* spin, like thermal ionization or phonon scattering, become ineffective at low temperature. As shown in [27], spin relaxation in this case is determined by intrinsic (i.e., originated from spin–spin interactions within the quantum dot) mechanisms, which form a hierarchy of relaxation times (see Fig. 11.1).

The initial stage of the electron spin relaxation in the long-correlation-time regime is characterized by the time $T_4 \approx 1/\Omega_{NF}$. During this time, the electron polarization in an ensemble of quantum dots is diminished three times as a result of precession of individual electron spins in Overhauser fields of nuclear spin fluctuations. This was most spectacularly demonstrated in a time-resolved experiment on an ensemble of p-InGaAs/GaAs quantum dots [28]. Using singly positively charged dots allowed the authors to get rid of the electron–hole exchange interaction that would otherwise mask weaker effects of hyperfine coupling (see inset in Fig. 11.2). The circular polarization of photoluminescence (proportional to the electron mean spin) decreased

---

[2] Further increase of the spin relaxation time is blocked by small anisotropy of the electron–electron exchange interaction, giving rise to another spin relaxation mechanism.

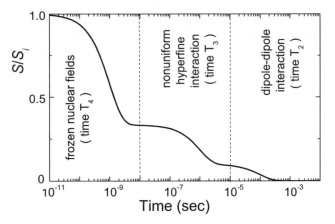

Fig. 11.1. Main stages of electron spin relaxation in an ensemble of quantum dots. The calculation is done for $N = 10^5$ nuclei per dot

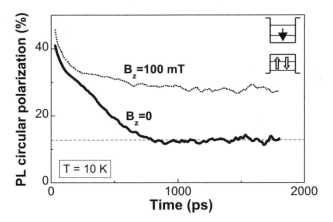

Fig. 11.2. Circular polarization dynamics of photoluminescence (PL) in singly positively charged InAs/GaAs quantum dots after a pump by 1.5 ps laser pulse for a longitudinal magnetic field $B_z = 0$ and $B_z = 100\,\text{mT}$ [28]. Insert: Energy diagram of the positively charged exciton X^+ formed by a photocreated electron–hole pair and a resident hole. Since unpolarized holes form a zero spin singlet, a spin-polarized electron is not coupled with the holes by the exchange interaction, which usually blocks the hyperfine interaction

down to about 30% of the initial value with the time constant very close to T_4 calculated for this type of quantum dots (Fig. 11.2). Application of the magnetic field, exceeding the nuclear fluctuation field, B_{NF}, and directed along the exciting light beam, suppressed spin relaxation, because the total field affecting the electron spin became nearly parallel to the initial spin direction.

The second stage of spin relaxation takes a much longer time T_3, which is of the order of the Larmor precession period of nuclear spins in the Knight field (i.e., $T_3 \approx T_4\sqrt{N}$). This stage originates from the fact that the electron coupling with different nuclei has different strength, because the electron density decreases from the central part of the quantum dot to the periphery. Those components of the Knight field, which are perpendicular to B_{NF}, oscillate at the frequency $1/T_4$ and for this reason produce almost no effect on the nuclear spins. But the nuclear spins are fully affected by the Knight-field component, *parallel* to B_{NF}. Since the nuclear spins are oriented randomly, each of them is directed at some angle to B_{NF}, but the contributions of nuclear spin components, perpendicular to B_{NF}, to the total Overhauser field exactly compensate each other. The rotation of nuclear spins around the Knight field with different frequencies destroys this compensation, resulting in slow rotation of the B_{NF} vector. Naturally, the direction of the Knight field follows the direction of B_{NF}. With thousands of nuclei each having its own Larmor frequency, the spin dynamics of the electron–nuclear spin system becomes very complex [29–33]. The electron polarization at times longer than T_3 can be assessed with a statistical approach [27], assuming that the only conserved quantities in the system are the total spin (due to rotational symmetry) and the amplitude of the Overhauser field B_{NF} (due to energy conservation). In that way, it was shown that the electron polarization is determined by the ratio $\langle a^2 \rangle / \langle a \rangle^2$, which characterizes the spread of the Knight field amplitudes over the quantum dot. For typical quantum dots, the second stage of spin relaxation diminishes the polarization 3–4 times, so that after the time T_3 it is reduced to approximately 0.1 of the initial value [27].

Finally, the third stage of spin relaxation is governed by the dipole–dipole interaction of nuclear spins, and characterized by the time T_2. This interaction does not conserve the total spin of the system, hence the electron polarization completely vanishes after the time T_2 (Fig. 11.1).

Experimental observation of the last two stages of spin relaxation requires using n-doped quantum dots, to avoid limitation by the short radiative lifetime of photoexcited electrons, which is of the order of one nanosecond in III–V semiconductors. So far, no decisive experiments have been reported. According to [34], the lifetime of the remaining (after the first stage) polarization in n-InP dots is longer than 12 ns but shorter than 1 μs. A complicated spin–echo experiment with gate-induced coupled quantum dots in a GaAs/AlGaAs heterostructure [35] gave the decoherence time of the electron spin precession in the Overhauser field of nuclear fluctuations, equal to 1.2 μs; it is however unclear whether this decoherence was due to intrinsic or extrinsic mechanisms.

11.3 Dynamic Polarization of Nuclear Spins

Because of the small value of the nuclear magnetic moment, a considerable equilibrium polarization of nuclear spins in constant magnetic fields, accessible in the laboratory (of the order of 10 Tesla), can be reached only at milliKelvin temperatures of the sample. However, a high polarization of nuclei can be realized at much higher temperatures (normally a few Kelvin, but in some cases up to the liquid nitrogen temperature) and much weaker magnetic fields (just a few Gauss), by using dynamic polarization of nuclear spins. The dynamic polarization is essentially a transfer of angular momentum from electrons to nuclei via the hyperfine interaction. For example, electron spin relaxation by nuclei, considered in the previous section, is accompanied by such a spin transfer, because the hyperfine interaction conserves the total spin of the interacting particles. Then the component of the nuclear spin, collinear to the magnetic field, will accumulate, since the longitudinal relaxation time of the nuclear spin, T_1, is very long (see Sect. 11.1.4). Now we see that there are certain necessary conditions for the dynamic polarization: (i) the nuclear spins should be placed in a magnetic field (either an external one, or the mean Knight field of spin-polarized electrons); (ii) if the dynamic polarization is induced by spin-polarized electrons,[3] the electron mean spin should not be perpendicular to the magnetic field. Obviously, the nuclear polarization and the Overhauser field will change their direction to the opposite, if the sign of the electron mean spin is inverted (for example, if the helicity of the excitation light is changed).

11.3.1 Electron Spin Splitting in the Overhauser Field

The optical dynamic nuclear polarization in semiconductors was first detected using the conventional nuclear magnetic resonance (NMR) in bulk silicon [1]. However, the sensitivity of conventional NMR, where the change of the absorption of radio-frequency power is measured, is usually insufficient to detect nuclear polarization in nanostructures.[4] Much more sensitive optical methods were developed, which use the effect of the polarized nuclei on electron spins. The most direct of them is spectroscopic observation of the splitting of electron spin levels by the Overhauser field. Because of its technical complexity, this method was realized only recently [5] (see Fig. 11.3); now it is widely used for studying nuclear effects in quantum dots.[5]

One of the advantages of this method is that it gives a possibility to detect the dynamic nuclear polarization in the Faraday geometry, where the external magnetic field is parallel to the mean spin of photogenerated electrons. Under these conditions, the nuclear polarization has its maximum value. For example, it reaches 65% in GaAs quantum dots [39].

This method also gives a possibility to measure the Knight field. As mentioned above, no dynamic polarization can develop if nuclear spins are not subjected to a magnetic field. If electrons are spin-polarized, the total magnetic field at the nucleus is a sum of the external magnetic field B and the mean Knight field B_e. It goes to zero if these two contributions compensate each other, i.e., $B = -B_e$. Under these conditions, the electron Zeeman splitting is minimal. The dips on the magnetic-field de-

---

[3] This is a common situation in semiconductors. However, there are also certain ways of dynamic polarization with unpolarized electrons, which will be briefly considered below.

[4] Multiple scanning of NMR spectra and other technical improvements used in [36, 37] have allowed to detect the conventional NMR in a sandwich of a few tens of quantum wells. This approach is practically impossible to apply to a single quantum dot because of small number of nuclei within the dot.

[5] The Overhauser field can also induce rotation of the polarization plane of linearly polarized light (Faraday effect), even in the absence of electrons, which was analyzed theoretically by Artemova and Merkulov and proposed as a method for detecting nuclear polarization [38].

318   V.K. Kalevich et al.

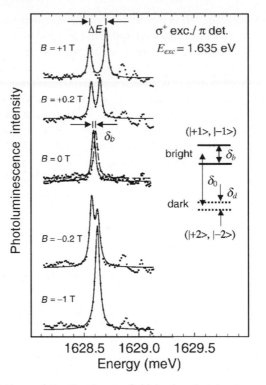

Fig. 11.3. Manifestations of the Overhauser field in the photoluminescence spectrum of a single quantum dot formed at the interface fluctuation in a GaAs/AlGaAs quantum well [39]. The splitting of the bright-exciton doublet in the sum of the external and nuclear magnetic fields under circularly polarized excitation changes its value when the external field is inverted. The spectrum at zero external field is obtained under linearly polarized excitation, so that the nuclear field is also zero. The small residual splitting is due to the exchange interaction of the electron and the hole

pendences of the electron spin splitting in single self-organized InGaAs/GaAs quantum dots, clearly seen in panel (a) in Fig. 11.4, indicate lower polarization of nuclei at $B = -B_e$ [6]. Note that the Zeeman splitting does not disappear entirely: because of inhomogeneity of the electron density, it is impossible to exactly compensate the Knight field for all the nuclei within the quantum dot. The diminishing of the nuclear polarization at $B = -B_e$ is reflected also in the polarization of photoluminescence (panel (b) in Fig. 11.4), because at lower Overhauser fields the electron spin relaxation becomes faster. This latter effect was earlier observed for donor-bound electrons in bulk semiconductors [20]. Other methods of measuring the Overhauser and Knight fields will be discussed in Sects. 11.4.2 and 11.5.3.

Even in this simplest experimental geometry, the behavior of the dynamic polarization is rather complex. The point is that the Overhauser field modifies the energy spectrum of electrons, and, as a result, the rate of the spin transfer from electrons to nuclei changes. The system thus develops a feedback, which may be either positive,

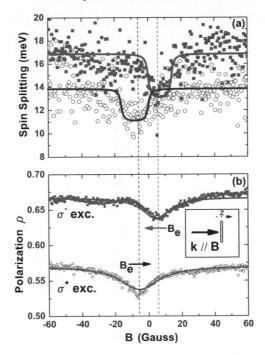

Fig. 11.4. Spin splitting (**a**) and polarization of photoluminescence (**b**) as a function of applied external magnetic field, measured in single self-organized InGaAs/GaAs quantum dots [6]. The minima on the curves correspond to compensation of the mean Knight field by the external magnetic fields. At these points, the nuclear polarization is close to zero

or negative, depending on the specific positions of electron energy levels, on the sign of the external magnetic field and on the helicity of the excitation light. If the feedback is positive, several stable states of the electron–nuclear spin system may exist. In some cases, the self-polarization, i.e., spontaneous development of a considerable nuclear polarization in the absence of optical spin orientation, may become possible. Nonlinear phenomena of this kind, intensively investigated in the last few years in low-dimensional structures, are considered in the next section.

11.3.2 Stationary States of the Electron–Nuclear Spin System in Faraday Geometry

The stationary states of the electron–nuclear spin system are determined by balance of several relaxation processes: (1) spin relaxation of electrons by nuclei, (2) spin relaxation of nuclei by electrons, (3) losses (due to, e.g., spin diffusion), (4) spin-dependent recombination, excitation, or capture of electrons to localized states, accompanied by flip–flop spin transitions, which change the spin projections of electrons and nuclei on the magnetic field by ± 1 and ∓ 1, respectively. The latter contribution is more typical for nanostructures than for bulk semiconductors and was not considered in first theoretical works [20, 22]. In the approximation of short correlation time, the rate equation for the nuclear mean spin, which would account for all the contributions but the last one, can be written in the following general form:

$$\langle \dot{I} \rangle = -W_0(B^*)\big[\langle I \rangle/Q - \big(S - S_T(B^*)\big)\big] - \langle I \rangle/T_N(B),   \tag{11.13}$$

where S is the electron mean spin, $S_T(B^*)$ is its thermally equilibrium value in the effective field $B^* = B + B_N$, the time constant $T_N(B)$ characterizes the losses of the nuclear spin due to dipole–dipole relaxation or diffusion,[6] $Q = I(I+1)/s(s+1)$, and $W_0 = 1/\tau_{eN}$ is the relaxation rate of electrons by nuclei [22]. To take into account the spin-dependent recombination and other contributions unrelated to the electron spin polarization, the right-hand side of this equation should be complemented by a phenomenological parameter $W_1(B^*)$. Spins, as well as magnetic fields, change sign under time inversion. For this reason, $W_1(B^*)$ should be an odd function of the magnetic field. Correspondingly, $W_1(0) = 0$, while its value at a finite magnetic field depends on the specifics of the energy spectrum and spin-dependent electron transitions involved. The stationary states of the electron–nuclear spin system are solutions of the algebraic equation obtained by putting the time derivative in (11.13) equal to zero:

$$W_1(B^*) - W_0(B^*)\big[\langle I \rangle/Q - \big(S - S_T(B^*)\big)\big] - \langle I \rangle/T_N(B) = 0.   \tag{11.14}$$

11.3.3 Dynamic Polarization by Localized Electrons

Probably the first theoretical investigation of the stationary states of the electron–nuclear spin system was performed in early 1970s by Dyakonov and Perel [22] for donor-bound electrons in a bulk semiconductor. Analogous situations are realized for resident electrons in singly negatively charged quantum dots [40], and for photoexcited electrons in singly positively charged quantum dots [8], where spins of the photoexcited and resident holes form a singlet and do not affect the electron spin. In this case, $W_1 = 0$. The thermally equilibrium electron polarization can usually be neglected also ($S_T = 0$). Still, (11.14) remains nonlinear because of the dependence of the spin relaxation rate W_0 on the precession frequency of the electron spin in a magnetic field [22] (see also (1.9) in Chap. 1):

$$W_0(B^*) = \frac{W(0)}{1 + (\mu_B g_e/\hbar)^2 (B^*)^2 \tau_c^2}.   \tag{11.15}$$

According to (11.14),

$$\langle I \rangle = \frac{QS}{1 + Q(W_0 T_N)^{-1}}.   \tag{11.16}$$

---

[6] In the domain of weak external fields, the dependence of T_N on the Knight field may become important. In this case B in the expression for T_N should be replaced with $B + B_e$.

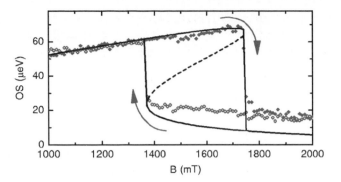

Fig. 11.5. A hysteresis of the Overhauser shift, OS (the contribution of the Overhauser field into the Zeeman splitting) as a function of the magnetic field, in negatively charged InAs quantum dots [40]

As seen from (11.15) and (11.16), electrons transfer their mean spin to nuclei most effectively, when B^* is close to zero. At low losses, $W_0 T_N \gg 1$, the nuclear mean spin is proportional to the electron mean spin[7]:

$$\langle I \rangle = QS = \frac{I(I+1)}{s(s+1)} S. \tag{11.17}$$

However, if $W_0 T_N \leq 1$, the dependence of $W_0 T_N$ on $\langle I \rangle$ through the nuclear field $B_N \propto \langle I \rangle$ becomes important, and (11.16) is nonlinear. With like signs of B_N and B, there is always only one solution. If the signs of B_N and B are different, the nuclear field can compensate the external field, making possible the existence of three solutions. Two of them, those with the largest and the smallest values of B^*, are stable, while the solution with the intermediate value of B^* is unstable. The stability of the two stable states can be qualitatively understood from the following considerations. The state with a high nuclear polarization corresponds to the effective field B^* close to zero, i.e., to the most favorable conditions for spin transfer from electrons to nuclei, which supports nuclei in the polarized state. The solution with low nuclear polarization corresponds to a high B^* (because the external field is not compensated by the Overhauser field), which suppresses the spin transfer to nuclei. At large external fields, exceeding the highest possible Overhauser field, only one state with low nuclear polarization exists.

Bistability of the electron–nuclear spin system may also result from the effect of Overhauser field on other (i.e., not caused by hyperfine interaction) channels of electron spin relaxation. Suppression of spin relaxation due to splitting of electron spin levels by the field B_N makes the electron mean spin S depend on $\langle I \rangle$. Consequently, (11.17) becomes a non-linear equation, having two stable solutions at certain values of parameters [22]. In this case, the nuclear polarization at large external fields is

---

[7] If nuclei experience quadrupole interaction, (11.17) is transformed into general expression $I = \hat{A}S$. The properties of the tensor $\hat{A}$ for crystals of the cubic symmetry are studied in [41], where its components for GaAlAs alloys are also calculated.

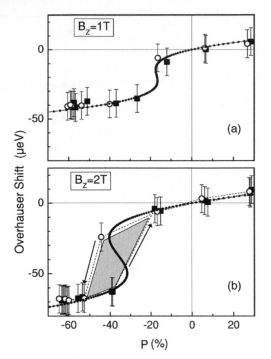

Fig. 11.6. A hysteresis of the Overhauser shift (the contribution of the Overhauser field into the Zeeman splitting) as a function of circular polarization of the excitation light, in a positively charged InAs quantum dot for two values of the external field B_z [8]

high, because the mean electron spin is the largest. Similar effects may also appear as a consequence of suppression of exciton spin relaxation by the Overhauser field [42].

By changing the external field, it is possible to pass the bistability region and to observe a hysteresis of the nuclear polarization (Fig. 11.5). The state of the electron–nuclear spin system can also be manipulated by changing the mean spin of optically oriented electrons (Fig. 11.6). Note that in the experiments shown in Figs. 11.5 and 11.6 the nuclear polarization in large external fields was low. This indicates the suppression of electron–nuclear spin transfer as the main physical reason for appearance of the observed hysteresises.

11.3.4 Cooling of the Nuclear Spin System

In weak magnetic fields $B \leq B_L \sim 1$ G, the polarization, brought by electrons into the nuclear spin system, relaxes as a result of nuclear dipole–dipole interaction with the characteristic time T_2. This does not, however, prevent the development of a high nuclear polarization in weak external fields or in the mean Knight field of spin-polarized electrons [6, 20, 43–45].

The answer to this apparent paradox is that it is energy conservation, not spin conservation, that matters. The spin transfer into the nuclear spin system in what-

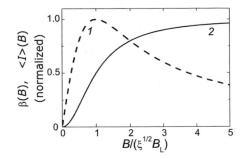

Fig. 11.7. Optical cooling of the nuclear spin system by oriented electrons. Magnetic field dependences of reciprocal spin-temperature (*1*) and mean spin of nuclei (*2*)

ever weak a magnetic field results in changing its Zeeman energy. The dipole–dipole interaction does not change the total energy of the nuclear spin system, which is well isolated from the lattice, but transforms the Zeeman energy into the energy of spin–spin interactions. As a result, the nuclear spin temperature Θ_N decreases. With low Θ_N, the magnetic field induces an equilibrium (within the nuclear spin system) nuclear polarization, which is not affected by the dipole–dipole relaxation and decays with the spin-lattice relaxation time T_1.

Theoretical calculations [46] (see also [20]), performed in the approximation of short correlation time, yield the following expression for the reciprocal nuclear spin temperature $\beta = (k_B \Theta_N)^{-1}$, where k_B is the Boltzmann constant:

$$\beta = \frac{3I}{\mu_n} \frac{f}{s(s+1)} \frac{(\boldsymbol{BS})}{B^2 + \xi B_L^2}. \tag{11.18}$$

Here ξ is a numerical factor of the order of one,[8] $f \leq 1$ is a phenomenological factor, accounting for losses of heat in the nuclear spin system.

The corresponding equilibrium value of the mean nuclear spin is[9]

$$\langle \boldsymbol{I} \rangle = \frac{I(I+1)f}{s(s+1)} \frac{(\boldsymbol{BS})\boldsymbol{B}}{B^2 + \xi B_L^2}. \tag{11.19}$$

The reciprocal temperature of cooled nuclei is the highest for $B = \sqrt{\xi} B_L$ and goes to zero at $B \to 0$ and $B \to \infty$ (see Fig. 11.7). Therefore, the spin temperature Θ_N is minimal for $B = \sqrt{\xi} B_L$ and can be as low as $\approx 10^{-7}$ K at $S = 0.25$ [46]. The minimal Θ_N reached experimentally is $\approx 10^{-6}$ K [45].

The mean spin of nuclei increases with increasing magnetic field, and at $B \gg \sqrt{\xi} B_L$ and zero losses becomes equal to $\langle \boldsymbol{I} \rangle = [I(I+1)/s(s+1)]S$, in accordance with (11.17).

---

[8] For magnetodipole interaction, $2 \leq \xi \leq 3$ depending on the correlation of Knight field on neighboring nuclei.

[9] To take into account the Knight field, $\boldsymbol{B}$ should be replaced by $(\boldsymbol{B} + \boldsymbol{B}_e)$ in (11.18) and (11.19).

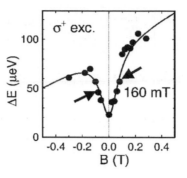

Fig. 11.8. The dependence of the splitting of exciton spin levels on the external magnetic field in a single GaAs quantum dot [39]. The contribution from the Overhauser field increases in the field range up to 0.16 T and then saturates

11.3.5 Polarization of Nuclei by Excitons in Neutral Quantum Dots

When an electron–hole pair is captured or excited by light into a neutral dot, the spin of the hole is not compensated by another hole, as in a positively charged dot, and participates in spin–spin interactions. Since holes do not interact with nuclear spins [19], the most important interaction is their spin exchange with the electron. In the most common case of the exciton formed by a heavy hole, this interaction is equivalent to application of an effective magnetic field to the electron spin. This exchange field is parallel to the structure axis, and may be as strong as a few Tesla [39]. Because of energy conservation, this field almost completely forbids the flip–flop transitions between electron and nuclear spins. They remain possible only as a part of a combined process of exciton capture or recombination, which provides the needed energy. Formally, this corresponds to an anomalously large denominator in (11.15), where the role of B^* is played by the exchange field of the hole. In the same way, the external magnetic field and the Knight field of the electron block the dipole–dipole relaxation of the nuclear polarization, if $|B + B_e| > B_L$. The nuclear spin relaxation is possible only when the Knight field abruptly changes at the moments of recombination and capture of the exciton. In this regime, the main term in the formula for the dipole–dipole relaxation of the nuclear spin is given by the expression [39]:

$$T_N(B) = T_N(0)\left(1 + \frac{B^2}{\xi B_L^2}\right), \tag{11.20}$$

where ξ is a numeric coefficient of order 1 (see Sect. 11.3.4). In weak magnetic fields, the rate of nuclear spin relaxation is many times larger than the rate of dynamic polarization ($W_0 T_N \ll 1$), and nuclear spins are practically unpolarized. An appreciable polarization of nuclei appears only in sufficiently strong external fields $B \geq 0.1\,\text{T} \gg B_L$, which suppress the dipole–dipole relaxation (Fig. 11.8).

In even stronger magnetic fields (when the Zeeman splitting becomes comparable to the exchange-induced energy separation of dark and bright exciton doublets) the transition rates between exciton levels start to change, modifying the coefficients W_0

and W_1 in (11.14). As shown in [47], this should lead to a bistability, similar to the one considered in Sect. 11.3.3. This bistability was indeed observed in [7].

11.3.6 Current-Induced Dynamic Polarization in Tunnel-Coupled Quantum Dots

Equation (11.14) does not forbid the appearance of the nonequilibrium nuclear polarization even if the electron mean spin is zero. In this case, the stationary value of the nuclear spin is

$$\langle I \rangle = \frac{QW_1 + W_0 S_T(B^*)}{W_0 + Q/T_N}. \tag{11.21}$$

The dynamic polarization of nuclei due to non-zero equilibrium value of the electron spin S_T in a magnetic field, while the electrons are intentionally kept unpolarized ($S = 0$), is the classical Overhauser effect [15]. At high temperatures, when S_T is negligibly small, the dynamic polarization is still possible if $W_1 \neq 0$. A typical example is the dynamic polarization of nuclei in the process of singlet–triplet relaxation, observed in many condensed matter systems [48, 49] under excitation by unpolarized light or in chemical reactions.

An analogous phenomenon has been demonstrated recently in semiconductors [50]. The nuclear polarization occurred when a current was passed through a pair of coupled quantum dots. The structure was arranged in such a way that the electron was entering one of the dots, tunneling to the other one, and then escaping to the drain lead. The adjustment of gate potentials ensured constant presence of a resident electron in the second dot. For this reason, tunneling was possible only if the spins of the traveling and resident electrons formed a singlet state. As the traveling electron was initially unpolarized, the probability of the singlet configuration was 1/4. Correspondingly, in 3/4 of cases the electrons appeared in a triplet state, and the tunneling was blocked until the spin state of the electrons changed due to interaction with nuclei. This situation is analogous to the spin-dependent recombination in optical experiments, and it also leads to a non-zero W_1. Physically, this is due to the fact that the application of a magnetic field results in the difference between flip–flop transition rates from triplet states having opposite electron spin projections $+1$ and -1. As the hyperfine interaction conserves the total spin of the electrons and nuclei, nuclear spins get polarized. The resulted Overhauser field was detected by its influence on the conductivity of the coupled quantum dots due to shifts of the electron energy levels, changing the tunneling probability. In this way, nuclear polarization of 40% was measured [50].

11.3.7 Self-Polarization of Nuclear Spins

The parameters S_T and W_1 in (11.21) can, generally speaking, not be equal to zero even in zero external field, if there already exists some nuclear field. The right-hand side of (11.21) is in this case an odd function of $\langle I \rangle$, and we again are dealing with a nonlinear equation. This equation always has a trivial solution, $\langle I \rangle = 0$. However, situations are possible when a solution with a nonzero $\langle I \rangle$ appears, while the solution with zero nuclear spin becomes unstable. This means that, under certain conditions, a nonequilibrium polarization of nuclei may arise spontaneously, in the absence of both optical orientation and magnetic field (strictly speaking, a weak magnetic field is still required to suppress the dipole–dipole nuclear relaxation; however, this field plays no role in the spin transfer into the nuclear system). This phenomenon is called the dynamic self-polarization of nuclei.

The possibility of the dynamic self-polarization was first pointed out by Dyakonov and Perel in 1972 [51]. They suggested a mechanism based on the Over-hauser effect, i.e., in terms of (11.21), it results from a non-zero value of the equilibrium electron spin in the nuclear field, $S_T(B_N)$. For this reason, the critical temperature for the appearance of this effect even in the absence of losses is about 1 K.

Another theoretical model of the dynamic self-polarization, based on spin-dependent recombination of excitons in quantum dots, was considered in [52]. The spontaneous development of nuclear polarization within this model is brought about by recombination of non-radiative excitonic states (dark excitons), which are transformed into radiative (bright) ones when the electron swaps its spin with one of the nuclei. The dark and bright doublets are split by the electron–hole exchange interaction. The Overhauser field splits both doublets, making the energy separation between the initial and the final state of the flip–flop transition larger for one spin direction and smaller for the other. The consequent difference of transition probabilities (formally expressed by the parameter $W_1(B_N)$ in (11.21)) gives rise to a spin influx into the nuclear spin system, leading to a further increase of the Overhauser field. This mechanism is virtually temperature-independent. It was estimated that the self-polarization can be easily realized in GaAs-based quantum dots.

Recently, an analogous mechanism has been theoretically considered for the system of coupled quantum dots, described in the previous subsection [53].

In spite of many optimistic estimates, the dynamic self-polarization of nuclear spins has not been so far experimentally observed. A possible reason is strong losses of nuclear polarization present in real structures.

11.4 Dynamic Nuclear Polarization in Oblique Magnetic Field

Before the advent of single-quantum-dot spectroscopy, most experiments on optical polarization of nuclear spins were performed in magnetic fields directed at an angle to the excitation light beam. This geometry is still widely used, both in continuous-wave and time-resolved experiments. The reason is that it provides a technically simple way of detecting the spin polarization of nuclei via its influence on the electron polarization. The projection of the mean spin of optically oriented electrons on the external magnetic field B gives rise to the dynamic polarization of nuclei. The resulting nuclear spin, parallel to B, produces the Overhauser field B_N. Larmor precession of electron spin about the total field $B^* = B + B_N$ changes the direction of the electron spin polarization. This change is detected by optical means: using either its influence on the polarization of a probe light beam, or on polarization of the photoluminescence. In anisotropic structures, the oblique-field geometry can bring about qualitatively new phenomena. In particular, the electron–nuclear spin system can demonstrate bistability even at low losses (see Sect. 11.4.3), i.e., under the conditions where it would not appear in the Faraday geometry (for comparison, see Sect. 11.3.3).

The behavior of the electron–nuclear spin system in an oblique magnetic field can often be described within a simple model assuming that the time derivative of the electron spin density s is a sum of three contributions: (1) generation of polarized electrons by absorbed circularly polarized light, (2) precession of their spins in the sum of the external magnetic field and mean nuclear field, and (3) spin relaxation and recombination of electrons[10]:

$$\frac{\partial s}{\partial t} = G S_i + [\boldsymbol{\Omega}^* \times s] - \frac{s}{T_{es}}, \tag{11.22}$$

where G and S_i are generation rate and the mean spin of photoexcited electrons; $\boldsymbol{\Omega}^* = \boldsymbol{\Omega}_B + \boldsymbol{\Omega}_N$ is the electron Larmor frequency in the sum of the external and Overhauser fields; $T_{es} = (\tau_e^{-1} + \tau_{es}^{-1})^{-1}$ is the lifetime of the electron's oriented state; τ_e and τ_{es} are the recombination lifetime and the spin relaxation time of electrons. Under excitation by short pulses of circularly polarized light, the generation term is zero between the pulses, and (11.22) yields damped oscillations of the spin components perpendicular to $\boldsymbol{\Omega}^*$—the so-called spin beats. The value of the mean spin $S = s/n$ under continuous wave excitation ($G S_i = $ const) is obtained from (11.22) by putting the time derivative equal to zero:

$$S = \frac{G T_{es}}{n} \frac{S_i + [\boldsymbol{\Omega}^* T_{es} \times S_i] + (\boldsymbol{\Omega}^* T_{es} \cdot S_i) \cdot \boldsymbol{\Omega}^* T_{es}}{1 + (\boldsymbol{\Omega}^* T_{es})^2}, \tag{11.23}$$

where n is the electron concentration.

When the mean spin of photoexcited electrons S_i is directed along the axis z, the spin projection S_z is

$$S_z = S_0 \left(\frac{\sin^2 \alpha}{1 + (\boldsymbol{\Omega}^* T_{es})^2} + \cos^2 \alpha \right), \tag{11.24}$$

where α is the angle between z-axis and the total magnetic field $\boldsymbol{\Omega}^*$. In strong magnetic fields (large Ω^*), S_z is equal to $S_0 \cos^2 \alpha$. Depolarization of electrons by the magnetic field is called the Hanle effect.

If the distribution of the electron polarization is spatially inhomogeneous, the electron spin diffusion may have a strong impact on the Hanle effect [54–56]. This is not the case for quantum dot structures where electrons are strongly localized.

11.4.1 Larmor Electron Spin Precession

Time-resolved polarization spectroscopy gives a possibility to register time evolution of the mean electron spin (see Chap. 5). This way, the Larmor frequency of the

---

[10] Here we neglect the equilibrium polarization of electrons in the external and nuclear magnetic fields, assuming that their temperature is sufficiently high.

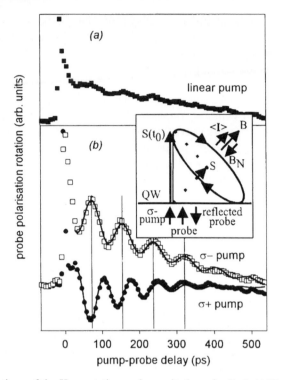

Fig. 11.9. Oscillations of the Kerr rotation under excitation of a GaAs/AlGaAs quantum well (QW) by circularly polarized light pulses in an oblique magnetic field (**b**) [58]. The oscillation frequency is the electron spin Larmor frequency in the sum of the external field and the Overhauser field. It changes its value when the helicity of the excitation light is changed. For comparison, the Kerr-rotation signal shows no oscillations at linearly polarized excitation (**a**)

electrons can be directly measured, from which the Overhauser field can be found [57, 58]. Figure 11.9 gives a nice example. Here, the oscillating projection of the electron spin on the structure axis was measured using Kerr rotation of linearly polarized probe pulse, time-delayed with respect to a circularly polarized pump pulse. The spin-related nature of the Kerr signal was proved by absence of oscillations under linearly polarized pump. The dependence of the Larmor frequency on the pump helicity revealed the contribution of the Overhauser field: it inverted direction when the pump polarization was changed to the opposite, while the external field did not, and as a result the total field changed its value. Further confirmation of the dynamic polarization of nuclei came from slow drift of the electron Larmor frequency after switching on the pump light, reflecting spin accumulation in the nuclear system and nuclear spin diffusion.

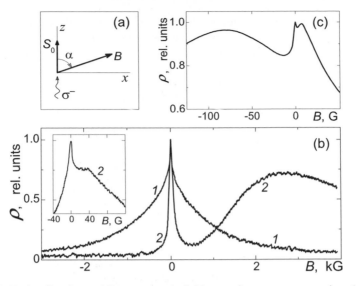

Fig. 11.10. Hanle effect in an oblique magnetic field at continuous-wave pumping. (**a**) Geometry of experiment with an oblique magnetic field. The excitation beam is directed along z-axis while luminescence is collected in the opposite direction. Here S_0 is a mean electron spin in a zero magnetic field. (**b**) Experimental Hanle curves for a single 100 Å-thick GaAs/Al$_{0.3}$Ga$_{0.7}$As (001) quantum well, measured on 1e–1hh luminescence line at alternating (34 kHz) (*curve 1*) and constant (*curve 2*) circular polarization of the exciting light. $\alpha = 85°$, $T = 2$ K [24]. (**c**) The Hanle curve measured in an epitaxial layer of Ga$_{0.5}$In$_{0.5}$P at a constant circular polarization of the pump, $\alpha = 65°$, $T = 77$ K [61]

11.4.2 Polarization of Electron–Nuclear Spin-System in an Oblique Magnetic Field

If polarized luminescence is used as a detector, the nuclear field B_N manifests itself most directly when the external magnetic field B is applied at an angle α to the excitation beam [59] (Fig. 11.10(a)). That is a result of the fact that, in accordance with (11.19), nuclear spins are oriented either along B or opposite to it. Therefore, the field B_N enhances or reduces (depending on its sign) the effect of the external field component transverse to the mean electron spin S.

Typical curves of the luminescence depolarization in an oblique magnetic field, measured in 100 Å-thick GaAs/Al$_{0.3}$Ga$_{0.7}$As quantum well at continuous wave pumping,[11] are shown in Fig. 11.10(b).

Curve 1 was obtained with the circular polarization of exciting light being alternated in sign at a high (34 kHz) frequency.[12] In this case, the Overhauser field is absent since the nuclear dynamic polarization can not follow the rapid alternation

---

[11] The degree of circular polarization of luminescence propagating along the z-axis $\rho \propto S_z$ [2].

[12] To do so, the exciting beam was passed trough a photoelastic polarization modulator [60] operating at 34 kHz.

of the mean electron spin. Therefore, the curve 1 is a purely electronic Hanle curve, which is symmetric in magnetic field and has a maximum at $B = 0$. In a strong field $B \gg B_{1/2}$ ($B_{1/2}$ is the half-width of the Hanle curve measured in perpendicular external field), this curve approaches the value $\rho(0) \cos^2 \alpha$, corresponding to the z-projection of the electron spin, $S_z = S_B \cos \alpha = S_0 \cos^2 \alpha$, conserved in the strong field.

Curve 2 in Fig. 11.10(b) was measured at constant circular polarization of the pump beam. Three conspicuous maxima, with the photoluminescence polarization close to $\rho(0)$, are seen in this curve. The central maximum is situated at $B = 0$, one of the additional maxima—at a relatively large field ($\approx 2.5\,\mathrm{kG}$), and the other one—at a much weaker field ($\approx 40\,\mathrm{G}$). In all the three maxima, the electron depolarization is slowed down, but for different reasons.

According to (11.6) and (11.19), the Overhauser field $\boldsymbol{B}_N$ is directed along the sum of the external and electron fields ($\boldsymbol{B} + \boldsymbol{B}_e$).

At zero external field, the polarization of nuclei can occur only in the Knight field. In this case the Knight field is directed along S_0, that is $\boldsymbol{B}_e \propto S_0$. Therefore, the nuclear field is also directed along S_0, and, for this reason, it cannot depolarize electrons.

When the external field is not equal to zero, the electrons are depolarized by the total field ($\boldsymbol{B} + \boldsymbol{B}_N$). As the nuclear field sharply rises with the increase of the external field, reaching large values already at $B \sim B_L$, it strongly enhances the depolarizing action of the weak external field. This explains the narrow central maximum.

The additional maximum at the relatively weak magnetic field in the inset in Fig. 11.10(b) arises from compensation of the longitudinal component of the external field by the electron field [61]. According to (11.19), almost no nuclear polarization occurs at this maximum because the nuclei are located in a total longitudinal field whose value is close to zero: $(B_z + B_{e0}) \approx 0$.[13] The field $B \approx -B_{e0}/\cos \alpha$, at which the maximum is observed, can be used to find the Knight field value. One can find from curve 2 in the insert in Fig. 11.10(b) that $B_{e0} = (4 \pm 1)\,\mathrm{G}$.

In a strong field $B \gg B_L, B_e$, the nuclear field is collinear to the external field. The condition for compensation of the external field by the nuclear field, $(B + B_N) = 0$, defines the position of the additional maximum on the Hanle curve at strong external fields [59]. Therefore, it can be used to measure the Overhauser field value. One can find from the curve 2 in Fig. 11.10(b) that $B_N \approx 2.5\,\mathrm{kG}$ at $\alpha = 85°$.

According to (11.6), the measurement of nuclear field, $B_N \propto S_B = S_0 \cos \alpha$, permits to find the nuclear polarization $\langle I \rangle / I$. Using the value $B_{N\,\mathrm{max}} = 56\,\mathrm{kG}$ for bulk GaAs [20, 54] and the measured value $B_N \approx 2.5\,\mathrm{kG}$, one can find that $\langle I \rangle / I \approx 4.5\%$ at $\alpha = 85°$. One can estimate that the nuclear polarization will reach $\approx 50\%$ in a longitudinal ($\alpha = 0°$) magnetic field. Note, that up to $T = 77\,\mathrm{K}$ the nuclear polarization retains a substantial value ($\sim 1\%$ at $\alpha = 60°$).

The direction of the nuclear field $\boldsymbol{B}_N$ is determined by the signs of both nuclear (g_n) and electron (g_e) g-factors (see (11.6)), while the Knight field $\boldsymbol{B}_e$ is always antiparallel to S (see (11.9)). It turns out in [61] that at $g_n > 0$ in crystals with

---

[13] $B_{e0} = B_e(S = S_0)$.

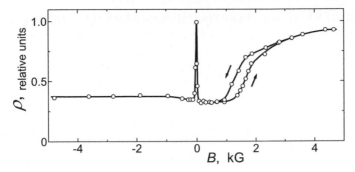

Fig. 11.11. Hysteresis on the experimental Hanle curve recorded for 80 Å-wide GaAs/Al$_{0.3}$Ga$_{0.7}$As (001) quantum well in an oblique magnetic field at a constant circular polarization of the exciting light, $\alpha = 60°$, $T = 2\,\mathrm{K}$ [63]

$g_e < 0$ both additional maxima are observed at the same field sign (Fig. 11.10(b)), whereas at $g_e > 0$ they appear at different signs[14] of the external field (Fig. 11.10(c)).

Thus, an analysis of luminescence depolarization in an oblique magnetic field allows to find the values of the Overhauser and Knight fields and the sign of electron g-factor.

11.4.3 Bistability of the Electron–Nuclear Spin System in Structures with Anisotropic Electron g-Factor and Spin Relaxation Time

The electron–nuclear spin system, coupled by hyperfine interaction, is an essentially nonlinear system. The manifestations of this nonlinearity, resulting from the dependence of relaxation rates on the Overhauser field, are considered in Sect. 11.3. The other group of non-linear phenomena is brought about by the anisotropy of nuclear or electron spin states. The effects of the nuclear anisotropy were first investigated in bulk crystals Al$_{0.26}$Ga$_{0.74}$As, where, under substitution of Ga atoms by Al atoms, a portion of As nuclei are subjected to a quadrupole perturbation, and their nuclear field, in the general case, does not coincide with the direction of external field. The most striking observations were bistabilities and self-sustained oscillations of electron and nuclear spin polarizations. These effects are described in detail in [16, 62]. Here, we concentrate at the nonlinear effects caused by the anisotropy of the electron g-factor and spin relaxation time, which often appears in semiconductor nanostructures.

---

[14] For a certain sign of g_e, the position of the strong-field maximum is reversed with respect to $B = 0$, if $g_n < 0$ [59]. The detailed analysis of the positions of the additional maxima depending on the g_e and g_n signs and the angle α (including the case of $\alpha = 90°$ when polarization of nuclei is possible solely in the Knight field) is done in [61], see also [16].

Bistability of Electron–Nuclear Spin System Induced by Anisotropy of Electron *g*-Factor

The bistability of electron–nuclear spin system can be induced by anisotropy of the electron *g*-factor, as it was found in quantum wells [63, 64]. Because the electron *g*-factor components along ($g_\parallel$) and across ($g_\perp$) the quantum well growth axis are different [65–67], the axis $\boldsymbol{\Omega}$ of Larmor precession of electron spins in the oblique external magnetic field $\boldsymbol{B}$ does not coincide with the field direction: $\boldsymbol{\Omega}$ is not parallel to $\boldsymbol{B}$. At the same time, due to isotropy of hyperfine interaction, the mean nuclear spin is directed along $\boldsymbol{B}$ (we do not consider the field range of $|B| \leq B_L, B_e$), and the precession axis of electron spins in the nuclear field $\boldsymbol{\Omega}_N = A\langle\boldsymbol{I}\rangle/\hbar$ is collinear to the external field: $\boldsymbol{\Omega}_N \parallel \boldsymbol{B}$. Thus, the electron depolarization is realized by the Overhauser field at $B \ll B_N$ and by the external field when $B \gg B_N$. As a consequence, the electron spin precession axis in the total field $(B + B_N)$ changes its direction under variation of the magnitude of the oblique external field.

The availability of two precession axes gives rise to bistability of the electron–nuclear spin system. Such a bistability was observed in quantum wells at continuous-wave [63] and pulse [64] pumping. It appears as a hysteresis on the experimental Hanle curve in oblique field as shown in Fig. 11.11. The dependence of $\rho(B)$ in Fig. 11.11 was recorded in GaAs/Al$_{0.3}$Ga$_{0.7}$As (001) quantum well of 80 Å width, where the $g_\parallel$ and $g_\perp$ are significantly different: $g_\parallel/g_\perp = 2.2$ and $g_\parallel < g_\perp < 0$ [63, 65, 66].

The calculation of the stationary states of the electron–nuclear spin system using (9.23) describes qualitatively the hysteresis in Fig. 11.11 for $g_\parallel/g_\perp = 2.2$ [63, 68]. The calculation also shows [63, 68] that in quantum wells of a smaller width, where the signs of $g_\parallel$ and $g_\perp$ are different [65–67], there are two regions of bistability, realized in external magnetic fields of different signs. Since in the absence of nuclear spin anisotropy the mean nuclear spin is always collinear to the external field, the spin dynamics is described by a single differential equation of the first order, and self-sustained oscillations cannot appear.

As the width of quantum well increases, the anisotropy of the electron *g*-factor decreases [65–67], and the properties of the electron–nuclear spin system approach those in a bulk crystal. The Hanle curves measured in quantum wells of 100 Å or wider do not differ qualitatively from the Hanle curves in bulk semiconductors, as shown in Sect. 11.4.2.

Bistability of the Electron–Nuclear Spin System, Induced by Anisotropy of Electron Spin Relaxation

The electron size quantization in quantum wells is accompanied by an anisotropy of their spin relaxation time τ_s [69]. The numerical calculation in [63] shows that this anisotropy can also give rise to a bistability. This bistability was not observed experimentally. This might be related to the fact that a strong τ_s anisotropy is predicted for free electrons, while nuclei are effectively polarized by localized electrons, which are less sensitive to size quantization effects. The conditions for observation of this bistability are most favorable in (110) quantum wells, where the anisotropy of τ_s is the highest [69], see also [64, 70, 71].

11.5 Optically Detected and Optically Induced Nuclear Magnetic Resonances

11.5.1 Optically Detected Nuclear Magnetic Resonance

A direct evidence for the nuclear polarization is the observation of nuclear magnetic resonance (NMR) under application of a radio-frequency field $B_{rf} = 2B_1 \cos \omega t$ perpendicular to the external field B. When a resonant condition $\omega = \gamma_n B$ is fulfilled, the nuclear polarization decreases, thus decreasing the Overhauser field B_N.

A commonly used way of detecting the NMR optically, demonstrated for the first time by Ekimov and Safarov [72], is the measurement of luminescence polarization ρ in a longitudinal [72–74] or in an oblique ([59] and [24]) magnetic field. In an oblique field, the resonant decrease of the Overhauser field is accompanied by a sharp change of ρ on the slopes of the additional maximum on the Hanle curve, where the external field is compensated by the nuclear field. The optically detected NMR spectra of the bulk GaAs obtained with a slow sweep of the radio-frequency field are shown in Fig. 11.12 (curve 1) and in Fig. 11.14(a). The resonance signals from all the isotopes (^{75}As, ^{69}Ga, ^{71}Ga) of the GaAs crystal lattice are clearly seen in these spectra.

The resonant decrease of the Overhauser field can be observed as a change of the electron spin splitting in a single quantum dot in longitudinal field [75] or through time-resolved Faraday rotation in oblique field [76].

11.5.2 Multispin and Multiquantum NMR

Large nuclear polarization realized in a weak magnetic field under optical pumping permits to observe multispin and multiquantum resonances [77–80] which are strongly forbidden in a conventional NMR [15, 81].

Multispin resonances are observed at the double ($\omega = 2\gamma_n B$) or triple ($\omega = 3\gamma_n B$) Larmor frequency and correspond to the simultaneous flip of two or three nuclear spins in the same direction induced by one radio-frequency quantum. Such resonances are permitted due to admixing of $|M \pm 1\rangle$ and $|M \pm 2\rangle$ spin states to $|M\rangle$ by operators $\hat{I}^{\pm}\hat{I}_z$ and $\hat{I}^{\pm}\hat{I}^{\pm}$ of the nonsecular part of the dipole–dipole interaction (11.10) [15, 81]. The admixture is proportional to B_L/B, and a probability of the multispin transition decreases with increasing B as B_L^2/B^2.

Figure 11.12 presents NMR spectra for GaAs crystal. Fundamental single-spin resonances of ^{75}As, ^{69}Ga, and ^{71}Ga nuclei are seen at small radio-frequency field amplitude $B_1 = 0.08$ G (curve 1). With increase in B_1 to 0.8 G, all six possible two-spin resonances appear (curve 2), including flip–flip resonances, both of one isotope [2(^{75}As), 2(^{69}Ga), 2(^{71}Ga)] and of different isotopes (^{75}As + ^{69}Ga, ^{75}As + ^{71}Ga,

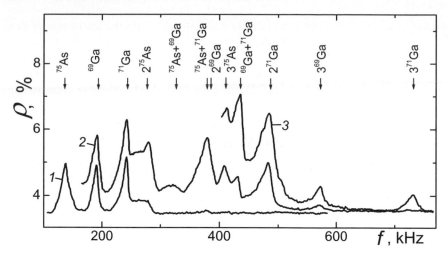

Fig. 11.12. Optically detected NMR spectra of bulk GaAs measured in an oblique field $B = 187\,$G at continuous-wave optical excitation with a constant circular polarization and radio-frequency field $B_{rf} = 2B_1 \cos 2\pi f t$ applied along the y-axis [80]. $T = 1.9\,$K, $\alpha = 84°$. B_1, G: 1—0.08, 2—0.8, 3—4.8. *Arrows* indicate NMR frequencies

^{69}Ga $+\,^{71}$Ga). Further increase in B_1 to 4.8 G induces flip–flip–flip transitions of one isotope [3(^{75}As), 3(^{69}Ga), and 3(^{71}Ga)] (curve 3). In a smaller magnetic field $B = 84\,$G where the probability of multispin transitions is higher, three-spin resonances [2(^{75}As)+^{71}Ga, ^{75}As+2(^{69}Ga), 2(^{75}As)+^{69}Ga, 2(^{69}Ga)+^{71}Ga], whose frequencies are equal to the sum of the double and single resonant frequencies of the different isotopes, are recorded at $B_1 = 5.6\,$G.

The multispin resonance of different kinds of nuclei does not disappear even if single-spin resonance of one of them is saturated. For example, two-spin resonance at the frequency $\omega_2(^{69}$Ga $+\,^{75}$As) is retained when the resonance at the frequency $\omega_1(^{69}$Ga) is saturated.[15] In other words, multispin transition can occur when only one of the species of nuclei is polarized while others belong to nonpolarized spin subsystems.

Multiquantum NMR is realized if several radio-frequency quanta are absorbed with the total energy equal to the energy of the NMR transition, besides the energies of these quanta can be different. As an example, two-photon resonances for the transition at the triple Larmor frequency of the ^{75}As nuclei (409 kHz in a 187-G field) are shown in Fig. 11.13. The resonances are recorded in the presence of the second radio-frequency field $B_{rf}^{(1)} = 2B_1^{(1)} \cos 2\pi f^{(1)}t$ with the fixed frequency $f^{(1)}$ equal to 196.5 kHz (curve 1) or 186.5 kHz (curve 2). Theory [82–84] confirms the high probability of the multiquantum multispin NMR.

---

[15] Two radio-frequency fields of different frequencies are applied to the sample in this case.

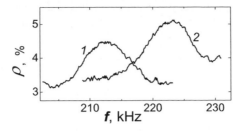

Fig. 11.13. Optically detected NMR signal in a GaAs crystal with the absorption of two radio-frequency quanta of different frequencies $f^{(1)}$ and f, the sum of which corresponds to resonance at triple the Larmor frequency of the ^{75}As nuclei in an oblique field $B = 187\,\text{G}$ ($\alpha = 84°$, $T = 1.9\,\text{K}$). The radio-frequency field $B_1^{(1)} = 2.9\,\text{G}$ is of the fixed frequency $f^{(1)} = 196.5\,\text{kHz}$ (*1*) or $186.5\,\text{kHz}$ (*2*). The radio-frequency field of varying frequency f is $B_1 = 0.8\,\text{G}$ [80]

11.5.3 Optically Induced NMR

Under optical pumping, the NMR in a semiconductor can be induced solely by light [85, 86], see also [57, 87–89]. The excitation light needs to be circularly polarized and partly modulated at the NMR frequency either in circular polarization or in intensity.

In an oblique magnetic field, the constant part of the circularly polarized excitation light creates dynamic polarization (cooling) of nuclei as described in Sect. 11.4.2. Resonant depolarization (i.e., resonant heating) of the dynamically polarized nuclei is caused by the oscillating part of the Knight field, whose transverse component plays the role of radio-frequency field. Since such a NMR is induced by purely optical means, without using radio-frequency field, it is called "optically induced NMR" [85] or "all-optical NMR" [57].

According to (11.9), for nuclei of one type, the Knight field averaged over the localization volume is

$$\boldsymbol{B}_{\text{e}}(r) = F b_{\text{e}} \boldsymbol{S}, \tag{11.25}$$

where S is the mean electron spin, $F = n_{\text{d}}/N_{\text{d}}$ is the localization center filling factor (N_{d} is the total density of centers, and n_{d} is the density of occupied centers), $b_{\text{e}}/2$ is the Knight field of fully polarized electrons at $F = 1$. In turn, S is governed by the degree of circular polarization P, whereas F is controlled by the intensity J of the incident light. Thus, the Knight field oscillating at some frequency can be produced by modulating either P or J at this frequency. Under continuous wave pumping, such modulation can be easily achieved by means of the linear electro-optic effect in a KDP crystal [85].

Figure 11.14 shows the NMR spectra for GaAs recorded at weak ($\sim$1%) modulation of the light polarization (Fig. 11.14(b)) and light intensity (Fig. 11.14(c)). Their comparison with the spectrum (Fig. 11.14(a)) induced by a weak radio-frequency field shows that the resonant depolarization of nuclei by the oscillating electron field is very effective.

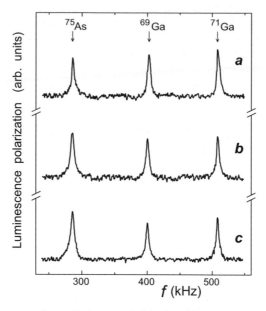

Fig. 11.14. NMR spectra for p-GaAs recorded in an oblique magnetic field B with slow variation of a frequency of the radio-frequency field (**a**), modulation of the circular polarization of light (**b**), and modulation of the intensity of light (**c**). The measurement is performed under continuous wave optical pumping. $B = 392\,\text{G}$, $\alpha = 77°$, $T = 1.9\,\text{K}$ [85]. (**a**) Radio-frequency field amplitude $2B_1 = 0.02\,\text{G}$, (**b**) polarization modulation depth $m_\text{P} = (P_\text{max} - P_\text{min})/(P_\text{max} + P_\text{min}) = 0.021$, (**c**) intensity modulation depth $m_\text{J} = (J_\text{max} - J_\text{min})/(J_\text{max} + J_\text{min}) = 0.021$

The comparison of the amplitudes of the NMR signals induced by modulated polarization with those induced by radio-frequency field allows to find the Knight field magnitude for each type of nuclei. Using this method, the field b_e for ^{71}Ga nuclei was found to be $(17 \pm 8)\,\text{G}$ in GaAs (at $T = 1.9\,\text{K}$) and $(59 \pm 29)\,\text{G}$ in $\text{Al}_{0.26}\text{Ga}_{0.74}\text{As}$ (at $T = 77\,\text{K}$) [85].

Deep ($\sim$100%) modulation of intensity can induce resonant heating of nuclei at the double Larmor frequency [87–89]. Such transitions with $\Delta m_\text{I} = \pm 2$ are due to coupling between the modulated electric field of photoexcited electrons and the nuclear quadrupole moment.

It is worth noting that modulation of pump polarization can also result in *resonant cooling* of nuclei. The cooling is accompanied by the resonant change of nuclear polarization and formally can be also classified as the all-optical NMR. However, the resonant cooling has quite different physical nature since it creates the dynamic nuclear polarization by oscillating Knight field. The resonant cooling was observed experimentally in bulk semiconductors [90, 91] and quantum wells [92, 93]. The theory was developed in [94] for bulk semiconductors but remains also valid for nanostructures. For a detailed review of the effect see [16]. Resonant cooling leads to the change in the photoluminescence polarization with a shape of dispersion curve unlike the NMR signals in Fig. 11.14, which reflect the heating of the nuclear spins and have a shape of the absorption curve. Resonant cooling is negligible for small (~1%) modulation of the Knight field. It becomes comparable with resonant heating for strong (~100%) modulation, at which the resultant change in polarization is described by superposition of absorption and dispersion signals [86].

11.6 Spin Conservation in the Electron–Nuclear Spin System of a Quantum Dot

Electron and nuclear spins in a semiconductor form a complex, nonlinear, strongly coupled system. The dynamics of this system is characterized by a variety of relaxation times, which are sometimes difficult to separate when analyzing experimental data. Though these issues have been mostly discussed in previous sections of this chapter, it is worthwhile to summarize the information for readers interested in the interpretation of "spin memory" experiments, i.e., measurements of the photoluminescence polarization, Faraday/Kerr rotation, splitting of electron spin levels or any other parameter sensitive to the state of the electron–nuclear spin system, as a function of time after the system has been prepared in a spin-polarized state. Such experiments, especially those revealing a "long-lived" (that can mean anything from microseconds to minutes and more) spin memory, currently attract a great interest. In this section, we discuss the time scales relevant to such experiments, and then propose a scheme of the analysis of experimental data that may help in separating different mechanisms of spin conservation.

11.6.1 Time Scales for Preservation of Spin Direction and Spin Temperature

The lifetime of nonequilibrium spin polarization in the nuclear spin system is characterized by the time T_2; this time also determines the decoherence of the spin components, perpendicular to the external magnetic field. The time T_2 is limited by the dipole–dipole interaction and is typically of the order of 10^{-4} s.

The time T_2 may become much longer if nuclear spins ($I > 1/2$) experience the quadrupole splitting in strained structures (for example, about 10 ms in InP quantum dots, as we have estimated from the data of [18]).

The nuclear spin temperature is conserved during the spin-lattice relaxation time T_1. The time T_1 is especially long if nuclei are not coupled with electrons by the hyperfine interaction. At low temperatures and in the absence of electrons, T_1 riches minutes (as in GaAs [44, 73]) or even days (as in Si [95]). If a magnetic field (or the mean Knight field of spin-polarized electrons) is applied to dynamically cooled nuclei, they develop a quasi-equilibrium spin polarization. This polarization can be erroneously interpreted as a manifestation of the spin conservation. In reality, this is *energy* conservation.

For nuclei in a quantum dot (or around a donor center), occupied by an electron, the coupling to the lattice is effectively mediated by the electron spin. For such nuclei, T_1 is much shorter and may even become comparable with T_2 [45, 96].

On the other hand, cooled nuclear spins in a quantum dot containing an electron may bind with the electron spin to form the *nuclear spin polaron*, as has been theoretically predicted in [97, 98]. In this case, T_2 is renormalized due to the influence of the Knight field of the single electron:

$$T_{2\text{pol}} \approx T_2 \frac{\langle I \rangle^2 N}{I(I+1)}, \tag{11.26}$$

where $\langle I \rangle$ is the mean nuclear spin in the polaron, and N is the number of nuclei in the quantum dot. The necessary condition for the polaron formation is a high (close to maximum) nuclear polarization. If nuclei are completely polarized, one can estimate $T_{2\text{pol}} \approx T_2 N \sim 10\,\text{s}$ for the dot containing $N \sim 10^5$ nuclei.[16]

The hierarchy of relaxation times of the mean electron spin coupled to unpolarized nuclei was discussed in detail in Sect. 11.2. If the nuclei are polarized, the electron spin relaxation requires changing its spin projection on the Overhauser field. This process may be induced by interaction either with other charge carriers, or, in their absence, with phonons. In the latter case, the electron relaxation times can be quite long (up to seconds at liquid-helium temperatures) [99].

The dynamics of spin temperature in a quantum dot or around a donor may be also controlled by nuclear spin diffusion. The diffusion time depends on the spatial pattern of the nuclear temperature and may range from T_2 to infinity [45, 73, 100].

11.6.2 A Guide to Interpretation of Experiments on "Spin Memory"

At present, four possible reasons are known for long-lived (up to minutes and more) memory of the electron–nuclear spin system after switching off the preliminary optical pumping. The first one is a long, with time T_1, relaxation of nuclear spin temperature to the lattice temperature [15]. The three other reasons are due to suppression of fast nuclear dipole relaxation. The dipole relaxation can be suppressed: (1) by the external magnetic field[17] [15], (2) by the quadrupole splitting of nuclear spin levels [15, 16, 18] and (3) by the formation of a nuclear spin polaron [97, 98]. The increase of relaxation time in cases (2) and (3) we will call the anomalous prolongation of time T_2.

Only three of these reasons hold in zero external magnetic field. We will consider the differences in their manifestations.[18] Let us analyze the common features of numerous measurements of slow spin relaxation, not paying much attention to details.

---

[16] Conservation of the polarization of resident electrons in singly negatively charged InGaAs dots during $\approx 0.2\,\text{s}$ after optical pumping in zero external magnetic field was attributed in [98] to formation of the nuclear polaron. However, the value of nuclear polarization was not measured directly in this experiment. Therefore, additional experimental and theoretical efforts are required to elucidate the reason of the observed long-lived spin memory.

[17] The dipole relaxation can also be suppressed by the mean Knight field of optically polarized electrons. However, this field is absent in the darkness after switching off the pump light.

[18] The consideration below is valid for the case of zero external magnetic field only.

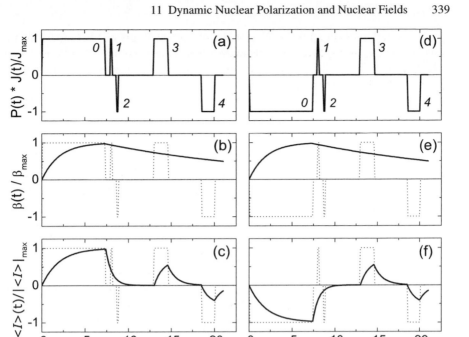

Fig. 11.15. Timing diagrams for qualitative analysis of slow spin relaxation in the electron–nuclear spin system of a quantum dot in a typical pump–probe optical experiment at zero external magnetic field. (**a**) and (**d**) are polarizations of pump and probe pulses, (**b**) and (**e**) are time dependences of reciprocal nuclear spin temperature for opposite helicities of the pump, (**c**) and (**f**) are time dependences of nuclear polarization for opposite helicities of the pump. In zero external magnetic field, nuclear spin temperature is not sensitive to the change of pump polarization sign, while nonequilibrium nuclear spin reverses its direction

Then, we will formulate the method for analysis of experimental results and apply it to the experiment of [96] as an example.

Let some parameter Z of the electron–nuclear spin system be measured, which is related to a nuclear polarization $\langle I \rangle$ and has different values at different signs of $\langle I \rangle$. For measurements in zero magnetic field, this allows us to separate long memory related to a long relaxation time T_1 of the spin temperature from the memory due to the anomalously long relaxation of nonequilibrium spin.

A sketch of typical optical measurement of polarization relaxation time in quantum dots without magnetic field is shown in Fig. 11.15. A dipole–dipole relaxation time T_2 is taken as the time unit. The sample is illuminated with a pump pulse "0" of circularly polarized light. (The curves (a), (b), (c) are plotted for polarization σ^+, while the curves (d), (e), (f) are drawn for polarization σ^-.) During the pump pulse, nuclei interact with electrons, which are oriented by light. The electrons transfer to the nuclei their polarization and create a mean Knight field, which stabilizes the polarization transferred to the nuclei.

The duration of pump pulse "0" (or the train of such pulses) is larger than the typical time T_{Ne} of polarization of nuclei by photoelectrons (curves in Fig. 11.15 are calculated for $T_{Ne} = 2T_2$). During the pulse, the nuclear polarization $\langle I \rangle$ and the reciprocal nuclear spin temperature β achieve saturation.

On completion of the pump pulse (in the darkness), the nuclear polarization and the spin temperature relax with characteristic times T_2 and T_1, respectively. Usually, $T_2 \ll T_1$, and only in anomalous cases the non-equilibrium nuclear spin can last for a long time. In a semiconductor of GaAs type without magnetic field and quadrupole splitting of nuclear spin levels, $T_2 \sim 10^{-4}$ s, and the value of $T_1 > 1$ min is not surprising after experiments [44, 45, 73]. Curves in Fig. 11.15 are drawn for $T_1 = 20T_2$.

The state of the nuclear spin system after a pump pulse is detected by a probe light pulse. The duration of probe pulse τ_{prob} should be short in comparison with T_{Ne}. Otherwise, this pulse will change the value of β. However, if $\tau_{prob} \ll T_2$ (pulses 1 and 2), then no noticeable changes of nuclear polarization occur during the probe pulse. If $\tau_{prob} \geq T_2$ (pulses 3 and 4), then under a mean Knight field the nuclear polarization begins to rise with a characteristic time T_2 and can reach its equilibrium value, which corresponds to the reciprocal temperature β conserved in the darkness up to the arrival of the probe pulse. Thus, both the short relaxation time of nonequilibrium spin and the long relaxation time of nuclear spin temperature can be experimentally observed.

One can separate these two completely different relaxation processes by changing the sign of the probe pulse polarization, as it is shown in Fig. 11.15. In this figure, the pulses 1 and 3 have polarization σ^+, and pulses 2 and 4 have polarization σ^-. If the mean nuclear spin $\langle I(t_{prob}) \rangle$, induced by the probe pulse (and the measured parameter Z, related to it) does not depend on the probe–pulse polarization sign, then it is a nonequilibrium nuclear spin, which has been conserved after pumping. If $\langle I(t_{prob}) \rangle$ changes its sign with the change of the probe–pulse polarization sign (as shown in Fig. 11.15), then we deal with a thermal mechanism of spin memory. In this case, the long relaxation times are quite natural.

The dependence $\langle I(t_{prob}) \rangle$ on the sign of the pump–pulse polarization has an opposite character. It follows from (11.18), that the spin temperature is controlled by a product of nonequilibrium electron spin and Knight field, that is $\beta_{pump} \propto \sigma^2_{pump}$. Therefore, in the absence of an external magnetic field the value and sign of β do not depend on the pump polarization sign (Fig. 11.15, panels (b) and (e)). At the same time, the nuclear polarization generated by the pump pulse changes sign together with the sign of pump polarization because, according to (11.19), $\langle I_{pump} \rangle \propto \sigma_{pump}$. Thus, if the nuclear polarization in probe pulse remembers the preliminary pumping, but does not remember its sign, we deal with long times of thermal relaxation.[19] Here, relaxation times of order of minutes are not surprising. If polarization in probe pulse remembers not only the pumping magnitude, but its sign also, then the traces

---

[19] In pump–probe Faraday or Kerr rotation measurements, where probe pulse is always linearly polarized, the effects of conservation of the nuclear polarization and nuclear spin temperature can be separated using their dependence on the sign of the pump circular polarization.

of nonequilibrium spin $\langle I_{pump} \rangle$ remain in the system. In this case, long (noticeably exceeding $T_2 \sim 10^{-4}$ s) times of spin relaxation demand taking into account suppression mechanisms of nuclear dipole–dipole interaction—quadrupole suppression [16, 18], or creation of a spin polaron [97, 98].

The dependence of spin memory on the presence of an electron in the quantum dot can be used to separate the quadrupole and polaron mechanisms of dipole-relaxation suppression. In the absence of electron, the polaron is not created, and only the quadrupole suppression mechanism exists.[20]

Let us apply the suggested above approach for analysis of long-living spin memory observed by Maletinsky et al. in self-organized InGaAs quantum dots [96]. The pump–probe microphotoluminescence technique was used to measure the dynamics of generation and decay of the nuclear spin polarization in charged single quantum dots. Both pump and probe pulses were circularly polarized. The optical polarization of nuclei was achieved via their hyperfine interaction with a resident electron in a singly negatively charged dot. During the pump pulse of width τ_{pump}, the dynamic nuclear polarization arises, reaching about 15% under optimal conditions [6]. The Overhauser shift of the photoluminescence line of negatively charged (X^-) exciton was measured. Its magnitude is directly proportional to the Overhauser field, which is conserved during the dark interval τ_{wait} before switching on the probe pulse.

Applying a proper electric bias to the structure (see the voltage diagram in Fig. 11.16(a)), the authors of [96] could quickly (during 30 μs) change the quantum dot charge. Let us analyze the results for those dots, which contain neither an electron nor a hole in the dark. To get the dots empty, the resident electron is removed from an initially singly-charged dot just after switching off the pump pulse and is injected back at the end of the dark period τ_{wait} before switching on the probe pulse. Figure 11.16 shows the Overhauser-shift decay, obtained in such an empty dot under σ^+ (empty squares) and σ^- (full squares) pumping. The exponential fit (solid curve in Fig. 11.16(c)) indicates a decay time constant of $\tau_{decay} \sim 2.3$ s.

Among three mentioned reasons for the long spin memory (conservation of a low nuclear temperature, polaron formation, or quadrupole suppression) the polaron formation should be excluded, since there is no electron in the dot during the dark interval, and the nuclear spin polaron cannot be created. The Overhauser shift does

---

[20] Another method of distinguishing between the polaron and quadrupole mechanisms, by the dependence of the memory effect on the crystal temperature, was proposed in [98]. Indeed, the polaron formation requires low temperature and is critically sensitive to a temperature increase [97], while the quadrupole suppression does not strongly depend on temperature. However, in [98], as well as in many other works, spin memory effects were detected by measuring the polarization of electrons, not nuclei. Electron and nuclear spins are decoupled at elevated temperature. As a result, any manifestation of the nuclear polarization, not only the polaron, may become unobservable. For this reason, if the electron polarization is used as a detector, the temperature dependence does not allow unambiguous separating of the polaron and quadrupole suppression of spin relaxation. It is noteworthy that this method may appear effective if the nuclear spin memory is detected by the splitting of electron energy levels in the Overhauser field [96] or from the absorption spectrum of radio-frequency power by the cooled nuclear spin system in zero external field [80].

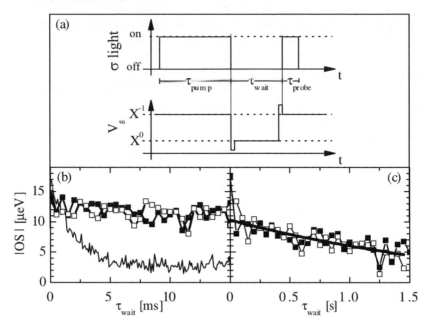

Fig. 11.16. Time decay of the Overhauser shift (OS) in a zero-charged single quantum dot recorded in zero external magnetic field after switching-off the optical pump [96]. (**a**) Timing diagram for the gate voltage switching experiment: During the dark period τ_{wait}, the gate voltage, V_g, is switched to a value where the quantum dot is empty (neutral, X^0, exciton is the stable charge complex). Using transient pulses, the switching time is $30\,\mu s$. (**b**) Time decay of nuclear polarization in the absence of the resident electron under σ^+ (*empty squares*) and σ^- (*full squares*) excitation. For comparison, *the solid curve* shows the mean of the data obtained in X^- exciton where resident electron is present in the quantum dot during the dark period. (**c**) Same measurement as in (**b**), but over a longer time scale. The exponential fit (*solid curve*) indicates a decay time constant of $\tau_{decay} \sim 2.3\,s$

not reverse its direction irrespective of the probe-pulse polarization (linear, σ^+ or σ^-) [101]. This rules out the conservation of a low spin temperature. Thus, we can conclude that the quadrupole suppression is responsible for the long spin memory observed in empty dots in [96].

11.7 Conclusions

Let us briefly summarize the issues discussed in this chapter.

The hyperfine coupling between electron and nuclear spins gives a possibility to polarize (to cool) nuclear spins and detect their polarization. The experimental manifestations of the nuclear polarization include the line splitting in optical spectra of quantum dots, the all-optical nuclear magnetic resonance, and many other bright effects.

The system of strongly coupled electron and nuclear spins is nonlinear. Under certain conditions, it develops multistability, self-sustained oscillations and hysteretic phenomena.

The theory based on the approximation of short correlation time is usually able to correctly describe the behavior of the electron–nuclear spin system. Nevertheless, some experiments performed on long-lived, well-isolated spin systems of quantum dots require approaches going beyond this approximation.

The unique physical properties of a localized electron interacting with nearby nuclei have been provoking questions, for many years, about a possible use of such physical systems in information technology. Lately, quantum computation with nuclear spins has been frequently discussed. So far, no practical results were achieved. However, the physical beauty of electron–nuclear spin phenomena, together with a remaining hope for revolutionary applications, fuel new experimental and theoretical research into this very attractive field.

Acknowledgements. This work was partly supported by the Russian Foundation for Basic Research and grants of the Russian Academy of Sciences. The work of IAM was partly conducted at the Center for Nanophase Materials Sciences, which is sponsored at Oak Ridge National Laboratory by the Division of Scientific User Facilities, U.S. Department of Energy.

References

[1] G. Lampel, Phys. Rev. Lett. **20**, 491 (1968)
[2] F. Meier, B.P. Zakharchenya (eds.), *Optical Orientation* (North-Holland, Amsterdam, 1984)
[3] S.A. Crooker, J.J. Baumberg, F. Flack, N. Samarth, D.D. Awschalom, Phys. Rev. Lett. **77**, 2814 (1996)
[4] R.E. Worsley, N.J. Traynor, T. Grevatt, R.T. Harley, Phys. Rev. Lett. **76**, 3224 (1996)
[5] S.W. Brown, T.A. Kennedy, D. Gammon, E.S. Snow, Phys. Rev. B **54**, R17339 (1996)
[6] C.W. Lai, P. Maletinsky, A. Badolato, A. Imamoglu, Phys. Rev. Lett. **96**, 167403 (2006)
[7] A.I. Tartakovskii, T. Wright, A. Russell, V.I. Fal'ko, A.B. Van'kov, J. Skiba-Szymanska, I. Drouzas, R.S. Kolodka, M.S. Skolnick, P.W. Fry, A. Tahraoui, H.-Y. Liu, M. Hopkinson, Phys. Rev. Lett. **98**, 26806 (2007)
[8] P.-F. Braun, B. Urbaszek, T. Amand, X. Marie, O. Krebs, B. Eble, A. Lemaitre, P. Voisin, Phys. Rev. B **74**, 245306 (2006)
[9] T. Takagahara (ed.), *Quantum Coherence, Correlation and Decoherence in Semiconductor Nanostructures* (Academic, London, 2003)
[10] D.D. Awschalom, N. Samarth, D. Loss (eds.), *Semiconductor Spintronics and Quantum Computation* (Springer, Berlin, 2002)
[11] D.K. Young, J.A. Gupta, E. Johnston-Halperin, R. Epstein, Y. Kato, D.D. Awschalom, Semicond. Sci. Technol. **17**, 275–284 (2002)
[12] I. Žutić, J. Fabian, S. Das Sarma, Rev. Mod. Phys. **76**, 323–410 (2004)
[13] R. Winkler, *Spin–Orbit Coupling Effects in Two-Dimensional Electron and Hole Systems* (Springer, Berlin, 2003)
[14] A. Abragam, B. Bliney, *Electron Paramagnetic Resonance of Transition Ions* (Clarendon Press, Oxford, 1970)

[15] A. Abragam, *The Principles of Nuclear Magnetism* (Clarendon Press, Oxford, 1961)
[16] V.G. Fleisher, I.A. Merkulov, in *Optical Orientation*, ed. by F. Meier, B.P. Zakharchenya (North-Holland, Amsterdam, 1984), pp. 173–258
[17] V.L. Berkovits, V.I. Safarov, Fiz. Tverd. Tela **20**, 2536 (1978); Sov. Phys. Solid State **20**, 1468 (1978)
[18] R.I. Dzhioev, V.L. Korenev, Phys. Rev. Lett. **99**, 37401 (2007)
[19] E.I. Gr'ncharova, V.I. Perel, Fiz. Tekhn. Poluprovodn. **11**, 1697 (1977); Sov. Phys. Semicond. **11**, 997 (1977)
[20] D. Paget, G. Lampel, B. Sapoval, V.I. Safarov, Phys. Rev. B **15**, 5780 (1977)
[21] G.R. Khutsishvili, Usp. Fiz. Nauk **87**, 211 (1965); Sov. Phys. Usp. **8**, 743 (1966)
[22] M.I. Dyakonov, V.I. Perel, Z. Eksp. Teor. Fiz. **65**, 362 (1973); Sov. Phys. JETP **38**, 177 (1974)
[23] I.D. Vagner, T. Maniv, Phys. Rev. Lett. **61**, 1400 (1988)
[24] V.K. Kalevich, V.L. Korenev, O.M. Fedorova, Pis'ma Z. Eksp. Teor. Fiz. **52**, 964 (1990); JETP Lett. **52**, 349 (1990)
[25] R.I. Dzhioev, K.V. Kavokin, V.L. Korenev, M.V. Lazarev, B.Ya. Meltser, M.N. Stepanova, B.P. Zakharchenya, D. Gammon, D.S. Katzer, Phys. Rev. B **66**, 245204 (2002)
[26] R.I. Dzhioev, V.L. Korenev, I.A. Merkulov, B.P. Zakharchenya, D. Gammon, Al.L. Efros, D.S. Katzer, Phys. Rev. Lett. **88**, 256801 (2002)
[27] I.A. Merkulov, Al.L. Efros, M. Rosen, Phys. Rev. B **65**, 205309 (2002)
[28] P.F. Braun, X. Marie, L. Lombez, B. Urbaszek, T. Amand, P. Renucci, V. Kalevich, K. Kavokin, O. Krebs, P. Voisin, Y. Masumoto, Phys. Rev. Lett. **94**, 116601 (2005)
[29] A.V. Khaetskii, D. Loss, L. Glazman, Phys. Rev. Lett. **88**, 186802 (2002)
[30] A.V. Khaetskii, D. Loss, L. Glazman, Phys. Rev. B **67**, 195329 (2003)
[31] E.A. Yuzbashyan, B.L. Altshuler, V.B. Kuznetsov, V.Z. Enolskii, J. Phys. A **38**, 7831 (2005)
[32] K.A. Al-Hassanieh, V.V. Dobrovitski, E. Dagotto, B.N. Harmon, Phys. Rev. Lett. **97**, 037204 (2006)
[33] G. Chen, D.L. Bergman, L. Balents, Phys. Rev. B **76**, 45312 (2007)
[34] B. Pal, S.Yu. Verbin, I.V. Ignatiev, M. Ikezawa, Y. Masumoto, Phys. Rev. B **75**, 125332 (2007)
[35] J.R. Petta, A.C. Johnson, J.M. Taylor, E.A. Laird, A. Yacoby, M.D. Lukin, C.M. Marcus, M.P. Hanson, A.C. Gossard, Science **309**, 2180 (2005)
[36] S.E. Barrett, R. Tycko, L.N. Pfeiffer, K.W. West, Phys. Rev. Lett. **72**, 1368 (1994)
[37] R. Tycko, J.A. Reimer, J. Phys. Chem. **100**, 13240 (1996)
[38] E.S. Artemova, I.A. Merkulov, Fiz. Tverd. Tela **27**, 1558 (1985); Sov. Phys. Solid State **27**, 941 (1985)
[39] D. Gammon, Al.L. Efros, T.A. Kennedy, M. Rosen, D.S. Katzer, D. Park, S.W. Brown, V.L. Korenev, I.A. Merkulov, Phys. Rev. Lett. **86**, 5176 (2001)
[40] P. Maletinsky, C.W. Lai, A. Badolato, A. Imamoglu, Phys. Rev. B **75**, 035409 (2007)
[41] M.I. Dyakonov, I.A. Merkulov, V.I. Perel, Z. Eksp. Teor. Fiz. **76**, 314 (1979); Sov. Phys. JETP **49**, 160 (1979)
[42] R.I. Dzhioev, B.P. Zakharchenya, V.L. Korenev, M.V. Lazarev, Fiz. Tverd. Tela **41**, 2193 (1999); Phys. Solid State **41**, 2014 (1999)
[43] R.I. Dzhioev, B.P. Zakharchenya, V.G. Fleisher, Pis'ma Z. Tekh. Fiz. **2**, 193 (1976); Sov. Tech. Phys. Lett. **2**, 73 (1976)
[44] V.K. Kalevich, V.D. Kulkov, V.G. Fleisher, Izv. Akad. Nauk SSSR Ser. Fiz. **46**, 492 (1982); Bull. Acad. Sci. USSR Phys. Ser. **46**, 70 (1982)

[45] V.K. Kalevich, V.D. Kulkov, V.G. Fleisher, Pis'ma Z. Eksp. Teor. Fiz. **35**, 17 (1982); JETP Lett. **35**, 20 (1982)
[46] M.I. Dyakonov, V.I. Perel, Z. Eksp. Teor. Fiz. **68**, 1514 (1975); Sov. Phys. JETP **41**, 759 (1975)
[47] I.A. Merkulov, Phys. Usp. **45**, 1293 (2002)
[48] V.A. Atsarkin, Sov. Phys. Usp. **21**, 725 (1978)
[49] K.M. Salikhov, Y.N. Molin, R.Z. Sagdeev, A.L. Buchachenko, *Spin Polarization and Magnetic Effects in Chemical Reactions* (Elsevier, Amsterdam, 1984)
[50] J. Baugh, Y. Kitamura, K. Ono, S. Tarucha, Phys. Rev. Lett. **99**, 96804 (2007)
[51] M.I. Dyakonov, V.I. Perel, Pis'ma Z. Eksp. Teor. Fiz. **16**, 563 (1972); JETP Lett. **16**, 398 (1972)
[52] V.L. Korenev, Pis'ma Z. Eksp. Teor. Fiz. **70**, 124 (1999); JETP Lett. **70**, 129 (1999)
[53] M.S. Rudner, L.S. Levitov, Phys. Rev. Lett. **99**, 36602 (2007)
[54] M.I. Dyakonov, V.I. Perel, in *Optical Orientation*, ed. by F. Meier, B.P. Zakharchenya (North-Holland, Amsterdam, 1984), pp. 11–71
[55] R.I. Dzhioev, B.P. Zakharchenya, R.R. Ichkitidze, K.V. Kavokin, P.E. Pak, Fiz. Tverd. Tela **35**, 2821 (1993); Phys. Solid State **35**, 1396 (1993)
[56] R.I. Dzhioev, B.P. Zakharchenya, V.L. Korenev, M.N. Stepanova, Fiz. Tverd. Tela **39**, 1975 (1997); Phys. Solid State **39**, 1765 (1997)
[57] J.M. Kikkawa, D.D. Awschalom, Science **287**, 473 (2000)
[58] A. Malinowski, R.T. Harley, Solid State Commun. **114**, 419 (2000)
[59] M.I. Dyakonov, V.I. Perel, V.L. Berkovits, V.I. Safarov, Z. Eksp. Teor. Fiz. **67**, 1912 (1974); Sov. Phys. JETP **40**, 950 (1975)
[60] S.N. Jasperson, S.F. Schnatterly, Rev. Sci. Instrum. **40**, 761 (1969)
[61] B.P. Zakharchenya, V.K. Kalevich, V.D. Kulkov, V.G. Fleisher, Fiz. Tverd. Tela **23**, 1387 (1981); Sov. Phys. Solid State **23**, 810 (1981)
[62] E.S. Artemova, E.V. Galaktionov, V.K. Kalevich, V.L. Korenev, I.A. Merkulov, A.S. Silbergleit, Nonlinearity **4**, 49 (1991)
[63] V.K. Kalevich, V.L. Korenev, Pis'ma Z. Eksp. Teor. Fiz. **56**, 257 (1992); JETP Lett. **56**, 253 (1992)
[64] H. Sanada, S. Matsuzaka, K. Morita, C.Y. Hu, Y. Ohno, H. Ohno, Phys. Rev. B **68**, 241303R (2003)
[65] E.L. Ivchenko, A.A. Kiselev, Fiz. Tekh. Poluprovodn. **26**, 1471 (1992); Sov. Phys. Semicond. **26**, 827 (1992)
[66] E.L. Ivchenko, A.A. Kiselev, Pis'ma Z. Eksp. Teor. Fiz. **67**, 41 (1998); JETP Lett. **67**, 43 (1998)
[67] P. Le Jeune, D. Robart, X. Marie, T. Amand, M. Brousseau, J. Barrau, V. Kalevich, D. Rodichev, Semicond. Sci. Technol. **12**, 380 (1997)
[68] A.A. Kiselev, Fiz. Tverd. Tela **35**, 219 (1993); Phys. Solid State **35**, 114 (1993)
[69] M.I. Dyakonov, V.Yu. Kachorovski, Fiz. Tekh. Poluprovodn. **20**, 178 (1986); Sov. Phys. Semicond. **20**, 110 (1986)
[70] Y. Ohno, R. Terauchi, T. Adachi, F. Matsukura, H. Ohno, Phys. Rev. Lett. **83**, 4196 (1999)
[71] S. Döhrmann, D. Hägele, J. Rudolph, M. Bichler, D. Schuh, M. Oestreich, Phys. Rev. Lett. **93**, 147405 (2004)
[72] A.I. Ekimov, V.I. Safarov, Pis'ma Z. Eksp. Teor. Fiz. **15**, 453 (1972); JETP Lett. **15**, 179 (1972)
[73] D. Paget, Phys. Rev. B **25**, 4444 (1982)
[74] G.P. Flinn, R.T. Harley, M.J. Snelling, A.C. Tropper, T.M. Kerr, J. Luminescence **45**, 218 (1990)

[75] D. Gammon, S.W. Brown, E.S. Snow, T.A. Kennedy, D.S. Katzer, D. Park, Science **277**, 85 (1997)

[76] M. Poggio, D.D. Awschalom, Appl. Phys. Lett. **86**, 182103 (2005)

[77] V.L. Berkovits, V.I. Safarov, Pis'ma Z. Eksp. Teor. Fiz. **26**, 377 (1977); JETP Lett. **26**, 256 (1977)

[78] V.A. Novikov, V.G. Fleisher, Z. Eksp. Teor. Fiz. **74**, 1026 (1978); Sov. Phys. JETP **47**, 539 (1978)

[79] V.K. Kalevich, V.D. Kulkov, I.A. Merkulov, V.G. Fleisher, Fiz. Tverd. Tela **24**, 2098 (1982); Sov. Phys. Solid State **24**, 1195 (1982)

[80] V.K. Kalevich, V.G. Fleisher, Izv. Akad. Nauk SSSR Ser. Fiz. **47**, 2294 (1983); Bull. Acad. Sci. USSR Phys. Ser. **47**, 5 (1983)

[81] C.P. Slichter, *Principles of Magnetic Resonance* (Springer, Berlin, 1980)

[82] J.R. Franz, C.P. Slichter, Phys. Rev. **148**, 287 (1966)

[83] Yu.G. Abov, M.I. Bulgakov, A.D. Gul'ko, F.S. Dzheparov, S.S. Trostin, S.P. Borovlev, V.M. Garochkin, Pis'ma Z. Eksp. Teor. Fiz. **35**, 344 (1982); JETP Lett. **35**, 424 (1982)

[84] T.Sh. Abesadze, L.L. Buishvili, M.G. Menabde, Z.G. Rostomashvili, *Radiospectroskopiya* (Perm' University, Perm', 1985), pp. 99–108

[85] V.K. Kalevich, Fiz. Tverd. Tela **28**, 3462 (1986); Sov. Phys. Solid State **28**, 1947 (1986)

[86] V.K. Kalevich, V.L. Korenev, V.G. Fleisher, Izv. Akad. Nauk SSSR Ser. Fiz. **52**, 434 (1988); Bull. Acad. Sci. USSR Phys. Ser. **52**, 16 (1988)

[87] G. Salis, D.T. Fuchs, J.M. Kikkawa, D.D. Awschalom, Y. Ohno, H. Ohno, Phys. Rev. Lett. **86**, 2677 (2001)

[88] G. Salis, D.D. Awschalom, Y. Ohno, H. Ohno, Phys. Rev. B **64**, 195304 (2001)

[89] M. Eickhoff, B. Lenzman, G. Flinn, D. Suter, Phys. Rev. B **65**, 125301 (2002)

[90] V.K. Kalevich, V.D. Kulkov, V.G. Fleisher, Fiz. Tverd. Tela **22**, 1208 (1980); Sov. Phys. Solid State **22**, 703 (1980)

[91] V.K. Kalevich, V.D. Kulkov, V.G. Fleisher, Fiz. Tverd. Tela **23**, 1524 (1981); Sov. Phys. Solid State **23**, 892 (1981)

[92] V.K. Kalevich, B.P. Zakharchenya, Fiz. Tverd. Tela **37**, 3525 (1995); Sov. Phys. Solid State **37**, 1938 (1995)

[93] V.K. Kalevich, B.P. Zakharchenya, *Proceedings 23th International Conf. on the Physics of Semiconductors*, Berlin, Germany, vol. 3 (World Scientific, Singapore, 1996), p. 2455

[94] I.A. Merkulov, M.N. Tkachuk, Z. Eksp. Teor. Fiz. **83**, 620 (1982); Sov. Phys. JETP **56**, 342 (1982)

[95] N.T. Bagraev, L.S. Vlasenko, Z. Eksp. Teor. Fiz. **83**, 2186 (1982); Sov. Phys. JETP **56**, 1266 (1982)

[96] P. Maletinsky, A. Badolato, A. Imamoglu, Phys. Rev. Lett. **99**, 56804 (2007)

[97] I.A. Merkulov, Fiz. Tverd. Tela **40**, 1018 (1998); Phys. Solid State **40**, 930 (1998)

[98] R. Oulton, A. Greilich, S.Yu. Verbin, R.V. Cherbunin, T. Auer, D.R. Yakovlev, M. Bayer, I.A. Merkulov, V. Stavarache, D. Reuter, A.D. Wieck, Phys. Rev. Lett. **98**, 107401 (2007)

[99] A.V. Khaetskii, Yu.V. Nazarov, Phys. Rev. B **61**, 12639 (2000)

[100] M.N. Makhonin, A.I. Tartakovskii, I. Drouzas, A. Van'kov, T. Wright, J. Skiba-Szymanska, A. Russell, V.I. Fal'ko, M.S. Skolnick, H.-Y. Liu, M. Hopkinson, cond-mat/0708.2792, 4 Oct. 2007

[101] P. Maletinsky, A. Imamoglu, private communication

Nuclear–Electron Spin Interactions in the Quantum Hall Regime

Y.Q. Li and J.H. Smet

The discoveries of the integer and fractional quantum Hall effects [1, 2] in the early 1980s opened an exciting field for condensed matter physics. The continuous effort in this area during the past quarter century has been particularly rewarding. The interplay between Landau quantization, disorder, electron–electron and spin interactions gives rise to a large variety of fascinating electronic states [3–6]. Some examples are incompressible quantum fluids with quasi-particles carrying fractional charge [7–11], compressible metallic states composed of composite particles [12–14], insulating Wigner crystal electron solids [15, 16], Bose–Einstein condensates [17], Skyrmion spin textures in quantum Hall ferromagnets [18, 19], non-Abelian states [20–22], electron liquid crystal phases [23–28], and so forth. Indeed, this list is incomplete.

In this chapter, we shall review some of the interesting physics related to the spin degrees of freedom. The focus will be on the spin interactions between electrons and nuclei in the quantum Hall regime. This review is not intended to be exhaustive, but rather attempts to give a flavor of the richness of the physics involved by describing some experiments that were carried out in Stuttgart and elsewhere. Here, only experiments on GaAs based two-dimensional electron systems will be addressed. They offer the highest carrier mobility, which is an important prerequisite to observe many of these interesting but fragile quantum Hall states. Two-dimensional hole systems as well as other two-dimensional electron systems which are based on AlAs, InAs, GaN, other III–V semiconductors, SiGe, Si/SiO_2, and II–VI semiconductors are beyond the scope of this chapter.

This chapter is organized as follows. In Sect. 12.1, we will give a brief introduction to the quantum Hall effects and the interaction between nuclear and electron spins. Subsequently, some experimental techniques to study the nuclear–electron spin interactions will be described in Sect. 12.2. It will be followed by a review of recent experimental work on several quantum Hall systems where the interaction between nuclear and electron spins is involved. These experiments not only shed light on the properties of the electronic states and their excitations, but also suggest novel means for manipulating and detecting nuclear spins.

12.1 Introduction

12.1.1 The Quantum Hall Effects in a Nutshell

When electrons are constrained to move in two dimensions only, they exhibit many interesting properties that do not exist in three-dimensional systems. The most celebrated examples are the integer and fractional quantum Hall effects (QHE) [1, 2]. They manifest in a quantization of the Hall resistance and the vanishing of the diagonal resistance. Figure 12.1 shows transport data recorded on a high mobility sample when a magnetic field is applied perpendicular to the plane in which the two-dimensional (2D) electrons are confined. The filling factor ν is defined as the ratio between the 2D carrier density n_s and the magnetic flux quantum density $n_B = B/\Phi_0$, i.e.,

$$\nu = \frac{n_s}{n_B} = \frac{n_s \Phi_0}{B},$$ (12.1)

where B is the magnitude of the perpendicular magnetic field. If ν is equal or close to either an integer or any of the magical rational numbers, ν_0, the Hall resistance exhibits a plateau and takes on a quantized value equal to

$$R_{xy} = \frac{1}{\nu_0} \frac{h}{e^2}.$$ (12.2)

Here, h is the Planck constant, $-e$ is the charge of an electron, and $\Phi_0 = h/e$ is the magnetic flux quantum. As seen in Fig. 12.1, plateaus in the Hall resistance are accompanied by vanishing of the longitudinal (or diagonal) resistance, i.e., $R_{xx} \to 0$ as the temperature $T \to 0$. These two hallmarks of the QHE are universal: independent of the sample geometry, host material, as well as the details of the disorder landscape seen by the electrons. The universality, high precision (better than one part in 10^9 [29] for the integer quantum Hall effect) and robustness of the Hall resistance quantization suggest that the underlying physics is highly nontrivial.

In 2D, the vanishing of R_{xx} implies that also the diagonal conductivity σ_{xx} drops to zero as $T \to 0$. At finite T, σ_{xx} exhibits an Arrhenius type of temperature dependence [15, 30, 31]

$$\sigma_{xx} \propto \exp\left(-\frac{\Delta}{2k_B T}\right).$$ (12.3)

This is similar to an insulator and a superconductor, in which the existence of an energy gap is responsible for an exponential drop of the conductivity or the resistivity, respectively. Here, the Arrhenius behavior of σ_{xx} also stems from the existence of a thermal activation gap which we will denote as $\Delta/2$.

The energy gaps in the integer quantum Hall effect regime can be attributed to the quantization of the cyclotron motion of electrons as well as the Zeeman splitting. The perpendicular magnetic field discretizes the single particle energy spectrum into a ladder of equidistant Landau levels with energies E_n

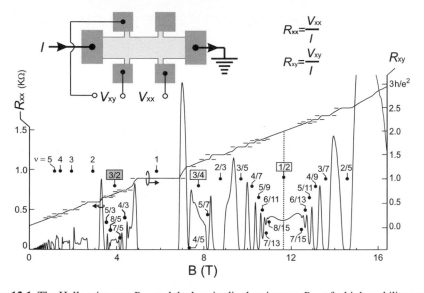

Fig. 12.1. The Hall resistance R_{xy} and the longitudinal resistance R_{xx} of a high mobility modulation doped GaAs/AlGaAs 2D electron system subjected to a perpendicular magnetic field of magnitude B. The numbers in the plot refer to either integer fillings or any of the rational values ν at which the integer or fractional quantum Hall effects are observed. Some even denominator fractional fillings have been highlighted. At these fillings a composite fermion Fermi sea forms (see below)

$$E_n = \left(n + \frac{1}{2}\right)\hbar\omega_c, \qquad (12.4)$$

with the Landau level or orbital index $n = 0, 1, 2, \ldots$ and the cyclotron frequency $w_c = eB/m$, where m is the effective mass of a conduction band electron in GaAs.[1] Each Landau level is split further into two sublevels with opposite spin orientation and separated by the Zeeman energy $E_Z = g^*\mu_B B$. Here, g^* is the exchange enhanced electron g-factor.[2] Hence, each level is characterized by two quantum numbers: the spin $s = \uparrow$ or $\downarrow$, and the orbital index n. Each of the Zeeman split Landau levels is macroscopically degenerate. The degeneracy is equal to the number of flux quanta that thread the sample, i.e., the magnetic flux density n_B times the sample area. As the magnetic field is increased, each Landau level can accommodate more

---

[1] SI units are used throughout this chapter.

[2] The electron g-factor in bulk GaAs is $g = -0.44$. In the quantum Hall regime, the Zeeman gap can be enhanced by an order of magnitude because of exchange interactions between electrons [32]. This exchange enhanced Zeeman gap ($E_Z = g^*\mu_B B$) is important for electron transport properties at small odd integer filling factors. Exchange enhancement appears in transport because it probes the energy required to flip the electron spin in the limit of infinite wave number q. In optical experiments, such as electron spin resonance, electron–electron interactions are not relevant since they probe the energy in the limit $q \to 0$. In such experiments, the single particle Zeeman gap is observed: $\Delta_Z = g\mu_B B$ [33] with μ_B the Bohr magneton.

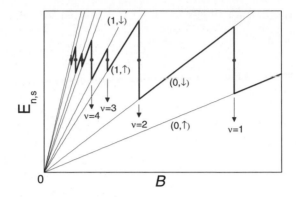

Fig. 12.2. The discrete single particle energy spectrum of a disorder-free 2D electron system with fixed density n_s in a perpendicular magnetic field. Each Zeeman split Landau level (LL) is characterized by an orbital index $n = 0, 1, 2, \ldots$ and a spin index $s = \uparrow, \downarrow$. Only the lowest two pairs of Zeeman split Landau levels ($n = 0, 2$) are labeled. As the magnetic field increases, the filling factor is reduced as a result of increasing LL degeneracy, and the chemical potential (*thick line*) jumps abruptly in the gap separating two adjacent Landau levels at integer filling factors. The size of the Zeeman gap is exaggerated for clarity

electrons and the topmost partially filled Landau level is gradually emptied at fixed carrier density. When it is entirely depopulated (corresponding to an integer filling), the chemical potential drops in between two Landau levels. Adding an additional electron requires a large energy to cross the gap and the system is said to be incompressible.[3] At low enough temperatures, no electronic states are available to scatter into and the conductivity (as well as the resistivity) drops to zero. The behavior of the chemical potential and the Landau levels as a function of applied magnetic field is schematically depicted in Fig. 12.2.

In order to account for the vanishing of R_{xx} over an extended range of the magnetic field and the appearance of a plateau in R_{xy}, it is important to include disorder. When potential fluctuations due to impurities occur on a length scale larger than the magnetic length $l_B = [\hbar/(eB)]^{1/2}$, the Landau levels follow the variations in the electrostatic potential. This broadens the Landau levels, and the states belonging to a Landau level need to be classified into two categories. At the center of the Landau levels, extended states form, which allow current transport from the source to the drain contact. The tails of the Landau levels comprise localized states, which correspond to electrons that either encircle potential hills or are trapped in a valley of the disorder landscape. The localized states do not contribute to current transport. As

---

[3] The compressibility is defined as $\kappa = -(1/A)(\partial A/\partial P)|_N$ with A the area of the sample, P the pressure, and N the total number of particles in the 2D system. For a 2D electron system, $\kappa^{-1} = n_s^2 \partial\mu/\partial n_s$ (See for example, [6]). When the 2D electron system condenses in an integer (or a fractional) quantum Hall state, it turns incompressible, i.e., $\kappa = 0$. The two hallmarks of the quantum Hall effects (vanishing R_{xx} and quantized R_{xy}) not only appear at exact integer (or rational) fillings as suggested by Fig. 12.2, but also persist over a finite range of magnetic fields due to disorder broadening. The same holds for the incompressibility.

long as the Fermi energy is pinned in the tail of a Landau level and these localized states are being emptied as the magnetic field is ramped up, the transport properties are frozen: R_{xx} remains zero and R_{xy} stays fixed at the plateau value. To account for the precise value of the Hall resistance plateaus a more extensive treatment is required [34].

In contrast to the integer quantum Hall effect, the fractional quantum Hall effect can not be understood from the single particle energy spectrum. Additional gaps do arise in the many body energy spectrum. They are the result of electron correlations in high magnetic fields [7], and hence more difficult to understand as it requires the analysis of many body wave functions. Fortunately in 1989, Jain succeeded in identifying suitable quasiparticles which allow to convert the system of strongly interacting 2D electrons into a system of noninteracting or weakly interacting quasiparticles. These quasiparticles are referred to as composite fermions. Each of them is assembled from one electron and two magnetic flux quanta, or more generally an even number, $2p$, of flux quanta [35]. This bond between electrons and flux quanta turns out to be a natural way for electrons to avoid one another. They, just as electrons, are sent onto circular orbits by a field. Unlike electrons, they do not experience the external applied magnetic field but only an effective field B^*, which is greatly reduced from the applied field by an amount equal to the field produced by all the flux quanta of the other composite fermions:

$$B^* = B - 2p\Phi_0 n_s. \tag{12.5}$$

The effective field vanishes at filling $1/2p$. The system is gapless and hence compressible. The composite fermions were predicted to form a Fermi sea with a well-defined Fermi surface [36]. This scenario was impressively confirmed in a series of ballistic transport experiments [37–41]. As one moves away from filling $\nu = 1/2p$, Landau quantization of the composite fermion cyclotron motion gives rise to a discrete spectrum of composite fermion Landau levels. The cyclotron energy gaps separating the composite Fermion Landau levels are no longer determined by the band mass of GaAs, but rather follow the Coulomb interaction strength. The successive depopulation of these Landau levels produces the integer quantum Hall effect of composite fermions. It is equivalent to the fractional quantum Hall effect of the original electrons. The fractional quantum Hall state with rational filling factor ν_0 can then be viewed as the integer quantum Hall state of composite fermions with integer filling factor $q = n_s \Phi_0 / |B^*|$. It satisfies

$$\nu_0 = \frac{q}{2pq \pm 1}. \tag{12.6}$$

The plus and minus signs correspond to B^* parallel and antiparallel to B.

So far we have ignored the finite size of the 2D electron system. However, any real sample has boundaries. These boundaries play an important role in the quantum Hall regime [42]. Moreover, they help in understanding the hallmarks of the quantum Hall effect. In the bulk of the sample, the Fermi level is pinned in a band of localized states and transport through bulk states can not take place. Near the boundary, the

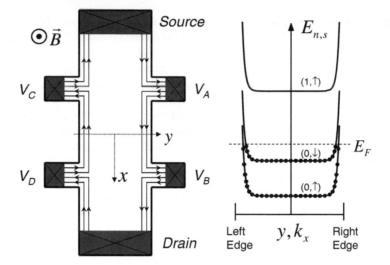

Fig. 12.3. Edge channels in a Hall bar with ideal contacts for bulk filling factor $v = 2$. At this bulk filling, two edge channels run along the boundaries of the sample. They belong to the two Zeeman split Landau levels $(0, \uparrow)$ and $(0, \downarrow)$. The applied electrochemical potential at the source and drain contact is equal to μ_1 and μ_2. Floating contacts acquire an electrochemical potential equal to the average electrochemical potential of the incoming edge channels. Therefore, $-eV_A = -eV_B = \mu_1$ for instance

electrostatic potential and also the Landau levels bend upwards due to depletion of the charge carriers. Where the Fermi energy crosses through the Landau level, there exists an extended state that runs all the way along the boundary. Classically, this so-called edge state corresponds to a skipping orbit of the electron. The number of edge states, or channels, is equal to the number of occupied spin split Landau levels in the bulk region. These edge states are chiral, since electrons propagate only in one direction, as indicated schematically in Fig. 12.3. This direction is determined by the sign of the applied field and is opposite for opposite sides of the sample. Since upward and downward moving electrons are located on opposite sides of the sample, back-scattering is strongly suppressed. The large spatial separation between the chiral edge states and the isolating or incompressible character of the bulk is what accounts for the dissipationless transport near integer fillings.

An ideal contact at equilibrium absorbs all incoming edge channels. While edge channels leaving an ideal contact are filled up to the electrochemical potential of the contact. If a contact is left floating (for instance, a potential probe), it acquires a potential equal to the average potential of all incoming edge channels. Potential probes located on one side of the sample will thus take on equal potential (for instance $-eV_A = -eV_B = \mu_1$) and the current flows despite the absence of a voltage drop, i.e., $R_{xx} = 0$. Dissipation only occurs at the other end of the sample, due to the applied voltage difference between the source and the drain contacts. Each edge channel may be viewed as a one-dimensional channel.

Irrespective of the details of their energy spectrum, 1D channels contribute a conductance equal to e^2/h [43]. The total net current supported by the sample is therefore equal to [43, 44]

$$I = N\frac{e}{h}(\mu_1 - \mu_2), \tag{12.7}$$

where N is the number of edge channels, and μ_1 and μ_2 are the electrochemical potentials of right and left edges (equivalently the source and drain contacts), respectively. For ideal contacts, we then obtain a Hall resistance $R_{xy} = V_{AC}/I = (\mu_1 - \mu_2)/(-eI) = -Nh/e^2$, which is exactly the quantization of the Hall resistance in the integer quantum Hall regime. The above treatment of edge channels is a single particle picture. A more accurate description requires a self-consistent treatment which includes electron screening. Edge channels then acquire a finite width and the boundary reconstructs into compressible and incompressible regions. The above conclusions remain however valid.

For non-ideal contacts, edge channels are partially transmitted or reflected. Quantum point contacts, narrow constrictions in the 2D electron system often defined electrostatically with metallic split gates deposited on top of the 2D electron system [45, 46], can be regarded as extreme cases of non-ideal contacts. Their properties have been exploited on a number of occasions in the context of the quantum Hall effect. For instance, van Wees et al. [47] demonstrated that edge channels can be selectively populated and detected with quantum point contacts. Furthermore, interedge channel scattering can be induced by applying voltage differences between edge channels [48]. As will be described in Sect. 12.3.2, scattering between spin-resolved edge channels can be exploited to dynamically polarize nuclear spins.

So far, we have laid out some of the basic concepts of the QHE. They form the basis for subsequent discussions of the interactions between the electron and nuclear spins. Readers are referred to any of the excellent textbooks [3–6, 13, 14, 49] and review articles [50–54] to obtain a more comprehensive overview of the subject.

12.1.2 Electron Spin Phenomena in the Quantum Hall Effects

The electron spin degree of freedom is not essential for the occurrence of the quantum Hall effects, but it brings additional richness into the quantum Hall physics. A school example is the degree of spin polarization of fractional quantum Hall states. At filling factor $\nu < 1$, where the most prominent fractional quantum Hall features occur, one may think that all electrons reside in the lowest Landau level and have their spins aligned in the direction of the external applied magnetic field. The spin degree of freedom—just like the orbital degree of freedom—would be entirely frozen out. However, for conduction band electrons in GaAs, the Zeeman energy E_Z is small. It is only about $1/70$ of the cyclotron energy $E_c = \hbar\omega_c$ due to the small electron g-factor and the small effective mass of electrons. The Coulomb energy $E_C = 1/(4\pi\epsilon\epsilon_0)(e^2/l_B)$ is also far larger.[4] In a typical magnetic field of $10\,\mathrm{T}$,

---

[4] For bulk GaAs, $g = -0.44$, $m = 0.067m_e$, $\epsilon = 12.9$. It is often convenient to write these energies in the following form: $E_Z \simeq 0.296B\,\mathrm{K}$, $E_C \simeq 50.8\sqrt{B}\,\mathrm{K}$, and $E_c \simeq 20.1B\,\mathrm{K}$.

$\Delta_Z \sim 3\,\mathrm{K}$, $E_c \sim 200\,\mathrm{K}$, and the Coulomb energy E_C is $\sim 160\,\mathrm{K}$. Halperin first pointed out that the small Zeeman energy invalidates the assumption that fractional quantum Hall states when $\nu < 1$ are fully spin-polarized [55]. The energy gap responsible for the appearance of the fractional quantum Hall effect is only a fraction of the Coulomb energy and comparable to the Zeeman energy. As a result, many fractional quantum Hall states are not necessarily fully spin-polarized. Chakraborty and Zhang [56, 57] showed that for the $\nu = 2/5$ fractional quantum Hall state, a spin unpolarized phase has lower energy than the polarized one in the limit of small Zeeman energy. The ground state of $\nu = 2/3$ was also predicted by Xie et al. [58] to be possibly spin-unpolarized. Transport measurements have confirmed that the electron spin degree of freedom can indeed not be ignored despite the high magnetic field. Transitions from spin unpolarized or partially polarized states to fully polarized ones have been observed at many different fractional filling factors, including 8/5 [59], 4/3 [60], 2/3 [61], 3/5 [62], 2/5 (under hydrostatic pressure) [63], and others [64].

The composite fermion theory provides an intuitive picture to understand these spin related phenomena ubiquitous in the fractional quantum Hall regime. Spin transitions occur whenever composite fermion Landau levels with different spin quantum numbers cross. The orbital Landau level splitting Δ_c^* of composite fermions is determined by the Coulomb energy and hence scales with the inverse of the average interparticle distance or $\sqrt{n_s}$. At fixed filling factor, this also implies that Δ_c^* grows with $\sqrt{B}$. On the other hand, the Zeeman splitting depends linearly on the applied magnetic field B. The different B-dependence of E_C and E_Z makes it possible to vary the ratio $\eta = E_Z/E_C$ simply by tuning the electron density n_s (while simultaneously sweeping B to keep the filling factor $\nu = \Phi_0 n_s/B$ fixed). Energy levels with different spin indices can then be brought into degeneracy. An example is shown in Fig 12.4 for filling factor 2/3 or 2/5. At these filling factors, two levels are completely full. Each level is denoted by its orbital index ($n = 0, 1, \ldots$) and spin ($\uparrow, \downarrow$). At low carrier density or magnetic field, the spin splitting is smaller than the Landau quantization energy. Levels ($0 \uparrow$) and ($0 \downarrow$) are filled and there is no net spin polarization. As the density or magnetic field is increased, levels ($0 \downarrow$) and ($1 \uparrow$) eventually cross since the Zeeman splitting rises more rapidly than the composite fermion cyclotron energy does. Level ($0 \downarrow$) is emptied and the electronic system becomes fully spin-polarized. The transition occurs at the transition field B_{tr} where η reaches a critical value. As the ($0 \downarrow$) and ($1 \uparrow$) levels approach, the energy gap separating the filled and empty levels diminishes. We anticipate the quantum Hall effect to disappear near the crossing. Indeed, in transport experiments, the Hall resistance is no longer quantized and the longitudinal resistance no longer vanishes [62, 65]. More direct evidence for changes in the spin polarization was obtained from circular polarization resolved photoluminescence experiments [66] as well as from resistively detected NMR studies [125]. Many of these transitions are also ac-

---

Similarly, $l_B \simeq 25.65/\sqrt{B}\,\mathrm{nm}$. The unit of B is Tesla in these equations. Here E_Z does not take into account exchange enhancement. For full spin polarization, for instance at $\nu = 1$, the Zeeman energy for transport can be one order of magnitude larger, which is still much smaller than other energy scales.

2D Electron Density

n_{tr}

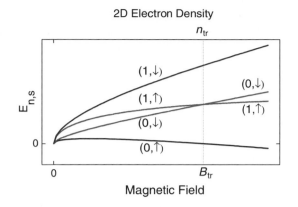

Fig. 12.4. The crossing of composite fermion Landau levels due to the competition between the Coulomb energy and the Zeeman energy. Shown here is the case for two filled composite fermion Landau levels, i.e., electron filling factor $2/(4p \pm 1)$, $p = 1, 2, \ldots$. At $B < B_{\mathrm{tr}}$ (or equivalently $n_s < n_{\mathrm{tr}}$), the electron spin polarization $\mathcal{P} = 0$ with levels $(0 \uparrow)$ and $(0 \downarrow)$ filled; whereas at $B > B_{\mathrm{tr}}$, $(0 \uparrow)$ and $(1 \uparrow)$ are occupied, and $\mathcal{P} = 1$

companied by hysteresis and other phenomena that appear in conventional magnetic materials. It has been shown that the physics at crossings of Landau levels with different quantum numbers can indeed be described in the language of quantum Hall ferromagnetism [69].

Landau level crossings for integer fillings are not possible by simply tuning the carrier density, because both the electron cyclotron energy and the Zeeman energy are proportional to B. However, by applying an in-plane magnetic field in addition to the perpendicular component B, it is possible to align two electron Landau levels with different spins in the integer quantum Hall regime. To first order the in-plane field leaves the electron cyclotron energies unaltered. However, the Zeeman energy increases, since it is determined by the total magnetic field B_{tot} rather than just the perpendicular field component B. Hence, if the field is not perpendicular to the 2D plane, E_Z should be written as $g^* \mu_B B_{\mathrm{tot}}$. Because the Zeeman splitting is two orders of magnitude smaller than the electron cyclotron energy in GaAs, large in-plane magnetic fields or tilt angles close to $90°$ are needed to produce a level coincidence (See, for instance, [69]). As a consequence, studies of the spin physics associated with level crossings in single layer GaAs based 2D electron system have been mainly restricted to fractional quantum Hall states.[5]

Despite the irrelevance of level coincidences, the spin degree of freedom does play an important role in GaAs even in the integer quantum Hall regime. Due to the exchange interaction between the electrons, energy gaps at odd integer fillings can be one order of magnitude larger compared to the single-particle Zeeman gap [32]. The ground state at filling $\nu = 1$ is a ferromagnet with all electron spins aligned

---

[5] In SiGe based 2D electron systems, such coincidences can be achieved at much smaller tilt angles due to the larger electron g-factor and effective mass [166].

with the magnetic field. This should come as no surprise. However, it has also been shown that all the spins would remain polarized by the exchange interaction even in the limit of vanishing Zeeman energy [18].[6] Like in conventional ferromagnets, spin wave excitations are present in such quantum Hall ferromagnets. As will be discussed in detail later, this ferromagnet also has gapless excitations that can lead to an enhanced interaction between the electron spins and the nuclear spin subsystem of the GaAs host.

Compressible states such as the $\nu = 1/2$ and $1/4$ composite fermion Fermi seas also exhibit rich spin physics [66, 70]. Depending on the relative strength of E_C and E_Z, the composite fermion Fermi sea at filling $\nu = 1/2$ and other even denominator fractional fillings can be either partially or fully spin polarized. This has important implications for the nuclear–electron spin interaction (see Sect. 12.3.6).

In short, phenomena related to the electron spin degree of freedom are abundant in the quantum Hall regime. This rich variety of spin related states is available in a single sample at the turn of a voltage knob. The gate voltage allows to tune the density or filling factor. We will demonstrate that the existence of gapless spin excitations or the presence of overlapping continuous energy spectra for spin-up and spin-down populations for some of these states provides ample opportunities to control and manipulate the interaction with the nuclei of the GaAs crystal hosting the 2D electron system.

12.1.3 Nuclear Spins in GaAs-Based 2D Electron Systems

The nuclei contained in the GaAs crystal which hosts the 2D electron system (two gallium isotopes ^{69}Ga and ^{71}Ga, and ^{75}As) all have nuclear spin $I = 3/2$ (see Table 12.1 for more information). Owing to their large masses, the magnetic moments of nuclei $\mu_N = \gamma_n \hbar I$ are typically about three orders of magnitude smaller than that of an electron. Nuclear spins interact with each other via magnetic dipole–dipole interactions or indirectly through the interaction mediated by surrounding electrons [71]. These interactions are, however, extremely weak, so that ferromagnetic or antiferromagnetic ordering of the nuclear spins is not possible unless cooling brings the spin temperature of the nuclei below $\sim 10^{-7}$ K [75]. In most cases, we only need to consider nuclear paramagnetism. At thermal equilibrium, the spin population follows the Boltzmann distribution. One finds that the nuclear spin polarization, defined as $\mathcal{P}_N = \langle I \rangle / I$, can be described by the Brillouin function $B_I(x)$, with $x = I \gamma_n \hbar B_{tot}/(k_B T)$. If the temperature is not too low ($k_B T \gg I \gamma_n \hbar B_{tot}$), $\mathcal{P}_N$ reduces to the Curie law of the following form:

$$\mathcal{P}_N = \frac{\langle I \rangle}{I} \cong \frac{\gamma_n \hbar (I+1) B_{tot}}{3 k_B T}, \qquad (12.8)$$

where k_B is the Boltzmann constant.

---

[6] Similar arguments can also be applied to fractional fillings $\nu = 1/m$, where m is an odd number. For example, at $\nu = 1/3$, the ground state is a quantum Hall ferromagnet with electron spin polarization $\mathcal{P} = 1$.

Table 12.1. Properties of the three types of nuclei in GaAs. The natural abundance and the gyromagnetic ratio data are cited from [76]. A_H and b_N are based on the evaluation of $|u(0)|^2$ by Paget et al. [77]

	^{69}Ga	^{71}Ga	^{75}As
Spin quantum number I	3/2	3/2	3/2
Natural abundance x_n	60.108%	39.892%	100%
Reduced gyromagnetic ratio $\frac{1}{2\pi}\gamma_n$ (MHz/T)	10.2478	13.0208	7.3150
$\gamma_n\hbar/\mu_B$ ($\times 10^{-3}$)	0.732	0.930	0.523
$\gamma_n\hbar/k_B$ (mK/T)	0.492	0.625	0.351
$|u(0)|^2/v_0$ (10^{25} cm^{-3})	5.8	5.8	9.8
Hyperfine constant A_H (μeV)	38	49	46
Full polarization nuclear field b_N (T)	−1.37	−1.17	−2.76

Hyperfine Coupling

Apart from interacting with their neighbors, nuclear spins also interact with electrons in their surrounding via the hyperfine interaction (see Chaps. 1 and 11).

For GaAs/AlGaAs based 2D electron systems, the hyperfine interaction is essentially just the Fermi contact interaction between the nuclei and the s-type conduction band electrons. The corresponding Hamiltonian is written as

$$H_{HF} = A_H |\Phi(r)|^2 v_0 I \cdot S, \qquad (12.9)$$

with

$$A_H \cong \frac{4\mu_0}{3} \frac{\mu_B \gamma_n \hbar}{v_0} |u(0)|^2. \qquad (12.10)$$

Here A_H is the hyperfine coupling constant, μ_0 the magnetic constant, $|\Phi(r)|^2$ the amplitude of the envelope of the electron wave function which satisfies $\int |\Phi(r)|^2 d^3r = 1$, v_0 the volume of the crystal unit cell, and $|u(0)|^2$ the dimensionless Bloch amplitude of the electron wave function at the site of the nucleus with the normalization condition $\int |u(r)|^2 d^3r = v_0$ [78]. For s-electrons, $|u(0)|^2$ is quite large (on the order of $\sim 10^3$ for Ga and As) due to the sharp maxima in the electron density at the nuclear sites. Consequently, A_H is about 40–50 μeV (see Table 12.1).

The hyperfine Hamiltonian can also be expressed in terms of raising and lowering operators:

$$H_{HF} \propto A_H I \cdot S = \frac{A_H}{2}(I^+ S^- + I^- S^+) + A_H I_z S_z, \qquad (12.11)$$

where the first term is called the spin flip–flop term. It describes spin transfer between the electron and the nucleus while conserving total spin. The electron flips its spin while the nuclear spin is reversed in the opposite direction. This dynamic term is responsible for many processes including electron spin decoherence [78], nuclear spin relaxation [75, 78], and dynamic nuclear spin polarization [78]. When nuclear spins are driven out of thermal equilibrium, the flip–flop process may provide a channel for nuclear spins to equilibrate provided total energy can be conserved. In a magnetic field, the large mismatch in the Zeeman energies of electrons and nuclei may prevent flip–flop processes because energy conservation is hard to fulfill. In many systems, such as normal metals, the hyperfine flip–flop process plays a dominant role in nuclear spin relaxation. Conversely, if nonequilibrium electron spins are generated, the spin flip–flop process can dynamically polarize the nuclear spins. Many techniques, such as optical pumping [79], electron spin resonance (ESR) [80], electrical injection from ferromagnets [81], etc., exploit flip–flop processes to dynamically polarize the nuclear spin subsystem.

The static part of the hyperfine interaction, described by the second term in (12.11), can be regarded as a change in the Zeeman energy of electrons when the nuclei are spin polarized, and vice versa. For polarized nuclei, the change in the electron Zeeman energy can be quantified in terms of an additional effective nuclear magnetic field $\mathbf{B}_N$ acting on the electron spin. The contribution to the electron Zeeman energy can then be written as $g\mu_B \mathbf{S} \cdot \mathbf{B}_N$. This is manifested in ESR experiments as an Overhauser shift in the electron spin precession frequency $\Delta f_{\text{Ovh.}} = g\mu_B B_N / h$. Summing up contributions from all three types of nuclei in contact with the electron wave function, we obtain

$$\mathbf{B}_N = \sum_{i=1}^{3} b_{N,i} \frac{\langle \mathbf{I}_i \rangle}{I}, \tag{12.12}$$

with

$$b_{N,i} = \frac{4\mu_0}{3} \frac{I_i}{g} \gamma_{n,i} \hbar \rho_{n,i} |u_i(0)|^2 = \frac{A_H I x_{n,i}}{g\mu_B}. \tag{12.13}$$

Here, $\rho_{n,i} = x_{n,i}/v_0$, and $x_{n,i}$ is the natural abundance of the ith type of nuclei. Because of the contact type interaction and the reduced electron g-factor, B_N can be as large as 5.3 T in GaAs (see Table 12.1).

Similarly, polarized electrons act on the nuclear spins. The corresponding effective magnetic field for the nuclei equals

$$\mathbf{B}_e = b_e \langle \mathbf{S} \rangle, \tag{12.14}$$

with

$$b_e = -\frac{4\mu_0}{3} n_e \mu_B |u(0)|^2, \tag{12.15}$$

where n_e is the 3D electron density. For a 2D electron system confined in a quantum well or a heterostructure, it is more convenient to write b_e as

$$b_e^{2D} = -\frac{4\mu_0}{3} n_s \mu_B |u(0)|^2 |\phi(z)|^2, \tag{12.16}$$

where $\int |\phi(z)|^2 \, dz = 1$ and $\phi(z)$ is the 1D envelope of the electron wave function of the lowest sub-band of the potential well forming the 2D electron system. As a first order approximation, $|\phi(z)|^2 \sim 1/w$, and w is the width of the envelope of the electron wave function in the growth direction. Consequently, B_e can be increased by using narrower quantum wells. For a typical 2D electron system, $n_s \sim 10^{11}$ cm^{-2}, B_e is quite small (on the order of 10^{-3} T) even for full electron spin polarization. Nevertheless, B_e can be detected and is referred to as the Knight shift $K_s = \gamma_n B_e/(2\pi)$ in nuclear magnetic resonance (NMR) experiments. A measurement of the Knight shift has turned into a very powerful method for determining the electron spin polarization.

Nuclear Spin Relaxation in High Magnetic Fields

As mentioned earlier, there is a large mismatch of a factor of $\sim 10^3$ between the Zeeman energies of electrons and nuclei. In order to satisfy energy conservation during spin flip–flop processes, this difference in the Zeeman energy must be compensated for by for instance a change in the kinetic energy of the electrons. In a 3D normal metal, the energy conservation requirement is easily fulfilled, since the energy spectrum of both spin-up and spin-down electrons is continuous. As a result, nuclear spin relaxation can take place efficiently via the Fermi contact flip–flop mechanism. The spin relaxation time T_1 is given by [82]

$$T_1^{-1} = \frac{\pi}{\hbar} A_H^2 v_0^2 |\Phi(r)|^4 \int D_\uparrow(\varepsilon) D_\downarrow(\varepsilon) f(\varepsilon) \big[1 - f(\varepsilon)\big] d\varepsilon, \tag{12.17}$$

where $f(\varepsilon)$ is the Fermi–Dirac distribution function, and $D_\uparrow(\varepsilon)$ and $D_\downarrow(\varepsilon)$ are the density of states for spin-up and spin-down electrons, respectively. In the low temperature limit ($k_B T \ll \varepsilon_F$), (12.17) simplifies into

$$T_1^{-1} = \frac{\pi}{\hbar} A_H^2 v_0^2 |\Phi(r)|^4 D_\uparrow(\varepsilon_F) D_\downarrow(\varepsilon_F) k_B T, \tag{12.18}$$

where ε_F is the Fermi energy. Korringa [83] found that T_1^{-1} for metals with noninteracting electrons can be rewritten in a more convenient form

$$T_1^{-1} = \frac{4\pi}{\hbar} \left(\frac{\gamma_n}{\gamma_e}\right)^2 \left(\frac{K_s}{f_0}\right)^2 k_B T, \tag{12.19}$$

with $\gamma_e = e/m_e$ the gyromagnetic ratio of the electrons, and $f_0 = \frac{1}{2\pi}\gamma_n B$ the resonance frequency of the nuclei. Equation (12.19) is often referred to as the Korringa relation. For a 2D electron system, T_1^{-1} can be written in an analogous form

$$T_1^{-1} = 16\pi^3 \hbar \left(\frac{K_s^{\max}}{n_s}\right)^2 D_\uparrow(\varepsilon_F) D_\downarrow(\varepsilon_F) k_B T, \tag{12.20}$$

where $K_s^{\max}$ is the Knight shift for a fully spin polarized 2D electron system. The influence of the envelope of the electron wave function $\phi(z)$ is included in $K_s^{\max}$.

For a 2D electron system subjected to a strong magnetic field, it is no longer obvious that energy conservation for the flip–flop processes can be fulfilled through a change in the electron's kinetic energy, because of the discretization of the single particle energy spectrum into a ladder of spin split Landau levels. The nuclear spin and electron spin subsystems become well decoupled. A strong suppression of

Korringa-like spin relaxation results and long T_1 times are anticipated. Nevertheless, the longest T_1 in the quantum Hall regime reported so far are only on the order of 10^3 s. It suggests any of the following possibilities: (1) There exist nuclear spin relaxation mechanisms that do not rely on the hyperfine flip–flop scattering but rather on electron spin–orbit coupling [84]; (2) Hyperfine flip–flops take place but are assisted by other mechanisms that can compensate for the Zeeman energy mismatch [73, 85, 86]; (3) The contribution of nuclear spin diffusion is not separately determined in the measurements of the nuclear spin relaxation rate, which causes an overestimation of T_1^{-1}.

Nuclear Spin Diffusion

In a sample containing a 2D electron system, the active device volume where the electron wave function is non-zero only amounts to a nanometer sized region. Therefore, the number of nuclei affected by the electron system is negligible in comparison with the nuclei in the surrounding bulk. These few nuclear spins do however interact with the nuclei in the bulk via the magnetic dipole–dipole interaction. As described in Chaps. 1 and 11, the dipolar interaction between nuclei is responsible for nuclear spin diffusion. In case of the 2D electron system, only those nuclei inside or near the active device region are of interest. Nuclear spin diffusion is then important if there exists a non-equilibrium nuclear spin polarization in the active region. This nuclear spin polarization may get diluted by sharing it with the vastly larger number of nuclei in the surrounding bulk. Nuclear spin diffusion has been observed in many systems where a nuclear spin polarization was generated only locally [88, 89].

12.2 Experimental Techniques

The technique most frequently used to investigate nuclear spin phenomena is nuclear magnetic resonance (NMR). The sample is exposed to a steady magnetic field B_z. In addition, a magnetic field alternating in the x–y plane at a radio frequency (RF) f is applied. When the splitting of the nuclear spin levels for any of the nuclear species coincides with the incident RF radiation, i.e., $f = \gamma_n B_z/(2\pi)$, resonant absorption occurs and can be detected by the NMR electronics. The hyperfine interaction between the nuclei and the surrounding electrons may cause shifts in the resonance frequency from the value expected of the bare Zeeman splitting. These shifts provide valuable information about the electronic and chemical structure of the material at hand. The sophisticated pulse sequences, which have been developed during the past decades, can be exploited to measure the nuclear spin dynamics [90], as well as to manipulate the nuclear spins for quantum information processing [91, 92].

Unfortunately, traditional NMR techniques still lack the required sensitivity to detect the small number of nuclei interacting with the itinerant electrons of a single 2D layer.

One possible approach to improve the sensitivity of conventional NMR is to increase the nuclear spin polarization away from its equilibrium value. This will enhance the free induction decay signal in the RF coils. Dynamical nuclear polarization has been accomplished via electron spin resonance (ESR) [80], through optical pumping of spin polarized electrons [79, 93], or electrically by inducing spin flip–flop scattering between edge channels [94] or in the fractional quantum Hall regime where two phases with different electron spin configurations co-exist (i.e., at the $\nu = 2/3$ spin phase transition; see Sect. 12.3.4) [95–97]). In all these cases, it has been demonstrated that a large nuclear spin polarization can be generated. Still, an intrinsic problem to conventional NMR is that it probes not only those nuclei interacting with the 2D electron layer, but also the vastly larger number of nuclei of the surrounding bulk.

An alternative with excellent sensitivity to *selectively* probe only those nuclear spins interacting with the 2D electron layer relies on the effective Zeeman effect these nuclear spins exert via the hyperfine interaction. As discussed in Sect. 12.1, the nuclear field modifies the Zeeman energy of the electrons. A change in the Zeeman energy may leave strong signatures in the measured resistivity of the 2D electron system [94–100]. Examples include the resistance in the flanks of the $\nu = 1$ quantum Hall state [98, 100], near the spin-transition at fractional filling factor 2/3 [95–97], and when edge channel scattering between edge channels of opposite spin occurs [94, 99]. Though electrical detection is convenient and sensitive, it should be kept in mind that, in contrast with traditional NMR, it does not measure the nuclear spin polarization directly.

In traditional NMR, determining T_1 is straightforward: One simply needs to fit the data to the following exponential decay function:

$$\mathcal{P}_N(t) = \mathcal{P}_N^0 + \Delta\mathcal{P}_N \exp\left(-\frac{t}{T_1}\right), \tag{12.21}$$

where $\mathcal{P}_N(t)$ is the time evolution of nuclear spin polarization, and $\mathcal{P}_N^0$ the equilibrium nuclear spin polarization. For a measurement of the nuclear spin relaxation time T_1 using resistive detection, complications may arise because the change in resistance (ΔR) is not necessarily proportional to the change in nuclear spin polarization ($\Delta\mathcal{P}_N$).

Nuclear spins can also be detected with optical means, either by extracting the spin polarization from circular polarization resolved photoluminescence [79] or by Faraday/Kerr rotation [90, 101]. These methods have been demonstrated for III–V semiconductors at liquid-helium temperatures, but have not yet been extended to quantum Hall systems at dilution refrigerator temperatures. Also the developments in magnetic force resonance microscopy [102] may eventually provide a valuable tool to study nuclear spins in quantum Hall regime with high sensitivity and spatial resolution.

12.3 Nuclear Spin Phenomena in the Quantum Hall Regime

One of the marvellous aspects of quantum Hall physics is that a large variety of spin related electronic states can be accessed in a single sample. This gives us some unique opportunities to tailor the spin interactions between the electrons and the nuclei and to develop novel recipes or procedures to manipulate and detect nuclear spins. A measurement of the nuclear spin polarization in turn provides a very useful probe of the spin polarization of the electronic states. Here we review some of the recent work on the interaction of the 2D electron system with the nuclear spins in the quantum Hall regime.

12.3.1 The Role of Disorder

As pointed out previously, in the quantum Hall regime the large mismatch between the Zeeman energy of the nuclei and the electrons must be overcome in order for flip–flop processes to occur. The Landau quantization of the kinetic energy of the electrons prevents this mismatch to be compensated for by a corresponding change in the electron's kinetic energy. As a consequence, nuclear spin relaxation via the contact hyperfine interaction would normally be entirely suppressed in an ideal 2D electron system, in which disorder is absent and the energy spectrum is discrete [103]. In real samples however, the electrostatic potential from the randomly distributed impurities broadens the Landau levels. In particular, at low magnetic fields, electron states with different spin indices may overlap and allow for Korringa like spin relaxation.

The first measurement of the nuclear spin relaxation time T_1 in a 2D electron system was reported by Berg and co-workers [73, 80]. In their experiments, the electron spin subsystem is driven out of equilibrium by illuminating the sample with microwave radiation whose frequency matches the energy spacing between the Zeeman split Landau levels, i.e., $\hbar\omega = g\mu_B(B_{tot} + B_N)$. Here, B_N is included to account for a possible non-zero nuclear spin polarization due to low temperatures or dynamic nuclear spin polarization. The excited electrons may relax back. The reversal of their spins to the original state can proceed via flip–flop processes with nuclei if energy conservation can be fulfilled for instance in the presence of disorder broadening. As a result the nuclear spin subsystem becomes dynamically polarized. A nuclear spin polarization as high as $\mathcal{P}_N \simeq -8\%$ has been achieved in this manner. This polarization corresponds to an effective nuclear field B_N of more than 0.4 T. It acts back on the electron spin subsystem and shifts the electron spin resonance frequency.

The resonant absorption of the incident microwave radiation heats up the electronic system. The increased temperature [104], as well as any changes in the Zeeman energy due to dynamic nuclear spin polarization, influence the resistance. This is the basis for resistively detected ESR. Close to odd-integer fillings, the change in the resistance can be understood at a qualitative level from the Arrhenius type T-dependence. These ESR experiments obtain a g-factor close to the value of bulk GaAs ($g = -0.44$). ESR involves no momentum transfer and hence probes the properties of spin-flip excitations in the limit $q \to 0$, where their energy is immune to Coulomb interaction phenomena. It therefore simply equals the bare Zeeman energy.

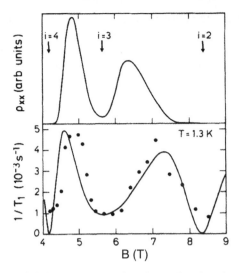

Fig. 12.5. Longitudinal resistivity ρ_{xx} (*top*) and nuclear spin relaxation rate $1/T_1$ (*bottom*) as a function of the applied magnetic field. The integer filling factors have been marked in *the top panel*. Relaxation rates in *the bottom panel* were obtained from resistively detected ESR measurements. The line is the theoretical curve describing the disorder mediated nuclear spin relaxation in the quantum Hall regime. Reprinted from [73]

In contrast, thermally activated transport studies may yield a strong filling factor dependent enhancement for states with large electron spin polarization. Transport requires charge separation and hence reveals the spin flip excitation energy in the limit $q \to \infty$ instead.

From the shift of the ESR-frequency as a function of time, it is possible to extract the time dependence of B_N (which is proportional to $\mathcal{P}_N$) and to determine the nuclear spin relaxation time T_1. Figure 12.5 summarizes the results obtained by Berg et al. [73]. The nuclear spin relaxation rate T_1^{-1} is plotted as a function of filling factor. The filling factor dependence is very pronounced and exhibits some similarity to the filling factor dependence of the longitudinal resistivity ρ_{xx}. These data suggest some correlation between the electronic states, which develop at different filling factors, and the interaction strength between the electron and the nuclear spins. The observed filling factor dependence of the relaxation rate can be explained at least at a qualitative level with a theoretical model which considers both disorder mediated flip–flop processes and the exchange enhanced Zeeman splitting [73, 105, 106]. As an illustration, consider integer filling $\nu = 3$. At this filling the relaxation rate reaches a minimum and takes on a value of $\sim 10^{-3}$ s^{-1}. The overlap between the disorder-broadened Landau levels ($1 \uparrow$) and ($1 \downarrow$) indeed reaches a minimum at filling 3 when the Fermi level lies in the middle of the Zeeman gap. This minimum would however not be observable for the disorder strength encountered in the experiment of Berg *et al.* if exchange enhancement of the Zeeman splitting were absent.

12.3.2 Edge Channel Scattering

In the previous section, we learned that the relaxation of a non-equilibrium electron spin population, which has been created by ESR, can dynamically polarize nuclear spins via the hyperfine flip–flop process. However, even without incident microwave radiation, it is possible to dynamically polarize the nuclei. Here an example will be discussed where only a dc-bias voltage generates nuclear spin polarization. It is based on electron spin flip scattering between spin-resolved edge channels [89, 94].

The device, which was used in [94], is depicted in Fig. 12.6. It consists of a rectangular shaped mesa of the 2D electron system, three ohmic contacts (denoted as electrodes 1, 2, and 3) and two quantum point contacts defined by split-gates A–B and A–C, respectively. The filling factor in the bulk is set to 2 by applying an appropriate perpendicular magnetic field B, so that in the bulk of the sample the two lowest Zeeman split Landau levels (0 ↑) and (0 ↓) are completely filled, and two edge channels with opposite spins run along the boundaries of the sample. The bias voltages applied to split gates A, B, and C are tuned such that the outer (spin-up) edge channel is fully transmitted by both quantum point contacts, while the inner (spin-down) edge channel is completely reflected.

Now let us consider the case where electrode 1 is biased with voltage V, whereas electrodes 2 and 3 are grounded. Electrode 3 is grounded via a transimpedance amplifier. It allows to measure the net current flowing through this contact. The net outgoing current from electrode 1 is carried by the outer edge channel with spin-up due to the filtering effect of quantum point contact A–B and this outer edge channel is filled up to the applied electrochemical potential $-eV$. It runs along the boundary of split gate A. Both edge channels emanating from electrode 3 are at zero bias voltage. The inner channel with downward spin is reflected at quantum point contact A–B and runs alongside the outer edge channel with opposite spin in the corner defined by split gate A. If no scattering occurs between the inner and outer channel running in the corner of split gate A, current originating from electrode 1 will be drained entirely through electrode 2 and no current is detected at electrode 3. If scattering does occur, a current will be detected at electrode 3. Scattering from one edge channel into the other edge channel requires a reversal of the electron spin. The spin reversal may take place with the assistance of nuclear spins and dynamic nuclear spin polarization is expected. The resulting nuclear spin polarization is confined locally to the region where the two edge channels are in close proximity. I–V characteristics indeed confirm this scenario. An example is shown at the bottom of Fig. 12.6 and is discussed below in some detail.

As mentioned in Sect. 12.1.1, edge channels acquire a finite width due to screening. For each edge channel a compressible region appears and two adjacent compressible regions are separated by an incompressible region. The situation at equilibrium, i.e., at zero bias voltage, is depicted in level diagram (a). Both compressible regions have the same electrochemical potential. The situation is very different if edge channels are filled up to different electrochemical potentials. In particular, an asymmetry develops between forward bias and reverse bias. For forward bias ($V > 0$), the electrochemical potential of the outer edge channel, $\varphi = -eV$, is lowered. Conse-

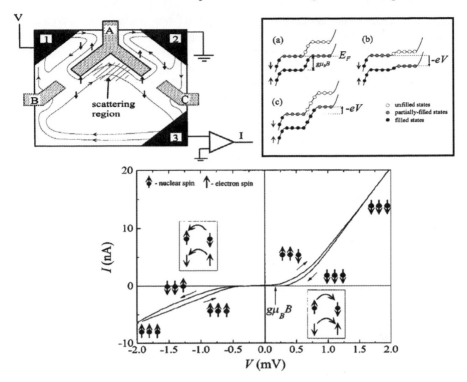

Fig. 12.6. *Upper left panel*: An edge channel spin diode. *Upper right panel*: energy level diagram for no bias (**a**), positive (**b**) and negative bias voltage V (**c**). *Bottom panel*: Current measured at electrode 3 as a function of the applied bias voltage. At zero bias voltage, an incompressible strip (*sloped region with filled states*) separates the two partially filled compressible regions (*flat regions with grey filled dots at the Fermi level E_F*) that belong to the levels with opposite spin orientation. For positive bias voltage (**b** in *right panel* at *the top*), the incompressible strip narrows and a large increase in the current is observed when eV exceeds the Zeeman energy. At reverse bias, the width of the incompressible strip increases and interedge scattering is initially suppressed. At sufficiently high voltage, however, electron tunneling between the two compressible regions takes place as the spatial separation between filled and empty states shrinks. Reprinted from [94]

quently, the incompressible strip becomes narrower. When $|\varphi|$ exceeds the Zeeman energy $E_Z = g^* \mu_B B_{tot}$, the incompressible strip is eliminated. States in the upper level with a strong spatial overlap with partially filled states in the lower level get filled. The overlap enhances the probability for interedge channel scattering. It results in a large increase in current. Conversely, for reverse bias voltages ($V < 0$), the width of the incompressible strip increases and edge channel scattering is suppressed for small reverse bias. At large reverse bias voltages, however, tunneling to empty states in the upper level may take place, since the spatial gap between states with the same energy, but belonging to different levels, shrinks. The energy level diagrams to some extent remind of a conventional p–n junction. The resulting I–V characteristic in Fig. 12.6 indeed resembles that of a p–n junction.[7] It is nonlinear and asymmetric with the polarity of the bias voltage. When recording the I–V characteristic, one can sweep the bias voltage either up or down. Comparing the data acquired during both sweep directions reveals strong hysteretic behavior. The appearance of hysteresis is an important piece of evidence for the involvement of nuclear spins.

The scattering of electrons between the two edge channels requires spin reversal. The hyperfine flip–flop process may help in accomplishing spin reversal. Nuclei become polarized and act back on the electron spin subsystem by modifying the electron Zeeman energy, $E_Z = g^* \mu_B (B_{tot} + B_N)$. The threshold voltage for conduction in forward bias direction will move to higher voltages for positive B_N and to smaller voltages otherwise. The resulting changes in the I–V characteristic may be exploited to detect the degree of nuclear spin polarization $\mathcal{P}_N$. For forward bias, scattering of the electrons from the (spin-down) inner channel into the (spin-up) outer channel causes a positive B_N (corresponding to $\mathcal{P}_N < 0$). The larger the forward bias, the higher the spin scattering rate will be, and thus a larger B_N is expected. At a certain bias voltage, the nuclear field B_N will be larger, if this bias voltage is approached by sweeping from positive to negative bias voltages than if the sweep started from negative values. This is true unless the voltage is swept extremely slowly so that either the nuclear spin polarization saturates or a stationary state is reached for every bias voltage. Because of the slow time scales on which the interactions with the nuclei occur, hysteresis is commonly observed. The different current observed during upward and downward sweeps can therefore be attributed to the difference in the Zeeman gap caused by B_N. Similar arguments can also to be applied to explain the hysteresis for reverse bias. Dixon et al. estimated a $\mathcal{P}_N$ as high as $\sim$85% (corresponding to $B_N \sim 4\,\mathrm{T}$) from the changes in the I–V characteristics in Fig. 12.6.

Würtz et al. [99] also studied the nuclear–electron spin interaction in a device with spin-resolved edge channels. A different device geometry enabled them not only to generate a large degree of nuclear spin polarization by creating a voltage difference between adjacent edge channels as Dixon and others did, but they were also able to detect the reverse effect. A voltage output appeared due to a local nonequilibrium in the nuclear spin polarization initially created by dynamical nuclear polarization. The device arrangement was coined a hyperfine battery. According to their model, the output voltage of the hyperfine battery is directly linked to the degree of nuclear spin polarization: $V_{out} = -g^* \mu_B B_N / e$. Hence, recording this voltage represents an alternative method to measure the local nuclear spin polarization. When optimizing the pump current, they were able to obtain an output voltage of $V_{out} = 0.32\,\mathrm{meV}$, from which they deduced an effective nuclear field of $B_N \approx 5.2\,\mathrm{T}$, or nearly full nuclear spin polarization.

Dynamic nuclear polarization induced by interedge channel scattering and the ability to detect the change in the nuclear spin polarization in edge channel transport

---

[7] An earlier version of the spin diode in the quantum Hall regime was reported by Kane et al. [107]. In that device, electron spin scattering between regions of different filling factors ($\nu > 1$ and $\nu < 1$) was held responsible for dynamic nuclear spin polarization and the accompanied observation of hysteresis in the I–V curves.

devices were exploited by Machida et al. to demonstrate coherence of the nuclear spins [108]. In their experiments, pulsed NMR sequences were applied locally with a strip-line geometry in the region where edge channel spin scattering occurred. The nuclear spin decoherence time T_2 was estimated to be 80 μs using a spin-echo measurement, in agreement with theory (see Chap. 1). Machida et al. [109] also extended the work to spin-resolved edge channels in the fractional quantum Hall regime. Based on the polarity of the nuclear polarization, the electron spin polarization of different fractional quantum Hall edge channels was inferred.

12.3.3 Skyrmions

At filling factor $\nu = 1$, the exchange interaction between electrons leads to full electron spin polarization even in the limit of vanishing Zeeman energy [18]. All electrons reside in the lowest spin-split LL (0 ↑) and their spins are aligned parallel to the magnetic field. This ground state is often referred to as a quantum Hall ferromagnet. In the single particle picture, an extra electron added to the system must reside in the next spin-split Landau level (0 ↓). A large exchange penalty has to be paid for having a single electron with opposite spin, since all available electron states are occupied in the (0 ↑) Landau level. The interacting 2D system is capable of lowering the required energy through a more complex rearrangement of the electrons among these two levels. A vortex-like spin texture is formed instead. The spin at its center is opposite to the applied magnetic field and gradually reverses towards the perimeter [18, 110–112]. This type of spin texture, which still carries a single unit of electron charge, $-e$, but may involve flipping more than one electron spin, is called a *skyrmion* due to its mathematical connection to a topological soliton solution of the Skyrme Lagrangian, which describes nuclear matter [113]. The energy required to form a skyrmion can be about half of the exchange enhanced Zeeman energy required to flip a single electron spin. Analogously, anti-skyrmions carrying a charge $+e$ also exist because of particle-hole symmetry.

The size of the skyrmions is determined by the competition between the Coulomb energy and the Zeeman energy. This competition is often characterized by the ratio $\eta = E_Z/E_C$. For large η, the Zeeman energy is dominant and the skyrmion reduces to a single spin flip as in the independent electron picture. In the opposite limit, i.e., $\eta \rightarrow 0$, the size of the skyrmions becomes infinite. In a typical GaAs based 2D electron system, the skyrmions spread over a limited range only, and a finite but small number of spins ($s > 1$) are reversed. The first evidence for the existence of skyrmions was provided by Barrett et al. using an optically pumped NMR technique [19]. These experiments were performed on multiple quantum well samples in order to be sufficiently sensitive to the physics of the 2D electron layers. The experiment is based on (12.14). The Knight shift of the NMR resonances can be used to deduce the spin polarization of the 2D electron system. As shown in Fig. 12.7, the experimentally observed Knight shift decreases much more rapidly as the filling factor deviates from $\nu = 1$ than expected for the independent electron picture ($s = 1$). Fitting the data near $\nu = 1$ yields $s = 3.6$, i.e., for each added electron charge, 3.6 spins are flipped instead of just one, consistent with theoretical calculations. In addition,

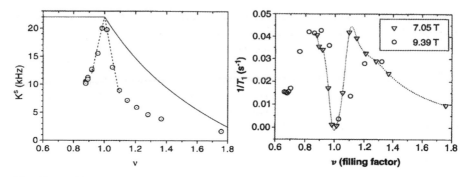

Fig. 12.7. Filling factor dependence of the Knight shift (*left*, *circles*) and the nuclear spin relaxation rate T_1^{-1} (*right*, *circles* and *triangles*) obtained with an optically pumped NMR technique. In the left panel, the solid line is calculated based on the independent electron picture ($s = 1$), and the dashed line corresponds to $s = 3.6$, i.e., the reversal of 3.6 spins per electron charge added or extracted from the $\nu = 1$ quantum Hall ground state. Reprinted from [19, 93]

temperature dependent transport studies in tilted magnetic fields [114], and polarization resolved optical absorption measurements [115] further support the existence of skyrmionic excitations in the 2D electron system near $\nu = 1$.

Besides their static Zeeman effect on nuclear spins, skyrmions also significantly alter the physics of the nuclear spin relaxation. As one moves away from filling $\nu = 1$, a skyrmion or anti-skyrmion forms for each electron charge added or removed. The density of skyrmions (or anti-skyrmions) is proportional to $|1 - \nu|$. At sufficiently large density and low temperature, skyrmions were predicted to organize into a crystal at low temperatures [86, 116, 117]. Isolated skyrmions possess a rotational symmetry associated with in-plane spin components. When forming a crystal, this symmetry is broken. It leads to a gapless Goldstone spin-wave mode with a linear dispersion in the long wavelength limit. Without this gapless mode, the hyperfine spin flip–flop process would remain suppressed, since the energy required to create a single skyrmion is still three orders of magnitude larger than the energy involved in flipping a nuclear spin. The gapless Goldstone mode of a Skyrme crystal may greatly enhance the nuclear spin relaxation as it overcomes the large energy mismatch for flipping an electron spin together with a nuclear spin. The resulting relaxation rate T_1^{-1} was predicted to have a Korringa type temperature dependence and should follow the same filling factor dependence as the skyrmion density: $T_1^{-1} \propto |1 - \nu|$ [86]. This filling factor dependence seems to agree qualitatively with the results of the optically pumped NMR measurements (up to $|1 - \nu| = 0.1$, see Fig. 12.7) [93]. Similar results were also obtained from transport measurements, in which the nuclear spin relaxation rate was detected resistively by using the properties of the spin transition near $\nu = 2/3$ [96, 97] (see Sect. 12.3.4 for further details). This is a beautiful example of how investigations of the nuclear–electron spin interaction may help in identifying low energy or gapless collective excitations in the 2D electron system.

Despite this qualitative agreement some experimental controversy still exists. The Skyrme crystal is likely to melt into a liquid state when either the density of the skyrmions or the temperature is raised sufficiently [117]. An enormous peak in the specific heat at a temperature of 42 mK was attributed to this solid-to-liquid transition [86, 118, 119]. The optically pumped NMR experiments were, however, carried out at $T = 1.5$–4.2 K [93]. In this temperature range, melting should have occurred and the skyrmions are anticipated to be in the liquid state. Yet, the relaxation rate indicates the presence of the gapless Goldstone mode. Presumably, the gapless spin-wave mode is still present as an over-damped mode [117]. The temperature dependence of T_1 also needs further investigation. Some have reported Korringa-like behavior [120], while other work has asserted that the temperature dependence does not obey the Korringa law [100]. New techniques that can measure T_1 at ultra-low temperatures, would be particularly helpful to clarify the issues regarding the Skyrme crystallization and melting, or other phase transitions that skyrmions may undergo.

12.3.4 Nuclear–Electron Spin Interactions at $v = 2/3$

Ising Ferromagnetism and Domains

Among the many fractional filling factors exhibiting a spin transition, filling factor $v = 2/3$ has attracted the largest attention. Within the composite fermion description of the fractional quantum Hall effect, $v = 2/3$ corresponds to having two filled spin split composite fermion Landau levels. The energy spacing between two adjacent Landau levels with the same spin is determined by the Coulomb interaction and hence follows $E_C \propto B^{1/2}$. The Zeeman splitting however is proportional with B_{tot}. As illustrated in Fig. 12.4, this difference in the functional dependence of E_C and E_Z may give rise to a level crossing of the $(0 \downarrow)$ and $(1 \uparrow)$ levels at the transition field B_{tr}. Depending on the ratio $\eta = E_Z/E_C$, two different ground states are conceivable: a spin-unpolarized ground state with the $(0 \uparrow)$ and $(0 \downarrow)$ levels completely filled when $\eta < \eta_{tr} = \eta(B_{tr})$, and a spin-polarized ground state when η exceeds η_{tr} for which the $(0 \uparrow)$ and $(1 \uparrow)$ levels are occupied.

Experimentally, η can be varied either by tilting the magnetic field or by changing the electron density while keeping the filling factor fixed. By altering η, the energy gap of the $v = 2/3$ fractional quantum Hall-state can be tuned. Figure 12.4 shows that the energy gap for the spin unpolarized state decreases with increasing η, while for the spin polarized state the gap grows with increasing η. The phase transition occurs at $\eta = \eta_{tr}$ when the gap vanishes. The fractional quantum Hall effect disappears. The longitudinal resistance acquires a finite, non-zero value and the Hall resistance is no longer quantized to the 2/3 plateau value. This has been confirmed in numerous transport experiments including thermal activation studies [61, 62, 65, 97, 121, 122]. An example is depicted in Fig. 12.8(a) [65]. It shows a plot of R_{xx} in the (n_s, v)-plane. Following a line of constant filling factor $v = 2/3$ from low to high density, the system first condenses in the spin unpolarized ground state $(\uparrow\downarrow)$, then shows non-zero resistance, and finally the polarized ground state $(\uparrow\uparrow)$ develops. This reentrant fractional quantum Hall behavior can also be clearly seen in panel (b)

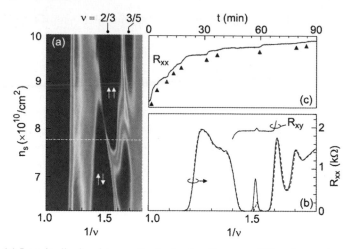

Fig. 12.8. (**a**) Longitudinal resistance R_{xx} in the density versus filling factor plane. The color scale corresponds to 0–2.5 kΩ. A non-zero R_{xx} at filling 2/3 signals the phase transition. The two phases with spin polarization $\mathcal{P} = 0$ and 1 are labeled with ($\uparrow\downarrow$) and ($\uparrow\uparrow$), respectively. (**b**) Hysteresis in R_{xx} for $n_s \approx 7.8 \times 10^{10}$ cm^{-2} (*dashed line sweep up; solid line sweep down*). The Hall resistance R_{xy} is shown for the sweep down only. (**c**) Time dependence of R_{xx} after interrupting the field sweep. Reprinted from [65]

where a cross-section through the data at fixed carrier density has been plotted. This graph also includes the Hall resistance. It deviates from its plateau value at the transition.

In [65], it was pointed out that the transport quantities in the vicinity of the phase transition show hysteresis, as well as time-dependent behavior. An example of hysteresis is shown in Fig. 12.8(b). The time dependent behavior is illustrated in panel (c). The time dependence is logarithmic and exhibits a number of sudden jumps reminiscent of the Barkhausen effect in conventional ferromagnets. These features can indeed be understood when describing the phase transition in terms of Ising ferromagnetism. In the integer quantum Hall regime, Jungwirth and co-workers [69] investigated the physics for the analogous problem when two electron Landau levels with different orbital indices and spins are brought into degeneracy. It was shown that the quantum ferromagnetism language provides a proper description of such a system. The transition is of first order nature and is accompanied by domain formation. Experimentally, transitions at $v = 2$ and 4 [123] in double layers systems were reported and successfully discussed within this framework. Even though little theoretical work has been performed in the fractional quantum Hall regime [124], it is believed that the basic conclusions remain valid in this regime [63, 65, 69]. Hysteresis as well as the Barkhausen jumps can be accounted for when the system breaks up into domains of different spin polarization.

Direct evidence for the co-existence of two types of domains was obtained from resistively detected NMR experiments combined with conventional NMR [125]. The details of how nuclear magnetic resonance can be detected in the resistivity of the

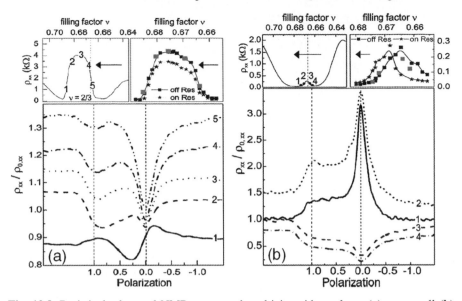

Fig. 12.9. Resistively detected NMR spectra when driving either a large (**a**) or a small (**b**) current through the sample. *The upper left panels* in (**a**) and (**b**) show the longitudinal resistance as a function of filling factor when a high (**a**) or small (**b**) current is imposed through the sample. *The arrows* indicate the direction of the filling factor sweep. The numbers indicate the positions where resistively detected NMR was performed. These numbers also serve as labels for the corresponding NMR spectra plotted in the bottom graphs. The abscissae of the NMR spectra do not show the NMR frequency itself, but rather the degree of electron spin polarization. It was possible to determine the degree of electron spin polarization by calibrating the NMR frequency in the absence of electron spin polarization (signal in conventional NMR from the GaAs bulk crystal) and for full electron spin polarization at filling $\nu = 1/2$ (with resistively detected NMR). Two resonance lines corresponding to $\mathcal{P} = 0$ and 1 are clearly resolved near filling factor 2/3 for both low and high sample current. The panels at the top on the right of (**a**) and (**b**) compare the ρ_{xx} value when the RF frequency is on and off-resonance. Reprinted from [125]

sample will be described in Sect. 12.3.5. Conventional NMR on the GaAs substrate was used to calibrate the resonance frequency in the absence of any electron spin polarization. Subsequently, the maximum Knight shift for full spin polarization of the 2D electron system was determined with resistively detected NMR by performing the experiment at a filling factor for which the degree of spin polarization is well understood. The authors choose filling factor $\nu = 1/2$ to calibrate the maximum Knight shift, where at sufficiently high values of the magnetic field or density, a sea of composite fermions with fully aligned spins forms (see Sect. 12.3.6). Finally, NMR spectra were recorded near filling factor 2/3 at densities where the spin transition was observed in transport. These NMR spectra are shown in Fig. 12.9. Two resonance lines were observed. Their frequencies correspond to the two extremes $\mathcal{P} = 0$ and 1. It represents unequivocal spectroscopic evidence for the co-existence of unpolarized and fully spin polarized domains at $\nu \sim 2/3$.

12.3.5 Resistively Detected NMR at $v = 2/3$

As seen in Fig. 12.9, it is possible to exploit the properties of the $v = 2/3$ spin transition for detecting the nuclear magnetic resonance in a straightforward resistance measurement. At temperatures below 250 mK, the thermal nuclear spin polarization $\mathcal{P}_N$ is no longer negligible. The effective nuclear field relocates the position of the phase transition, because it alters the ratio $\eta = E_Z/E_C$. The electron Zeeman energy is $E_Z = g^*\mu_B(B_{tot} + B_N)$, while the Coulomb energy is unaffected by B_N. For small B_N, the shift in the transition field, ΔB_{tr} (with $\eta(B_{tr}) = \eta_{tr}$), is approximately $-2B_N$.[8] Cooling the sample, increases $\mathcal{P}_N$ and hence shifts the transition to higher values of the applied magnetic field (for fixed v). Conversely, exposure of the sample to RF radiation in resonance with any of the nuclear spin precession frequencies, scrambles the nuclear spin polarization and shifts the phase transition in the opposite direction. Both cases have been observed in experiment [125, 126]. The peaked behavior of the resistance near the transition allows resistive detection of the NMR. If the incident RF radiation is on-resonance, the phase transition is relocated. When monitoring the resistance at fixed field and density near the initial location of the transition, the resistance will drop as the spin transition is relocated to lower values of the magnetic field. Under off-resonance conditions, the transition remains at the same density or field and the resistance remains unaltered. A number of NMR experiments relying on this resistive detection scheme have been reported [65, 95, 97, 125] and have demonstrated a response for the ^{69}Ga, ^{71}Ga, or ^{75}As nuclei. The ^{27}Al nuclei, which reside in the barrier confining the 2D electron system, have not been detected, presumably due to the small overlap of the electron wave function with the barrier region.

Current Induced Nuclear Spin Polarization

Transport through the landscape of domains with different spin polarization near the spin transition at filling $v = 2/3$ poses an intriguing problem. It has only been addressed at a qualitative, hand-waving level. Disorder promotes the 2D electron system to break up into domains with different spin polarization. Naively, one may anticipate that the current carrying quasi-particles are forced into transiting the two types of domains. A transition from one domain to the other requires the reversal of the electron spin. In the vicinity of the level crossing, the energy involved for spin reversal may be small. Provided spin–orbit coupling and acoustic phonon emission do not dominate, spin reversal may take place with the help of the hyperfine interaction and a flop of a nuclear spin. Current flow may then polarize the nuclear spins in a dynamic fashion. Presumably the nuclear spin polarization is initially confined

---

[8] The composite fermion Landau levels $(0 \downarrow)$ and $(1 \uparrow)$ cross each other when $B_{tot} = B_{tr}$ with B_{tr} satisfying $\alpha B^{1/2} = \beta(B_{tot} + B_N)$. Here B and B_{tot} are the perpendicular and total magnetic field, respectively. It follows that $B_{tr} = \frac{1}{2}(B_{tr}^0 - 2B_N + [(B_{tr}^0)^2 - 4B_N B_{tr}^0]^{1/2})$, where $B_{tr}^0 = (\alpha/\beta)^2 \cos\theta$. For $B_N \ll B_{tr}^0$, $\Delta B_{tr} = B_{tr}(B_N) - B_{tr}^0 \simeq -2B_N$. θ is the angle between B_{tot} and the sample surface normal.

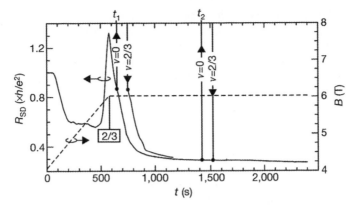

Fig. 12.10. Time dependence of the resistance (R_{SD}) near filling factor $\nu = 2/3$ after a field sweep in a two-terminal device. The magnetic field sweep starts from $\nu = 1$ and is interrupted at $\nu = 2/3$. R_{SD} continues to change. The experiment is then repeated, but at times t_1 and t_2 all charge carriers are removed for 90 s by applying a suitable gate voltage. When the original carrier density is restored, the resistance continues its descent from nearly the same value as if the carrier depletion had never taken place. Reprinted from [96]

to the domain boundaries. The polarized nuclear spins act back on the electron spin subsystem. The ratio η drifts away as the nuclear spins are dynamically polarized. First indications for the importance of dynamic nuclear spin polarization near the spin transition were obtained by Kronmüller et al. [127]. These authors observed a large enhancement of the longitudinal resistance peak near $\nu = 2/3$ when applying a large current and using a low magnetic field sweep rate. The resistance peak was accompanied by hysteresis and was time dependent on a slow time scale typical for nuclear spin interactions. Subsequently, Kraus et al. [126] found that these features remain observable even at $T = 250$ mK, at which the thermal nuclear spin polarization can be ignored. Hashimoto et al. [97] also studied the influence of current on the transport near $\nu \sim 2/3$, and also observed a similar enhancement in R_{xx}. The enhancement of R_{xx} was attributed to local dynamic nuclear spin polarization at the domain walls following the scenario described above. The resulting *inhomogeneously* polarized nuclear spins act as an additional source of disorder [125, 126]. Since the current induced local nuclear spin polarization modifies the local value of η, the domain pattern may change as the nuclear polarization builds up. A microscopic picture based on transport theory has however not been developed.

Storage Capability of Nuclear Spins

Carrier depletion experiments carried out in [96] and [97] have demonstrated that the local nuclear spin polarization may store the domain configuration. Figure 12.10 shows an example. The resistance of the sample is plotted as a function of time. Initially, the sample is at filling factor $\nu = 1$. The magnetic field is swept up to filling factor 2/3. Even though the field sweep is interrupted at filling 2/3, the resistance continues to change. The same experiment is then repeated, but at times t_1 and t_2 all charge carriers are removed from the sample with a top gate. After a time interval of 90 s, the original electron density is restored. The resistance returns to almost exactly the same value as before depletion and then continues its descent as if depletion had never taken place. Only the nuclear spins can be invoked as the storage medium for the domain configuration in the absence of charge carriers. During depletion, electron mediated nuclear spin relaxation is interrupted because conduction electrons are absent. This is a remarkable illustration of the storage capability of nuclear spins. When the sample is refilled with electrons, the local nuclear field *restores* the domain morphology, which existed prior to depletion.

Nuclear Magnetometry Based on the $\nu \sim 2/3$ Spin Transition

Techniques to study the nuclear spin dynamics comprise two crucial ingredients: a scheme to detect the degree of nuclear spin polarization and an elegant way of disturbing the nuclear spin subsystem reproducibly from its equilibrium or stationary state so as to elicit a time-dependent response. The subsequent recovery as a function of time discloses the sought-after information. The transport properties at the $\nu \sim 2/3$ spin transition lend themselves for accomplishing both tasks. A gradual change in the polarization of the nuclear spins relocates the 2/3 spin transition. This shift leaves a clear signature in the resistance when choosing a fixed working point near the transition where a small change in E_Z/E_C induces a dramatic resistance variation. Hence, a straightforward resistance measurement reveals information about gradual changes in the degree of nuclear spin polarization.

The current induced nuclear spin polarization discussed above can be used to generate a nonequilibrium nuclear spin population, whose recovery can subsequently be monitored. Some examples of how to combine these effects into a powerful method to study the nuclear spin relaxation are discussed below.

Example 1: Filling Factor Dependence of the Nuclear Spin Relaxation Rate

The nuclear spin relaxation rate as a function of filling factor was investigated across a large filling range in [96] and [97]. Figure 12.11 summarizes the results of experiments performed in [97]. By imposing a current at the $\nu \sim 2/3$ spin transition, the nuclear spin subsystem was polarized until R_{xx} reached saturation. Subsequently, the filling factor was set to a different value ν_{temp} by varying the electron density with the help of a gate. This filling factor was maintained for some time τ. The original filling factor ν close to 2/3 was then restored, and the resulting change in the resistance, ΔR_{xx}, is measured. The time constant T_r extracted from fitting ΔR_{xx} as a function of τ to an exponential decay function $\Delta R_{xx}(\tau) = \Delta R_{xx}^0 [1 - \exp(-\tau/T_r)]$ is then taken *equal* to the nuclear spin relaxation time T_1. The relaxation rate obtained in this manner is shown in Fig. 12.11 as a function of ν_{temp}. The filling factor dependence of T_1^{-1} is very pronounced and qualitatively in agreement with the results from optically pumped NMR experiments [93]. At $\nu = 1$, T_1^{-1} reaches a minimum. This is expected from the suppression of the hyperfine flip–flop process due to the presence

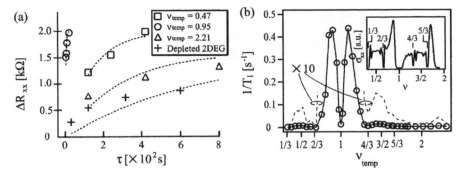

Fig. 12.11. Filling factor dependence of the spin relaxation rate T_1^{-1} obtained as described in the text. In this experiment, nuclear spins are first dynamically polarized by driving a large current through the sample at $v = 0.69$. Subsequently, the nuclear spin subsystem is allowed to relax at filling factor v_{temp} for a time τ before the original filling factor $v = 0.69$ is restored. The change in R_{xx} at $v = 0.69$ after the excursion to filling v_{temp} for a time τ (ΔR_{xx}) is measured. Some examples for four different v_{temp} are shown in the left panel. The time dependence $\Delta R_{xx}(\tau)$ reflects the nuclear spin relaxation. Fitting $\Delta R_{xx}(\tau)$ to an exponential decay function $\Delta R_{xx}^0 [1 - \exp(-\tau/T_r)]$ yields the time constant T_r (*left panel*: symbols are experimental data points, while *dotted lines* are the fits). The nuclear spin relaxation time T_1 is assumed equal to T_r and is plotted in *the right panel* as a function of v_{temp}. Reprinted from [97]

of a large energy gap for electron spin reversal. The two maxima, which appear in the vicinity of—but away from—$v = 1$, can be attributed to the gapless low energy excitations of the skyrmion crystal, as discussed in Sect. 12.3.3. Similar results were also reported in [96].

Example 2: Suppression of Skyrmion Enhanced Nuclear Spin Relaxation

Near filling 1, the exchange interaction favors the formation of skyrmion spin textures over single electron spin flips. The size of a skyrmion depends on the ratio $\eta = E_Z/E_C$ [111]. The Coulomb interaction attempts to spread the charge of a skyrmion across an area as large as possible, but the Zeeman energy prefers to keep the area small in order to reduce the number of spins flipped. In the limit of vanishing Zeeman energy ($\eta \to 0$), the skyrmion size diverges and an infinite number of spins is flipped. As η increases, skyrmions shrink in size and fewer number of spins are reversed. In the limit of large Zeeman energy (large η), a single electron spin flip is energetically more favorable. It is anticipated that the gapless Goldstone mode associated with the Skyrme crystal, which forms due to interactions among the skyrmions, will then disappear. As a result, the nuclear spin relaxation facilitated by this gapless mode should be suppressed [86]. This has been confirmed experimentally in [96]. The experiment is briefly described here.

The nuclear spin relaxation in this experiment is detected by measuring the resistance at $v = 0.65$. This resistance will be denoted as $R_{0.65}$. η is varied by tilting the sample, so the magnetic field is no longer perpendicular to the sample surface. The

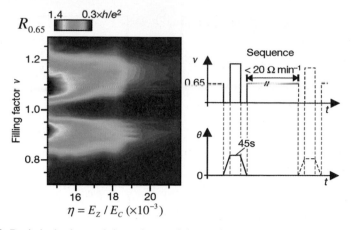

Fig. 12.12. Resistively detected dependence of the nuclear spin relaxation on η and ν. Here $R_{0.65}$ is the resistance at filling factor $\nu = 0.65$ of a two-terminal device. The ratio $\eta = E_Z/E_C$ is changed by tilting the sample in situ to an angle θ with respect to the axis of a superconducting magnet. The nuclear spin relaxation is investigated as a function of η and ν by executing the measurement sequence depicted in *the right panel*. See the text for details. Reprinted from [96]

measurement sequence and the results are depicted in Fig. 12.12. $R_{0.65}$ is plotted as a function of (ν, η). The following procedure was executed for each data point. The electron system is allowed to relax at $\nu = 0.65$ and zero tilt angle until the resistance $R_{0.65}$ has saturated. The saturation value is approximately equal to $0.3h/e^2$. The sample is rotated to reach a certain value η. To avoid changes in the nuclear spin polarization during rotation, the sample is depleted while rotating. After rotation, the filling factor is then set to values between 0.7 and 1.4 for 45 s. The sample is depleted again, rotated back and the original filling factor 0.65 is restored. The resistance value is recorded immediately upon return to $\nu = 0.65$. This value is plotted in Fig. 12.12. The same experiment is repeated for different angles, i.e., values of η.

Nuclear spin relaxation during the excursion changes $\mathcal{P}_N$ for some values of ν and causes an increase in the resistance $R_{0.65}$ upon return to $\nu = 0.65$. The highest resistance value ($\sim 1.4h/e^2$) is obtained for filling factors ν near 0.9 and 1.1 for the smallest values of η. This behavior can be attributed to the rapid nuclear spin relaxation mediated by the gapless Goldstone mode of a Skyrme crystal. The position of the maximum nuclear spin relaxation rate is close to $|1 - \nu| = 0.1$, in agreement with the results from optically pumped NMR [93] (even though the experiments here were carried out at $T \leq 50$ mK, much lower than the temperatures in the NMR experiments). As η increases, the filling factor range with high resistance values shrinks. Eventually, $R_{0.65}$ no longer exhibits any change for the filling factors covered by the experiment. It suggests that the nuclear spin relaxation rate is no longer enhanced for large η at any filling factor around $\nu = 1$. At large η, the size of the skyrmions shrinks and the interaction among skyrmions is too weak to bring about a Skyrme crystal. The associated gapless Goldstone mode which helps nuclear spin relaxation vanishes [86].

Example 3: The Filling Factor Dependence of the Nuclear Spin Polarization

The previous examples focused on the nuclear spin relaxation as a function of filling factor. The time dependent behavior was monitored by measuring the resistance at some filling factor close to the $\nu \sim 2/3$ spin transition. The resistance serves as the detector for nuclear spin polarization $\mathcal{P}_N$ even though the relationship between the resistance and $\mathcal{P}_N$ is not known at the quantitative level. Since the effective nuclear field B_N shifts the phase transition to a different transition density n_{tr} or magnetic field B_{tr}, an alternative (but more time consuming) approach for detecting $\mathcal{P}_N$ consists in determining the exact location of the transition. In contrast with the resistive detection schemes, it has the advantage that the change in $\mathcal{P}_N$ can be extracted directly from B_{tr} or equivalently n_{tr}.

This method was first applied in [131]. The main result of that work is shown in Fig. 12.13. This graph illustrates the location of the phase transition as a function of the filling factor at which the 2D electron system is allowed to equilibrate or rest for 180 s. The sophisticated measurement sequence is schematically depicted in the inset on the left. It is explained in the figure caption. In the main panel, the magnetic field at which the $\nu = 2/3$-transition takes place is plotted.

The result is astonishing. The transition field is strongly filling factor dependent. For the two extreme cases, $\nu_{rest} = 1/2$ and 0.9, the difference in B_{tr} is about 3 T. This corresponds to a change in the effective nuclear field ΔB_N of ~ 1.5 T, or a $\Delta \mathcal{P}_N$ of approximately 20%. Since no current is flowing through the sample when it rests at $\nu = \nu_{rest}$ (this makes up 99.3% of measurement time), dynamic nuclear polarization induced by externally imposed current can be excluded for such a large ΔB_N. At thermal equilibrium, we expect virtually no dependence of the degree of nuclear spin polarization on the filling factor. The effective magnetic field B_e of the electron spins is at least three orders of magnitude smaller than the external magnetic field B. Therefore, the thermal distribution of the nuclei should essentially remain unperturbed, even if the electron spin polarization varies drastically with filling factor. Does some mechanism drive the nuclear spin subsystem out of equilibrium? If so, it has not been identified so far. Even if it is accepted that dynamic nuclear polarization takes place, it remains to be understood what the final state of the nuclear spin subsystem should be as a function of the rest filling factor. These questions require further investigation.

Despite a number of open questions, these results offer a straightforward recipe to manipulate the degree of nuclear spin polarization and elicit a time-dependent response without the need for incident microwave radiation, optical pumping, or large currents. The latter approaches inevitable increase the electron temperature. According to Fig. 12.13, it is sufficient to let the 2D electron system rest at a properly chosen ν_{rest}. This method allows for investigating the nuclear–electron spin interactions at the lowest available temperatures. Such studies are bound to be very fruitful in view of the many intriguing but fragile fractional quantum Hall states which demand ultra-low temperatures.

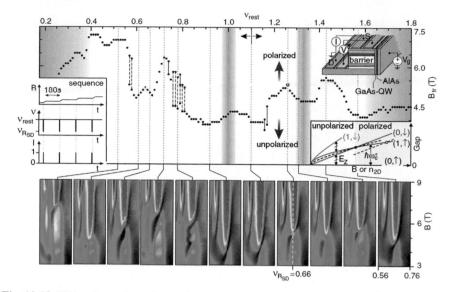

Fig. 12.13. Filling factor dependence of the location of the $\nu \sim 2/3$ spin transition. The sample (*upper-right inset*) is a two terminal device in which the 2D electron system is confined to a 20 nm wide quantum well. The quantum well is grown onto a cleaved edge of a GaAs(001) substrate. The 2D electron system channel is 250 μm wide and 3 μm long. It is contacted with two n-doped GaAs layers (source and drain). The carrier density n_s can be varied with gate voltage V_g. *The left inset* shows the time sequence of the magnetic field B, filling factor ν and the current I (in nA) imposed through the sample. For each value of ν_{rest}, B is swept from 3 to 9 T in 0.1 T steps. The gate voltage tracks B during the B-sweep to maintain fixed filling factor ν_{rest}. After a new B value has been reached, the sample is left to relax for 180 s. Subsequently, V_g is adjusted for a short excursion to $\nu_{RSD} = 0.66$ and simultaneously the current is turned on. The source drain resistance R_{SD} is recorded after a 2 s wait in order to account for time constants of the signal acquisition system. Then the current is turned off again and the original filling ν_{rest} restored. B is swept to the next set point and the entire procedure is repeated up to 9 T. The phase transition field B_{tr}, at which $R_{SD}(\nu_{RSD} = 0.66)$ reaches a minimum, is plotted as a function of ν_{rest} (*dotted curve in the upper panel*). *The bottom panels* are the outcome of similar measurements but with ν_{RSD} varied from 0.56 to 0.76. They visualize the spin transition for selected values of ν_{rest}. Reprinted from [131]

Other Examples

The behavior and properties of the resistance near the $\nu \sim 2/3$ spin transition were also exploited to demonstrate quantum coherence of the nuclear spins [128]. Kumada and co-workers [129, 130] recently applied a similar detection method for the study of nuclear spin relaxation in bilayer 2D systems. Despite this success, we would like to stress that no systematic study or theory has been carried out or developed to determine a quantitative relationship between the resistance at $\nu \sim 2/3$ and the degree of nuclear spin polarization $\mathcal{P}_N$. The complicated nature of the $\nu \sim 2/3$ transition suggests that the resistance may not simply depend linearly on $\mathcal{P}_N$. A non-linear dependence of the resistance on $\mathcal{P}_N$ may introduce some error in extracting T_1.

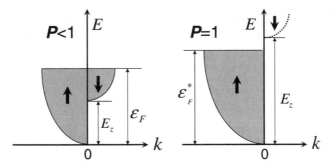

Fig. 12.14. Spin polarization of a noninteracting composite fermion Fermi sea. Full spin polarization ($\mathcal{P} = 1$) occurs when $E_Z > \varepsilon_F^* = \hbar^2(k_F^*)^2/2m^*$ with $k_F^* = \sqrt{4\pi n_s}$

These errors will be more severe for large changes in $\mathcal{P}_N$. Nevertheless, this resistive method offers unparalleled sensitivity and a very convenient way to obtain qualitative information about the interaction between the nuclear spins and the 2D electron spins. This information would be difficult to acquire with other techniques.

12.3.6 Composite Fermion Fermi Sea at $\nu = 1/2$

According to the composite fermion model, the strongly interacting 2D electron system at filling factor $\nu = 1/2$ can be viewed as a Fermi liquid of weakly interacting composite fermions. The composite fermions no longer experience the external applied magnetic field, but rather an effective magnetic field ($\langle B^* \rangle$) which vanishes at exact half-filling. Similar to electrons in zero magnetic field, composite fermions form a Fermi sea at $\nu = 1/2$ [36, 132] with a well-defined Fermi wave-vector. Away from half-filling, they are sent onto circular cyclotron orbits with a diameter determined by the non-zero effective magnetic field. This was impressively confirmed in surface acoustic wave experiments [38], ballistic transport measurements in periodically modulated structures such as antidot arrays [37] or 1D density modulations [133, 134], as well as in magnetic focusing experiments [39, 40].

Spin Polarization of the Composite Fermion Fermi Sea

The effective magnetic field only controls the orbital degree of freedom of composite fermions. The spin degree of freedom still listens to the external applied magnetic field. In view of the large external fields at which composite fermions form, it was often assumed in early studies that the composite fermion Fermi sea is fully spin polarized. However, later work revealed that this is not always the case [64, 66]. Under the assumption of non-interacting composite fermions and a parabolic energy dispersion $E = (\hbar^2 k^2)/(2m^*)$, the density of states is constant $D_\uparrow(E) = D_\downarrow(E) = m^*/(2\pi\hbar^2)$ and the Fermi sea is fully polarized (see Fig. 12.14) when the Zeeman energy $E_Z = g^* \mu_B B_{tot}$[9] exceeds $\varepsilon_F^* = (\hbar^2 k_F^2)/(2m^*)$ with $k_F^* = \sqrt{4\pi n_s}$. When $E_Z < \varepsilon_F^*$, the composite fermions are partially polarized and the spin polarization is equal to

$$\mathcal{P} = \frac{k_{F\uparrow}^2 - k_{F\downarrow}^2}{k_{F\uparrow}^2 + k_{F\downarrow}^2} = \frac{E_Z}{\varepsilon_F^*}. \tag{12.22}$$

The composite fermion effective mass m^* plays a crucial in determining the spin polarization and requires some discussion. At $\nu = 1/2$, all electrons are accommodated in the lowest Landau level and their kinetic energy is quenched.[10] The kinetic energy and the effective mass of composite fermions is therefore no longer related to the effective mass m of the GaAs conduction band electrons. It originates solely from the Coulomb interaction between electrons, and hence scales with the Coulomb energy E_C. For typical magnetic field strengths, m^* is much larger than m and comparable to the free electron mass m_e instead [66]. For the sake of completeness we point out that different effective masses have to be distinguished depending on the physical context [14]. For instance, the effective mass relevant for describing thermal activation studies ($m_a \sim 0.079\sqrt{B}m_e$, with B in Tesla) [36, 135, 136], is different from the mass m^* which enters in the description of the spin polarization. According to theory, this so-called polarization mass is given by $m^* = \xi\sqrt{B}m_e$. Calculations by Park and Jain yield $\xi = 0.60$ [137]. Note that $\varepsilon_F^* \propto E_C$. The spin polarization can then be rewritten in a form which underlines the importance of the competition between E_C and E_Z:

$$\mathcal{P} = \frac{1}{\eta_c} \frac{E_Z}{E_C} = \frac{\eta}{\eta_c} \propto \frac{B_{tot}}{\sqrt{B}}. \tag{12.23}$$

Here, $\eta_c \simeq 0.022$ based on the value of m^* put forward by Park and Jain [137]. It is straightforward to show that the spin transition from $\mathcal{P} < 1$ to $\mathcal{P} = 1$ then takes place at $B_{tot} = (g^*\xi)^{-2} \cos\theta$.

This noninteracting composite fermion picture appears to capture the essential features of the experimental results on the spin polarization at $\nu = 1/2$. Kukushkin et al. [66] determined the degree of spin polarization at filling $\nu = 1/2$ directly by measuring the circular polarization of photoluminescence. The deduced composite fermion spin polarization has a transition from $\mathcal{P} < 1$ to $\mathcal{P} = 1$ at $B \simeq 9.3$ T, corresponding to $\xi \simeq 0.75$. The discrepancy with the predicted value by Jain and Park [137] can presumably be accounted for by finite width effects. Spin transitions of composite fermions at $\nu = 1/2$ were also observed in NMR experiments in which $\mathcal{P}$ was obtained by measuring the Knight shift [67, 68, 138]. Many of these experiments were carried out at $T > 0.3$ K, so the thermal energy $k_B T$ is no longer negligible in comparison to ε_F^*. Thermal smearing of the Fermi surface must be taken into account. Here we limit the discussion to the results obtained at $T < 0.1$ K in [138],

---

[9] According to an NMR experiment [138], the g^*-factor at $\nu = 1/2$ is close to the value of bulk GaAs ($g = -0.44$).

[10] This statement strictly speaking is an approximation. The kinetic energy of electrons becomes relevant when the mixing from higher Landau levels is considered. For magnetic fields sufficiently high, the large cyclotron gap $\hbar\omega_c$ makes constraining the electrons to the lowest Landau level a good approximation [12, 14].

so the finite temperature effects can largely be ignored. A standard spin-echo technique was used to measure the Knight shift of a multiquantum well sample. The ratio $\eta = E_Z/E_C$ was varied by tilting the sample relative to B_{tot} while keeping B fixed. For $\eta < 0.022$, $\mathcal{P}$ is proportional to η, while for larger η, $\mathcal{P} = 1$. This observation is in excellent agreement with the prediction of Park and Jain [137] (see (12.23)). Freytag et al. [138] also compared their experimental results with the Shankar–Murthy theory [139–141], which gives

$$\mathcal{P} = 0.117\lambda^{7/4}\frac{B_{tot}}{\sqrt{B}},\tag{12.24}$$

where λ is the finite thickness parameter in the Zhang–Das Sarma potential [142]. $\lambda = 1.6$ appears to provide good fits to the experimental data at various tilt angles and temperatures. The Shankar–Murthy theory is also able to fit the optically pumped NMR data recorded by Dementyev et al. [67] with a single fit parameter $\lambda = 1.75$ [140].

Nuclear Spin Relaxation at $\nu = 1/2$

For a partially polarized composite fermion Fermi sea, the hyperfine spin flip–flop process can take place. The conservation of energy is easily fulfilled due to the presence of gapless states for both spin directions. Therefore, the Korringa type spin relaxation is expected to play a dominant role in the nuclear spin relaxation if the temperature is not too low. For noninteracting composite fermions, the nuclear spin relaxation rate at $T \ll \varepsilon_F/k_B$ (given in (12.20)) can be written as

$$T_1^{-1} = \frac{4\pi(m^*)}{\hbar^3}\left(\frac{K_s^{max}}{n_s}\right)^2 k_B T.\tag{12.25}$$

Dementyev et al. [67] measured T_1 at $T = 0.3–1$ K with optically pumped NMR. They observed that the temperature dependence of T_1 follows the basic trend of the Korringa relation. A two parameter (m^*, J) fit was used in their data analysis. Here J is an interaction parameter introduced to account for the interactions between composite fermions. However, no single pair of m^* and J could be found to satisfactorily fit the T_1 and $\mathcal{P}$ data at various tilt angles. In contrast, the Shankar–Murthy theory was shown to provide fair agreement with the experiments [140, 143]. Freytag et al. [138] also compared T_1 data with theory. They found that the noninteracting composite fermion model underestimates T_1, while the Shankar–Murthy theory overestimates T_1.

Recently several experiments demonstrated that NMR can be electrically detected at $\nu = 1/2$ [120, 125, 144, 145]. In these experiments, RF radiation from an NMR coil surrounding the sample was used to depolarize the thermal nuclear spin polarization (which is sizable at $T < 0.2$ K). The reduction in nuclear spin polarization increases the Zeeman energy of the composite fermions. This causes an increase in the composite fermion spin polarization, which in turn manifests as a change in the longitudinal resistance R_{xx}. The spin polarization dependence of R_{xx} at $\nu = 1/2$, however, remains to be understood as so far theories have only been developed to treat electron transport for the fully polarized composite fermion Fermi sea [36, 146]. Nevertheless, the observed $\mathcal{P}$-dependence of R_{xx} allows nuclear spin relaxation to be measured at temperatures lower than those in previous experiments using traditional NMR approaches. Tracy et al. [145] measured T_1 down to $T \sim 35$ mK and observed that T_1^{-1} increases linearly with temperature for partial spin polarization of the composite fermion sea. Extrapolating the data to $T = 0$ yields $T_1^{-1} \sim 10^{-3}$ s^{-1}. This temperature independent offset in the nuclear spin relaxation was attributed to nuclear spin diffusion. They also measured T_1 at various magnetic fields by tuning the carrier density with a front gate while maintaining $\nu = 1/2$. It was found that T_1 is nearly B-independent for $\mathcal{P} < 1$. This is in contradiction with $T_1 \propto B^{-5/3}$ law expected from the noninteracting composite fermion picture [147]. Murthy and Shankar recently incorporated disorder into their theory, and their calculations seem to provide a better agreement with the experiment [147].

12.3.7 Other Cases

The work described in the previous sections is only a partial representation of the enormous experimental effort on the nuclear–electron spin interactions in the quantum Hall regime. This chapter is not intended to be exhaustive. Here we briefly discuss some work on other quantum Hall systems. We would like to emphasize that theses nuclear spin phenomena are by no means less interesting or less important than those mentioned in the previous subsections.

The Breakdown Regime of the Quantum Hall Effect

When the current applied to a 2D electron system exceeds a critical value, I_c, the quantum Hall effect breaks down [148, 149]. Several mechanisms [150], including electron heating [148, 151] and inter-Landau level scattering [152, 153], have been proposed to explain the breakdown process. Song and Omling [154] recently showed the relevance of nuclear spins in the electron transport near the onset of breakdown. Subsequently, Kawamura et al. [155] demonstrated that nuclear spins can be polarized in the breakdown regime of odd-integer filling factors. The excitation of electrons to the upper Zeeman-split Landau level with opposite spin dynamically polarizes the nuclear spins via the hyperfine flip–flop process.

The Wigner Crystal Phase of the 2D Electron System

At sufficiently high magnetic fields, the dominance of the Coulomb energy over the electron's kinetic energy will ultimately result in the formation of a Wigner crystal. It was predicted that the Wigner crystal will be energetically more favorable in comparison with the incompressible fractional quantum Hall liquids around filling $\nu = 1/5$ [156–158]. Transport measurements [15, 16] showed that the fractional quantum Hall series indeed terminates at $\nu = 1/5$. Microwave experiments on samples with lower disorder suggested that the electron solid is pinned by disorder [159], and that in very high magnetic fields there may exist two types of electron solid phases [160]. Gervais et al. [161] recently carried out resistively detected NMR experiments at the onset of the electron solid phase near $\nu = 2/9$ and $1/5$. It was found that T_1 is quite long ($\sim$350–1000 s) and also did not show any obvious dependence on the filling factor, contrary to what was observed at filling factors near $\nu \sim 1$.

Two Sub-Band Systems

In the previous discussions, the carrier density in the two-dimensional electron system was sufficiently low to ensure that only the lowest sub-band is populated by the electrons. Zhang et al. [163] studied the electron transport in a 2D electron system with larger carrier density so that two sub-bands are occupied. It is then possible to bring two levels with opposite spins and different sub-band indices into degeneracy. It gives rise to quantum Hall ferromagnetism. Resistively detected NMR was used to find evidence for the ferromagnetic nature of some quantum Hall states [164].

Bilayer Systems

In a bilayer two-dimensional electron system, the interplay between the inter-layer and intralayer Coulomb interactions may give rise to exotic quantum Hall fluids, which do not occur in single layer systems. The most intriguing phase reported to date occurs at total filling factor 1. Experiments suggest that this phase is a Bose condensate of excitons [17]. As the layer separation is decreased, the bilayer system evolves from a system of weakly coupled compressible states each with filling factor $1/2$ into an interlayer phase coherent quantum Hall state. Several groups have demonstrated [129, 144] that nuclear spin relaxation can be detected resistively also in these bilayer systems. Spielman et al. [144] used a combination of heat pulse and NMR techniques. These authors observed that the transition to the phase coherent quantum Hall state is accompanied by a rapid change in T_1. The experimental results suggest that the electron spin polarization decreases as the transition from the excitonic quantum Hall phase to the weakly coupled metallic phase takes place.

Also at other filling factors, bilayers exhibit rich spin physics. For instance at total filling factor 2, the bilayer can be treated as a quantum Hall ferromagnet in which the layer index acts as a pseudospin [112]. Kumada et al. [130] measured T_1 of this quantum Hall ferromagnet for two bilayers with different tunneling barriers. Current induced nuclear spin polarization when the system is reduced to a single layer near $\nu = 2/3$ with the help of the front and back gates was used to drive the nuclear spins out of equilibrium, and the change in nuclear spin polarization was monitored by measuring R_{xx} near $\nu = 2/3$. The observed temperature dependence of T_1 was argued to be evidence for a so-called canted antiferromagnet. Additional evidence for the canted nature of this phase was recently collected from Knight shift measurements [165].

12.4 Summary and Outlook

In conclusion, we have highlighted the rich family of spin related electronic states in the quantum Hall regime. They provide novel ways to manipulate nuclear spins via the hyperfine interaction. Various mechanisms for dynamical nuclear spin polarization were demonstrated experimentally. These include spin flip–flop induced by electron spin resonance, by interedge channel scattering, by interlevel scattering in the breakdown regime of the integer quantum Hall effect, and through spin flip scattering at domain boundaries, which form near spin phase transitions of some fractional quantum Hall states such as $\nu \sim 2/3$. Gate voltage control over nuclear spins appears feasible following the observation of a filling factor dependent nuclear spin polarization even in the absence of current flow. This phenomenon still remains to be understood.

Studies of the spin interactions between electrons and nuclei in the quantum Hall regime have led to new, all-electrical methods for selectively detecting nuclear spins residing in the region of the 2D electron system. For example, the longitudinal resistance R_{xx} at $\nu = 1/2, 2/3$ and other filing factors may serve as a sensitive detector for nuclear spin polarization. Quantum coherence of nuclear spins was demonstrated by combining miniaturized NMR pulse techniques with electrical manipulation and detection of nuclear spins.

Conversely, the dynamics of the nuclear spin polarization and the precession frequency of nuclear spins have been exploited as unique probes of the electron spin polarization of fragile electronic states. The information obtained from the Knight shift and the spin relaxation rate has deepened our understanding of spin related phenomena in quantum Hall systems. Despite the tremendous progress over the past two decades, we believe that the study of nuclear spins will continue to bring us additional and more profound insight into the remarkably rich quantum Hall effects. New approaches to manipulate and detect nuclear spins are bound to emerge as we gain more knowledge about the nuclear–electron spin interactions in these fascinating 2D electron systems.

References

[1] K. von Klitzing, G. Dorda, M. Pepper, Phys. Rev. Lett. **45**, 494 (1980)
[2] D.C. Tsui, H.L. Stormer, A.C. Gossard, Phys. Rev. Lett. **48**, 1559 (1982)
[3] R.E. Prange, S.M. Girvin (eds.), *The Quantum Hall Effect* (Springer, New York, 1987)
[4] T. Chakraborty, *The Quantum Hall Effects* (Springer, Berlin, 1995)
[5] S. Das Sarma (ed.), *Perspectives in Quantum Hall Effects* (Wiley, New York, 1997)
[6] Z.F. Ezawa, *Quantum Hall Effects—Field Effect Approach and Related Topics* (World Scientific, Singapore, 2000)
[7] R.B. Laughlin, Phys. Rev. Lett. **50**, 1395 (1983)
[8] R. dePicciotto et al., Nature **389**, 162 (1997)
[9] L. Saminadayar et al., Phys. Rev. Lett. **79**, 2526 (1997)
[10] M. Reznikov et al., Nature **399**, 328 (1999)
[11] J. Martin et al., Science **305**, 980 (2004)

[12] J.K. Jain, Phys. Today **53**(4), 39 (2000)
[13] O. Heinonen (ed.), *Composite Fermions* (World Scientific, Singapore, 1998)
[14] J.K. Jain, *Composite Fermions* (Cambridge University Press, Cambridge, 2007)
[15] R.L. Willett et al., Phys. Rev. B **37**, 8476 (1988)
[16] H.W. Jiang et al., Phys. Rev. Lett. **65**, 633 (1990)
[17] J.P. Eisenstein, A.H. MacDonald, Nature **432**, 691 (2004)
[18] S.L. Sondhi et al., Phys. Rev. B **47**, 16419 (1993)
[19] S.E. Barrett et al., Phys. Rev. Lett. **74**, 5112 (1995)
[20] G. Moore, N. Read, Nucl. Phys. B **360**, 362 (1991)
[21] M. Greiter, X.G. Wen, F. Wilczek, Nucl. Phys. B **374**, 567 (1992)
[22] S. Das Sarma, M. Freedman, C. Nayak, Phys. Today **59**, 32 (2006)
[23] A.A. Koulakov, M.M. Fogler, B.I. Shklovskii, Phys. Rev. Lett. **76**, 499 (1996)
[24] R. Moessner, J.T. Chalker, Phys. Rev. B **54**, 5006 (1996)
[25] S.A. Kivelson, E. Fradkin, V.J. Emery, Nature **393**, 550 (1998)
[26] R.R. Du et al., Solid State Commun. **109**, 389 (1999)
[27] E. Fradkin, S.A. Kivelson, Phys. Rev. B **59**, 8065 (1999)
[28] E. Fradkin et al., Phys. Rev. Lett. **84**, 1982 (2000)
[29] A. Hartland et al., Phys. Rev. Lett. **66**, 969 (1991)
[30] A.M. Chang et al., Phys. Rev. B **28**, 6133 (1983)
[31] G.S. Boebinger et al., Phys. Rev. B **36**, 7919 (1987)
[32] A. Usher et al., Phys. Rev. B **41**, 1129 (1990)
[33] M. Dobers, K. von Klitzing, G. Weimann, Phys. Rev. B **38**, 5453 (1988)
[34] R.B. Laughlin, Phys. Rev. B **23**, 5632 (1981)
[35] S.H. Simon, in *Composite Fermions*, ed. by O. Heinonen (World Scientific, Singapore, 1998)
[36] B.I. Halperin, P.A. Lee, N. Read, Phys. Rev. B **47**, 7312 (1993)
[37] W. Kang et al., Phys. Rev. Lett. **71**, 3850 (1993)
[38] R.L. Willett et al., Phys. Rev. Lett. **71**, 3846 (1993)
[39] V.J. Goldman, B. Su, J.K. Jain, Phys. Rev. Lett. **72**, 2065 (1994)
[40] J.H. Smet et al., Phys. Rev. Lett. **77**, 2272 (1996)
[41] I.V. Kukushkin et al., Nature **415**, 409 (2002)
[42] B.I. Halperin, Phys. Rev. B **25**, 2185 (1982)
[43] M. Büttiker, Phys. Rev. B **38**, 9375 (1988)
[44] A.H. MacDonald, P. Streda, Phys. Rev. B **29**, 1616 (1984)
[45] T.J. Thornton et al., Phys. Rev. Lett. **56**, 1198 (1986)
[46] H.Z. Zheng et al., Phys. Rev. B **34**, 5635 (1986)
[47] B.J. van Wees et al., Phys. Rev. Lett. **62**, 1181 (1989)
[48] B.J. van Wees et al., Phys. Rev. B **43**, 12431 (1991)
[49] D. Yoshioka, *The Quantum Hall Effect* (Springer, Berlin, 2002)
[50] K. von Klitzing, *25 Years of Quantum Hall Effects*. Séminar Poincaré (Vol. 2, p. 1) (2004)
[51] H.L. Stormer, Rev. Mod. Phys. **71**, 875 (1999)
[52] D.C. Tsui, Rev. Mod. Phys. **71**, 891 (1999)
[53] R.B. Laughlin, Rev. Mod. Phys. **71**, 863 (1999)
[54] S.M. Girvin, *Introduction to the Fractional Quantum Hall Effect*. Séminar Poincaré (Vol. 2, p. 53) (2004)
[55] B.I. Halperin, Helv. Phys. Acta **56**, 75 (1983)
[56] T. Chakraborty, F.C. Zhang, Phys. Rev. B **29**, 7032 (1984)
[57] F.C. Zhang, T. Chakraborty, Phys. Rev. B **30**, 7320 (1984)
[58] X.C. Xie, Y. Guo, F.C. Zhang, Phys. Rev. B **40**, 3487 (1989)

[59] J.P. Eisenstein et al., Phys. Rev. Lett. **62**, 1540 (1989)
[60] R.G. Clark et al., Phys. Rev. Lett. **62**, 1536 (1989)
[61] J.P. Eisenstein et al., Phys. Rev. B **41**, 7910 (1990)
[62] L.W. Engel et al., Phys. Rev. B **45**, 3418 (1992)
[63] H. Cho, J.B. Young, W. Kang, K.L. Campman, A.C. Gossard, M. Bichler, W. Wegscheider, Phys. Rev. Lett. **81**, 2522 (1998)
[64] R.R. Du et al., Phys. Rev. Lett. **75**, 3926 (1995)
[65] J.H. Smet et al., Phys. Rev. Lett. **86**, 2412 (2001)
[66] I.V. Kukushkin, K. von Klitzing, K. Eberl, Phys. Rev. Lett. **82**, 3665 (1999)
[67] A.E. Dementyev et al., Phys. Rev. Lett. **83**, 5074 (1999)
[68] S. Melinte et al., Phys. Rev. Lett. **84**, 354 (2000)
[69] T. Jungwirth et al., Phys. Rev. Lett. **81**, 2328 (1998)
[70] A.S. Yeh, H.L. Stormer, D.C. Tsui, Phys. Rev. Lett. **82**, 592 (1999)
[71] M.A. Ruderman, C. Kittel, Phys. Rev. **96**, 99 (1954)
[72] T. Machida et al., Physica E **21**, 921 (2004)
[73] A. Berg et al., Phys. Rev. Lett. **64**, 2563 (1990)
[74] O.V. Lounasmaa, Phys. Today **42**(10), 26 (1989)
[75] A. Abragam, *The Principles of Nuclear Magnetism* (Oxford University Press, London, 1961)
[76] D.R. Lide (ed.), *CRC Handbook of Chemistry and Physics* (CRC Press, Boca Raton, 2008)
[77] D. Paget, G. Lampel, B. Sapoval, V.I. Safarov, Phys. Rev. B **15**, 5780 (1977)
[78] M.I. Dyakonov, V.I. Perel, in *Optical Orientation*, ed. by F. Meier, B.P. Zakharchenya (Elsvier, New York, 1984)
[79] F. Meier, B.P. Zakharchenya (eds.), *Optical Orientation* (Elsevier, New York, 1984)
[80] M. Dobers et al., Phys. Rev. Lett. **61**, 1650 (1988)
[81] J. Strand et al., Phys. Rev. Lett. **91**, 036602 (2003)
[82] C.P. Slichter, *Principles of Magnetic Resonance* (Harper & Row Publishers, New York, 1963)
[83] J. Korringa, Physica **16**, 601 (1950)
[84] K. Hashimoto et al., Phys. Rev. Lett. **94**, 146601 (2005)
[85] J.H. Kim, I.D. Vagner, L. Xing, Phys. Rev. B **49**, 16777 (1994)
[86] R. Côté et al., Phys. Rev. Lett. **78**, 4825 (1997)
[87] C.X. Deng, X.D. Hu, Phys. Rev. B **72**, 165333 (2005)
[88] D. Paget, Phys. Rev. B **25**, 4444 (1982)
[89] K.R. Wald et al., Phys. Rev. Lett. **73**, 1011 (1994)
[90] H. Sanada et al., Phys. Rev. Lett. **96**, 067602 (2006)
[91] I.L. Chuang et al., Nature **393**, 143 (1998)
[92] L.M.K. Vandersypen et al., Nature **414**, 883 (2001)
[93] R. Tycko et al., Science **268**, 1460 (1995)
[94] D.C. Dixon et al., Phys. Rev. B **56**, 4743 (1997)
[95] S. Kronmüller et al., Phys. Rev. Lett. **82**, 4070 (1999)
[96] J.H. Smet et al., Nature **415**, 281 (2002)
[97] K. Hashimoto et al., Phys. Rev. Lett. **88**, 176601 (2002)
[98] W. Desrat et al., Phys. Rev. Lett. **88**, 256807 (2002)
[99] A. Würtz et al., Phys. Rev. Lett. **95**, 056802 (2005)
[100] G. Gervais et al., Phys. Rev. Lett. **94**, 196803 (2005)
[101] J.M. Kikkawa, D.D. Awschalom, Science **287**, 473 (2000)
[102] D. Rugar et al., Nature **430**, 329 (2004)
[103] I.D. Vagner, T. Maniv, Phys. Rev. Lett. **61**, 1400 (1988)

[104] E. Olshanetsky et al., Phys. Rev. B **67**, 165325 (2003)
[105] S.V. Iordanskii, S.V. Meshkov, I.D. Vagner, Phys. Rev. B **44**, 6554 (1991)
[106] D. Antoniou, A.H. MacDonald, Phys. Rev. B **43**, 11686 (1991)
[107] B.E. Kane, L.N. Pfeiffer, K.W. West, Phys. Rev. B **46**, 7264 (1992)
[108] T. Machida et al., Physica E **25**, 142 (2004)
[109] T. Machida et al., Phys. Rev. B **65**, 233304 (2002)
[110] H.A. Fertig et al., Phys. Rev. B **50**, 11018 (1994)
[111] X.C. Xie, S. He, Phys. Rev. B **53**, 1046 (1996)
[112] S.M. Girvin, Phys. Today **53**(6), 39 (2000)
[113] T.H.R. Skyrme, Nucl. Phys. **31**, 556 (1962)
[114] A. Schmeller et al., Phys. Rev. Lett. **75**, 4290 (1995)
[115] E.H. Aifer, B.B. Goldberg, D.A. Broido, Phys. Rev. Lett. **76**, 680 (1996)
[116] L. Brey et al., Phys. Rev. Lett. **75**, 2562 (1995)
[117] C. Timm, S.M. Girvin, H.A. Fertig, Phys. Rev. B **58**, 10634 (1998)
[118] V. Bayot et al., Phys. Rev. Lett. **76**, 4584 (1996)
[119] V. Bayot et al., Phys. Rev. Lett. **79**, 1718 (1997)
[120] L.A. Tracy et al., Phys. Rev. B **73**, 121306 (2006)
[121] R.G. Clark et al., Surf. Sci. **229**, 25 (1990)
[122] K. Hashimoto, T. Saku, Y. Hirayama, Phys. Rev. B **69**, 153306 (2004)
[123] V. Piazza et al., Nature **402**, 638 (1999)
[124] K. Vyborny et al., Phys. Rev. B **75**, 045434 (2007)
[125] O. Stern et al., Phys. Rev. B **70**, 075318 (2004)
[126] S. Kraus et al., Phys. Rev. Lett. **89**, 266801 (2002)
[127] S. Kronmüller et al., Phys. Rev. Lett. **81**, 2526 (1998)
[128] G. Yusa et al., Nature **434**, 1001 (2005)
[129] N. Kumada et al., Phys. Rev. Lett. **94**, 096802 (2005)
[130] N. Kumada, K. Muraki, Y. Hirayama, Science **313**, 329 (2006)
[131] J.H. Smet et al., Phys. Rev. Lett. **92**, 086802 (2004)
[132] V. Kalmeyer, S.C. Zhang, Phys. Rev. B **46**, 9889 (1992)
[133] J.H. Smet et al., Phys. Rev. Lett. **83**, 2620 (1999)
[134] R.L. Willett, K.W. West, L.N. Pfeiffer, Phys. Rev. Lett. **83**, 2624 (1999)
[135] J.K. Jain, R.K. Kamilla, Phys. Rev. B **55**, R4895 (1997)
[136] J.K. Jain, R.K. Kamilla, Intern. J. Mod. Phys. B **11**, 2621 (1997)
[137] K. Park, J.K. Jain, Phys. Rev. Lett. **80**, 4237 (1998)
[138] N. Freytag et al., Phys. Rev. Lett. **89**, 246804 (2002)
[139] R. Shankar, G. Murthy, Phys. Rev. Lett. **79**, 4437 (1997)
[140] R. Shankar, Phys. Rev. B **63**, 085322 (2001)
[141] G. Murthy, R. Shankar, Rev. Mod. Phys. **75**, 1101 (2003)
[142] F.C. Zhang, S. Das Sarma, Phys. Rev. B **33**, 2903 (1986)
[143] R. Shankar, Phys. Rev. Lett. **84**, 3946 (2000)
[144] I.B. Spielman et al., Phys. Rev. Lett. **94**, 076803 (2005)
[145] L.A. Tracy et al., Phys. Rev. Lett. **98**, 086801 (2007)
[146] A.D. Mirlin, D.G. Polyakov, P. Wölfle, Phys. Rev. Lett. **80**, 2429 (1998)
[147] G. Murthy, R. Shankar, Phys. Rev. B **76**, 075341 (2007)
[148] G. Ebert, K. von Klitzing, K. Ploog, G. Weimann, J. Phys. C **16**, 5441 (1983)
[149] M.E. Cage et al., Phys. Rev. Lett. **51**, 1374 (1983)
[150] A.M. Martin et al., Phys. Rev. Lett. **91**, 126803 (2003)
[151] S. Komiyama et al., Solid State Commun. **54**, 479 (1985)
[152] O. Heinonen, P.L. Taylor, S.M. Girvin, Phys. Rev. B **30**, 3016 (1984)
[153] L. Eaves, F.W. Sheard, Semicond. Sci. Technol. **1**, 346 (1986)

[154] A.M. Song, P. Omling, Phys. Rev. Lett. **84**, 3145 (2000)
[155] M. Kawamura, H. Takahashi et al., Appl. Phys. Lett. **90**, 022102 (2007)
[156] P.K. Lam, S.M. Girvin, Phys. Rev. B **30**, 473 (1984)
[157] D. Levesque, J.J. Weis, A.H. MacDonald, Phys. Rev. B **30**, 1056 (1984)
[158] X.J. Zhu, S.G. Louie, Phys. Rev. B **52**, 5863 (1995)
[159] P.D. Ye et al., Phys. Rev. Lett. **89**, 176802 (2002)
[160] Y.P. Chen et al., Phys. Rev. Lett. **93**, 206805 (2007)
[161] G. Gervais et al., Phys. Rev. B **72**, 041310 (2005)
[162] Y.B. Lyanda-Geller, I.L. Aleiner, B.L. Altshuler, Phys. Rev. Lett. **89**, 107602 (2002)
[163] X.C. Zhang, D.R. Faulhaber, H.W. Jiang, Phys. Rev. Lett. **95**, 216801 (2005)
[164] X.C. Zhang, G.D. Scott, H.W. Jiang, Phys. Rev. Lett. **98**, 246802 (2007)
[165] N. Kumada, K. Muraki, Y. Hirayama, Phys. Rev. Lett. **99**, 076805 (2007)
[166] P. Weitz et al., Surf. Sci. **362**, 542 (1996)

13

Diluted Magnetic Semiconductors: Basic Physics and Optical Properties

J. Cibert and D. Scalbert

13.1 Introduction

Diluted Magnetic Semiconductors (DMS) form a new class of magnetic materials, which fill the gap between ferromagnets and semiconductors [1]. In the early literature these DMS were often named semimagnetic semiconductors, because they are midway between nonmagnetic and magnetic materials. DMS are semiconductor compounds $(A_{1-x}M_xB)$ in which a fraction x of the cations is substituted by magnetic impurities, thereby introducing magnetic properties into the host semiconductor AB. This makes a great difference with semiconducting ferromagnets, i.e., ferromagnetic materials exhibiting semiconductor-like transport properties, which have been known for some time (see a review in [2]). A DMS is expected to retain most of its classical semiconducting properties, and to offer the opportunity of a full integration into heterostructures, including heterostructures with the host material. The great challenge and ultimate goal of the research in this field is to obtain DMS ferromagnetic at room temperature, which can be integrated in semiconductor heterostructures for electronic or optoelectronic applications. This is one of the key issues for the advent of spintronics devices.

Among the principal DMS families, II–VI and, to a less extent, III–V based DMS, with Mn as the magnetic impurity, are best understood. For this reason the present chapter will be mainly based on these compounds to introduce the well-established basic physics of DMS. More details can be found in review papers such as [3–5]. Some issues related to work in progress, generally on novel materials, will be also discussed but only briefly.

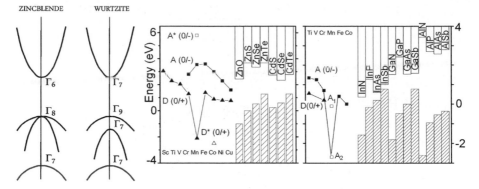

Fig. 13.1. *Left*: schematic of zinc-blende and wurtzite band structures near the Brillouin zone center. *Right*: approximate positions of transition metals levels relative to the conduction and valence band edges of II–VI and III–V compounds; *triangles* note the d^N/d^{N-1} donor and *squares* the d^N/d^{N+1} acceptor states [6]

13.2 Band Structure of II–VI and III–V DMS

The band structures of DMS are quite similar to those of their host II–VI or III–V compounds. They exhibit the band structures of zinc-blende or wurtzite semiconductors, and, except for some III–V compounds, they possess a direct band gap (Fig. 13.1). But in addition, the d-states of the Mn atoms, with a more or less localized character, contribute to the total density of states, and they are responsible for the important magnetic properties of DMS, as will be seen in Sect. 13.3.

In II–VI based DMS, Mn atoms behave as isoelectronic impurities, and in general do not introduce bound states. The two $4s$ electrons of Mn atoms participate to the covalent bonding, while the Mn d shell remains relatively inert. Hence, in zero magnetic field, the semiconducting properties of DMS look like those of non magnetic II–VI alloys. As in other standard alloys, DMS exhibit a shift of the energy gap with Mn concentration, and alloy fluctuations responsible for potential fluctuations, and eventually tails of localized states at the band edges.

In III–V DMS, Mn atoms introduce energy levels in the gap of the semiconductor. In the best understood case of antimonides and arsenides, Mn behaves as a shallow acceptor. It keeps its d^5 configuration and is surrounded by a weakly bound hole [7]. Therefore, at relatively small Mn concentrations, III–V DMS undergo an insulator to metal transition (although the critical density can be higher in the DMS than in the host material [8, 9]).

When a magnetic field is applied, or a spontaneous magnetization appears, magnetic properties of DMS come into play. The p–d hybridization is essential to understanding magnetic and magneto-optical properties of DMS. As will be seen in the next section, this leads to a strong exchange interaction between holes in the valence band and Mn atoms. Therefore the magnetic properties of DMS depend critically on p–d hybridization and on the positions of d-levels in the host band structure (Figs. 13.1 and 13.2). These positions determine the energy needed to promote

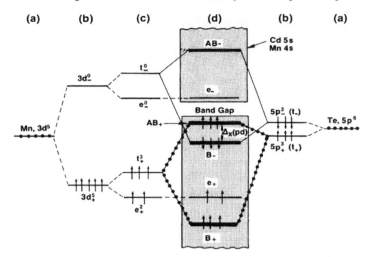

Fig. 13.2. Schematic diagram of the p–d hybridization in CdMnTe after Wei and Zunger [10]. (**a**) Atomic unpolarized levels, (**b**) exchange-split atomic levels, (**c**) crystal-field split levels, (**d**) final interacting states. Level repulsion with occupied spin-up and unoccupied spin-down d-states determines the sign of p–d exchange interaction (see Sect. 13.3)

an electron from the occupied d-level to the top of the valence band (d^5/d^4 donor level), or to promote an electron from the top of the valence band to the unoccupied d-level (d^5/d^6 acceptor level). In the latter case there is an extra energy cost due to the intra-d-shell Coulomb energy (the energy to be paid to add an electron on the Mn d orbitals), U_{eff}, which is particularly large in the case of the very stable d^5 configuration, see Fig. 13.1. In the one-electron representation the d-orbitals are split by the tetrahedral crystal field into doubly degenerate e_g states, which do not mix with the anions p-orbitals due to symmetry, and triply degenerate t_{2g} states, which mix with the p-orbitals.

Experimentally, let us recall the existence of spin–flip excitations within the Mn d-shell, observed at 2.2 eV and above in CdMnTe. The ground state of Mn in the d^5 configuration has a total spin $S = 5/2$, following Hund's rule, and zero orbital angular momentum. Spin–flip of one d electron gives excited states of the d-shell with total spin $S = 3/2$ and non-zero total angular momentum. Contrary to expectation based on parity rule and spin conservation, optical transitions between these excited states and the ground state are electric dipole active, as a result of the lack of inversion symmetry of the tetrahedral crystal field and mixing of $S = 5/2$ and $S = 3/2$ states by spin–orbit interaction. Although relatively weak,[1] these transitions cannot be ignored in the optical properties of wide gap DMS [11–13]. The same mixing is also responsible for spin–lattice relaxation of isolated Mn atoms in II–VI DMS, which will be discussed in Sect. 13.6.

---

[1] More intense, spin-allowed transitions are observed for magnetic ions with a configuration other than d^5.

13.3 Exchange Interactions in DMS

The dominant exchange interactions in DMS are fairly well established. The sp–d exchange interactions between band states and localized Mn d orbitals are mainly responsible for the enhanced magneto optical properties of DMS. The optical transitions involved in most of experiments are near the direct energy gap of the semiconductor, therefore we will be mainly concerned with exchange interactions at the Brillouin zone center.

13.3.1 s, p–d Exchange Interaction

s–d Exchange Interaction

The s–d exchange interaction is a simple example of direct exchange interaction. Due to Pauli exclusion principle, two electrons with the same spin avoid each other, thus their (repulsive) Coulomb interaction energy is reduced, while electrons with opposite spins can approach each other. The s–d exchange interaction is thus "ferromagnetic".[2] It may be described by the phenomenological Kondo-like Hamiltonian $H_K = -\sum_i J_{s-d}(\mathbf{r} - \mathbf{R}_i)\mathbf{S}_i \cdot \mathbf{s}$, with a positive exchange coupling constant J_{s-d}. Here and in the following, $\mathbf{s}$ and $\mathbf{r}$ are the spin and position (= argument of the wave function) of the carrier, $\mathbf{S}_i$ and $\mathbf{R}_i$ the spin and position (fixed) of the magnetic impurity labeled i. Exchange interaction is short range because it exists only in the region where the s and d orbitals overlap.

The matrix elements must be evaluated between Bloch states $|\mathbf{k}, m_s\rangle$, of wave function $\langle \mathbf{r}|\mathbf{k}\rangle = u_{c,\mathbf{k}}(\mathbf{r}) \exp(i\mathbf{k}.\mathbf{r})$, so that [14]

$$\langle \mathbf{k}', m'_s|H_K|\mathbf{k}, m_s\rangle = -\sum_i J_{\mathbf{k}',\mathbf{k}} e^{i(\mathbf{k}-\mathbf{k}').\mathbf{R}_i} \mathbf{S}_i \cdot \langle m'_s|\mathbf{s}|m_s\rangle. \tag{13.1}$$

In particular, the exchange integral for a Bloch state $u_{c,0}$ at the bottom of conduction band is generally denoted $N_0\alpha$ and is defined as

$$N_0\alpha = J_{0,0} = \frac{\int |u_{c,0}(\mathbf{r})|^2 J_{s-d}(\mathbf{r}) \, d^3\mathbf{r}}{\int |u_{c,0}(\mathbf{r})|^2 \, d^3\mathbf{r}}, \tag{13.2}$$

where the integrals can be calculated over the unit cell volume $v_0 = N_0^{-1}$ (=1/4 of the cubic cell in the zinc-blende structure). This parameter gives the magnitude of the giant Zeeman effect described in Sect. 13.5. It is the same exchange integral $N_0\alpha$ which appears when confinement effects are described within the effective mass approximation: in its simplest form, the effective Hamiltonian is written neglecting any $\mathbf{k}$-dependance of the exchange integral, so that it reads

$$H = -\alpha \sum_i \mathbf{S}_i \cdot \mathbf{s}\delta(\mathbf{r} - \mathbf{R}_i). \tag{13.3}$$

---

[2] This can be misleading. Since Landé factors have different signs for s and d electrons in (Cd, MnTe), parallel spins means antiparallel magnetic moments. A consequence is the field-dependance of Larmor frequency of conduction electrons in a sample with a low Mn content, see Fig. 13.17.

p–d Exchange Interaction

When the *d* levels fall within the valence band, as is the case in II–VI based DMS, *p–d* hybridization leads to an additional exchange interaction mechanism called kinetic exchange, with "antiferromagnetic" sign [15]. Kinetic exchange dominates over the usual direct exchange and the resulting exchange integral denoted $N_0\beta$ becomes negative.[3]

The mechanism of kinetic exchange can be understood most simply by looking at the position of energy levels of the hybridized Mn and valence band states, assuming that the Mn spins (up) are fully polarized. Figure 13.2 shows that close to the valence band maximum spin up antibonding states are shifted to high energy, while spin down bonding states are shifted to low energy. To second order in perturbation these shifts are expressed as

$$\delta E_\uparrow \simeq \frac{(V_{pd})^2}{\epsilon_v - \epsilon_d}, \tag{13.4}$$

$$\delta E_\downarrow \simeq -\frac{(V_{pd})^2}{\epsilon_d + U_{\text{eff}} - \epsilon_v}, \tag{13.5}$$

where V_{pd} is the *p–d* hybridization parameter, ϵ_v and ϵ_d are the positions of the valence band maximum and of the Mn t_{2g} levels, respectively, and U_{eff} is the intra *d*-shell Coulomb energy (the energy to be paid to add a sixth electron on the Mn *d* orbitals). The energy difference between spin up and spin down states is related to the *p–d* exchange integral as $\delta E_\uparrow - \delta E_\downarrow = -N_0\beta S$. Apart from a numerical factor the expression of $N_0\beta$ derived from the Schrieffer–Wolff transformation [16] is recovered in a physically transparent way

$$N_0\beta \simeq -\frac{(4V_{pd})^2}{S}\left[\frac{1}{\epsilon_v - \epsilon_d} + \frac{1}{\epsilon_d + U_{\text{eff}} - \epsilon_v}\right]. \tag{13.6}$$

Deviations from Local Exchange

s, p–d exchange interactions being short range, a contact interaction is often assumed as in (13.3). However, in some cases it is necessary to go beyond this approximation, in particular when the *k*-dependance of exchange integrals is to be considered. This happens for example in a quantum well of thickness *L* in which the confinement of the carrier imposes a finite momentum $p \sim h/2L$. In the case of electrons in the conduction band this reduces the *s–d* exchange because kinetic exchange (with sign opposite to direct exchange) becomes allowed at finite *k* [17]. This can be easily understood in a *k–p* approximation, as a consequence of the mixing of

---

[3] A negative β means also an "antiferromagnetic" coupling between the Mn spins and a gas of holes occupying the upper levels of the valence band (both spin and energy are reversed by switching from electrons to holes). But the magnetic moments of antiparallel spins can be parallel if the effective Landé factors have opposite signs, as in (Cd, Mn)Te. Note that the exchange parameter β applies to the spin of the hole, not its total angular momentum.

conduction and valence band states at $k \neq 0$ (in other words s–d hybridization becomes allowed away from the zone center). The effect in the valence bands was also considered in [14], which predicts even larger effects in quantum dots.

Also, the contact interaction is unable to account for the eventual existence of bound states induced by exchange interaction. This situation may arise when the ratio of the exchange integral to the effective mass of the carrier is large enough, in this case the gain in exchange energy is larger than the increase of kinetic energy due to carrier localization, and it is energetically more favorable for the carrier to be localized around the magnetic atom. This corresponds to the formation of the so-called Zhang-Rice polaron. This configuration is described in Sect. 13.5.1.

13.3.2 d–d Exchange Interactions

Contrary to transition metals, in which the overlap between d orbitals is sizable and leads to formation of d bands, direct exchange interactions between Mn atoms are negligible in DMS because d-orbitals do not overlap. Therefore the most important d–d exchange interactions are mediated by valence band or conduction band states. Because the carrier density can be easily controlled in semiconductors, DMS exhibit different exchange interactions typical for insulators (superexchange, double-exchange, Bloembergen–Rowland) or metals (RKKY, for Ruderman–Kittel–Kasuya–Yosida). The accepted terminology distinguishes between indirect exchange, which is mediated by the polarization of the intervening medium (electrons or holes in semiconductors), and superexchange, which is due to covalent mixing of magnetic (d) and non magnetic (s–p) orbitals. RKKY and Bloembergen–Rowland are familiar examples of indirect interactions.

Superexchange

In undoped DMS, and in most magnetic insulators, superexchange is the dominant interaction mechanism. It is also a consequence of hybridization of the d levels with the occupied sp orbitals. In order to understand the physical basis of superexchange it is enough to consider a simple three-site picture in which two nearest-neighbors Mn moments interact via a common anion [18]. Due to hybridization, d electrons with opposite spins tend to slightly delocalize on the anion, thus lowering their kinetic energy, while d electrons with same spin cannot delocalize due to Pauli exclusion (see Fig. 13.3). Thus, following Anderson [19], superexchange is a consequence of "kinetic exchange", similarly to p–d exchange discussed above, and generally this is the determining factor that leads to antiferromagnetic coupling between local spins in non-metals. An approximate expression for the superexchange coupling parameter can be obtained in second order of perturbation as $J \sim \delta E_{\uparrow\downarrow} = -b^2/U_{\text{eff}}$ [18], where b represents the hopping integral between the Mn atoms, and U_{eff} is the intra d-shell Coulomb energy. Of course, since hopping proceeds via the p orbital of the anion, it is already a second order process in the hybridization potential, so that superexchange can be viewed as a fourth order process shown in the right part of Fig. 13.3.

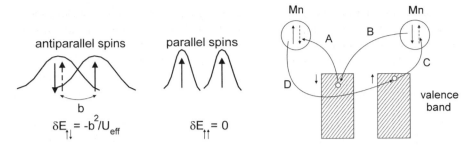

Fig. 13.3. Superexchange interaction. *Left*: wave function for antiparallel and parallel spin configurations. Hopping is possible only for antiparallel configuration and thereby reduces the kinetic energy. *Right*: fourth-order process enabling spin exchange between two Mn atoms via virtual creation of two holes in the valence band. *Solid (dashed) arrows* represent spins in the initial (final) state

Double Exchange

The double exchange mechanism, first introduced by Zener, can be found in mixed-valence compounds, in which magnetic ions exist in different charge states. This mechanism was considered to explain the magnetic properties of manganites and perovskites [20], but could be also important in wide gap DMS such as ZnMnO and GaMnN in which $Mn^{2+}(d^5)$ and $Mn^{3+}(d^4)$ could coexist. Double exchange proceeds via virtual hopping of a d electron from one Mn to another, according to the transition $(d^4 - d^5) \rightarrow (d^5 - d^4)$. If Mn spins are parallel this virtual transition does not cost any energy (in real cases the two Mn sites may be inequivalent, they may feel for example different built-in electric fields). Hence, the gain in kinetic energy is simply given by the hopping integral b introduced above for the superexchange mechanism. If the spins are antiparallel the virtual transition involves an excited state of the Mn d-shell in which the fifth electron has an opposite spin, which costs energy according to Hund's rule. Therefore the gain in kinetic energy is much less for the antiparallel spin configuration, and this results in a ferromagnetic interaction.

RKKY

The d–d exchange interactions considered until now are relevant for insulators. On the contrary, RKKY mechanism becomes efficient when a high density of free carriers is present in the sample, as typical in metals or in highly degenerate semiconductors. It is at the origin of carrier induced ferromagnetism in several DMS. This is an indirect coupling in which a magnetic moment interacts with the spin polarization of the carriers induced by another nearby magnetic moment. Thus is can be treated in second order of perturbation of the s–d (or p–d) exchange interaction. A carrier in the Fermi sea is scattered to an empty state ($k > k_F$) by exchange interaction with the first magnetic atom, and then it is scattered back to its initial state by the second magnetic atom. The calculation was done originally by Rudermann and Kittel who considered the indirect interaction between two nuclei mediated by the hyperfine interaction with free electrons [21]. This is completely analogous to the problem of indirect exchange interaction between two magnetic atoms. The indirect exchange coupling takes the form of the Heisenberg Hamiltonian with $H = -J_{ij}\mathbf{S}_i \cdot \mathbf{S}_j$ [4]

$$J_{ij} = \frac{\rho(\epsilon_F)k_F^3 J_{sp-d}^2}{2\pi} f(2k_F R_{ij}), \quad \text{for the 3D case } f(x) = \frac{\sin(x) - x\cos(x)}{x^4}.$$
(13.7)

J_{ij} is positive at small distances and oscillates with the distance R_{ij} between Mn atoms. These oscillations have the same origin as the Friedel oscillations: they come from the cutoff at around $2k_F$ for the maximum change of carrier wave vector in the exchange scattering with magnetic atoms.

An expression of $f(x)$ has been derived for 2D and 1D cases [22], which are also important because many recent advances in the physics of DMS concern structures of reduced dimensionality.

13.4 Magnetic Properties

Magnetic properties of DMS are governed by the exchange interactions between the local moments introduced by magnetic atoms. In Sect. 13.3.2 we have seen that these interactions can be strongly modified by the presence of free carriers, which introduce a ferromagnetic coupling between local moments, hence the magnetic properties of undoped and doped DMS are discussed separately. We focus on materials containing Mn in the d^5 configuration: the ground state is derived from the ^6S atomic state, so that in the crystal the Mn spin g-factor is highly isotropic due to the lack of orbital momentum.

13.4.1 Undoped DMS

Paramagnetism and the Brillouin Function

At relatively low Mn concentration, or high temperature, Mn spins behave more or less as if they were independent from each other. It means that in a first approximation interactions between Mn spins can be neglected, the orientation of spins in a magnetic field being only limited by thermal activation. The average spin is given by $\langle S_z \rangle = -S\mathbf{B}_S(\xi)$, where $S = 5/2$ and $\mathbf{B}_S$ is the well-known Brillouin function for spin S

$$\mathbf{B}_S(\xi) = \frac{2S+1}{2}\coth\left(\frac{2S+1}{2S}\xi\right) - \frac{1}{2}\coth\left(\frac{\xi}{2S}\right),$$
(13.8)

and the argument ξ is the ratio of Zeeman to thermal energy $\xi = g\mu_B S B/k_B T$. For $\xi \ll 1$, i.e., low field or high temperature, Taylor expansion gives $\mathbf{B}_S \simeq \xi(S+1)/3$.

---

[4] Note that in literature the definition of J_{ij} is not the same everywhere (sign and factor 2 which comes from a double summation over i and j on an ensemble of sites).

The magnetization M is linear in B, $M = \chi_{Mn}B$, where the static magnetic susceptibility χ_{Mn} has the Curie form

$$\chi_{Mn} = \frac{N_0 x (g\mu_B)^2 S(S+1)}{3k_B T} = \frac{C}{T}. \tag{13.9}$$

Antiferromagnetism and the Modified Brillouin Function

Unless in the very dilute limit, when Mn–Mn interactions can be neglected, the Curie law is not sufficient to describe a real DMS. This is fortunate because interesting and non trivial magnetic properties are a consequence of these interactions.[5] Mn–Mn interactions give rise to deviations from the Brillouin function, described as a "modified Brillouin function" and "magnetization steps".[6]

Mn–Mn interactions can be included in a simple mean-field approach considering that, in addition to the external magnetic field, each moment feels the molecular (exchange) field B_m created by all other moments and proportional to their magnetization $B_m = \lambda M$. This gives the well-known Curie–Weiss law $\chi_{Mn} = C/(T - \theta)$, where the Curie–Weiss temperature $\theta = \lambda C$ is negative (positive) if dominant Mn–Mn interactions are antiferromagnetic (ferromagnetic). Using the Heisenberg Hamiltonian it is very easy to calculate B_m and to get the expression for the Curie–Weiss temperature

$$\theta = S(S+1) \sum_j J_{ij}/3k_B, \tag{13.10}$$

where the summation runs over all lattice sites occupied by magnetic atoms, except the "central" site i.

Experimentally the Curie–Weiss law is well verified in DMS at temperatures higher than the typical temperature of the Mn–Mn interactions. Otherwise, higher-order corrections to the Curie–Weiss law can be obtained by expanding the partition function of the Heisenberg Hamiltonian for randomly diluted Mn atoms in powers of $(k_B T)^{-1}$ [27].

At lower temperature, another Curie–Weiss law is observed, with different parameters. Although there is no general expression for the magnetization at arbitrary x, B and T, in this temperature range and for moderate fields, an empirical modified Brillouin function introduced by Gaj, Planel, and Fishman [28] accurately describes magnetization data. In the modified Brillouin function, T and x are replaced respectively by $T + T_0$, where $T_0 > 0$, and $x_{eff} < x$, in order to take into account the presence of antiferromagnetic interactions:

---

[5] Antiferromagnetic and spin–glass phases are observed in DMS with high Mn content (or very low temperatures [23, 24]). The magnetic properties of DMS in these regimes are quite interesting, but beyond the scope of this chapter. The interested reader may refer to [25, 26].

[6] The modified Curie–Weiss law and the magnetization steps described in this section are caused by the interactions between two Mn spins having isotropic g-factor. Very similar deviations from the Brillouin function are also observed for isolated spins with anisotropic g-factor (non-d^5 ions), which also exhibit level crossings already for a single spin.

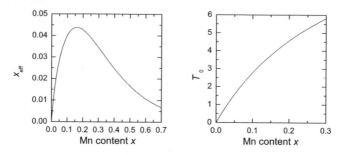

Fig. 13.4. Parameters of the modified Brillouin function, fitted to (CdMn)Te data

$$M = N_0 x_{\text{eff}} g \mu_B S \mathbf{B}_S \big[g \mu_B S B / k_B (T + T_0) \big]. \tag{13.11}$$

Accordingly, the susceptibility may be written as

$$\chi_{\text{Mn}} = \frac{N_0 x_{\text{eff}} (g \mu_B)^2 S(S+1)}{3 k_B T} = \frac{C_{\text{eff}}}{T + T_0}. \tag{13.12}$$

The model has been extensively tested and the values of the parameters have been accurately determined for several materials. Figure 13.4 shows the fit suggested in [29] for (Cd, Mn)Te: $x_{\text{eff}} = x[0.265 \exp(-43.34x) + 0.735 \exp(-6.19x)]$, and $T_0 = (35.37\,\text{K})x/(1 + 2.752x)$. Although essentially phenomenological, this extremely useful model has received an appealing interpretation in terms of a formation of antiferromagnetically coupled nearest-neighbor pairs. In the relevant field and temperature ranges, say, 0 to 5 T and 1.5 to 20 K in (Cd, Mn)Te, x_{eff} closely matches the number of "free spins", i.e., the number of Mn atoms which remain when the nearest-neighbor pairs have been deducted. The number of nearest-neighbors on the metal sublattice in the zinc-blende and the wurtzite structures is 12, so that in the low-x limit and for a random distribution of Mn atoms, one simply expects $x_{\text{eff}} = x(1 - x)^{12}$. Analytical expressions have been given in order to take into account small clusters, which form already for small values of x, for bulk material where they apply up to $x \simeq 0.1$ [30], but also at an interface between the DMS and its non-magnetic host [31]. A Monte Carlo calculation has further shown the validity of the model for higher values of the Mn content, up to $x \simeq 0.8$ [32]. Note that the formation of nearest-neighbor pairs puts a severe limitation to the magnetic properties of a DMS in the field and temperature range where this applies, since the maximum x_{eff} is less than 0.05.

The same mechanism of blocking of the nearest-neighbor pairs gives rise to magnetization steps observed at high field and low temperature, a regime where the modified Brillouin function is not valid (Fig. 13.5). These magnetization steps occur when the magnetic field becomes strong enough to unlock nearest-neighbors Mn pairs, and give a precise measurement of the nearest-neighbors d–d exchange integral [30, 33–37]. The positions of magnetization steps are easily deduced from the nearest-neighbors pair energy levels (Fig. 13.5)

$$E = -J_1 \big[S(S+1) - 35/2 \big] + g \mu_B S_z B, \tag{13.13}$$

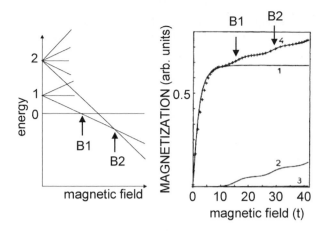

Fig. 13.5. *Left*: lowest energy levels of nearest-neighbors Mn pairs (labeled by their total spin *S*) and two first level crossings (*left*). At each level crossing there is a step in magnetization (*right*) which adds a component (*2*), (*3*) to the modified Brillouin function (*1*). From [35]

where J_1 is the nearest-neighbors exchange constant, S is the total spin of the pair, and S_z its projection along the magnetic field.[7]

13.4.2 Carrier-Induced Ferromagnetism

In Sect. 13.3.2 it was shown that, in a metal containing disordered magnetic impurities or in a degenerate DMS, the sign of the RKKY coupling is oscillating with the distance between the local moments. In a metal at low temperature, it gives rise to spin-glass phases or to complex ordered phases. In a doped DMS, the free carrier density is usually lower than the magnetic atom density. This means that the average distance between magnetic atoms is less than the inverse Fermi wave vector, so that the dominating interaction between nearest magnetic atoms is expected to be ferromagnetic. The Curie–Weiss temperature can be calculated using (13.7) and (13.10) and replacing the summation over the magnetic ion position by an integration; one gets the Curie–Weiss temperature [22]

$$\theta = S(S+1)N_0 x \rho(\epsilon_F) I^2 / 12 k_B, \tag{13.14}$$

with $I = \alpha$ for electrons and $I = \beta$ for holes. As θ is positive, a paramagnetic to ferromagnetic transition is expected at the Curie–Weiss temperature, with appearance of a spontaneous magnetization below θ.

Zener Model

It is also instructive, and convenient for calculations (see [7]), to adopt the point of view of Zener [38, 39] and consider that the stable magnetic phase is that which min-

---

[7] The exchange interactions with distant Mn atoms are weak, but nevertheless they slightly shift the magnetization steps and must be taken into account for a precise determination of J_1.

imizes the total free energy of the system. The Zener model assumes uniform spin polarization (neglects Friedel oscillations). The free energy functional[8] of the interacting carriers and Mn can be expanded in powers of Mn and carrier magnetizations, M and m, respectively, up to second order, as

$$F(M, m) = \frac{1}{2\chi_{Mn}} M^2 - MB + \frac{1}{2\chi_c} m^2 - mB - \frac{I}{(g\mu_B)(g_c\mu_B)} Mm. \quad (13.15)$$

The first and second set of two terms represent the free energy of Mn and carriers, respectively. The form of these terms, which depends on the magnetic susceptibilities χ_{Mn} and χ_c, is dictated by the requirement that minimizing the free energy functional in an external magnetic field B should yield the correct value of the magnetization for each subsystem (Mn or carriers). The physical meaning of these terms is that the more it costs (free) energy to align spins, the smallest is the susceptibility. Finally, the last term stands for the Mn–carrier exchange energy. We ignore for the moment the vectorial character of all quantities. Minimizing $F(M, m)$ with respect to m leads to

$$m = \chi_c I M / (g\mu_B g_e \mu_B), \quad (13.16)$$

which is nothing but the spin polarization of the carrier gas resulting from the giant Zeeman splitting to be described in Sect. 13.5.1, with $I = \alpha$ for electrons and $I = \beta$ for holes. In a symmetric way, minimizing $F(M, m)$ with respect to M leads to

$$M = \chi_{Mn} I m / (g\mu_B g_c \mu_B). \quad (13.17)$$

The key parameter is thus

$$I^2 \frac{\chi_{Mn}}{(g\mu_B)^2} \frac{\chi_c}{(g_c\mu_B)^2}. \quad (13.18)$$

If this is less than unity, the equilibrium magnetization is zero in the absence of an applied field: this is the high temperature, paramagnetic phase. If the temperature is lowered, χ_{Mn} increases and the characteristic parameter can reach unity: this sets the Curie temperature T_C, which can be calculated by inserting the expressions for the Curie law (13.9), and Pauli [$\chi_c = 1/4(g_c\mu_B)^2\rho(\epsilon_F)$] susceptibility in this expression. The result is the same as (13.14). Antiferromagnetic Mn–Mn interactions can be easily taken into account by use of the Curie–Weiss susceptibility, (13.12). On the one hand, this reduces χ_{Mn} and thus T_C. On the other hand, carrier–carrier exchange interactions enhance χ_c (by a factor A_F called the Fermi liquid parameter) and thus increase T_C.

Equations (13.16) and (13.17) describe carrier induced ferromagnetism as a mutual polarization, of the carrier gas by the localized spins (through the giant Zeeman effect), and of the localized spins by the carriers (by an effect similar to the Knight shift). The two effects can be measured separately, see Sects. 13.5.1 and 13.6.3. Below T_C, a finite, spontaneous magnetization is expected (which can be calculated using the complete expressions for the magnetization of each system, not the linearized ones involving the susceptibility). In regular metals, the magnetization is locally saturated and the spin polarization of the carriers is only partial and follows the

---

[8] An equivalent expression of the free energy is given in [40].

magnetization: this configuration may be achieved in strongly doped $(Ga, Mn)As$, see Sect. 13.5.5. If the carrier density is low, carriers will be fully spin-polarized (a so-called "half-metal" is realized) and one expects the magnetization of the Mn system to be weak [22] and to keep increasing when lowering further the temperature. This gives rise to an unusual, characteristic upward curvature of the $M(T)$ curve: both the complete polarization of the carriers and the upward curvature of the $M(T)$ curve are observed in $(Cd, Mn)Te$ quantum wells [41].

Role of the Valence Band

The Curie temperature predicted by (13.14) is proportional to the square of the carrier–Mn exchange coupling I, and through the Pauli susceptibility to the density of states at Fermi energy. Therefore holes are much more efficient than electrons in mediating ferromagnetism in DMS, because both exchange coupling (see Sect. 13.5.1 below) and effective mass are larger. However, there is much more information contained in the susceptibility of holes, and spin–orbit interaction induces a strong valence band anisotropy, which in general cannot be neglected. For strong spin–orbit coupling (typical for tellurides) the heavy-hole spin must be aligned along its k-vector close to the center of the Brillouin zone. This reduces the exchange spin splitting by a factor $|\cos(\theta)|$, where θ is the angle between k and the magnetization [42]. Since the spin projection enters the susceptibility to the square, after averaging $\cos^2(\theta)$ over all k-vector directions, one gets the hole spin susceptibility reduced by a factor of 3. Actually, the reduction is a little bit less when light-hole/heavy-holes terms are taken into account [8]. This was applied to p-doped $(Zn, Mn)Te$. In $(Ga, Mn)As$, with a smaller spin–orbit coupling and a larger Fermi energy (i.e., holes further apart from the center of the Brillouin zone), a more detailed calculation must be performed taking into account the nonparabolic structure of the valence band, and this gives rise to a very complex anisotropy [43]. A very simple case is that of a two-dimensional hole gas such as realized in $(Cd, Mn)Te$ quantum wells, see below Sect. 13.5.2. It is a great advantage of the Zener model that it allows such a straightforward and hand-waving incorporation of all the complexity of the valence band structure.

Disorder

The basic assumption of the Zener model is that disorder can be neglected, and doped DMS offer many sources of disorder (random distribution of spins, random distribution of acceptors, localization of carriers and metal-insulator transition, many types of defects...). Many theoretical studies deal with the problem of disorder in doped DMS. A preliminary review is [44]. According to an extreme assumption, carriers are fully localized, and form magnetic polarons by aligning the Mn spins in their neighborhood. The size of the magnetic polaron increases when the temperature is lowered, thus increasing the polaron–polaron interaction: two interacting polarons tend to align parallel to each other in order to avoid frustration amongst their common spins. A ferromagnetic phase occurs when the interaction becomes strong enough (see, e.g., [45]).

More sophisticated models have been proposed which rely upon an ab initio calculation of the Mn–Mn interaction in various configurations; these data are then used in analytical models aimed at describing disorder properly without involving lengthy computations [46]. A general feature of these models is that the Curie temperature calculated for a random distribution of Mn impurities is generally lower than in the mean field model. However more recent analyses suggest that a intentional inhomogeneous distribution of the magnetic impurities in a DMS could give rise to both a high Curie temperature while preserving strong effects in magnetotransport or magneto-optics [7].

The Zener model has lead to the prediction that p-doped Ga(Mn)N and Zn(Mn)O could be ferromagnetic at room temperature, which triggers considerable effort in growing such structures. The main assumption is that one can incorporate Mn impurities, up to $x_{\text{eff}} = 0.05$, into ZnO or GaN with a strong p-type doping (3.5×10^{20} holes per cm^3), and that it retains the same d^5 configuration. While the incorporation of Mn and doping are technological challenges left to crystal growers, the question of the electronic configuration gave rise to many debates amongst both theoreticians and experimentalists, with many different methods employed and compared.

13.5 Basic Optical Properties

13.5.1 Giant Zeeman Effect

Linear Approximation for the Spin–Carrier Interaction

Let us first focus onto the simplest case, that of a II–VI semiconductor with zinc-blende structure, a moderate band gap, and Mn impurities in the d^5 configuration, i.e., in the isotropic $^6A_1$ ground state with zero orbital momentum and no electrical activity.[9] (Cd, Mn)Te is one possible example. Vertical interband transitions with selection rules described in Chap. 1 give rise to excitons. They strongly dominate photoluminescence, photoluminescence excitation, reflectivity, and transmission spectra at photon energies close to the band gap. More precisely, they are shifted by the exciton binding energy, E_B, with respect to interband transitions depicted in Fig. 13.6(a).

The unique feature in a DMS is the so-called *giant Zeeman effect*, which directly results from the s–d and p–d interactions described in Sect. 13.3.1. Calculating the expectation value of the s–d exchange Hamiltonian, cast in the form of (13.3), we obtain a spin splitting of the conduction band equal to $\alpha \langle \sum_i S_i \delta(\mathbf{r} - \mathbf{R}_i) \rangle$. If we do not worry too much about the exact meaning of the average in this expression, we can recognize the definition of the magnetization of the Mn spins: *this is the key feature*

---

[9] Magnetic impurities with a non-d^5 configuration have a non-zero orbital momentum. This usually gives rise to a more complex, anisotropic behavior of the magnetization of the magnetic impurity, and also to a complex, but different, behavior of the giant Zeeman effect [47, 48]. The Jahn–Teller effect can significantly alter this anisotropy [49].

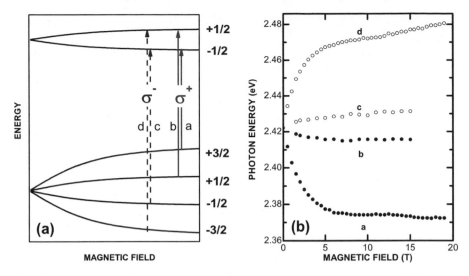

Fig. 13.6. (a) Scheme of band to band transitions close to band gap energy, and their polarization, in the Faraday configuration. The spin states of the electron in the conduction band, and hole in the valence band, are indicated. One assumes $\alpha > 0$ and $\beta < 0$ as observed in tellurides and selenides. (b) Experimental transition energies in a $Zn_{0.95}Mn_{0.05}Te$ sample, as a function of the applied field, in the Faraday configuration [3]

of a DMS, the bands split proportionally to the Mn magnetization [28]. The correct method to evaluate the average is to use the virtual-crystal approximation over the Cd and Mn distribution, and a mean-field approximation for the Mn spins. Writing a similar expression, with α replaced by β, for the valence band, we obtain a simple description of the energies of the transitions shown in Fig. 13.6. In the presence of a magnetic field applied in the z-direction, they are given by

$$E_{\pm} = E_G - E_B \pm \frac{1}{2}N_0(\alpha - \beta)x_{\text{eff}}\langle S_z \rangle, \tag{13.19}$$

for the heavy-holes excitons with σ^+ and σ^- circular polarizations, respectively, and

$$E_{\pm} = E_G - E_B \mp \frac{1}{2}N_0\left(\alpha + \frac{\beta}{3}\right)x_{\text{eff}}\langle S_z \rangle, \tag{13.20}$$

for the light-hole excitons. Here E_G is the band gap energy, E_B the binding energy of the exciton, and the average spin $\langle S_z \rangle$ is given by the modified Brillouin function. In order to give orders of magnitude, let us evaluate the giant Zeeman splitting for $Cd_{1-x}Mn_xTe$: $E_G = (1606+1750x)$ eV, $E_B = 10$ meV, $N_0\alpha = 0.22$ eV and $N_0\beta = -0.88$ eV. That means that at low temperature (2 K) and fields around a few Teslas so that $\langle S_z \rangle = -\frac{5}{2}$, and for $x_{\text{eff}} = 0.04$, the giant Zeeman splitting easily reaches

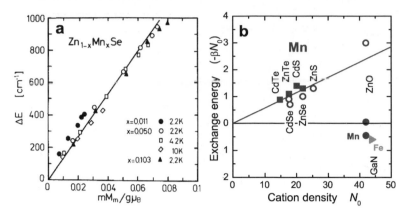

Fig. 13.7. (**a**) Experimental check of the proportionality between the giant Zeeman effect and the Mn magnetization [52]. (**b**) Exchange parameter in different II–VI DMS with Mn, and in GaN:Fe which has the same d^5 configuration, determined either from magnetooptics (*closed symbols*) or from X-ray spectroscopy (*open symbols*) [43, 53]

100 meV—a significant fraction of the band gap.[10] Another example is shown in Fig. 13.6.

Determination of Exchange Integrals

Direct checks of the proportionality between the Zeeman shifts of the four lines and magnetization have been performed in most DMS for different Mn content, temperature and applied field. An example is shown in Fig. 13.7(a). The presence of both the light-hole and heavy-hole excitons, which split proportionally to $N_0(\alpha + \beta/3)$ and $N_0(\alpha - \beta)$, respectively, allows one to determine both $N_0\alpha$ and $N_0\beta$. A general trend is that $N_0\alpha$ assumes the same positive value, $N_0\alpha \simeq 0.2$ meV, in all II–VI DMS, while $N_0\beta$ is large, negative, and proportional to N_0, see Fig. 13.7.

The other method of determining $N_0\beta$ is to use (13.6) with parameters deduced from X-ray spectroscopy [6]. The values found are generally very close to those determined from magneto-optical spectroscopy (Fig. 13.7), which gives some confidence in the model. This method is now commonly used to predict values of the exchange integral in materials, where the magneto-optical determination has not been achieved [6, 50, 51].

Deviations from the Simple Model

The previous description must be altered in many occasions: in semiconductors with wurtzite structure, when the exciton binding energy is large, in the case of a strong

---

[10] These large values of the giant spin splitting are often cast into an appealing but sometimes misleading form, that of an effective Landé factor which can be up to several hundreds. However one should keep in mind that this effective g-factor is strongly temperature dependent, and that the giant Zeeman effect is highly non-linear in field.

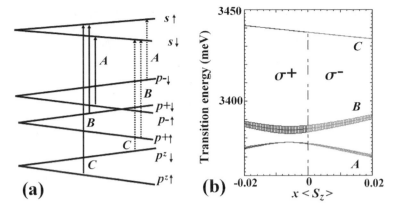

Fig. 13.8. (a) Giant Zeeman shifts of the conduction and valence bands in a wurtzite semiconductor, drawn with α and $\beta > 0$, and optical transitions in σ^+ (*solid arrows*) and σ^- (*dotted*)
polarizations. (b) The anticrossing of the corresponding excitons in a ZnO-based DMS [55]

p–d exchange, in the case of magnetic impurities with a non-zero orbital momentum.
We will only briefly address these issues.

Semiconductors with light anions (some selenides, sulfides, ZnO, GaN) usually
have the wurtzite structure. Samples of best quality are often epitaxial layers with
the hexagonal c-axis as the growth axis. We will focus on the optical properties in
the Faraday configuration, with the directions of light propagation and magnetic field
parallel to the c-axis, which is then the natural quantization axis (z in the following).
In addition, these semiconductors feature a small spin–orbit coupling. As a result
(Fig. 13.8(a)), the p-like states at the top of the valence band are split by the hexago-
nal anisotropy, into a ($|p^z\uparrow\rangle$, $|p^z\downarrow\rangle$) doublet, inactive in σ polarization (which gives
rise to the so-called C-exciton), and states formed on the $|p^+\rangle\,|p^-\rangle$ orbital states (or
$|\pm 1\rangle$ states, which behave as $x \pm iy$) which are optically active in σ polarization. The
spin–orbit coupling further splits these states into two doublets, ($|p^+\uparrow\rangle$, $|p^-\downarrow\rangle$) and
($|p^+\downarrow\rangle$, $|p^-\uparrow\rangle$), thus forming the so-called A and B excitons. The electron–spin is
conserved in these optical transitions, so that opposite giant Zeeman shifts apply to A
and B, proportional to $N_0|\alpha - \beta|$. Exciton C acquires a non-zero oscillator strength
through spin–orbit coupling, so that the electron spin is flipped during the transition
and the shift is proportional to $N_0|\alpha + \beta|$. However if the oscillator strength of C
remains too weak, the straightforward determination of $N_0\alpha$ and $N_0\beta$, separately, is
not possible.

In addition, other parameters have to be taken into account properly, so that the
simple description, where excitons closely follow the band-to-band transitions, completely fails. This is particularly true in ZnO and GaN: the electron–hole exchange
interaction which occurs within the tightly bound excitons (exciton binding energies
are as high as 25 meV in GaN and 60 meV in ZnO), strongly mixes the original states.
This causes strong anticrossings when excitons approach each other as a result of the

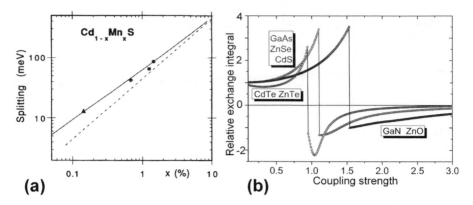

Fig. 13.9. (**a**) Giant Zeeman splitting in (Cd, Mn)S. *Symbols* are experimental data from litera-ture, *the dotted line* is the mean field model, and *the solid line* the nonperturbative model [56]. (**b**) Effective exchange integral describing the giant Zeeman splitting in wide band gap DMS based on GaN or ZnO, as seen in magneto-optics, relatively to the actual exchange integral, as a function of the depth of the local potential (adapted from [58])

giant Zeeman effect [54], see Fig. 13.8(b), so that the shift of the exciton is far from being proportional to the magnetization.

The picture is also significantly altered in the case of a strong p–d exchange, so that the perturbation scheme no more holds. The effect of a single magnetic im-purity on the carrier can be described by a local potential with two contributions: the chemical shift, which does not depend on the spin of the carrier as in all ternary semiconductors, and the spin-dependant s–d or p–d exchange interaction. Rough estimates for holes in (Cd, Mn)Te are provided by the values of the valence band off-set (about 0.5 eV) and the p–d exchange parameter $N_0\beta = -0.88$ eV, respectively. For holes with spin parallel to the Mn spin, the two contributions add, giving rise to a repulsive potential. For holes with antiparallel spin, the two contributions partly cancel so that the potential is slightly attractive, but too weak to form a bound state (remember that a quantum dot with a too small height and extension does not create a bound state). In this case, the mean field model and virtual crystal approximation hold.

In a DMS with a larger p–d exchange interaction, the local attractive potential for a hole with spin antiparallel, is close to the limit where a bound state is formed. A nonlinear behavior then sets up. For instance, it was argued that the effective width of the bound state is determined by the Mn–Mn distance, hence by x_{eff}, so that the spin splitting is strongly enhanced at low Mn content. Experimental deviations from the mean field prediction were observed in (Cd, Mn)S (Fig. 13.9), and explained by such a nonperturbative scheme [56], with the same value of the band offset but with $N_0\beta = -1.5$ eV. The potential seen by the carrier is in fact due to a random distribution of localized dots, with variable depth determined by the spin state of the Mn atom: alloy theory was applied later to calculate the whole giant Zeeman splitting [57].

Finally, in the case of an even stronger p–d exchange interaction, a bound state is expected. A comparison between the results of X-ray spectroscopy and magneto-optics strongly suggests that this is the case in ZnO and GaN: while $N_0\beta$ calculated from X-ray spectroscopy data and (13.6) keeps to be proportional to N_0 even for these large band gap semiconductors [51], magneto-optics demonstrates a surprisingly small, positive value of $N_0\beta$ (Fig. 13.7). It has been proposed recently [58] that this can result from the formation of a bound state, see Fig. 13.9, with a transfer of the oscillator strength to the anti-bonding states. Experimental results on GaN:Fe [59] support this analysis, which could apply to GaN:Mn and ZnO-based DMS as well.

Other Magneto-Optical Spectroscopic Techniques

The giant Zeeman effect described in the previous section gives rise to optical magnetic circular dichroism $\rho = [I(\sigma^+) - I(\sigma^-)]/[I(\sigma^+) + I(\sigma^-)]$, where $I(\sigma^\pm)$ is the intensity transmitted for incident light with the $\sigma^\pm$ circular polarization. In its simple form, one assumes that the absorption curves for σ^+ and σ^- polarizations rigidly shift in opposite directions. Then the dichroism can be expressed as [60]

$$\rho = -\frac{\text{dln}(T(E))}{\text{d}E} \frac{\Delta E}{2},$$

(13.21)

with the zero-field transmission, $T(E)$, measured with non-polarized light at energy E, and $\Delta E = E(\sigma^+) - E(\sigma^-)$ the Zeeman splitting, a formula which is routinely used to test the intrinsic character of carrier induced ferromagnetism in a DMS. If ΔE is the giant Zeeman splitting, and if it is proportional to the magnetization, then indeed the dichroism can be used as a measure of that magnetization.

Figures 13.6(a) and 13.8(a), show that the dichroism at the band gap is related to the presence of spin–orbit coupling in the valence band, which splits sub-bands with different signs of the spin splitting (Γ_8 and Γ_7 in zinc-blende semiconductors, Γ_9 and the topmost Γ_7 in wurtzite semiconductors, see Fig. 13.1). If the different excitons giving rise to the dichroism are not clearly resolved, the dichroism is reduced and some care must be taken in the interpretation of a magnetic circular dichroism spectrum. This is the case in wide band gap DMS, which have a small spin–orbit coupling.

Also, the presence of a large density of carriers in the band with the heavier mass can change the sign of ρ due to the so-called Moss–Burstein shift. This is exemplified for a CdMnTe quantum well in Fig. 13.10. For the populated spin sub-band the absorption edge corresponds to vertical transitions at the Fermi wave vector. Hence, the absorption edge energy is increased by the kinetic energy of the excited electron–hole pair, which can be larger than the giant Zeeman splitting: the decisive parameter is the value, for each band, of the product of the Zeeman shift by the effective mass. As a result, the presence of a hole gas changes the sign of the shift of the absorption edge, Fig. 13.10(b). A change of sign of ρ follows. Similar effects have been invoked to explain an unexpected sign of ρ in GaMnAs [61], see Sect. 13.5.5.

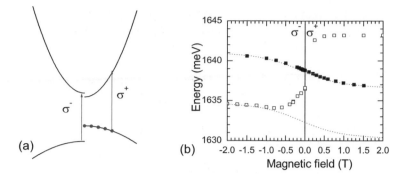

Fig. 13.10. (a) Moss–Burstein shift in the presence of a spin-polarized hole gas. With respect to the transition in σ^- polarization at $k = 0$, the transition in σ^+ polarization is at smaller energy if it occurs at $k = 0$ (no carriers) and at higher energy if it occurs at k non-zero (in the presence of carriers). (b) Position of the absorption in a CdMnTe quantum well with a hole gas ($p \simeq 2 \times 10^{11}$ cm^{-2}, *open squares*) and depleted (*closed squares*) [62]

Magnetic circular dichroism is the signature of a different absorption in the two circular polarizations. Kramers–Krönig relations imply that there is a corresponding difference in the σ^+/σ^- optical indices. This gives rise to the Faraday effect: a linearly polarized light transmitted along the direction of the magnetization experiences a rotation of its polarization proportional to the giant Zeeman splitting. Similarly, a linearly polarized light reflected at the surface of the material experiences a rotation of its polarization proportional to the giant Zeeman splitting (magneto-optical Kerr effect). Both methods are very useful to measure the magnetization of a DMS, in particular with spatial resolution or time resolution.

13.5.2 Optically Detected Ferromagnetism in II–VI DMS

The presence of a two-dimensional hole gas in a (Cd, Mn)Te quantum well, obtained either by remote nitrogen acceptors or by surface electron traps, induces a ferromagnetic transition between the localized Mn spins. This can be viewed as a two-dimensional version of carrier induced ferromagnetism, as described in Sect. 13.4.2. All observations have been done through spectroscopy, and particularly from the giant Zeeman splitting of the photoluminescence line, which is a measure of the magnetization of the Mn system. Basic features are a strong enhancement of the susceptibility (i.e., the slope of the giant Zeeman splitting induced by an applied field) in the high-temperature, paramagnetic phase, and a zero-field splitting in the low temperature, ordered phase, which demonstrates the presence of a spontaneous magnetization [63]. An attractive feature is that the low density of the carrier gas allows one to drive the system, isothermally and reversibly, from the paramagnetic to the ordered phase, just by changing the carrier density by an applied electric field (bias across a p–i–n structure) or by above-barrier illumination, Fig. 13.11 [64].

Another consequence of the low carrier density is that the Fermi wave vector is small enough to use a parabolic description of the valence band. In usual sam-

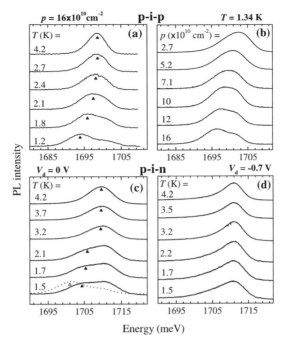

Fig. 13.11. Photoluminescence spectra for a modulation-doped p-type $Cd_{0.96}Mn_{0.04}Te$ quantum well located in a p–i–p structure and a modulation-doped p-type $Cd_{0.95}Mn_{0.05}Te$ quantum well in a p–i–n diode; (a) p–i–p structure at constant hole density and various temperatures; (b) same p–i–p structure with above barrier illumination to control the hole density, at fixed temperature; (c) p–i–n structure without bias at various temperatures: the hole density is constant (*solid lines*) or increased (*dotted line*) by additional Ar-ion laser illumination; (d) p–i–n structure with a -0.7 V bias (depleted quantum well) at various temperatures. Splitting and shift of the lines mark the transition to the ferromagnetic phase [64]

ples, grown coherently on a $(Cd, Zn)Te$ substrate with a smaller lattice parameter, the strain in the quantum well is compressive and its effect on the valence band adds to that of confinement: as a result, the two-dimensional hole gas is made only of heavy holes close to the center of the Brillouin zone (the Fermi energy is a few meV). As described in Chap. 1, these heavy holes form a doublet, $|3/2\rangle = |p^+\uparrow\rangle$ and $|-3/2\rangle = |p^-\downarrow\rangle$, with the quantization axis z normal to the quantum well, $|\uparrow\downarrow\rangle$ $(=|\pm1/2\rangle)$ labeling the spin states and $|p^\pm\rangle$ $(=|\pm1\rangle)$ the orbital moment. Thus this doublet is highly anisotropic, with $\pm1/2$ spin values along the z-axis but vanishing spin components in the x, y plane, hence with a usual Pauli spin susceptibility χ_{zz} but vanishing χ_{xx} and χ_{yy}. The spin susceptibility of the hole gas directly enters the spin–carrier coupling which governs carrier-induced ferromagnetism (see Sect. 13.4.2). Hence, a positive Curie–Weiss temperature is expected only when mea-

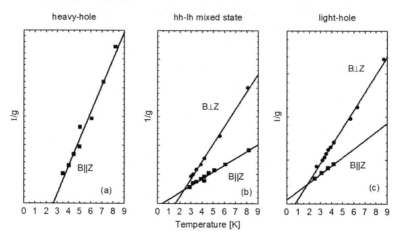

Fig. 13.12. The Curie–Weiss behavior obtained from photoluminescence spectroscopy in the paramagnetic phase of a p-doped (Cd, Mn)Te quantum well, with magnetic field parallel ($B\|z$) and perpendicular ($B\perp z$) to the normal to the well. The Landé factor is proportional to the Mn susceptibility. Sample (**a**) contains heavy holes, sample (**c**) light holes, and (**b**) is intermediate [65]

suring the susceptibility of the coupled system with the field applied normal to the quantum well.[11] This is checked in Fig. 13.12 [65].

In addition, in a broader quantum well grown on a CdTe substrate, which results in a smaller confinement energy overcompensated by the effect of the tensile strain, one forms a two-dimensional gas of *light* holes, with the opposite anisotropy. In this case the positive Curie–Weiss temperature is observed with the field applied in-plane.

Hence band-engineering techniques can be used to control the magnetic anisotropy, and the properties of (Cd, Mn)Te quantum wells are very well understood. The same mechanism gives rise to a complex behavior in (Ga, Mn)As where the Fermi energy is large, so that high-energy parts of the valence band are involved [43].

13.5.3 Quantum Dots

Quantum dots incorporating magnetic ions have revealed specific features—such as a broadening and a progressive redshift of the line—related to the dynamics of the spin–carrier interaction and to the formation of a magnetic polaron [66, 67]. An ultimate case of spin–carrier coupling is that of a single Mn spin, with a small, controlled number of carriers, as realized in CdTe quantum dots and observed by optical spectroscopy [68]. The single Mn impurity has been incorporated in the CdTe quantum dot during its growth. Electrons and holes can be introduced in the quantum dot by adjusting an electrical bias, electron–hole pairs are introduced by optical excitation either in the barrier or (better) by selective excitation in the dot. Thus three different

---

[11] This can be shown by introducing the three inequivalent directions in the expression of the free energy given by (13.15).

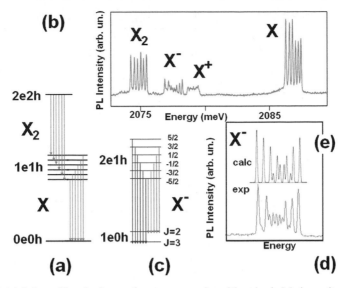

Fig. 13.13. (**a**) Selected level schemes for a quantum dot with a single Mn impurity, and neutral excitons. (**b**) Photoluminescence spectra showing the multiplets due to the so-called biexciton, positively charged exciton, negatively charged exciton, exciton. (**c**) Scheme of levels, and experimental (**d**) and theoretical (**e**) spectra, for the "charged exciton" X^- [68, 69]

types of objects interact: the 5/2 spin of Mn, the 1/2 spin of electrons, both with isotropic g-factor, and holes. As the dots are rather flat in shape, these are mostly of heavy-hole character, with a strongly anisotropic g-factor as described earlier for quantum wells.

With no carrier in the dot, the Mn spin gives rise to a six-fold degeneracy. With a single electron in the dot, the ferromagnetic exchange coupling with the Mn impurity results in two levels: a ground level where the two spins are parallel, with total spin $5/2 + 1/2 = 3$; an excited level where they are antiparallel with total spin $5/2 - 1/2 = 2$. The isotropy is preserved.

With a single hole in the dot, the anisotropy of the heavy hole, with $|\pm3/2\rangle$ states along the normal to the rather flat dot, imposes the axis of quantization: the antiferromagnetic coupling between the hole and the Mn spin can be viewed as an exchange field, along the growth axis, which lifts the six-fold degeneracy of the Mn spin. We thus expect six sublevels, each with a two-fold degeneracy, from $(|5/2\rangle|-3/2\rangle)$, $|-5/2\rangle|3/2\rangle$ to $(|5/2\rangle|3/2\rangle, |-5/2\rangle|-3/2\rangle)$. A similar scheme is expected in the presence of an optically active electron–hole pair (where the electron spin is antiparallel to the hole spin). Finally, adding two electrons (or two holes) does not change the picture since they automatically form a spin singlet. Thus the level scheme corresponding to one Mn spin, with 0, 1, 2 electrons and/or 0, 1, 2, holes, is shown in Fig. 13.13(a) and (c).

The first experimental evidence came from the photoluminescence of an electron–hole pair. Six sharp, intense, well-resolved lines are observed [68] for each circular polarization, as shown in the right part of the spectrum in Fig. 13.13(b). They replace the single sharp line observed on a nonmagnetic CdTe dot, and correspond to the six sublevels in the initial state and the full degeneracy in the final state. A rough estimation of the splitting for a Mn spin at the center of a dot of volume V, $|\alpha - \beta|/2V$, gives the correct order of magnitude $\sim 100\,\mu\mathrm{eV}$ for $V = 3 \times 10 \times 10\,\mathrm{nm}$.

Other features are observed in the presence of additional carriers. For instance, the X^- photoluminescence (Fig. 13.13(d)) can be understood as composed of two series of lines (Fig. 13.13(e)), with six sublevels in the initial state which contains an unpaired hole and two electrons forming a spin singlet, and two sublevels in the final state which contains an unpaired electron (Fig. 13.13(c)) [69]. The intensity of each line is simply given by the overlap of the Mn spin states in the initial and final states.

The oversimplified model described here accounts for the most simple spectra. More complex spectra have been attributed to asymmetric structures, to the anticrossing of dark excitons and bright ones, to a spatial dependance of the extension of the carrier wave function on the number of carriers, and so on.

13.5.4 Spin-Light Emitting Diodes

The idea of the spin-light emitting diodes gave a new impulse to efforts towards spin injection in semiconductors, as it provides an easy way to measure it in GaAs-based structures. Recent successes in spin-injection in GaAs are for a good part due to the application of this easy test. In the most straightforward configuration, a quantum well is inserted between a source of electrons spin-polarized along the normal to the quantum well, and a source of nonpolarized holes. The electroluminescence is measured in the direction normal to the quantum well, and its helicity is measured. If only heavy holes are involved, photons with σ^+ helicity are due to the recombination of a $|3/2\rangle$ hole and a $|-1/2\rangle$ electron, and photons with σ^- helicity are due to the recombination of a $|-3/2\rangle$ hole and a $|1/2\rangle$ electron. If the hole source is unpolarized, it provides an equal flux of $|3/2\rangle$ and $|-3/2\rangle$ holes, and the helicity of light is a direct measure of the spin-polarization of the electron arriving to the quantum well. If light holes are also involved, a correction factor equal to 2 has to be applied to take into account the optical selection rules. This configuration was invoked in [70] and more quantitatively assessed in [71], with in both cases a II–VI DMS as the source of spin-polarized electrons. Following this pioneering work, other spin injectors have been demonstrated, including ferromagnetic DMS and ferromagnetic metals. Note that, due to the anisotropy of holes, the correct configuration of a spin-LED using a quantum well requires that the spin polarization of the injected electrons be along the normal to the quantum well.

13.5.5 III–V Diluted Magnetic Semiconductors

To date, $Ga_{1-x}Mn_xAs$ is certainly the most studied diluted magnetic semiconductor. This is due to the presence of carrier induced ferromagnetism, to attractive magnetotransport properties, and to the many opportunities to fabricate heterostructures and even basic devices based on these magneto-transport characteristics. However, we will only give a short overview of this field, for two reasons.

First, since Mn behaves as an acceptor in GaAs, only p-doped samples (possibly compensated) are available. Therefore the resulting "dilute ferromagnetic semiconductor" must be taken as a whole, with less opportunities to identify basic mechanisms, which could be described analytically and thus lead to a more pedagogical presentation of the matter, in line with the scope of this book. In particular, optical spectroscopy is certainly not the easiest way to study such a highly doped semiconductor. Macroscopic tools such as magnetotransport and, to a lesser extent, magnetometry, are used instead.

The second reason is that our current understanding of this material still rapidly evolves: the model of ferromagnetism due to free holes [43, 72] is by far the one which explains most of the experimental findings, however how much disorder alters this oversimplified description, particularly at low carrier density, is still a matter of debate.

Disorder in (Ga, Mn)As is invoked because ferromagnetism is observed on both sides of the metal-insulator transition, and for theoretical reasons (see references in Sect. 13.4.2). Infrared spectroscopy can provide the relevant information by detecting the presence of an impurity band: absorption should be observed when an electron is excited from the valence band to the partly empty impurity band. However, transitions from the light-hole to the heavy-hole band should produce absorption in the same spectral range. A recent analysis of data available in the literature [9] concludes that the first mechanism applies to $Ga_{1-x}Mn_xAs$ with <1% Mn, while the latter one is more likely above 2% Mn.

In addition, the growth of $Ga_{1-x}Mn_xAs$ is still a challenge: best samples were obtained by molecular beam epitaxy at low temperature, with a strict control of the As incorporation, and subsequent low temperature annealing to eliminate Mn interstitials. Up to now, the record is $T_C = 173\,K$ [73], several groups having achieved similar values in very thin layers. But the incorporation of substitutional Mn remains below the values needed to achieve room temperature ferromagnetism—even according to the most optimistic estimates [74].

Particularly noteworthy is the strong anomalous Hall effect, which is observed in a p-doped DMS such as p-doped $Zn_{1-x}Mn_xTe$ [8], $Ga_{1-x}Mn_xAs$ [75] or $Ge_{1-x}Mn_x$ [76, 77]. Anomalous Hall effect is well known in magnetic metals, where it is considered as an easy way to measure the magnetization. This method was extensively used in $Ga_{1-x}Mn_xAs$ also, although it has been known for a long time that it is actually related to the spin polarization of the carriers: it is the spin–orbit coupling of the carrier itself which causes an asymmetric, spin dependant behavior of the scattering on nonmagnetic impurities. As a result, such a "spin dependant Hall effect" was measured in InSb, a semiconductor with a large g-factor, which allows to strongly polarize the electrons in the absence of any magnetic impurities [78]. In the case of a DMS with a large carrier density, the spin polarization of the carriers (and hence the transverse bias due to the anomalous Hall effect) may be expected to be proportional to the Mn magnetization: a recent theory [79] favorably compares to experimental data in GaMnAs [80].

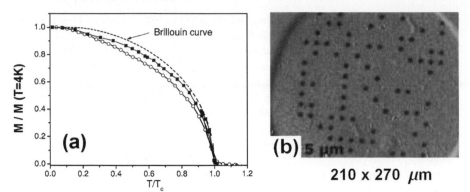

Fig. 13.14. (**a**) Temperature dependence of the magnetization (*open circles*) of an annealed GaMnAs sample with perpendicular magnetization, measured by SQUID magnetometry under $H = 250$ Oe applied along the growth axis, compared to that of the remnant PMOKE signal (*closed squares*). For comparison, the mean-field $S = 5/2$ Brillouin function is also plotted (*dashed line*) [83]. (**b**) Kerr microscopy snapshot of the magnetization reversal of dots patterned by hydrogen passivation [84]

As shown above for II–VI DMS, magneto-optics constitutes another class of methods bringing information on the spin–carrier coupling and on the spin-polarization of the carriers. Magnetic circular dichroism is observed in (Ga, Mn)As at photon energies close and below the band gap [81]. It was explained by invoking the effect of Moss–Burstein shift (which changes the sign of the dichroism) and disorder [61]. However, more recent studies over a wide range of composition conclude that there is probably still more to understand [82]. Faraday rotation and magneto-optical Kerr effect are closely related methods: magneto-optical Kerr effect is routinely used to get information on the magnetization of layers with perpendicular anisotropy (both effects are proportional to each other, at least in layers of good quality, see Fig. 13.14). Magneto-optical Kerr microscopy can be used to image magnetic domains in a uniform layer [83] and micrometer-sized dots obtained by hydrogenation [84].

13.6 Spin Dynamics

In a paramagnetic material spin dynamics is essentially reduced to single particle spin relaxation, of either carrier or magnetic atom, while in a ferromagnetic material one deals with collective spin dynamics of a large number of coupled carrier and Mn spins, such as magnetization precession and spin waves. Interesting collective effects may also exist in paramagnetic material provided the coupling between carriers and Mn spins is strong enough. Soft mode of precession in the vicinity of the ferromagnetic transition, mixed electron–Mn collective spin excitation in *n*-type CdMnTe quantum wells, and magnetic polarons formation, are examples of such collective spin dynamics.

13.6.1 Electron Spin Relaxation Induced by s–d Exchange

Carrier spin relaxation has been investigated for many years in semiconductors, and has lead to in depth understanding of the various underlying relaxation mechanisms and how they can be eventually reduced, in particular in structures of reduced dimensionality. This subject is covered in Chaps. 1, 2 and 6. Quite generally, spin relaxation is a result of existence of fluctuating effective fields (ω in frequency units) acting on the carrier spin during a correlation time τ_c. When $\omega\tau_c \ll 1$, the evolution of the electron spin looks like a random walk [85, 86] and the dynamic averaging formula can be applied to calculate the electron spin relaxation time as $1/\tau \sim \omega^2\tau_c$.

Even in very diluted DMS, with Mn concentration down to less than 1%, the s, p–d exchange interaction produces strong fluctuating fields, which are responsible for fast spin relaxation in these materials. This relaxation mechanism is much more efficient than usual mechanisms, such as Dyakonov–Perel, existing in non-magnetic semiconductors. In order to illustrate how it works let us derive an approximate expression for this spin relaxation mechanism for the case of non-degenerate electrons, which experience the exchange field of Mn atoms embedded in a quantum well.

In the absence of any localization effects, at temperature T, the electron wave function extends over the de Bröglie wavelength $\lambda \simeq h/(mkT)^{1/2}$ in the direction of the quantum well plane, and occupies a volume $V \sim L\lambda^2$ with L the quantum well width. This volume contains $N = N_0 x V$ Mn spins, with a total spin at zero field of the order of $N^{1/2}S$ producing an exchange field in frequency units

$$\omega \sim \frac{\alpha}{\hbar}\frac{\sqrt{N}S}{V}, \tag{13.22}$$

created by the *static* Mn spins. The electron experiences this field during the time $\tau_c \sim \hbar/kT$ it needs to travel the distance λ. The condition $\omega\tau_c \ll 1$ is generally well justified, so that we use the dynamic averaging formula to calculate the electron spin relaxation time as

$$\frac{1}{\tau_{e-Mn}} \sim \omega^2\tau_c = \frac{\alpha^2 m N_0 x}{\hbar^3 L}S^2. \tag{13.23}$$

Apart from a numerical factor this is the correct expression for the zero-field electron spin relaxation time. Numerical estimate for CdMnTe quantum wells gives spin relaxation times of the order of few picoseconds, much shorter than in nonmagnetic semiconductors, and close to the experimental values [87]. The general expression for the electron spin relaxation, including the field dependence of the longitudinal and transverse spin relaxation times has been derived by Semenov [88].

13.6.2 Mn Spin Relaxation

When considering the relaxation of Mn spins, one should distinguish between nonadiabatic relaxation, which involves a transfer of energy to a reservoir, and adiabatic relaxation, which is usually faster [89]. In the presence of an applied field, longitudinal spin relaxation is a non-adiabatic process since it requires dissipation of the

Zeeman energy accompanying the change of longitudinal spin component. Hence, it is generally much longer than transverse spin relaxation. This energy dissipation can be accommodated in part by the internal energy of the spin–spin interaction, while the remaining part will involve coupling to the lattice at longer times (and therefore generally this relaxation is not exponential [89]). Faster relaxation times can be measured in zero field [90].

Spin–Lattice Relaxation of Isolated Mn Spins

Spin relaxation of isolated Mn atoms in a lattice proceeds via phonon absorption or emission. It corresponds to the mechanism originally proposed by Heitler–Teller. According to this mechanism, spin and phonon can exchange energy via the orbit–lattice interaction. The phonons modulate the electric crystal field surrounding the magnetic atom. This field interacts with the orbital momentum of the ion, which in turn interacts with the spin via the spin–orbit coupling.

In the transition metal series Mn atom represents a special case because in its ground state configuration it has no angular momentum (it is a so-called S-state atom). However this is true only for a free Mn atom. When a Mn atom is placed in the crystal field of the lattice, spin–orbit coupling introduces small admixture of excited states, with non-zero angular momentum. This makes possible the spin–lattice relaxation of isolated Mn ions [91]. Direct absorption or emission of one phonon dominates at low temperature, while Raman two-phonon processes are more efficient at higher temperatures [92]. At liquid helium temperature this spin–lattice relaxation time τ_{SL} becomes very long (a fraction of second) because there is no more phonons available to induce a transition between different spin levels.

Spin–Lattice Relaxation via Mn Spin Clusters

Mn ions cannot be considered as isolated: one observes a strong acceleration of Mn spin relaxation measured under applied field when the Mn concentration increases [89] (see Fig. 13.15). This is explained by a relaxation of the "isolated" Mn spins, which proceeds via fast spin diffusion to "spin killing" centers such as Mn clusters [93]. Due to the random character of the distribution of Mn atoms on the cation sublattice, there is a finite probability to find nearest-neighbors or next-nearest-neighbors Mn ions forming pairs or larger clusters (see Sect. 13.4.1). These clusters most probably play the role of fast relaxing centers. The modulation of the anisotropic superexchange interactions between Mn spins by lattice vibrations makes possible transitions between different spin states in the clusters, and two-phonon Orbach processes (a kind of a resonant Raman process) dominate. This is mainly because energy level separation in the clusters falls into a region of high phonon density of states. For temperatures below 10 K, the strong dependance on x is still observed, however spin relaxation of clusters becomes inefficient and the detailed understanding of Mn spin relaxation mechanism in this temperature range is still lacking.

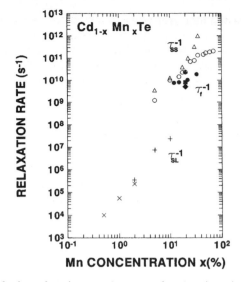

Fig. 13.15. Mn spin–lattice relaxation rate (*crosses*, *pluses*), spin–spin relaxation rate at 5 K (*empty circles*, *triangles*) compared to polaron formation rate (*full symbols*) [94]

Spin–Spin Relaxation

Coupling to the lattice is not mandatory for Mn spin relaxation. Anisotropic spin–spin interactions, which are at the origin of spin–lattice relaxation of Mn clusters, also provide a direct efficient spin relaxation channel.[12] The physical picture is similar to that of nuclear spin relaxation, although the orders of magnitude are quite different.

Nuclear spins are coupled by dipole–dipole interactions. This produces a so-called local field (of the order of one Gauss), which has different values at different lattice sites, and therefore results in a spread in precession frequency and a broadening of the nuclear magnetic resonance line [95]. This amounts to saying that there is a finite transverse spin relaxation time τ_{2s} of the order of 1 ms.

More precisely only the anisotropic part of the spin–spin interactions contributes to the line broadening [18]. In DMS, anisotropic interactions much larger than dipole–dipole coupling result from superexchange mediated by anions with spin–orbit coupling. This is the so-called Dzialoshinski–Moriya [96, 97] interaction. The broadening of Mn EPR line can be very large and correspondingly τ_{2s} is very short, of the order of 10^{-10} s (see Fig. 13.15). In addition, motional narrowing due to isotropic exchange interaction, the so-called exchange-narrowing, must be taken into account to explain the temperature dependance of the line width, and of τ_{2s} [98].

---

[12] Anisotropic spin–spin interactions do not conserve the total spin of the Mn atoms.

Spin Relaxation Assisted by Carriers

If carriers, either localized or delocalized, are present in the sample, they create an exchange field acting on the Mn spins. Fluctuations of this field lead to Mn spin relaxation. This is similar to the nuclear spin relaxation due to hyperfine interaction with electrons [95, 99], but again the orders of magnitude are quite different. The hyperfine coupling constant in III–V compounds for example is typically 3 to 4 orders of magnitude smaller than the exchange integrals in DMS. Since the spin relaxation rate associated with this mechanism varies as the square of the coupling constant, spin relaxation times of Mn are expected to be roughly up to 10^8 times shorter than those of nuclei.

Approximate formula can be obtained by considering that an equal number of mutual spin–flip determines Mn-induced carrier–spin relaxation and carrier-induced Mn–spin relaxation. In the case of a two-dimensional electron gas in a quantum well this gives the relation $n_e \tau_{Mn-e} = N_0 x L \tau_{e-Mn}$. For nondegenerate electrons we can use (13.23) and obtain

$$\frac{1}{T_{Mn-e}} \sim \frac{\alpha^2 m n}{\hbar^3 L^2}. \tag{13.24}$$

The corresponding expression for bulk DMS is easily derived along the same lines by simply using $V \sim \lambda^3$ as the typical volume of the electron wave function and the three-dimensional carrier density n instead of n/L in the quantum well (where n is the two-dimensional density). Considering the case of holes (holes are more effective in Mn spin relaxation than electrons because of the higher exchange integral and effective mass) the relaxation time is estimated to be

$$\frac{1}{T_{Mn-h}} \sim \frac{\beta^2 m^{3/2} p}{\hbar^4} (k_B T)^{1/2}. \tag{13.25}$$

For CdMnTe with hole density $p = 10^{14}\,cm^{-3}$, carrier temperature $T = 2\,K$, heavy-hole effective mass $m = 0.4\,m_0$, one finds $T_{Mn-h} \sim 10\,\mu s$ [100], which in general will not be the dominant contribution to Mn spin relaxation.

The general problem of Mn spin relaxation induced by exchange interaction with two-dimensional electrons, which can be far from thermal equilibrium, has been addressed in [101]. Such a situation may occur when the electrons are heated by optical excitation [101] or by heat pulses [102]. For zero field and close to thermal equilibrium, the general expression for the Mn relaxation time reduces to[13]

$$\frac{1}{\tau_{Mn-e}} = \frac{S(S+1)}{8\pi} \frac{\alpha^2 m^2}{\hbar^5 L^2} \left[1 - \exp\left(-\frac{\pi \hbar^2 n}{m k_B T} \right) \right] k_B T. \tag{13.26}$$

Equation (13.24) is recovered in the case of a nondegenerate gas. Figure 13.16 shows that τ_{Mn-e} decreases with increasing electron density, and saturates at a constant value at high electron density, when the electrons become degenerate, due to the constant 2D density of states. It also shows that at low Mn concentration and high

---

[13] Assuming that the electron gas is confined in a quantum well with infinite barriers.

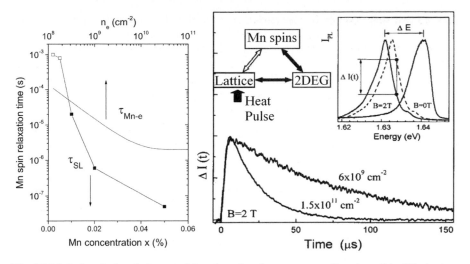

Fig. 13.16. *Left*: calculated electron–Mn relaxation time τ_{Mn-e} as a function of the 2D electron density (*upper scale*), and measured spin–lattice relaxation time τ_{SL} as a function of Mn content (*lower scale*) (after [101]). *Right*: experimental evidence of acceleration of Mn spin relaxation by a 2D electron gas detected by a change of photoluminescence intensity after a heat pulse [102]

enough electron density τ_{Mn-e} becomes shorter than τ_{SL}, thereby accelerating the Mn spin–lattice relaxation via the 2DEG shortcut [102].

Following the analogy with nuclear spins one could be tempted to say that localized carriers should be more effective in Mn spin relaxation than free carriers. However in the former case one has to deal with the problem of bound magnetic polaron, in which electron and Mn spins are strongly correlated (for a review see [103]) and an additional characteristic time, the polaron formation time τ_F, must be introduced to describe the evolution of the local magnetization after trapping of the carrier. It was shown that τ_F is closely related to the spin–spin relaxation time [94] (see Fig. 13.15).

13.6.3 Collective Spin Excitations in CdMnTe Quantum Wells

In doped quantum wells the collective motion of Mn spins may be strongly affected by the exchange interaction with the carriers. Obviously the Mn spin precession frequency is shifted by the exchange field B_C created by the spin polarized carriers. This effect is quite similar to the Knight shift of nuclear magnetic resonance in metals, which is a consequence of the hyperfine interaction, and has been first demonstrated in PbMnTe by Story et al. [104]. For this reason it was named EPR-Knight shift in order to distinguish this shift from the usual NMR-Knight shift. One might say that the Mn spins precess in the *static* total field $B + B_C$. This simple picture may fail to describe situations in which the coupling between carriers and Mn becomes strong enough, so that carriers and Mn spins behave collectively. Below two examples of these collective modes are given, along with a description based on the same simplifying assumptions as in the expression of the free energy (13.15).[14]

Soft Precession Mode in p-Doped Quantum Wells

This happens for example in the vicinity of the paramagnetic to ferromagnetic phase transition in p-doped quantum wells, where a soft precession mode of magnetization has been predicted [106], and observed in time-resolved Kerr rotation experiments [107]. The existence of a soft mode means that magnetization precession slows down as temperature is lowered down to the Curie temperature. Soft oscillation modes are typical for systems undergoing a phase transition. Here the soft precession mode is a consequence of spin–orbit interaction, which is a necessary condition for ferromagnetism to exist in two-dimensional systems, and at the same time is responsible for the strong anisotropy of the heavy-hole spin. Due to this anisotropy, the heavy-hole gas has a large susceptibility only in the direction perpendicular to the quantum well plane (see Sect. 13.4.2).

The simplest way to figure out how the softening emerges was given by Kavokin in [106]. As the magnetization M is tilted by a small angle θ from the direction of the external field B applied parallel to the quantum well plane, the Zeeman energy increases as $MB\theta^2/2$: the Larmor precession can be considered as an oscillation in a one-dimensional parabolic potential. In the presence of holes the energy required to tilt the magnetization is reduced by the exchange interaction. This lowers the rigidity of the system more and more as the temperature is lowered down to T_C, so that the Larmor frequency is expected to decrease with decreasing temperature.[15] This phenomenon has been observed in a p-doped CdMnTe quantum well (see Fig. 13.17), after tilting the magnetization by a short optical pulse (see Sect. 13.7.2). This could also explain the variation of the Larmor frequency around T_C in GaMnAs epilayers [108].

---

[14] In the description of the collective modes only small tilt angles of the magnetization with respect of the magnetic field will be considered, so that one can define the conjugated operators $\hat{X} = M_x/(g\mu_B M)^{1/2}$ and $\hat{P} = M_y/(g\mu_B M)^{1/2}$ which satisfy $[\hat{X}, \hat{P}] = i$ [105]. Similarly for the electrons one defines conjugated operators $\hat{x}$ and $\hat{p}$.

[15] This is a classical effect, which can be described from the Bloch equations of the coupled hole and Mn spins. One can also start from the expression of the free energy (13.15) for Mn coupled to heavy-holes

$$F(M, m) = \frac{M^2}{2\chi_{Mn}} - M \cdot B + \frac{m_x^2}{2\chi_h} - \frac{I M_x m_x}{(g\mu_B)(g_e\mu_B)},$$ (13.27)

and calculating m_x so as to minimize the free energy (i.e., assuming a collective Mn–carrier behavior). For small tilt angle θ the energy of the Mn system can then be expressed quantum mechanically using the creation and annihilation operators $\hat{A}^\dagger = [(1 - \zeta)^{1/2}\hat{X} + i(1 - \zeta)^{-1/2}\hat{P}]/\sqrt{2}$ and $\hat{A} = [(1 - \zeta)^{1/2}\hat{X} - i(1 - \zeta)^{-1/2}\hat{P}]/\sqrt{2}$, with $\zeta = I^2\chi_{Mn}\chi_h/(g\mu_B)^2(g_h\mu_B)^2$ the key parameter which goes to unity at the ferromagnetic transition. One finds the familiar expression for the quantum harmonic oscillator $H = g\mu_B B(1 - \zeta)^{1/2} \times (\hat{A}^\dagger\hat{A} + 1/2)$, the frequency being renormalized by $(1 - \zeta)^{1/2}$.

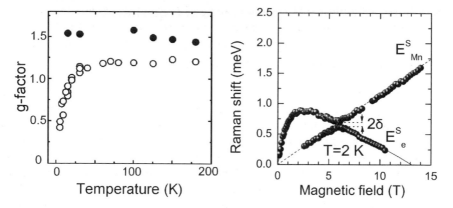

Fig. 13.17. *Left*: strong reduction of the effective *g*-factor attributed to the soft mode (*open circles*) at low temperature in a *p*-type CdMnTe quantum well where $T_C \approx 1.5\,\text{K}$ (from [107]). *Right*: anticrossing of the Mn and conduction electron spin excitations in an *n*-type CdMnTe quantum well (from [110])

Mixed Modes in *n*-Doped Quantum Wells

Another interesting situation occurs in *n*-doped quantum wells, when electron and Mn spin flip energies become comparable. This may happen for example in very diluted CdMnTe quantum wells [109, 110]: the Mn spin excitation corresponds to the normal Zeeman effect with $g = 2$ while the electron spin excitation first increases due to the giant Zeeman effect, then decreases because its normal Zeeman effect tends to align the spin in the opposite direction (g_e is negative).[16] Due to the *s–d* exchange interaction, the electron and Mn can mutually flip their spins, while conserving the total energy. In this case the spin–flip excitation is neither that of electron or Mn but a new collective spin excitation, which involves both electron and Mn.[17] The existence of these collective spin excitations has been evidenced as a shift in the Mn EPR line, and an anticrossing of electron and Mn spin–flip excitations observed in Raman scattering experiments, see Fig. 13.17 [110].

---

[16] This is a direct indication that the positive sign of the exchange integral α results into an antiparallel configuration of the *s* and *d* magnetic moments in CdMnTe. At even larger field, the Larmor frequency of the conduction electrons vanishes: this is not a soft mode, but it provides a good configuration for a study of skyrmions.

[17] The free energy reads

$$F(\boldsymbol{M}, \boldsymbol{m}) = -(\boldsymbol{M} + \boldsymbol{m}) \cdot \boldsymbol{B} - \frac{I \boldsymbol{M} \cdot \boldsymbol{m}}{(g\mu_B)(g_e\mu_B)}. \tag{13.28}$$

For small tilt angles of $\boldsymbol{M}$ and $\boldsymbol{m}$, the energy can be expanded up to quadratic terms using the conjugated operators introduced before $H = \frac{1}{2}(g_e\mu_B B + \Delta)(\hat{x}^2 + \hat{p}^2) + \frac{1}{2}(g\mu_B B + K)(\hat{X}^2 + \hat{P}^2) - (K\Delta)^{1/2}(\hat{X}\hat{x} + \hat{P}\hat{p})$ with the notations $\Delta = I M/g\mu_B$ and $K = Im/g_e\mu_B$ for the Overhauser and (EPR) Knight shifts respectively. In terms of annihilation and creation operators this becomes $H = \hbar\omega_e(\hat{a}\hat{a}^\dagger + 1/2) + \hbar\omega_{\text{Mn}}(\hat{A}\hat{A}^\dagger + 1/2) - \sqrt{K\Delta}(\hat{A}\hat{a}^\dagger + \hat{A}^\dagger\hat{a}) + \cdots$

13.7 Advanced Time-Resolved Optical Experiments

There is a great interest in dynamical spin processes in DMS both from scientific and technological viewpoints. Time-resolved optical experiments have been developed for many years to investigate these processes.

In direct band gap semiconductors such as III–V and II–VI compounds cw- and time-resolved photoluminescence have been widely used to study the spin relaxation of photoexcited carriers and excitons [99, 111]. These experiments are based on the fact that the degree of circular polarization of the emitted light is directly related to the degree of spin polarization of the recombining carriers.

Time-resolved photoluminescence can also be used to investigate the magnetization dynamics because the energy of optical transitions depends on magnetization (see Sect. 13.5). Most studies devoted to the dynamics of magnetic polaron formation rely on this technique. However spin-splittings of conduction and valence bands, which are proportional to magnetization, are generally not measurable in photoluminescence because carriers thermalize in the lowest spin levels. Only the energy shift of these lowest spin levels are measured, but these shifts are not due solely to the evolution of magnetization but also to spectral diffusion in the potential fluctuations due to disorder. Detailed studies, including resonant optical pumping below the mobility edge, are necessary to extract the correct magnetic polaron energy and formation time [112, 113]. Another disadvantage of time-resolved photoluminescence is that magnetization dynamics at time scales longer than photoexcited carriers lifetime is not easily accessible.

To circumvent these drawbacks one can use some very well known magneto-optical effects, such as Faraday rotation or, for nontransparent media, magneto-optical Kerr rotation, which are sensitive probes of magnetization (Sect. 13.5.1). If magnetization evolves in time after being perturbed, this evolution can be measured by time-resolved Faraday or Kerr rotation. Typically this is done by using a pump–probe configuration. An ultrafast laser pulse may alter in various ways, which will be examined below, the magnetization of the sample. The complete dynamics of magnetization, which includes excitation by the laser pulse and recovery of thermal equilibrium, is triggered by a linearly polarized laser pulse, which probes the Faraday or Kerr rotation at different time delays. This constitutes the principle of the so-called time-resolved magneto-optical Kerr rotation and time-resolved Faraday rotation.

The total magnetization of the sample contains two distinct contributions from magnetic atoms and from carriers, and their spin dynamics can be both studied by time-resolved magneto-optical Kerr rotation or Faraday rotation, as shown below.[18]

---

with $\hbar\omega_e = g_e\mu_B B + \Delta$ and $\hbar\omega_{Mn} = g\mu_B B + K$. This expression of two coupled harmonic oscillators shows that the electron and Mn spin–flip excitations are coupled and the anticrossing energy is given by $2\sqrt{K\Delta}$ [109]. Note that, as for the soft mode, a classical description in terms of Bloch equations give the same result.

[18] Care must be taken in the interpretation of Kerr or Faraday signal in magnetic materials, due complications introduced by the dichroic bleaching effect. Considering for example the Faraday rotation one can write the Faraday angle as $\theta_F \propto QM$, where Q is a magneto-optical coefficient. After excitation by a laser pulse the total variation of Faraday rotation contains

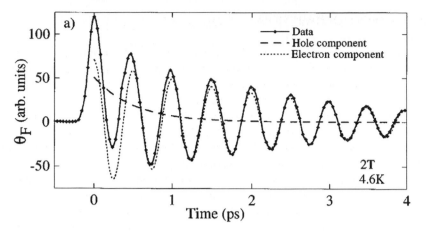

Fig. 13.18. (a) Induced Faraday rotation (*solid line*) in ZnCdMnSe/ZnSe quantum well, showing the rapid exponential decay of the hole spin superimposed on the electron precession (from [119])

13.7.1 Carrier Spin Dynamics

Carrier spin dynamics under the influence of an external magnetic field may be studied by time-resolved Kerr or Faraday rotation in usual nonmagnetic semiconductors and in DMS as well. Although Faraday and Voigt geometries are both eligible, Voigt geometry is generally used. In this geometry the laser light propagates perpendicular to the magnetic field and the photoexcited carriers have their spins initially perpendicular to the field. Electron spin precesses at terahertz frequencies in DMS and produces pronounced oscillations in the Kerr or Faraday signal, while the heavy-hole spin does not precess and contributes to an exponential decay in the signal (Fig. 13.18).

From these measurements electron and hole (transverse) spin relaxation times are deduced. In nonmagnetic semiconductors the electron spin relaxation can be due to spin precession about random internal magnetic fields (as in the Dyakonov–Perel mechanism). However, in addition to spin relaxation, spin dephasing due for example to a distribution of g-factors [117], may contribute to the decay of the transverse electron spin polarization. In II–VI DMS electron spin relaxation is much faster than in the corresponding nonmagnetic host, as expected because of the strong random exchange fields created by the magnetic atoms (see Sect. 13.6) [87, 118]. Note that in Voigt geometry the measured precession frequency gives a clear signature of the spins which are being observed, provided that their effective g-factors are known.

---

two terms $\delta\theta_F \propto M\delta Q + Q\delta M$, where the first term corresponds to the dichroic bleaching, i.e., a change of magneto-optical properties due to photoexcited carriers, and the second term contains the relevant information on magnetization dynamics [114–116].

13.7.2 Magnetization Dynamics

Magnetization Precession Induced by an Exchange Field

In semiconductors circularly polarized ultrafast laser pulses can generate transient effective magnetic fields acting on spins of carriers, nuclei, or magnetic atoms [119–124].

In magnetic materials magnetization precession can be induced by a fast transient effective field noncollinear with the initial magnetization direction. If the effective field changes slowly in comparison to the Larmor period, the magnetization follows adiabatically the total field direction and does not precess. But if the rise time of the transient field is much shorter than the Larmor period the magnetization experiences a torque, which triggers the coherent rotation of magnetic moments.

This effect has been studied in details in II–VI DMS quantum wells. Coherent rotation of magnetic ion spins has been demonstrated experimentally by Crooker et al. in ZnCdMnSe quantum wells with ZnSe barriers using time-resolved Faraday rotation [119]. ZnCdMnSe is a diluted magnetic semiconductor with no spontaneous magnetic ordering. The initial magnetization of Mn atoms is aligned parallel to the quantum well plane by an external magnetic field, and the exchange field perpendicular to the plane is induced by an ultrafast circularly-polarized laser pulse via the photoexcited spin-polarized carriers. The rise time of this field is only limited by the optical pulse width, while its decay is limited by the hole spin relaxation time at the picosecond time scale. After magnetization precession has been launched by the exchange field, free precession can persist for hundreds of picoseconds (Fig. 13.19).

It is important to note that in magnetic quantum wells the coherent rotation is mainly induced by photocreated holes. The hole spin is generally locked along the growth axis due to confinement and strain, while the electron spin precesses rapidly. It is therefore possible to cancel the electron spin polarization using a second pump pulse delayed with respect to the first one by a half of the electron Larmor period. The amplitude of the tilted magnetization appears to be insensitive to the relative delay between the two pump pulses, indicating that the coherent rotation is induced mainly by the hole spin polarization [125].

In addition, a careful analysis of the time-resolved Faraday rotation data reveals a dephasing of the oscillations associated with Mn spin precession [126]. This dephasing suggests the non-instantaneous decay of the transient field created by the holes [127] and allows for an indirect determination of the hole spin relaxation time [87].

The experimental demonstration of optically induced magnetization precession in ferromagnetic GaMnAs is more difficult because of the strong damping of this precession, and poor optical properties of this material. Magnetization rotation induced by photogenerated holes, as in II–VI DMS, was observed in ferromagnetic GaMnAs [108, 128], but the mechanism of photoinduced magnetization rotation is still not very clear [129, 130]. Magnetization rotation may also have a thermal origin, as first identified in ferromagnetic metals [131]. The effect relies on the temperature dependence of the anisotropy axis [132, 133]. The absorption of an ultrashort laser pulse increases locally the sample temperature, thus changing suddenly the anisotropy axis. Then the magnetization precesses coherently about the new anisotropy axis [134].

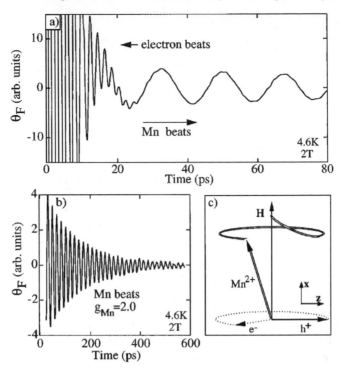

Fig. 13.19. (a), (b) Observation of electron and Mn spin precession (spin beats) via time-resolved Faraday rotation. (c) Mechanism for the coherent magnetization rotation initiated by the exchange field of the holes (after [119])

Demagnetization by Hot Carriers

Magnetization dynamics includes the coherent regime, such as magnetization precession addressed above, and incoherent processes such as demagnetization involving heating of the spin subsystem. Ultrafast demagnetization by intense laser pulses finds applications to magneto-optical recording and has been mainly investigated in metals [136]. In the standard description of the demagnetization process three thermal reservoirs are involved, the carriers, spins and lattice reservoirs as shown in Fig. 13.16.[19] The laser pulse primarily excites the carriers, which may then transfer the excess energy directly to the spins via carrier–spin coupling, or indirectly through the lattice. In DMS these two regimes of demagnetization can be observed depending on the density of excited carriers. At high excitation density the direct carrier–spin coupling dominates, while at low excitation density spins are heated indirectly by the lattice. The later process is much slower, since it is limited by the spin–lattice relaxation (see Sect. 13.6).

---

[19] In itinerant ferromagnets the distinction between carriers and spins is rather artificial as itinerant electrons contribute both to transport and magnetism (see [130]).

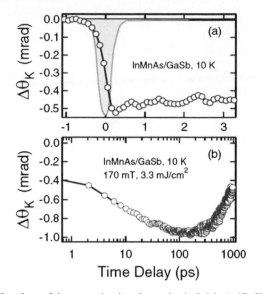

Fig. 13.20. (**a**) The first 3 ps of demagnetization dynamics in InMnAs/GaSb. Also shown is the cross correlation between the pump and probe pulses. (**b**) Demagnetization dynamics covering the entire time range of the experiment (up to ∼1 ns). There is a slow demagnetization process, which follows the fast component shown in (**a**) and completes only after ∼ 100 ps (from [135])

In ferromagnetic p-type InMnAs and GaMnAs ultrafast demagnetization is observed by TRKR on the timescale of several hundreds of femtoseconds (Fig. 13.20). It is interpreted as a consequence of transfer of polarization from Mn spins to holes via spin–flips. The holes become dynamically polarized at the expense of localized spins, and the dissipation of magnetization occurs through hole spin relaxation [130, 135].

Similarly, in paramagnetic undoped CdMnTe quantum wells direct heating of Mn spins through spin–flips with high density photoexcited carriers has been observed on the nanosecond timescale [137–139], while at low excitation density the indirect coupling via the lattice always dominates [100]. These observations are well explained by the competition between the Mn–carrier and Mn–lattice spin relaxation mechanism. Interestingly, the Mn spin heating induced by excited carriers is not homogeneous and lead to formation of hot and cold Mn spin domains [139, 140]. Time-resolved Kerr rotation experiments in Voigt geometry are well adapted to demonstrate this effect, since the electron spin Larmor frequency depends on the magnetization. The onset of two (or even three) different electron Larmor frequencies is a direct evidence of formation of Mn spins domains (Fig. 13.21). It was shown that in paramagnetic DMS, the injection of nonequilibrium electron spins in the presence of magnetic field may induce a spontaneous magnetization patterning. The observed bifurcation from homogeneous to inhomogeneous magnetization as the field increases is well reproduced by the theory [141].

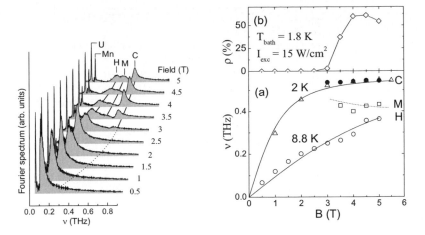

Fig. 13.21. *Left*: Fourier spectra of the TRKR signal $\theta_K(t)$ measured at 1.8 K on CdMnTe quantum wells. At low field a single line corresponding to electron spin precession shows up, which splits into lines H, M, and C above 2.5 T. These lines correspond to electron spins precessing in spatial regions of the quantum well with different Mn spin temperatures. Lines Mn and U correspond respectively to magnetization precession induced by the holes, and electron precession in a non magnetic layer. *Right*: (**a**) Frequencies of the lines H, M, and C, as a function of the field intensity. *Open triangles* show the electron–spin precession frequency under low excitation density when no spin instability occurs. Temperatures of hot and cold domains are deduced by fitting the Larmor frequencies with a Brillouin function. (**b**) Normalized spectral weight of the line C gives an estimate of the fraction of quantum well area occupied by the cold domains. (From [140])

References

[1] R.R. Galazka, Inst. Phys. Conf. Ser. **43**, 133 (1979)

[2] E.L. Nagaev, Phys. Rep. **346**, 387 (2001)

[3] J. Furdyna, J. Appl. Phys. **64**, R29 (1988)

[4] J.K. Furdyna, J. Kossut (eds.), *Diluted Magnetic Semiconductors*. Semiconductors and Semimetals, vol. 25 (Academic, New York, 1988)

[5] T. Dietl, in *Handbook on Semiconductors*, vol. 3b, ed. by T.S. Moss (North-Holland, Amsterdam, 1994), p. 1251

[6] J. Blinowski, P. Kacman, T. Dietl, in *Mat. Res. Soc. Symp. Proc.*, vol. 690, F6.9, ed. by T.J. Klemmer, J.Z. Sun, A. Fert (MRS, 2002); cond-mat/0201012

[7] T. Dietl, Semicond. Sci. Technol. **17**, 377 (2002)

[8] D. Ferrand, J. Cibert, A. Wasiela, C. Bourgognon, S. Tatarenko, G. Fishman, T. Andrearczyk, J. Jaroszynski, S. Kolesnik, T. Dietl, B. Barbara, D. Dufeu, Phys. Rev. B **63**, 85201 (2001)

[9] T. Jungwirth, J. Sinova, A.H. MacDonald, B.L. Gallagher, V. Novak, K.W. Edmonds, A.W. Rushforth, R.P. Campion, C.T. Foxon, L. Eaves, K. Olejnik, J. Masek, S.-R. Eric Yang, J. Wunderlich, C. Gould, L.W. Molenkamp, T. Dietl, H. Ohno, arXiv:0707.0665

[10] S.-H. Wei, A. Zunger, Phys. Rev. B **35**, 2340 (1987)

[11] J.P. Lascaray, J. Diouri, M. El Amrani et al., Sol. Stat. Commun. **47**, 709 (1983)

[12] Y.R. Lee, A.K. Ramdas, Sol. Stat. Commun. **51**, 861 (1984)

[13] Y.R. Lee, A.K. Ramdas, R.L. Aggarwal, Phys. Rev. B **33**, 7383 (1986)
[14] A.K. Bhattacharjee, Phys. Rev. B **58**, 15660 (1998)
[15] A.K. Bhattacharjee, G. Fishman, B. Coqblin, Physica B **117–118**, 449 (1983)
[16] J.R. Schrieffer, P.A. Wolff, Phys. Rev. **149**, 491 (1966)
[17] I.A. Merkulov, D.R. Yakovlev, A. Keller et al., Phys. Rev. Lett. **83**, 1431 (1999)
[18] B.E. Larson, H. Ehrenreich, J. Appl. Phys. **67**, 5084 (1990)
[19] P.W. Anderson, in *Solid States Physics*, vol. 14, ed. by F. Seitz, D. Turnbull (Academic, New York, 1963)
[20] P.W. Anderson, H. Hasegawa, Phys. Rev. **100**, 675 (1955)
[21] M.A. Rudermann, C. Kittel, Phys. Rev. **96**, 99 (1954)
[22] T. Dietl, A. Haury, Y. Merle d'Aubigné, Phys. Rev. B **55**, R3347 (1997)
[23] M.A. Novak, O.G. Symok, D.J. Zheng et al., J. Appl. Phys. **57**, 3418 (1985)
[24] M.A. Novak, O.G. Symko, D.J. Zheng et al., Phys. Rev. B **33**, 6391 (1986)
[25] B. Leclercq, C. Rigaux, Phys. Rev. B **48**, 13573 (1993)
[26] S.B. Oseroff, Phys. Rev. B **25**, 6584 (1982)
[27] J. Spalek, A. Lewicki, Z. Tarnawski et al., Phys. Rev. B **33**, 3407 (1986)
[28] J.A. Gaj, R. Planel, G. Fishman, Sol. Stat. Commun. **29**, 435 (1979)
[29] J.A. Gaj, W. Grieshaber, C. Bodin, J. Cibert, G. Feuillet, Y. Merle d'Aubigné, A. Wasiela, Phys. Rev. B **50**, 5512 (1994)
[30] Y. Shapira, S. Foner, D.H. Ridgley, K. Dwight, A. Wold, Phys. Rev. B **30**, 4021 (1984)
[31] W. Grieshaber, A. Haury, J. Cibert, Y. Merle d'Aubigné, A. Wasiela, J.A. Gaj, Phys. Rev. B **53**, 4891 (1996)
[32] J.M. Fatah, T. Piorek, P. Harrison, T. Stirner, W.E. Hagston, Phys. Rev. B **49**, 10341 (1994)
[33] R.L. Aggarwal, S.N. Jasperson, P. Becla et al., Phys. Rev. B **32**, 5132 (1985)
[34] R.R. Galazka, W. Dobrowolski, J.P. Lascaray et al., J. Mag. Mag. Mat. **72**, 174 (1988)
[35] J.P. Lascaray, A. Bruno, M. Nawrocki et al., Phys. Rev. B **35**, 6860 (1987)
[36] J.P. Lascaray, M. Nawrocki, J.M. Broto et al., Sol. State Commun. **61**, 401 (1987)
[37] Y. Shapira, S. Foner, P. Becla et al., Phys. Rev. B **33**, 356 (1986)
[38] C. Zener, Phys. Rev. **81**, 440 (1951)
[39] C. Zener, Phys. Rev. **83**, 299 (1951)
[40] T. Dietl, *Semimagnetic Semiconductors and Diluted Magnetic Semiconductors*, ed. by M. Averous, M. Balkanski. Ettore Majorana International Science Series (1990)
[41] P. Kossacki, D. Ferrand, A. Arnoult, J. Cibert, S. Tatarenko, A. Wasiela, Y. Merle d'Aubigné, J.-L. Staehli, J.-D. Ganière, W. Bardyszewski, K. Świątek, M. Sawicki, J. Wróbel, T. Dietl, Physica E **6**, 709 (2000)
[42] J.A. Gaj, J. Ginter, R.R. Gałazka, Phys. Stat. Sol. B **89**, 655 (1978)
[43] T. Dietl, H. Ohno, F. Matsukura, Phys. Rev. B **63**, 195205 (2001)
[44] C. Timm, J. Phys. Condens. Matter **15**, R1865 (2003)
[45] A. Kaminski, S. Das Sarma, Phys. Rev. Lett. **88**, 247202 (2002)
[46] G. Bouzerar, T. Ziman, J. Kudrnovský, Europhys. Lett. **69**, 812 (2005)
[47] A.K. Bhattacharjee, Phys. Rev. B **46**, 5266 (1992)
[48] J. Blinowski, P. Kacman, Phys. Rev. B **46**, 12298 (1992)
[49] S. Marcet, D. Ferrand, D. Halley, S. Kuroda, H. Mariette, E. Gheeraert, F.J. Teran, M.L. Sadowski, R.M. Galera, J. Cibert, Phys. Rev. B **74**, 125201 (2006)
[50] T. Mizokawa, A. Fujimori, Phys. Rev. B **56**, 6669 (1997)
[51] T. Mizokawa, T. Nambu, A. Fujimori, T. Fukumura, M. Kawasaki, Phys. Rev. B **65**, 085209 (2002)
[52] A. Twardowski, M. von Ortenberg, M. Demianiuk, R. Pauthenet, Sol. Stat. Commun. **51**, 849 (1984)

[53] W. Pacuski et al., APS March Meeting 2007
[54] W. Pacuski, D. Ferrand, J. Cibert, C. Deparis, J.A. Gaj, P. Kossacki, C. Morhain, Phys. Rev. B **73**, 035214 (2006)
[55] W. Pacuski, D. Ferrand, P. Kossacki, S. Marcet, J. Cibert, J.A. Gaj, A. Golnik, Acta Phys. Pol. A **110**, 303 (2006)
[56] C. Benoît à la Guillaume, D. Scalbert, T. Dietl, Phys. Rev. B **46**, 9853 (1992)
[57] J. Tworzydło, Phys. Rev. B **50**, 14591 (1994)
[58] T. Dietl, cond-mat/0703278
[59] W. Pacuski, P. Kossacki, D. Ferrand, A. Golnik, J. Cibert, M. Wegscheider, A. Navarro-Quezada, A. Bonanni, M. Kiecana, M. Sawicki, T. Dietl, Phys. Rev. Lett. (2007)
[60] K. Ando, Appl. Phys. Lett. **82**, 100 (2003)
[61] J. Szczytko, W. Bardyszewski, A. Twardowski, Phys. Rev. B **64**, 075306 (2001)
[62] P. Kossacki, J. Cibert, D. Ferrand, Y. Merle d'Aubigné, A. Arnoult, A. Wasiela, S. Tatarenko, J. Gaj, Phys. Rev. B **60**, 16018 (1999)
[63] A. Haury, A. Wasiela, A. Arnoult, J. Cibert, T. Dietl, Y. Merle d'Aubigné, S. Tatarenko, Phys. Rev. Lett. **79**, 511 (1997)
[64] H. Boukari, P. Kossacki, M. Bertolini, D. Ferrand, J. Cibert, S. Tatarenko, A. Wasiela, J.A. Gaj, T. Dietl, Phys. Rev. Lett. **88**, 207204 (2002)
[65] P. Kossacki, W. Pacuski, W. Maslana, J.A. Gaj, M. Bertolini, D. Ferrand, J. Bleuse, S. Tatarenko, J. Cibert, Physica E **21**, 943 (2004)
[66] A.A. Maksimov, G. Bacher, A. McDonald, V.D. Kulakovskii, A. Forchel, C.R. Becker, G. Landwehr, L.W. Molenkamp, Phys. Rev. B **62**, 7767 (2000)
[67] J. Seufert, G. Bacher, M. Scheibner, A. Forchel, S. Lee, M. Dobrowolska, J.K. Furdyna, Phys. Rev. Lett. **88**, 027402 (2002)
[68] L. Besombes, Y. Leger, L. Maingault, D. Ferrand, H. Mariette, J. Cibert, Phys. Rev. Lett. **93**, 207403 (2004)
[69] Y. Léger, L. Besombes, J. Fernández-Rossier, L. Maingault, H. Mariette, Phys. Rev. Lett. **97**, 107401 (2006)
[70] R. Fiederling, M. Keim, G. Reuscher, W. Ossau, G. Schmidt, A. Waag, L.W. Molenkamp, Nature **402**, 787 (1999)
[71] B.T. Jonker, Y.D. Park, B.R. Bennett, H.D. Cheong, G. Kioseoglou, A. Petrou, Phys. Rev. B **62**, 8180 (2000)
[72] T. Jungwirth, J. Sinova, J. Masek, J. Kucera, A.H. MacDonald, Rev. Mod. Phys. **78**, 809 (2006)
[73] K.Y. Wang, R.P. Campion, K.W. Edmonds, M. Sawicki, T. Dietl, C.T. Foxon, B.L. Gallagher, in *27th International Conference on the Physics of Semiconductors*, Flagstaff, July 2004, ed. by J. Mendez, C. Van de Walle (2005), p. 333
[74] T. Jungwirth, K.Y. Wang, J. Masek, K.W. Edmonds, J. König, J. Sinova, M. Polini, N.A. Goncharuk, A.H. MacDonald, M. Sawicki, R.P. Campion, L.X. Zhao, C.T. Foxon, B.L. Gallagher, Phys. Rev. B **72**, 165204 (2005)
[75] H. Ohno, Science **281**, 951 (1998)
[76] M. Jamet, A. Barski, T. Devillers, V. Poydenot, R. Dujardin, P. Bayle-Guillemaud, J. Rothman, E. Bellet-Amalric, A. Marty, J. Cibert, R. Mattana, S. Tatarenko, Nat. Mater. **5**, 653 (2006)
[77] Y.D. Park, A.T. Hanbicki, S.C. Erwin, C.S. Hellberg, J.M. Sullivan, J.E. Mattson, T.F. Ambrose, A. Wilson, G. Spanos, B.T. Jonker, Science **295**, 651 (2002)
[78] J.N. Chazalviel, Phys. Rev. B **11**, 3918 (1975)
[79] T. Jungwirth, Q. Niu, A.H. MacDonald, Phys. Rev. Lett. **88**, 207208 (2002)
[80] T. Jungwirth, J. Sinovaa, K.Y. Wang, K.W. Edmonds, R.P. Campion, B.L. Gallagher, C.T. Foxon, Q. Niu, A.H. MacDonald, Appl. Phys. Lett. **83**, 320 (2004)

[81] B. Beschoten, P.A. Crowell1, I. Malajovich, D.D. Awschalom, F. Matsukura, A. Shen, H. Ohno, Phys. Rev. Lett. **83**, 3073 (1999)
[82] R. Chakarvorty, K.J. Yee, X. Liu, P. Redlinski, M. Kutrowski, L.V. Titova, T. Wojtowicz, J.K. Furdyna, B. Janko, M. Dobrowolska, in *27th Internat. Conf. on the Physics of Semiconductors, AIP Conference Proceedings,* June 30, 2005, vol. 772, pp. 1337–1338
[83] L. Thevenard, L. Largeau, O. Mauguin, G. Patriarche, A. Lemaître, N. Vernier, J. Ferré, Phys. Rev. B **73**, 195331 (2006)
[84] L. Thevenard, A. Miard, L. Vila, G. Faini, A. Lemaître, N. Vernier, J. Ferré, S. Fusil, Appl. Phys. Lett. **91**, 142511 (2007)
[85] G. Fishman, G. Lampel, Phys. Rev. B **16**, 820 (1977)
[86] D. Pines, C.P. Slichter, Phys. Rev. **100**, 1014 (1955)
[87] C. Camilleri, F. Teppe, D. Scalbert et al., Phys. Rev. B **64**, 085331 (2001)
[88] Y.G. Semenov, Phys. Rev. B **67**, 115319 (2003)
[89] D. Scalbert, J. Cernogora, C.B. à La Guillaume, Sol. Stat. Commun. **66**, 571 (1988)
[90] M. Goryca, D. Ferrand, P. Kossacki, M. Nawrocki, W. Pacuski, W. Maslana, S. Tatarenko, J. Cibert, Phys. Stat. Sol. (b) **243**, 882 (2006)
[91] M. Blume, R. Orbach, Phys. Rev. **127**, 1587 (1962)
[92] V.Y. Bratus, I.M. Zaritskii, A.A. Konchits, G.S. Pekar, B.D. Shanina, Sov. Phys. Solid State **18**, 1348 (1976)
[93] D. Scalbert, Phys. Stat. Sol. (b) **189**, 193 (1996)
[94] T. Dietl, P. Peyla, W. Grieshaber et al., Phys. Rev. Lett. **74**, 474 (1995)
[95] A. Abragam, *The Principles of Nuclear Magnetism* (Oxford, 1961)
[96] I. Dzialoshinski, J. Phys. Chem. Solids **4**, 241 (1958)
[97] T. Moriya, Phys. Rev. Lett. **4**, 228 (1960)
[98] B.E. Larson, H. Ehrenreich, Phys. Rev. B **39**, 1747 (1989)
[99] M.I. Dyakonov, V.I. Perel, in *Optical Orientation*, ed. by F. Meier, B.P. Zakharchenya (North-Holland, Amsterdam, 1984)
[100] W. Farah, D. Scalbert, M. Nawrocki, Phys. Rev. B **53**, R10461 (1996)
[101] B. König, I.A. Merkulov, D.R. Yakovlev et al., Phys. Rev. B **61**, 16870 (2000)
[102] A.V. Scherbakov, D.R. Yakovlev, A.V. Akimov et al., Phys. Rev. B **64**, 155205 (2001)
[103] S. Takeyama, in *Magneto-optics*, ed. by S. Sugano, N. Kojima. Springer Series in Solid State Science (Springer, Berlin, 2000)
[104] T. Story, C.H.W. Swüste, P.J.T. Eggenkamp et al., Phys. Rev. Lett. **77**, 2802 (1996)
[105] K. Kavokin, I.A. Merkulov, Phys. Rev. B **55**, 7371 (1997)
[106] K. Kavokin, Phys. Rev. B **59**, 9822 (1999)
[107] D. Scalbert, F. Teppe, M. Vladimirova et al., Phys. Rev. B **70**, 245304 (2004)
[108] B. Sun, D. Jiang, Z. Sun et al., J. Appl. Phys. **100**, 083104 (2006)
[109] J. König, A.H. MacDonald, Phys. Rev. Lett. **91**, 077202 (2003)
[110] F.J. Teran, M. Potemski, D.K. Maude et al., Phys. Rev. Lett. **91**, 077201 (2003)
[111] L. Viña, J. Phys. Condens. Matter **11**, 5929 (1999)
[112] G. Mackh, W. Ossau, D.R. Yakovlev et al., Phys. Rev. B **49**, 10248 (1994)
[113] D.R. Yakovlev, K.V. Kavokin, I.A. Merkulov et al., Phys. Rev. B **56**, 9782 (1997)
[114] J.-Y. Bigot, L. Guidoni, E. Beaurepaire et al., Phys. Rev. Lett. **93**, 077401 (2004)
[115] E. Kojima, R. Shimano, Y. Hashimoto et al., Phys. Rev. B **68**, 193203 (2003)
[116] B. Koopmans, M. van Kampen, J.T. Kohlhepp et al., Phys. Rev. Lett. **85**, 844 (2000)
[117] J.M. Kikkawa, D.D. Awschalom, Phys. Rev. Lett. **80**, 4313 (1998)
[118] R. Akimoto, K. Ando, F. Sasaki et al., Phys. Rev. B **56**, 9726 (1997)
[119] S.A. Crooker, J.J. Baumberg, F. Flack et al., Phys. Rev. Lett. **77**, 2814 (1996)
[120] J.A. Gupta, R. Knobel, N. Samarth, D.D. Awschalom, Science **292**, 2458 (2001)
[121] J.M. Kikkawa, D.D. Awschalom, Science **287**, 473 (2000)

[122] A. Malinowski, R.T. Harley, Sol. Stat. Commun. **114**, 419 (2000)
[123] A. Malinowski, M.A. Brand, R.T. Harley, Physica E **10**, 13 (2001)
[124] G. Salis, D.T. Fuchs, J.M. Kikkawa et al., Phys. Rev. Lett. **86**, 2677 (2001)
[125] R. Akimoto, K. Ando, F. Sasaki et al., J. Appl. Phys. **84**, 6318 (1998)
[126] S.A. Crooker, D.D. Awschalom, J.J. Baumberg et al., Phys. Rev. B **56**, 7574 (1997)
[127] R. Akimoto, K. Ando, F. Sasaki et al., Phys. Rev. B **57**, 7208 (1998)
[128] Y. Mitsumori, A. Oiwa, T. Slupinski et al., Phys. Rev. B **69**, 033203 (2004)
[129] A.V. Kimel, G.V. Astakhov, G.M. Schott et al., Phys. Rev. Lett. **92**, 237203 (2004)
[130] J. Wang, C. Sun, Y. Hashimoto et al., J. Phys. Condens. Matter **18**, R501 (2006)
[131] M. van Kampen, C. Jozsa, J.T. Kohlhepp et al., Phys. Rev. Lett. **88**, 227201 (2002)
[132] K.-Y. Wang, M. Sawicki, K.W. Edmonds et al., Phys. Rev. Lett. **95**, 217204 (2005)
[133] U. Welp, V.K. Vlasko-Vlasov, X. Liu et al., Phys. Rev. Lett. **90**, 167206 (2003)
[134] D.M. Wang, Y.H. Ren, X. Liu et al., Phys. Rev. B **75**, 233308 (2007)
[135] J. Wang, C. Sun, J. Kono et al., Phys. Rev. Lett. **95**, 167401 (2005)
[136] M. Kaneko, in *Magnetooptics*, ed. by S. Suganom, N. Kojim (Springer, Berlin, 2000), pp. 271–315
[137] M.K. Kneip, D.R. Yakovlev, M. Bayer et al., Phys. Rev. B **73**, 035306 (2006)
[138] V.D. Kulakovskii, M.G. Tyazhlov, A.I. Filin et al., Phys. Rev. B **54**, R8333 (1996)
[139] M.G. Tyazhlov, V.D. Kulakovskii, A.I. Filin et al., Phys. Rev. B **59**, 2050 (1999)
[140] F. Teppe, M. Vladimirova, D. Scalbert et al., Phys. Rev. B **67**, 033304 (2003)
[141] M. Vladimirova, D. Scalbert, C. Misbah, Phys. Rev. B **71**, 233203 (2005)

Index

absorption
  bleaching, 256
    spin sensitive, 255
  resonant, 253
acceptors, 7
additional terms, 225
affinity, 286
angular dependent linewidth, 190
angular momentum, 12
anisotropic exchange splitting, 82
anomalous Hall effect, 22, 215, 239, 413
anticrossing, 421

balanced receiver, 125, 130
ballistic regime, 233
band structure of GaAs, 11
Berry phase, 231
biexciton, 99
binding energy, 7
Bir–Aronov–Pikus mechanism, 17, 18, 32, 35, 186, 205
bistability, 322
Bloch equations, 191, 201
Bloch waves, 203
Bloembergen N., 23
Bohr radius, 7, 180
Boltzmann equation, 239
Born approximation, 225, 230
Born parameter, 226
boundary conditions, 218
Brillouin function, 356, 397
Brillouin zone, 5
Brossel J., 2
bulk inversion asymmetry, 33, 40, 41, 183
Bychkov–Rashba field, 184
Bychkov–Rashba splitting, 20, 182

carrier induced ferromagnetism, 399, 408
circular polarized light, 30

circularly polarized light, 15
coherent spin dynamics, 25
collision integral, 239
common-atom system, 43
composite fermion, 351, 354, 379
compressibility, 350
compressible states, 356
conductance mismatch, 291
conduction band, 5, 40, 41
confinement energy, 39
continuous waveexcitation, 29
core states, 179
correlation time, 16, 192, 316
Coulomb energy, 354, 368, 380
critical density (of excitons), 83
Curie law, 356
Curie–Weiss law, 397
Curie–Weiss temperature, 397, 399
current induced spin rotation, 196
cyclotron energy, 354
cyclotron frequency, 348
cyclotron motion, 187, 190
cyclotron resonance
  plasma shift, 191

d-states, 390
Datta–Das device, 296
decoherence time
  intrinsic, 203
deep levels, 180, 202
density matrix, 232
dephasing, 193
dichroism
  circular, 155
  linear, 155
diffusion equation, 218, 287
diluted magnetic semiconductor, 298, 389
diluted magnetic semiconductors, 26
dipole–dipole interaction, 4, 23, 313, 360

direct and exchange interaction, 56
disorder, 350, 362
domain, 371, 373, 374
donor spin coherence, 204
donor wave function, 203
donors, 7
donors in Silicon, 203
doping, 6
    symmetrical, 195
double dots, 205
double exchange, 395
Dresselhaus coefficient, 33
    confinement energy dependence, 45
Dresselhaus splitting, 17, 18, 33, 183
Dyakonov–Perel mechanism, 17, 18, 20, 32,
        35, 39, 187, 204, 258, 259, 265
    strong scattering regime, 45
    weak scattering regime, 45
dynamic magnetic susceptibility, 187
dynamic nuclear polarization, 4, 26, 317
dynamical Kerr (Faraday) rotation, 78
dynamical nuclear polarization, 357, 363,
        364, 366, 373
    reverse effect, 366
Dysonian line shape, 199

eddy currents, 201
edge channel, 351
    scattering, 353, 364
edge state, 351
effective magnetic field, 16, 284
effective mass, 6, 380
effective mass approximation, 6
effective nuclear field, 358
electric dipole transitions, 200
electrical spin injection, 357
electro-optic modulator, 118
electron g-factor, 75
    anisotropy, 157
electron spin resonance, 357, 358, 360
    resistively detected, 362
electron spin transport, 36
electron wavefunction, 358
electron–electron scattering, 32, 45, 187
electron–hole exchange, 56
electron–hole interaction, 205
electron–hole spin interaction, 122
electron–nuclear spin system, 310

Elliott–Yafet mechanism, 17, 32, 35, 186,
        194, 204, 224, 258, 259, 263
elliptical exciton states, 83, 86
ENDOR, 203
energy spectrum, 5
ESR
    current induced, 199
    electric dipole, 198
    line shape, 188
    linewidth, 181
    photo, 180
    sensitivity, 181, 198
    transmission, 280
exchange assisted mutual exciton
        scattering, 86
exchange integral, 392, 404
exchange interaction, 3, 94, 392
exciton, 8, 38, 56, 93, 325
    fine structure, 56, 94
    positively charged, 71
exciton exchange energy, 73, 76
exciton exchange splitting, 81
exciton formation (bimolecular,
        geminate), 62
exciton g-factor, 73, 76
exciton longitudinal spin relaxation time, 68
exciton mutual exchange, 83
exciton quantum-beats spectroscopy, 76
exciton spin dynamics, 60
exciton transverse spin relaxation
        time, 68, 78
exciton–polaritons, 86
extended states, 350

Faraday effect, 30, 408
Faraday rotation, 25, 41, 123, 266, 319, 422
    time-resolved, 138
Faraday signals, 45
Fermi contact interaction, 4, 180, 357
Fermi level, 31
Fermi sea, 31, 356, 379
filling factor, 348, 351
fine structure, 180
forbidden gap, 5

g-factor, 4, 348
    anisotropy, 184, 331
    current-induced shift, 196
    in silicon, 179

GaAs, bulk, 35
GaAs/AlGaAs, 38, 47, 50
    modulation doped, 45
GaAs/AlGaAs, quantum well, 41
GaN, bulk, 36
gapless semiconductor, 5, 10, 14, 232, 239
GaSb, bulk, 35
GeMn, 413
giant magnetoresistance, 281
giant Zeeman effect, 392, 402
graded interface, 41
growth
    seeded, 205
    Stranski–Krastanow, 205
gyromagnetic ratio, 311
gyrotropic crystals, 22, 222

Hall resistance
    quantization of, 351
Hanle effect, 1, 21, 23, 30, 40, 41, 47, 196,
        260, 280
    oblique, 327
Hanle spectroscopy, 116
Hanle W., 1
Hartree–Fock, 45
heavy-, light-hole exciton, 59
helicity, 10
heterostructure, 14
hole g-factor, 75
hot luminescence, 11
hyperfine broadening, 206
hyperfine decoherence, 204
hyperfine interaction, 4, 18, 23, 108, 180,
        202, 203, 311, 313, 357
    in quantum dots, 169
    ligand, 180
hyperfine splitting, 202
hysteresis, 23, 322

impurities, 5
impurity scattering, 362
InAs, bulk, 35
InAs/GaSb, 44
indirect exchange, 394
indirect gap, 180
InGaAs/InP, 44
inhomogeneous broadening, 252
InSb, bulk, 35
intersubband scattering, 119

intrinsic mechanism, 223, 231
inverse scaling, 284
inverse spin Hall effect, 22, 215
inversion asymmetry
    bulk, 251
    structure, 251, 254, 263
inversion symmetry, 213
Ising ferromagnetism, 369
isoelectronic, 390
isotopes, 180

$k-p$ method, 32
Kastler A., 2
Kerr effect, 30
Kerr measurements, 38
Kerr rotation, 25, 123, 235, 328
    time-resolved, 138
kinetic coefficients, 214, 286
kinetic equation, 225
kinetic exchange, 393
Knight shift, 313, 358, 381
Korringa relation, 359
Kramers degeneracy, 94
Kramers–Kronig relation, 124

Lampel G., 2
Landau level, 348
Landé g-factor, 30
Larmor frequency, 30, 136, 192, 260, 267
Larmor precession, 30, 81, 118, 327
Larmor vector, 32
lateral confinement, 202
level crossing, 354, 355, 370
light and heavy holes, 8, 14, 56, 232
light polarization, 106
linear response, 286
liquid crystal retarder, 118
local nuclear field, 313
localized states, 350
long-range many-body interactions, 86
longitudinal-transverse splitting, 57
luminescence, 11
Luttinger Hamiltonian, 8, 9
    parameters, 59

magnetic
    random access memory, 301
magnetic circular dichroism, 407
magnetic dipole transitions, 200
magnetic flux quantum, 348

magnetic impurity, 389
magnetic interaction, 5
magnetic length, 351
magnetic moment, 2
magnetic polaron, 410, 419
magnetic resonance
  optically detected, 180
magnetic susceptibility, 181
magnetization patterning, 426
magnetization steps, 397
magneto-optical Kerr effect, 408, 422
magneto-plasma effects, 181
magnetoresistance, 280
Magnus effect, 213
metal insulator transition, 36
microwave absorption, 187
microwave cavity, 200
momentum relaxation, 188
momentum relaxation time, 252
momentum scattering time, 66, 68
Moss–Burstein shift, 407
motional narrowing, 188
motional slowing, 32, 45
Mott effect, 223

natural interface asymmetry, 33, 42
NMR, 203, 319, 358, 371, 381
  all-optical, 337
  multi-quantum, 333
  multi-spin, 333
  optically induced, 334
  optically pumped, 360, 381
  resistively detected, 361, 372, 382
  traditional, 360, 381
no-common-atom system, 43
nonequilibrium magnetization, 283
nonequilibrium thermodynamics, 280
nuclear field, 372
  compensation, 331
nuclear magnetometry, 377
nuclear magneton, 311
nuclear paramagnetism, 356
nuclear self-polarization, 326
nuclear spin, 4, 180
nuclear spin decoherence, 367
nuclear spin diffusion, 23, 360
nuclear spin memory, 374
nuclear spin polarization, 356, 361, 363,
    372, 377

nuclear spin relaxation, 316, 357, 359–363,
    368, 375
  Korringa type, 359, 381
  resistively detected, 375
nuclear spin system, 23
  cooling, 323
  resonant cooling, 337
nuclear spin temperature, 323
nuclear spins, 125

optical orientation of exciton, 59
optical pumping, 2, 357
  selection rules, 59
optical selection rules, 117
optical spin orientation, 15
optical transitions, 251
  interband, 253
  intersubband, 249, 251, 261
  intrasubband, 261
Orbach relaxation, 204
Overhauser field, 312
Overhauser shift, 313, 358, 363

Pauli blocking, 131
Pauli principle, 2, 30, 86
Pauli susceptibility, 400
phase space filling, 30, 83
phase synchronization, 162
phase transition, 354, 369
phonon LO, 97
photo-generation, 11
photoelastic modulator, 118
photogalvanic effect, 245
  circular, 246, 257
    interband, 253
  magneto-gyrotropic, 273
  magneto-photogalvanic, 247
  saturation, 256
  spin–galvanic, 257, 260
photogalvanic effects
  circular, 274
  magneto-photogalvanic, 273
photoluminescence, 29, 30
  time-resolved, 117
photoluminescence excitation, 31
polarization mass, 380
polaron, 338
potentiometric measurement, 293
power absorption, 201

precession frequency, 16
precession vector, 42
pump and probe methods, 30
pump–probe techniques, 123

quadrupole splitting, 311
quantum dot, 13, 202, 310, 410
    III–V, 205
quantum dots, 15
    natural, 122
    singly charged, 153
quantum Hall effect, 26, 347, 348, 350, 351
    fractional, 351
    integer, 348
quantum Hall ferromagnet, 355, 367
quantum point contact, 353
quantum well, 13, 39, 297
    [001]-oriented, 38
    [110]-oriented, 44, 119
    multiple, 59
quantum well structure
    multiple, 56
quantum wire, 13
quasimomentum, 5

Rabi oscillations, 158
random field, 16
random walk, 293
Rashba coefficient, 34
Rashba field, 20
Rashba splitting, 33
recombination, 8, 11
recombination time, 30
relaxation time, 252
    energy, 256
    momentum, 257, 264, 269
    spin, 256–258
resistance
    Hall, 348
    longitudinal, 348
resistance mismatch, 286
RKKY, 395

scattering amplitude, 225
Schrieffer–Wolff transformation, 393
screening, 351
second order effects, 219
selection rules, 12
self-sustained oscillations, 23

semimagnetic semiconductor, 389
Sentfleben effect, 211
shallow defects, 202
short-range electron–hole exchange, 75
short-range exchange, 56, 57
Shubnikov–deHaas effect, 181
side jump, 223, 230
side jump mechanism, 226
Silicon
    isotopically enriched, 203
single electron transistor, 205
single spin measurements, 131
singlet state, 325
skew scattering, 223, 224
Skyrme crystal, 368, 375
skyrmion, 367, 375
    charge of, 367
    collective excitation, 368
    Goldstone mode, 368
    size of, 375
    spin of, 368
soft mode, 420
spin accumulation, 22, 216, 233, 279, 280
spin amplification, 127
    resonant, 144
spin and charge transport, 22
spin asymmetry in scattering, 223
spin band splitting, 222, 233, 239
spin beats, 136
spin coherence
    frequency focusing, 169
    generation of, 137
    in quantum dots, 153
    in quantum wells, 140
    mode locking of, 160
    of electrons, 141
    of holes, 151
spin coherence time, 136
spin current, 22, 211, 247
    pure, 247, 268, 270, 273
spin density, 22
spin density matrix, 263
spin dependent Joule heating, 201
spin dephasing, 45
spin dephasing time, 136
spin diffusion length, 216, 233, 280
spin diode, 364
spin dynamics, 25, 414, 423, 425
    (111) quantum wells, 50

exciton-bound electron, 66
exciton-bound hole, 62
spin echo, 181, 204
spin excitation, 419
spin flip time, 283
spin flip–flop, 106, 357, 366
spin glass, 397
spin grating technique, 51
spin Hall effect, 22, 216, 268
spin Hamiltonian, 182
spin injection, 26, 279
spin layer, 216, 233
spin memory, 337
spin noise spectroscopy, 25, 129
spin orientation
    by current, 263, 266
    optical, 246, 257
        monopolar, 261
spin polarization, 354, 381
    of electrons, 369
spin quantum beats, 119
spin relaxation, 16, 32, 39, 415
    (110) quantum wells, 50
    dependence on spin polarization, 45
    electric field dependence, 50
    holes in bulk, 47
    holes in quantum wells, 47
    longitudinal, 188, 260
    Maille, de Andrada e Silva, and Sham, 68
    suppression in (110) quantum wells, 50
    suppression in (111) quantum wells, 50
    transversal, 260
    transverse time, 128
spin relaxation of holes, 20
    uniaxial stress, 47
spin relaxation time, 218, 283
spin separation, 268, 270
spin splitting, 40, 41, 252, 254, 259, 319
    at Fermi level, 45
spin splitting, anisotropy, 41
spin splitting in 2D, 20
spin synchronization, 162
spin texture, 367
spin transition, 372
spin valve, 281
    lateral, 289
spin–galvanic effect
    inverse, 263, 267
spin–lattice relaxation, 416

spin–orbit coupling, 32, 33
spin–orbit interaction, 4, 22, 239, 301, 360
    in silicon, 179
    quenching by localization, 204
spin–orbit splitting, 9
spin–spin correlations, 131
spin–spin relaxation, 417
spin-charge current coupling, 213
spin-dependent collision broadening, 86
spin-dependent energy shift, 86
spin-dependent scattering, 237
spin-LED, 412
spin-optoelectronic devices, 123
spin-phonon coupling, 198
spinning tennis ball, 227
split-off band, 11
Stark tuning, 204
stochastic spin polarization, 130
Stokes shift, 31
strain, 205
streak camera system, 118
structural inversion asymmetry, 33, 40,
    41, 183
subbands, 13
superexchange, 394
surface states, 14

thermal activation mass, 380
thermal dissociation, 38
thermalization, 11
time inversion, 211
time resolved pump–probe experiment, 78
time resolved spectroscopy, 422
time-resolved photoluminescence, 59, 60,
    81, 116
time-resolved polarized
    photoluminescence, 86
time-resolved pump–probe
    experiments, 71, 86
time-resolved techniques, 29
transresistance, 284
transverse relaxation time, 193
trion, 103
triplet state, 325
tunnel barrier, 286
tunnel junction
    magnetic, 280
tunnel magnetoresistance, 280

two-dimensional electrons, 13, 20, 45,
    183, 295
    Kerr data, 45
two-dimensional electrons and holes, 233
two-photon absorption, 65
type I quantum wells, 57
type II quantum wells, 82

ultrafast demagnetization, 425

valence band, 5, 8
valley degeneracy, 180
valley splitting, 205

van der Waals interaction, 86
vertical transitions, 11

warping of iso-energy surface, 10
Wigner density matrix, 239
Wollaston prism, 125, 130
Wood R., 1

Zeeman effect, 273
Zeeman energy, 23, 284, 311, 354, 368, 380
Zeeman splitting, 348
    exchange enhanced, 363
Zener model, 399

Springer Series in
SOLID-STATE SCIENCES

Series Editors:
M. Cardona   P. Fulde   K. von Klitzing   R. Merlin   H.-J. Queisser   H. Störmer

100 **Electron Correlations in Molecules and Solids**
3rd Edition By P. Fulde

101 **High Magnetic Fields in Semiconductor Physics III**
Quantum Hall Effect, Transport and Optics By G. Landwehr

102 **Conjugated Conducting Polymers**
Editor: H. Kiess

103 **Molecular Dynamics Simulations**
Editor: F. Yonezawa

104 **Products of Random Matrices**
in Statistical Physics By A. Crisanti, G. Paladin, and A. Vulpiani

105 **Self-Trapped Excitons**
2nd Edition
By K.S. Song and R.T. Williams

106 **Physics of High-Temperature Superconductors**
Editors: S. Maekawa and M. Sato

107 **Electronic Properties of Polymers**
Orientation and Dimensionality of Conjugated Systems
Editors: H. Kuzmany, M. Mehring, and S. Roth

108 **Site Symmetry in Crystals**
Theory and Applications
2nd Edition
By R.A. Evarestov and V.P. Smirnov

109 **Transport Phenomena in Mesoscopic Systems**
Editors: H. Fukuyama and T. Ando

110 **Superlattices and Other Heterostructures**
Symmetry and Optical Phenomena
2nd Edition
By E.L. Ivchenko and G.E. Pikus

111 **Low-Dimensional Electronic Systems**
New Concepts
Editors: G. Bauer, F. Kuchar, and H. Heinrich

112 **Phonon Scattering in Condensed Matter VII**
Editors: M. Meissner and R.O. Pohl

113 **Electronic Properties of High-T_c Superconductors**
Editors: H. Kuzmany, M. Mehring, and J. Fink

114 **Interatomic Potential and Structural Stability**
Editors: K. Terakura and H. Akai

115 **Ultrafast Spectroscopy of Semiconductors and Semiconductor Nanostructures**
By J. Shah

116 **Electron Spectrum of Gapless Semiconductors**
By J.M. Tsidilkovski

117 **Electronic Properties of Fullerenes**
Editors: H. Kuzmany, J. Fink, M. Mehring, and S. Roth

118 **Correlation Effects in Low-Dimensional Electron Systems**
Editors: A. Okiji and N. Kawakami

119 **Spectroscopy of Mott Insulators and Correlated Metals**
Editors: A. Fujimori and Y. Tokura

120 **Optical Properties of III–V Semiconductors**
The Influence of Multi-Valley Band Structures By H. Kalt

121 **Elementary Processes in Excitations and Reactions on Solid Surfaces**
Editors: A. Okiji, H. Kasai, and K. Makoshi

122 **Theory of Magnetism**
By K. Yosida

123 **Quantum Kinetics in Transport and Optics of Semiconductors**
By H. Haug and A.-P. Jauho

Springer Series in
SOLID-STATE SCIENCES

Series Editors:
M. Cardona   P. Fulde   K. von Klitzing   R. Merlin   H. J. Queisser   H. Störmer

124 **Relaxations of Excited States and Photo-Induced Structural Phase Transitions**
Editor: K. Nasu

125 **Physics and Chemistry of Transition-Metal Oxides**
Editors: H. Fukuyama and N. Nagaosa

126 **Physical Properties of Quasicrystals**
Editor: Z.M. Stadnik

127 **Positron Annihilation in Semiconductors**
Defect Studies
By R. Krause-Rehberg and H.S. Leipner

128 **Magneto-Optics**
Editors: S. Sugano and N. Kojima

129 **Computational Materials Science**
From Ab Initio to Monte Carlo
Methods. By K. Ohno, K. Esfarjani, and Y. Kawazoe

130 **Contact, Adhesion and Rupture of Elastic Solids**
By D. Maugis

131 **Field Theories for Low-Dimensional Condensed Matter Systems**
Spin Systems
and Strongly Correlated Electrons
By G. Morandi, P. Sodano,
A. Tagliacozzo, and V. Tognetti

132 **Vortices in Unconventional Superconductors and Superfluids**
Editors: R.P. Huebener, N. Schopohl, and G.E. Volovik

133 **The Quantum Hall Effect**
By D. Yoshioka

134 **Magnetism in the Solid State**
By P. Mohn

135 **Electrodynamics of Magnetoactive Media**
By I. Vagner, B.I. Lembrikov, and P. Wyder

136 **Nanoscale Phase Separation and Colossal Magnetoresistance**
The Physics of Manganites
and Related Compounds
By E. Dagotto

137 **Quantum Transport in Submicron Devices**
A Theoretical Introduction
By W. Magnus and W. Schoenmaker

138 **Phase Separation in Soft Matter Physics**
Micellar Solutions, Microemulsions,
Critical Phenomena
By P.K. Khabibullaev and A.A. Saidov

139 **Optical Response of Nanostructures**
Microscopic Nonlocal Theory
By K. Cho

140 **Fractal Concepts in Condensed Matter Physics**
By T. Nakayama and K. Yakubo

141 **Excitons in Low-Dimensional Semiconductors**
Theory, Numerical Methods,
Applications
By S. Glutsch

142 **Two-Dimensional Coulomb Liquids and Solids**
By Y. Monarkha and K. Kono

143 **X-Ray Multiple-Wave Diffraction**
Theory and Application
By S.-L. Chang

144 **Physics of Transition Metal Oxides**
By S. Maekawa, T. Tohyama,
S.E. Barnes, S. Ishihara,
W. Koshibae, and G. Khaliullin

145 **Point-Contact Spectroscopy**
By Y.G. Naidyuk and I.K. Yanson

146 **Optics of Semiconductors and Their Nanostructures**
Editors: H. Kalt and M. Hetterich

Printing: Krips bv, Meppel, The Netherlands
Binding: Stürtz, Würzburg, Germany